D1687699

*Edited by
Richard Dronskowski,
Shinichi Kikkawa, and
Andreas Stein*

**Handbook of
Solid State Chemistry**

Edited by
Richard Dronskowski,
Shinichi Kikkawa, and
Andreas Stein

Handbook of Solid State Chemistry

Volume 4: Nano and Hybrid Materials

WILEY-VCH

WILEY-VCH Verlag GmbH & Co. KGaA

Editors

Richard Dronskowski
RWTH Aachen
Institute of Inorganic Chemistry
Landoltweg 1
52056 Aachen
Germany

Shinichi Kikkawa
Hokkaido University
Faculty of Engineering
N13 W8, Kita-ku
060-8628 Sapporo
Japan

Andreas Stein
University of Minnesota
Department of Chemistry
207 Pleasant St. SE
Minneapolis, MN 55455
USA

Cover Credit: Sven Lidin, Arndt Simon and Franck Tessier

All books published by **Wiley-VCH** are carefully produced. Nevertheless, authors, editors, and publisher do not warrant the information contained in these books, including this book, to be free of errors. Readers are advised to keep in mind that statements, data, illustrations, procedural details or other items may inadvertently be inaccurate.

Library of Congress Card No.: applied for

British Library Cataloguing-in-Publication Data
A catalogue record for this book is available from the British Library.

Bibliographic information published by the Deutsche Nationalbibliothek
The Deutsche Nationalbibliothek lists this publication in the Deutsche Nationalbibliografie; detailed bibliographic data are available on the Internet at http://dnb.d-nb.de.

© 2017 Wiley-VCH Verlag GmbH & Co. KGaA, Boschstr. 12, 69469 Weinheim, Germany

All rights reserved (including those of translation into other languages). No part of this book may be reproduced in any form – by photoprinting, microfilm, or any other means – nor transmitted or translated into a machine language without written permission from the publishers. Registered names, trademarks, etc. used in this book, even when not specifically marked as such, are not to be considered unprotected by law.

Print ISBN: 978-3-527-32587-0
oBook ISBN: 978-3-527-69103-6

Cover Design   Formgeber
Typesetting   Thomson Digital, Noida, India
Printing and Binding   Markono Print Media Pte Ltd, Singapore

Printed on acid-free paper

Preface

When you do great science, you do not have to make a lot of fuss. This oft-forgotten saying from the twentieth century has served these editors pretty well, so the foreword to this definitive six-volume *Handbook of Solid-State Chemistry* in the early twenty-first century will be brief. After all, is there any real need to highlight the paramount successes of solid-state chemistry in the last half century? – Successes that have led to novel magnets, solid-state lighting, dielectrics, phase-change materials, batteries, superconducting compounds, and a lot more? Probably not, but we should stress that many of these exciting matters were derived from curiosity-driven research — work that many practitioners of our beloved branch of chemistry truly appreciate, and this is exactly why they do it. Our objects of study may be immensely important for various applications but, first of all, they are interesting to us; that is, how chemistry defines and challenges itself. Let us also not forget that solid-state chemistry is a neighbor to physics, crystallography, materials science, and other fields, so there is plenty of room at the border, to paraphrase another important quote from a courageous physicist.

Given the incredibly rich heritage of solid-state chemistry, it is probably hard for a newcomer (a young doctoral student, for example) to see the forest for all the trees. In other words, there is a real need to cover solid-state chemistry in its entirety, but only if it is conveniently grouped into digestible categories. Because such an endeavor is not possible in introductory textbooks, this is what we have tried to put together here. The compendium starts with an overview of materials and of the structure of solids. Not too surprisingly, the next volume deals with synthetic techniques, followed by another volume on various ways of (structural) characterization. Being a timely handbook, the fourth volume touches upon nano and hybrid materials, while volume V introduces the reader to the theoretical description of the solid state. Finally, the sixth volume reaches into the real world by focusing on functional materials. Should we have considered more volumes? Yes, probably, but life is short, dear friends.

This handbook would have been impossible to compile for three authors, let alone a single one. Instead, the editors take enormous pride in saying that they managed to motivate more than a hundred first-class scientists living across the globe, each of them specializing in (and sometimes even shaping) a subfield of solid-state chemistry or a related discipline, and all of these wonderful colleagues did their very best to make our dream come true. Thanks to all of you; we sincerely appreciate your contributions. Thank you, once again, on behalf of solid-state chemistry. The editors also would like to thank Wiley-VCH, in particular Dr. Waltraud Wüst and also Dr. Frank Otmar Weinreich, for spiritually (and practically) accompanying us over a few years, and for reminding us here and there that there must be a final deadline. That being said, it is up to the reader to judge whether the tremendous effort was justified. We sincerely hope that this is the case.

A toast to our wonderful science! Long live solid-state chemistry!

Richard Dronskowski
RWTH Aachen, Aachen, Germany

Shinichi Kikkawa
Hokkaido University, Sapporo, Japan

Andreas Stein
University of Minnesota, Minneapolis, USA

Contents

Volume 1: Materials and Structure of Solids

1 **Intermetallic Compounds and Alloy Bonding Theory Derived from Quantum Mechanical One-Electron Models** *1*
Stephen Lee and Daniel C. Fredrickson

2 **Quasicrystal Approximants** *73*
Sven Lidin

3 **Medium-Range Order in Oxide Glasses** *93*
Hellmut Eckert

4 **Suboxides and Other Low-Valent Species** *139*
Arndt Simon

5 **Introduction to the Crystal Chemistry of Transition Metal Oxides** *161*
J.E. Greedan

6 **Perovskite Structure Compounds** *221*
Yuichi Shimakawa

7 **Nitrides of Non-Main Group Elements** *251*
P. Höhn and R. Niewa

8 **Fluorite-Type Transition Metal Oxynitrides** *361*
Franck Tessier

9 **Mechanochemical Synthesis, Vacancy-Ordered Structures and Low-Dimensional Properties of Transition Metal Chalcogenides** *383*
Yutaka Ueda and Tsukio Ohtani

10	**Metal Borides: Versatile Structures and Properties** *435*	
	Barbara Albert and Kathrin Hofmann	
11	**Metal Pnictides: Structures and Thermoelectric Properties** *455*	
	Abdeljalil Assoud and Holger Kleinke	
12	**Metal Hydrides** *477*	
	Yaoqing Zhang, Maarten C. Verbraeken, Cédric Tassel, and Hiroshi Kageyama	
13	**Local Atomic Order in Intermetallics and Alloys** *521*	
	Frank Haarmann	
14	**Layered Double Hydroxides: Structure–Property Relationships** *541*	
	Shan He, Jingbin Han, Mingfei Shao, Ruizheng Liang, Min Wei, David G. Evans, and Xue Duan	
15	**Structural Diversity in Complex Layered Oxides** *571*	
	S. Uma	
16	**Magnetoresistance Materials** *595*	
	Ichiro Terasaki	
17	**Magnetic Frustration in Spinels, Spin Ice Compounds, $A_3B_5O_{12}$ Garnet, and Multiferroic Materials** *617*	
	Hongyang Zhao, Hideo Kimura, Zhenxiang Cheng, and Tingting Jia	
18	**Structures and Properties of Dielectrics and Ferroelectrics** *643*	
	Mitsuru Itoh	
19	**Defect Chemistry and Its Relevance for Ionic Conduction and Reactivity** *665*	
	Joachim Maier	
20	**Molecular Magnets** *703*	
	J.V. Yakhmi	
21	**Ge–Sb–Te Phase-Change Materials** *735*	
	Volker L. Deringer and Matthias Wuttig	

Index *751*

Volume 2: Synthesis

1 **High-Temperature Methods** *1*
Rainer Pöttgen and Oliver Janka

2 **High-Pressure Methods in Solid-State Chemistry** *23*
Hubert Huppertz, Gunter Heymann, Ulrich Schwarz, and Marcus R. Schwarz

3 **High-Pressure Perovskite: Synthesis, Structure, and Phase Relation** *49*
Yoshiyuki Inaguma

4 **Solvothermal Methods** *107*
Nobuhiro Kumada

5 **High-Throughput Synthesis Under Hydrothermal Conditions** *123*
Nobuaki Aoki, Gimyeong Seong, Tsutomu Aida, Daisuke Hojo,
Seiichi Takami, and Tadafumi Adschiri

6 **Particle-Mediated Crystal Growth** *155*
R. Lee Penn

7 **Sol–Gel Synthesis of Solid-State Materials** *179*
Guido Kickelbick and Patrick Wenderoth

8 **Templated Synthesis for Nanostructured Materials** *201*
Yoshiyuki Kuroda and Kazuyuki Kuroda

9 **Bio-Inspired Synthesis and Application of Functional Inorganic Materials by Polymer-Controlled Crystallization** *233*
Lei Liu and Shu-Hong Yu

10 **Reactive Fluxes** *275*

11 **Glass Formation and Crystallization** *287*
T. Komatsu

12 **Glass-Forming Ability, Recent Trends, and Synthesis Methods of Metallic Glasses** *319*
Hidemi Kato, Takeshi Wada, Rui Yamada, and Junji Saida

13 **Crystal Growth Via the Gas Phase by Chemical Vapor Transport Reactions** *351*
Michael Binnewies, Robert Glaum, Marcus Schmidt, and Peer Schmidt

14	Thermodynamic and Kinetic Aspects of Crystal Growth *375*
	Detlef Klimm

15	Chemical Vapor Deposition *399*
	Takashi Goto and Hirokazu Katsui

16	Growth of Wide Bandgap Semiconductors by Halide Vapor Phase Epitaxy *429*
	Yuichi Oshima, Encarnación G. Víllora, and Kiyoshi Shimamura

17	Growth of Silicon Nanowires *467*
	Fengji Li and Sam Zhang

18	Chemical Patterning on Surfaces and in Bulk Gels *539*
	Olaf Karthaus

19	Microcontact Printing *563*
	Kiyoshi Yase

20	Nanolithography Based on Surface Plasmon *573*
	Kosei Ueno and Hiroaki Misawa

Index *589*

Volume 3: Characterization

1	Single-Crystal X-Ray Diffraction *1*
	Ulli Englert

2	Laboratory and Synchrotron Powder Diffraction *29*
	R. E. Dinnebier, M. Etter, and T. Runcevski

3	Neutron Diffraction *77*
	Martin Meven and Georg Roth

4	Modulated Crystal Structures *109*
	Sander van Smaalen

5	Characterization of Quasicrystals *131*
	Walter Steurer

6	Transmission Electron Microscopy *155*
	Krumeich Frank

7	Scanning Probe Microscopy *183* *Marek Nowicki and Klaus Wandelt*
8	Solid-State NMR Spectroscopy: Introduction for Solid-State Chemists *245* *Christoph S. Zehe, Renée Siegel, and Jürgen Senker*
9	Modern Electron Paramagnetic Resonance Techniques and Their Applications to Magnetic Systems *279* *Andrej Zorko, Matej Pregelj, and Denis Arčon*
10	Photoelectron Spectroscopy *311* *Stephan Breuer and Klaus Wandelt*
11	Recent Developments in Soft X-Ray Absorption Spectroscopy *361* *Alexander Moewes*
12	Vibrational Spectroscopy *393* *Götz Eckold and Helmut Schober*
13	Mößbauer Spectroscopy *443* *Hermann Raphael*
14	Macroscopic Magnetic Behavior: Spontaneous Magnetic Ordering *485* *Heiko Lueken and Manfred Speldrich*
15	Dielectric Properties *523* *Rainer Waser and Susanne Hoffmann-Eifert*
16	Mechanical Properties *561* *Volker Schnabel, Moritz to Baben, Denis Music, William J. Clegg, and Jochen M. Schneider*
17	Calorimetry *589* *Hitoshi Kawaji*
	Index *615*

Volume 4: Nano and Hybrid Materials

1	Self-Assembly of Molecular Metal Oxide Nanoclusters *1* *Laia Vilà-Nadal and Leroy Cronin*
1.1	Introduction to Self-Assembly *1*
1.2	Molecular Metal Oxides: Polyoxometalates *3*

1.3	Mechanisms of Cluster Formation	6
1.4	Isomerism in Polyoxometalates	8
1.5	Building Blocks	12
1.6	Classic POM Synthesis	13
1.7	Novel Synthetic Approaches Using Flow Systems	14
1.8	Conclusions	16
	Acknowledgments	16
	References	17
2	**Inorganic Nanotubes and Fullerene–Like Nanoparticles from Layered (2D) Compounds** 21	
	L. Yadgarov, R. Popovitz-Biro, and R. Tenne	
2.1	Introduction	21
2.2	Recent Developments in Synthetic Methods	26
2.2.1	Nanotubes from Misfit Layer Compounds (MLC)	26
2.2.2	Doping of Inorganic Nanotubes and Fullerene-Like Nanoparticles	31
2.2.3	Core–Shell Inorganic Nanotube Superstructures	33
2.2.4	Single-to-Triple Wall Inorganic Nanotubes	33
2.3	Properties	35
2.3.1	General Outlook	35
2.3.2	The Effect of the Doping on the Properties of the Nanoparticles	38
2.3.2.1	Tribological Properties of the Re-Doped IF-MoS_2	38
2.3.2.2	Optical Properties of the Re-Doped IF-MoS_2	39
2.3.3	Properties of Individual WS_2 Nanotubes	41
2.3.3.1	Field-Effect Transistors Based on INT-WS_2	41
2.3.3.2	Electromechanical Properties of INT-WS_2	42
2.4	Applications	43
2.5	Conclusions	45
	References	46
3	**Layered Materials: Oxides and Hydroxides** 53	
	Ida Shintaro	
3.1	Layered Perovskite Oxides	53
3.1.1	The (100)-Layered Perovskite Oxides	53
3.1.1.1	Ruddlesden–Popper Phase ($A_{n+1}B_nO_{3n+1}$)	53
3.1.1.2	Dion–Jacobson Phase ($A'A_{n-1}B_nO_{3n+1}$)	55
3.1.1.3	Aurivillius phase (Bi_2O_2-$A_{n-1}B_nO_{3n+1}$)	56
3.1.2	The (110)-Layered Perovskite Oxides	57
3.1.3	Intercalation Properties of the Layered Perovskite Oxides	58
3.1.4	Conversion from 2D Perovskite to 3D Perovskite	58
3.1.5	Reaction with Other Chemical Reagents	59
3.2	Layered Metal Oxides	60
3.3	Layered Co Oxides	61
3.3.1	Li_xCoO_2	62
3.3.2	Na_xCoO_2	63

3.3.3	Other Layered Co Oxides 65	
3.4	Layered Manganese Oxides 66	
3.4.1	Intercalation Reaction of Layered Manganese Oxides 67	
3.4.2	Tunnel Structure Manganese Oxides 68	
3.5	Layered Copper Oxides 68	
3.5.1	Crystal Structure of Layered Copper Oxides 69	
3.5.2	$La_{2x}Ba_xCuO_4$ 69	
3.5.3	$YBa_2Cu_3O_{7-\delta}$ 70	
3.6	Layered Titanium Oxide and Niobium Oxide 71	
3.7	Layered Double Hydroxides 72	
3.8	Exfoliation of Layered Structures 73	
	References 74	

4 Organoclays and Polymer-Clay Nanocomposites 79
M.A. Vicente and A. Gil

4.1	Introduction 79	
4.2	Organophilization of Clay Minerals 81	
4.2.1	Organophilization of Smectites 82	
4.2.2	Organophilization of Kaolinite 84	
4.2.3	Organophilization of Fibrous Clays 84	
4.2.4	Organophilization of Other Clays 85	
4.3	Synthesis, Structures, and Physicochemical Characterizations 85	
4.4	Clay-Polymer Nanocomposites: Properties and Applications 88	
4.4.1	Mechanical Properties 88	
4.4.2	Thermal Properties and Fire Retardance 89	
4.4.3	Electrical and Electrochemical Properties 90	
4.4.4	Gas Permeation 90	
4.5	Future Applications 92	
	Acknowledgments 92	
	References 93	

5 Zeolite and Zeolite-Like Materials 97
Watcharop Chaikittisilp and Tatsuya Okubo

5.1	Introduction 97	
5.2	Structure and Classification 98	
5.3	Zeolite Synthesis 101	
5.3.1	Historical and Fundamental Views 101	
5.3.2	Recent Developments 105	
5.3.2.1	OSDAs: From Design of Efficient OSDAs to OSDA-Free Synthesis 105	
5.3.2.2	Synthesis of Zeolites from Layered Silicates 108	
5.3.2.3	Hierarchically Porous Zeolites 110	
5.4	Summary and Outlook 112	
	References 115	

6 Ordered Mesoporous Materials *121*
Michal Kruk

6.1 Mesoporous Materials *121*
6.2 Surfactants *122*
6.3 Micelle-Templated Ordered Mesoporous Materials *123*
6.4 Nonionic Poly(Ethylene Oxide)-Based Surfactants as Templates for OMMs *125*
6.5 Structure Control *127*
6.6 Pore Size Control *129*
6.7 Pore Connectivity *132*
References *135*

7 Porous Coordination Polymers/Metal–Organic Frameworks *141*
Ohtani Ryo and Kitagawa Susumu

7.1 Introduction and Fundamentals of PCPs/MOFs *141*
7.2 Synthetic Procedures *143*
7.3 How to use "Pores" in PCPs/MOFs *144*
7.3.1 Adsorption *144*
7.3.2 Release *147*
7.3.3 Conversion *148*
7.3.4 Visualization of Guest Species *150*
7.4 Functional Framework of PCPs/MOFs *151*
7.4.1 Magnetism *151*
7.4.2 Luminescence *152*
7.4.3 Electrical Conductivity *153*
7.4.4 Dielectric Properties *154*
7.5 Guests in the Pores of PCPs/MOFs *155*
7.6 Crystal Engineering of PCPs/MOFs *157*
7.7 Physical Chemistry in PCPs/MOFs *159*
7.8 Outlook *160*
References *160*

8 Metal–Organic Frameworks: An Emerging Class of Solid-State Materials *165*
Joseph E. Mondloch, Rachel C. Klet, Ashlee J. Howarth, Joseph T. Hupp, and Omar K. Farha

8.1 Introduction *165*
8.2 Synthesis of MOFs *166*
8.2.1 De Novo Synthesis *166*
8.2.2 Post-Synthesis Modification *168*
8.2.2.1 Metal-Based Node Modification *168*
8.2.2.2 Organic-Based Node Modification *171*
8.2.2.3 Linker Modification *172*
8.2.3 Building Block Replacement *173*
8.3 Activation of MOFs *174*

8.4	The Quest for Increasingly "Stable" MOFs *175*	
8.4.1	MOFs Containing M–N Bonds *176*	
8.4.2	MOFs Containing Metal–Carboxylate Bonds *177*	
8.5	Select Potential Applications of MOFs *179*	
8.5.1	Nerve Agent Degradation *179*	
8.5.2	Gas-Phase Catalysis with Alkenes *180*	
8.5.3	Environmental Pollution Remediation *182*	
8.6	Conclusions *183*	
	References *183*	

9 Sol–Gel Processing of Porous Materials *195*
Kazuki Nakanishi, Kazuyoshi Kanamori, Yasuaki Tokudome, George Hasegawa, and Yang Zhu

9.1	Introduction *195*	
9.2	Background and Concepts *196*	
9.2.1	Polymerization-Induced Phase Separation in Oxide Sol–Gels *196*	
9.2.2	Structure Formation in Parallel with Sol–Gel Transition *199*	
9.2.3	Macropore Control *200*	
9.2.4	Mesopore Control *200*	
9.3	Silica *201*	
9.3.1	Typical Synthesis Conditions *201*	
9.3.2	Additional Mesopore Formation by Aging *202*	
9.3.3	Hierarchically Porous Monoliths *203*	
9.3.4	Supramolecular Templating of Mesopores *204*	
9.3.5	Applications *205*	
9.4	Silsesquioxane and Other Silicone-Like Systems: Hybrid Aerogels and Low-Density Materials *206*	
9.4.1	Network Formation and Pore Control in Methylsilsesquioxane (MSQ) Systems *206*	
9.4.2	Methylsilsesquioxane Aerogels and Xerogels *209*	
9.4.3	Marshmallow-Like Gels *210*	
9.5	Titania and Zirconia *212*	
9.5.1	Choice of Starting Compounds *212*	
9.5.2	Control over Reactivity *214*	
9.5.3	Applications *215*	
9.6	Epoxide-Mediated System: (Oxy)hydroxides from Metal Salts *216*	
9.6.1	Gelation from Metal Salts and Acids *216*	
9.6.2	Hierarchically Porous Aluminum Hydroxide *218*	
9.6.3	Double Hydroxides, Mixed Metal Oxides, and Others *220*	
9.7	Metal Phosphate Systems: Layered Crystalline Phosphates and Related Compounds *222*	
9.7.1	Calcium Phosphate *222*	
9.7.2	Zirconium Phosphate and Its NaSICON-Type Derivatives *223*	
9.7.3	Titanium Phosphate *225*	

9.8	Functionalization by Postreduction: Carbon, Reduced Oxides, Carbides, and Nitrides *226*	
9.8.1	Carbon *226*	
9.8.2	Silicon Oxycarbide (SiO_xC_y) and Silicon Carbide (SiC) *229*	
9.8.3	Reduced Titanium Oxides (Ti_nO_{2n-1}) and Titanium Nitride (TiN) *231*	
9.8.4	Other Metal Carbides and Nitrides via "Urea Glass Route" *233*	
9.9	Summary *233*	
	Acknowledgments *234*	
	References *234*	

10 Macroporous Materials Synthesized by Colloidal Crystal Templating *243*
Jinbo Hu and Andreas Stein

10.1 Introduction *243*
10.2 Structure *244*
10.3 Synthesis *246*
10.3.1 Colloidal Spheres for CCTs *247*
10.3.2 Colloidal Crystal Assembly *247*
10.3.3 The Templating Process and Synthetic Alternatives *249*
10.4 Applications *258*
10.4.1 Optical Applications *258*
10.4.2 Catalytic Applications *260*
10.4.3 Electrochemical Energy Storage *260*
10.4.4 Electrochemical Sensing *263*
10.4.5 Fuel Cells *264*
10.4.6 Solar Cells *266*
10.4.7 Bioactive Materials and Tissue Engineering *267*
10.5 Conclusions and Outlook *268*
References *269*

11 Optical Properties of Hybrid Organic–Inorganic Materials and their Applications – Part I: Luminescence and Photochromism *275*
Stephane Parola, Beatriz Julián-López, Luís D. Carlos, and Clément Sanchez

11.1 Introduction *275*
11.2 Light-Emitting Hybrid Materials *276*
11.2.1 Introduction to Luminescence *276*
11.2.2 White Light Emission and LEDs *280*
11.2.3 Random and Feedback Lasers *286*
11.2.4 Luminescent Solar Concentrators *287*
11.2.5 Luminescent Thermometers *289*
11.3 Photochromic Hybrid Materials *291*
11.3.1 Introduction to Photochromism *291*
11.3.2 Organic Photochromism in Hybrids *293*
11.3.3 Inorganic and Organometallic Photochromism in Hybrids *304*
References *309*

12	**Optical Properties of Hybrid Organic–inorganic Materials and their Applications – Part II: Nonlinear Optics and Plasmonics** *317*

Stephane Parola, Beatriz Julián-López, Luís D. Carlos, and Clément Sanchez

12.1	Hybrid Materials for Nonlinear Optics *317*
12.1.1	Introduction to Nonlinear Optics *317*
12.1.2	Second-Order Nonlinear Materials *318*
12.1.2.1	Dye-Doped Inorganic Matrices *318*
12.1.2.2	Coordination and Organometallic Compounds Based Hybrid Systems *323*
12.1.3	Third-Order Nonlinear Materials *324*
12.1.3.1	Dispersion of Dyes in Sol–Gel or Organic Materials *324*
12.1.3.2	Polysilsesquioxanes Hybrids for Nonlinear Absorption *330*
12.1.3.3	Graphene Based Hybrid Materials for Nonlinear Absorption *330*
12.2	Plasmonic Hybrid Materials *333*
12.2.1	Optical Properties of Metal Nanoparticles *333*
12.2.2	Hybrids with Dyes and Plasmonic Nanostructures: Luminescence and Nonlinear Optical Properties *334*
12.2.2.1	Surface Functionalization of Metal Nanoparticles *335*
12.2.2.2	Encapsulation of Dyes in Metals *341*
12.2.2.3	Composite Materials and Thin Films *343*
12.3	General Conclusion and Perspectives (Parts I and II) *346*
	References *347*

13	**Bioactive Glasses** *357*

Hirotaka Maeda and Toshihiro Kasuga

13.1	Introduction *357*
13.2	Silicate Glasses *358*
13.2.1	Melt-Quenched Derived Glasses *358*
13.2.2	Sol–Gel-Derived Glasses *366*
13.2.3	Glass-Ceramics *366*
13.2.4	Functionalization of Bioactive Glasses *368*
13.3	Borate Glasses *369*
13.4	Phosphate Glasses *370*
13.4.1	Conventional Phosphate Glasses *370*
13.4.2	Phosphate Invert Glasses *374*
13.4.3	Sulfophosphate Glasses *376*
13.5	Summary *376*
	References *377*

14	**Materials for Tissue Engineering** *383*

María Vallet-Regí and Antonio J. Salinas

14.1	Tissue Engineering: General Concepts *383*
14.2	What can be Regenerated? *386*
14.3	Tissue Engineering in Bone *389*
14.3.1	Scaffolds *390*

14.3.1.1 Bioceramic Scaffolds 392
14.3.1.2 Polymeric Scaffolds 395
14.3.1.3 Metallic Scaffolds 397
14.3.1.4 Composite Scaffolds 399
14.3.2 The Cells 400
14.3.3 The Signals 402
14.4 Achievements in this Area 403
14.5 Where Are We Going? 404
Acknowledgments 405
References 405

Index 411

Volume 5: Theoretical Description

1 **Density Functional Theory** 1
Michael Springborg and Yi Dong

2 **Eliminating Core Electrons in Electronic Structure Calculations: Pseudopotentials and PAW Potentials** 29
Stefan Goedecker and Santanu Saha

3 **Periodic Local Møller–Plesset Perturbation Theory of Second Order for Solids** 59
Denis Usvyat, Lorenzo Maschio, and Martin Schütz

4 **Resonating Valence Bonds in Chemistry and Solid State** 87
Evgeny A. Plekhanov and Andrei L. Tchougréeff

5 **Many Body Perturbation Theory, Dynamical Mean Field Theory and All That** 119
Silke Biermann and Alexander Lichtenstein

6 **Semiempirical Molecular Orbital Methods** 159
Thomas Bredow and Karl Jug

7 **Tight-Binding Density Functional Theory: DFTB** 203
Gotthard Seifert

8 **DFT Calculations for Real Solids** 227
Karlheinz Schwarz and Peter Blaha

9 **Spin Polarization** 261
Dong-Kyun Seo

10   Magnetic Properties from the Perspectives of Electronic Hamiltonian:
     Spin Exchange Parameters, Spin Orientation, and Spin-Half
     Misconception   *285*
     Myung-Hwan Whangbo and Hongjun Xiang

11   Basic Properties of Well-Known Intermetallics and Some New Complex
     Magnetic Intermetallics   *345*
     Peter Entel

12   Chemical Bonding in Solids   *405*
     Gordon J. Miller, Yuemei Zhang, and Frank R. Wagner

13   Lattice Dynamics and Thermochemistry of Solid-State Materials from
     First-Principles Quantum-Chemical Calculations   *491*
     Ralf Peter Stoffel and Richard Dronskowski

14   Predicting the Structure and Chemistry of Low-Dimensional
     Materials   *527*
     *Xiaohu Yu, Artem R. Oganov, Zhenhai Wang, Gabriele Saleh, Vinit Sharma,
     Qiang Zhu, Qinggao Wang, Xiang-Feng Zhou, Ivan A. Popov,
     Alexander I. Boldyrev, Vladimir S. Baturin, and Sergey V. Lepeshkin*

15   The Pressing Role of Theory in Studies of Compressed Matter   *571*
     Eva Zurek

16   First-Principles Computation of NMR Parameters in
     Solid-State Chemistry   *607*
     Jérôme Cuny, Régis Gautier, and Jean-François Halet

17   Quantum Mechanical/Molecular Mechanical (QM/MM) Approaches   *647*
     *C. Richard A. Catlow, John Buckeridge, Matthew R. Farrow,
     Andrew J. Logsdail, and Alexey A. Sokol*

18   Modeling Crystal Nucleation and Growth and Polymorphic
     Transitions   *681*
     Dirk Zahn

     Index   *701*

Volume 6: Functional Materials

1   Electrical Energy Storage: Batteries   *1*
    Eric McCalla

2   Electrical Energy Storage: Supercapacitors   *25*
    Enbo Zhao, Wentian Gu, and Gleb Yushin

3	**Dye-Sensitized Solar Cells** *61*	
	Anna Nikolskaia and Oleg Shevaleevskiy	
4	**Electronics and Bioelectronic Interfaces** *75*	
	Seong-Min Kim, Sungjun Park, Won-June Lee, and Myung-Han Yoon	
5	**Designing Thermoelectric Materials Using 2D Layers** *93*	
	Sage R. Bauers and David C. Johnson	
6	**Magnetically Responsive Photonic Nanostructures for Display Applications** *123*	
	Mingsheng Wang and Yadong Yin	
7	**Functional Materials: For Sensing/Diagnostics** *151*	
	Rujuta D. Munje, Shalini Prasad, and Edward Graef	
8	**Superhard Materials** *175*	
	Ralf Riedel, Leonore Wiehl, Andreas Zerr, Pavel Zinin, and Peter Kroll	
9	**Self-healing Materials** *201*	
	Martin D. Hager	
10	**Functional Surfaces for Biomaterials** *227*	
	Akiko Nagai, Naohiro Horiuchi, Miho Nakamura, Norio Wada, and Kimihiro Yamashita	
11	**Functional Materials for Gas Storage. Part I: Carbon Dioxide and Toxic Compounds** *249*	
	L. Reguera and E. Reguera	
12	**Functional Materials for Gas Storage. Part II: Hydrogen and Methane** *281*	
	L. Reguera and E. Reguera	
13	**Supported Catalysts** *313*	
	Isao Ogino, Pedro Serna, and Bruce C. Gates	
14	**Hydrogenation by Metals** *339*	
	Xin Jin and Raghunath V. Chaudhari	
15	**Catalysis/Selective Oxidation by Metal Oxides** *393*	
	Wataru Ueda	
16	**Activity of Zeolitic Catalysts** *417*	
	Xiangju Meng, Liang Wang, and Feng-Shou Xiao	

17	**Nanocatalysis: Catalysis with Nanoscale Materials** *443*
	Tewodros Asefa and Xiaoxi Huang

18	**Heterogeneous Asymmetric Catalysis** *479*
	Ágnes Mastalir and Mihály Bartók

19	**Catalysis by Metal Carbides and Nitrides** *511*
	Connor Nash, Matt Yung, Yuan Chen, Sarah Carl, Levi Thompson, and Josh Schaidle

20	**Combinatorial Approaches for Bulk Solid-State Synthesis of Oxides** *553*
	Paul J. McGinn

Index *573*

1
Self-Assembly of Molecular Metal Oxide Nanoclusters

Laia Vilà-Nadal and Leroy Cronin

University of Glasgow, WestCHEM, School of Chemistry, University Avenue, Glasgow G12 8QQ, UK

1.1
Introduction to Self-Assembly

"[. . .] Verily at the first Chaos came to be, but next wide-bosomed Earth, the ever-sure foundations of all [. . .]." This sentence is extracted from Hesiod's Theogony composed around 700 BC, describing the origins and genealogies of the Greek gods [1]. About 300 years later, Democritus formulated atomistic theory and imagined all matter in the universe evolving from its atomistic components to form everything from the Earth to the stars [2]. In exploring this assembly and organization, the ancient questions have barely changed: how an ordered structure such as a natural beehive emerged? Or how to explain the geometrical beauty and the symmetry of a seashell? Complex organized patterns appear to be a feature of living and technologically produced systems.

Descartes published in 1644 the Principles of Philosophy in which he envisioned an ordered universe coming out of chaos obeying nature's laws through the organization of small objects into larger assemblage; it appears that Descartes himself was envisaging questions that are also relevant to nanoscience [3]. Indeed, the emergence of order from disorder has captivated the imagination of curious individuals for millennia. In our quest to understand, and perhaps control nature, we realized that breaking down and partitioning problems into their smallest components is useful to understand the whole picture, but complex dynamics and stochastic processes are prevalent at these scales.

A classic analogy to explain self-assembly is the following question: put the different parts of a car in a big box, and shake the whole, will you get a car? [4]. If that was even possible, the consequences would be mind-blowing, and have been often explored in fiction literature and films. Self-assembly is not a human creation, it is a spontaneous process and is well studied at smaller scale from atoms and molecules up to the millimeter scale and beyond [5]. The process by which an organized structure spontaneously forms from individual components,

---

Handbook of Solid State Chemistry, First Edition. Edited by Richard Dronskowski, Shinichi Kikkawa, and Andreas Stein.
© 2017 Wiley-VCH Verlag GmbH & Co. KGaA. Published 2017 by Wiley-VCH Verlag GmbH & Co. KGaA.

as a result of specific, local interactions among its components, is one of the most widely used definitions of self-assembly [6]. More importantly, at the primal level, the spontaneous organization of complex objects does not appear possible without an energy-consuming machinery, cf. the ribosome.

The concept of self-assembly emerged in the early 80s in organic chemistry. Self-assembly and self-organization are often used interchangeably but they refer to two very different processes, and hence have different meanings [7]. In order to clarify their meanings, we must introduce the concept of equilibrium. A chemical reaction has reached equilibrium when the concentrations of both reactants and products do not change over time. Chemical reactions in which the reactants convert to products and where the products *cannot* convert back to the reactants are an example of an irreversible process that leads to entropy production [8]. Therefore, a self-assembly reaction leading towards an equilibrium state, in thermodynamic terms, is due to the minimization of free energy in a closed system. Whereas, self-organization occurs far from equilibrium in an open system, since it requires an external energy source. For instance, the Belousov–Zhabotinsky (BZ) reaction, is a classic example of a nonequilibrium, dissipative reaction [9]. Nonequilibrium and dissipative systems have fascinated scientists for decades. Ilya Prigogine was awarded the 1977 Nobel Prize in Chemistry for his contribution to the field [10]. In essence, he discovered that by importing and dissipating energy into a chemical system, the rule of maximization of entropy, imposed by the second law of thermodynamics, was reversed. Simply put, it takes energy to create improbable configurations from disordered ones [11].

Beyond this, we all agree that there are chemical processes which, at equilibrium, do not dissipate energy, and others which are dissipative and far from equilibrium. It has been a challenging mission for the scientific community to establish "what self-assembly is" since the terminology and practice is multidisciplinary, crossing multiple length scales and spanning a range of forces and fields as diverse as cosmology and biology – length scales from nanoscopic to macroscopic and forces from "weak" to "strong" [12]. Whitesides and Grzybowsky have defined two types of self-assembly: (a) the static self-assembly resulting in an equilibrium state and energy minimization and (b) the dynamical self-assembly with energy dissipation [5]. A few examples of each type, see also Figure 1.1, are as follows. Static self-assembly: molecular crystals [13], lipid bilayers [14], and polymers [15]. Dynamic self-assembly: oscillating reactions [16] and reaction diffusion reactions [17].

Self-assembly in molecular systems can be determined by a number of characteristics such as the components (building blocks), interactions (balance between attractive and repulsive forces), reversibility (or adjustability), its environment (promotes the motion of its components), and finally mass transport and agitation (for instance, thermal motion to assure molecular mobility) [18]. Whether self-assembling molecular building blocks can form into well-organized structures depends on the ability to control their size, shape, and surface properties. Therefore, one of the main goals of

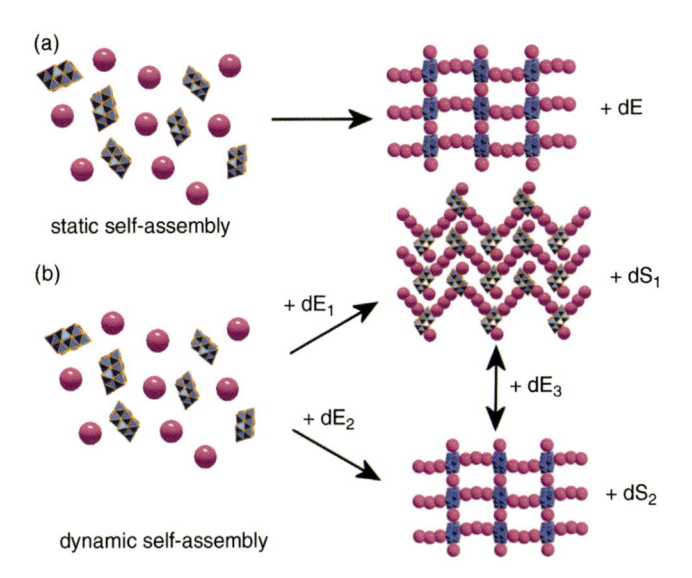

Figure 1.1 Graphical representation of static self-assembly (a) in which self-organization proceeds to an energetic minimum after an exothermic process releasing energy dE; (b) describes a situation out of equilibrium, dynamic self-assembly that requires a constant flow of energy, dE_1, dE_2, and dE_3, into the system to be sustainable, generating entropy dS_1 and dS_2.

self-assembly is being able synthesize building blocks with specified dimensions and forms, and chemically control their surface properties (e.g., charge, hydrophobicity, hydrophilicity, functionality) [19].

Self-assembly also has profound implications in life and origin of life research. One can imagine life as a dynamic self-organized system moving toward, but never reaching, equilibrium; since things at equilibrium are dead as they are not able to process information or self-organize. For example, the organelles of a cell can also be described as a collection of independent parts (or building blocks), each of them designed to perform a specific task. Hence, in our postindustrial society, the assembly line has long been considered one of the greatest innovations of the twentieth century. We can now envisage the power of directed self-assembly at small scales [20].

1.2 Molecular Metal Oxides: Polyoxometalates

Metal–oxygen anionic clusters of early transition metals (V, Nb, Ta, Mo, W) in their highest oxidation states constitute a family of compounds known as polyoxometalates (POMs) [21]. One of the founders of modern chemistry Berzelius, in 1826, described the formation of yellow compounds when phosphate and arsenate salts were mixed with molybdic acid [22]. It was not until 1934, when

Keggin carried out the first detailed crystallographic characterization of a POM salt [23]. POMs are still of great interest and are studied by many groups around the world. This is not surprising since POMs are structurally diverse, and their interesting properties have been exploited in multiple scientific fields, for example, catalysis, medicine, and materials science [24–26]. A seminal work in the classification and study of oxoanions and their salts was presented in the 1983 book by Pope [27]. Thanks to the recent technical advances in spectroscopic methods (IR, resonance Raman, visible/near-IR) as well as single-crystal X-ray structure analysis and electrospray ionization mass spectroscopy (ESI-MS) the number of characterized structures of POMs has increased enormously [28]. Understanding their mechanisms of assembly and the development of new applications has required a great effort for several research groups [29,30]. In addition, exploring the organizational aspects of polyoxometalate structures from small fragments to large species has allowed the synthesis of gigantic molybdenum structures [31]. In fact, the category of metal oxide clusters or POMs includes an unprecedented number of anionic multinuclear species that display a large number of structures, compositions, and sizes ranging from 1 to 5.6 nm, see Figure 1.2 [32]. With only six metal units, the smallest cluster is the so-called Lindqvist anion $[M_6O_{19}]^{n-}$ (in which M = Mo, W, etc.) [33]. The other two iconic structures are known as Keggin (12 metal centered) and Wells–Dawson (18 metal centered) clusters, both members of the heteropolyanions (HPAs) family: metal oxide clusters that encapsulate heteroanions, such as $[SO_4]^{2-}$, $[PO_4]^{3-}$, and so on [34]. Another example of a HPA is the Preyssler anion – this is a cluster with 30 W atoms surrounding a large internal cavity [35]. Finally, perhaps some of the most remarkable molecular structures known are those based upon polyoxomolybdates. For instance, the inorganic fullerene $\{Mo_{132}\}$ belongs to the family of the Keplerate [36] clusters reported by Müller and coworkers. Molybdenum oxide-based fragments, known as Mo-blue or Mo-brown, are reduced Mo-based POMs. Mo-based clusters have a rich structural flexibility under reducing conditions as shown by the isolation of the largest nonbiologically derived molecule to date, a lemon shaped cluster containing 368 molybdenum atoms ($\{Mo_{368}\}$, the "Blue Lemon") [32].

Polyoxometalates can be classified according to their structural characteristics. They are mainly divided into the following three categories:

1) *Heteropolyanions (HPAs)* are tungsten, molybdenum, or vanadium metal oxide clusters that encapsulate heteroanions such as $[SO_4]^{2-}$, $[PO_4]^{3-}$, $[AsO_4]^{3-}$, and $[SiO_4]^{4-}$. The incorporation of a heteroanion in the compound offers a great degree of structural stability in the cluster. For this reason, these are the most explored subset of POM clusters, mostly exploiting their catalytic properties. As stated earlier, the Keggin $[XM_{12}O_{40}]^{n-}$ and the Wells–Dawson $[X_2M_{18}O_{62}]^{n-}$ anions (where M = W or Mo; X is a tetrahedral template) are two of the most easily identified structures within the POM family, see Figure 1.2. Tungsten-based structures are the most robust, hence we have exploited their rigidity to develop lacunary derivatives, that is,

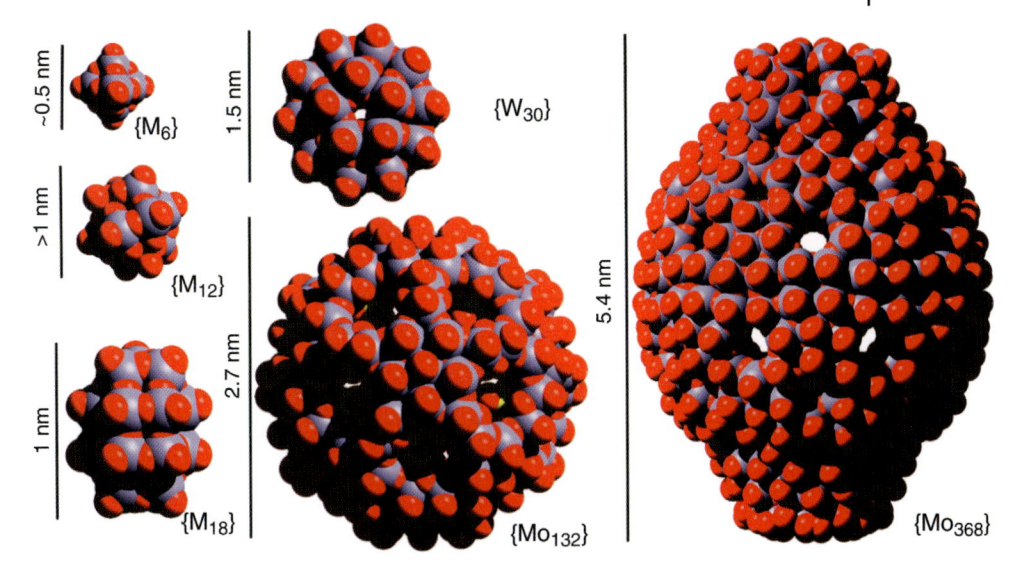

Figure 1.2 Structures of some POM clusters (space filling models M: blue, O: red, S: yellow). The {M_6} Lindqvist anion [M_6O_{19}]$^{n-}$ formed by a compact arrangement of six edge-shared MO$_6$ octahedra. The {M_{12}} Keggin structure [{XO$_4$}M$_{12}$O$_{36}$]$^{n-}$ composed of four M$_3$O$_{13}$ groups of three edge-shared MO$_6$ octahedra that are linked sharing corners to each other and to the central XO$_4$ tetrahedron. The {M_{18}} Wells–Dawson structure [{XO$_4$}$_2$M$_{18}$O$_{54}$]$^{n-}$ can be seen as two fused Keggin fragments. The {W_{30}} Preyssler anion [X^{n+}P$_5$W$_{30}$O$_{110}$]$^{(15-n)-}$ with an internal cavity that can be occupied by different cations (Na, Mn, Eu). The Keplerate-type structure {Mo_{132}}, resulting in "spherical disposition" of pentagonal {(Mo)Mo$_5$} building blocks. The largest cluster to date is the lemon shaped {Mo_{368}} containing 368 metal (1880 nonhydrogen) atoms formed by the linking of 64 {Mo$_1$}-, 32 {Mo$_2$}-, and 40 {Mo-(Mo$_5$)}-type units. The structures of the clusters are compared (to scale) to illustrate the wide range of sizes that POMs can achieve.

Keggin and Dawson anions with vacancies (most commonly with one, two, or three vacancies) that can be linked using electrophiles to larger aggregates in a predictable manner. The development of lacunary polyoxometalates based upon Keggin {M_{12-n}} and Dawson {M_{18-n}} is a large research area [37].

2) *Isopolyanions (IPAs)* are also composed of a metal oxide framework, but in this case without the heteroatom/heteroanion that makes them less stable. However, they are often used as building blocks to construct larger structures, together with their physical properties and high charges and strongly basic oxygen surfaces [38].

3) *Mo-blue and Mo-brown*-reduced nanosized POM clusters were historically described by Scheele in 1783. Their composition was unknown until Müller *et al.* reported the synthesis and structural characterization in 1995 of a very high-nuclearity cluster {Mo_{154}}, which exhibits ring topology, that crystallized from a solution of molybdenum blue [39]. Learning to control the experimental variables of these systems led to the discovery and proper characterization of the first member of the Mo-brown species that exhibited a

porous spherical topology $\{Mo_{132}\}$ [40]. This cluster can be formulated as $[\{Mo_2VO_4(CH_3OO)\}_{30}\{(Mo)Mo_5O_{21}\text{-}(H_2O)\}_{12}]^{42-}$ and is able to encapsulate over 100 water molecules inside its quasispherical polyoxomolybdate nanocapsule cluster, forming a water nanodrop [36]. The structuring of the water inside the $\{Mo_{132}\}$ nanocapsule resembles that of the C_{60} fullerene.

1.3
Mechanisms of Cluster Formation

As we have described, POMs are a diverse set of metal oxide compounds with nucleation number ranging from 6 to 368. The most common synthetic method is a "one-pot" synthesis, but this term only describes the final product of the reaction and the complicated network of interactive chemical processes present in the synthesis is obviated. In fact, this "one-pot" approach is misleading since it seems to indicate that POM systems are directed by simple rules, instead of being the result of self-assembly processes [37]. Synthetic routes to the most common structures are well known, mostly involving an acidification of alkaline aqueous solutions of simple oxoanions, controlling the pH and temperature. After the experience gained by unveiling new structures and optimizing their synthetic methods, it became apparent that POMs are the result of complex networks of chemical reactions governed by several variables. These self-assembly reaction networks allow to establish several equilibria that create a pool of available building blocks to assemble into more complex architectures, controlled by a long list of experimental variables such as: (1) concentration/type of metal oxide anion [38], (2) pH, (3) ionic strength, (4) heteroatom type/concentration, (5) presence of additional ligands, (6) reducing environment, (7) temperature and pressure of reaction (e.g., microwave, hydrothermal, refluxing), (8) counterion and metal-ion effect, and (9) processing methodology (one-pot, continuous flow conditions, 3D printing of reaction ware) [39–42]. New clusters are discovered, and up to some extent even "designed" by controlling the already listed experimental variables. Synthetic experience and careful variable control also helped to generate a vast library of building blocks, and to attempt the design of new POM-based materials.

Despite this synthetic knowledge, few studies have been dedicated to analyze the formation mechanisms of POMs. In fact, the processes of aggregation involved in the formation of these molecules are still poorly understood. Compared with other widely studied characteristics of POMs, such as electronic structure, redox behavior, and magnetic and nonlinear optical properties, the assembly mechanism of POMs still represents a challenge. Recently, the groups of Cronin and Poblet described the nucleation mechanisms of POMs with low nuclearities, by combining theoretical calculations and data from ESI-MS [43]. Standard density functional theory (DFT) calculations combined with molecular dynamics (MD) simulations demonstrate that once the dinuclear species are formed, consecutive steps of protonation and water condensation followed by aggregation take place. In these studies, we proposed two possible mechanisms involving successive protonation and water condensation steps, both thermodynamically favorable and with effectively no barrier at room temperature.

These results agree with the fragments observed in ESI-MS experiments for Lindqvist anions. These "soft-ionization" approaches are unique since they allow for well-defined resolution of multiple closely related species (in contrast to other common spectroscopic techniques such as UV–vis or IR) and can provide clear, well-resolved "snapshots" of a given system on a time scale down to within tens of seconds, if necessary. In this initial study, we concluded that the Lindqvist $[M_6O_{19}]^{n-}$ (M = Mo, W) anion formed in six consecutive steps that incorporate one metal unit at a time, and globally is an exothermic process [43]. We have also identified a common planar $[M_3O_{10}]^{2-}$ motif in the most stable tetra- and pentanuclear intermediate clusters that gives a greater stability to these intermediate clusters, see Figure 1.3. A detailed study by Lang et al. expanded the understanding of the self-assembly mechanism for the $[W_6O_{19}]^{2-}$ identifying relative Gibbs free energies and transition states (at 298 K and 1 atm) for the formation process of all the intermediates [44].

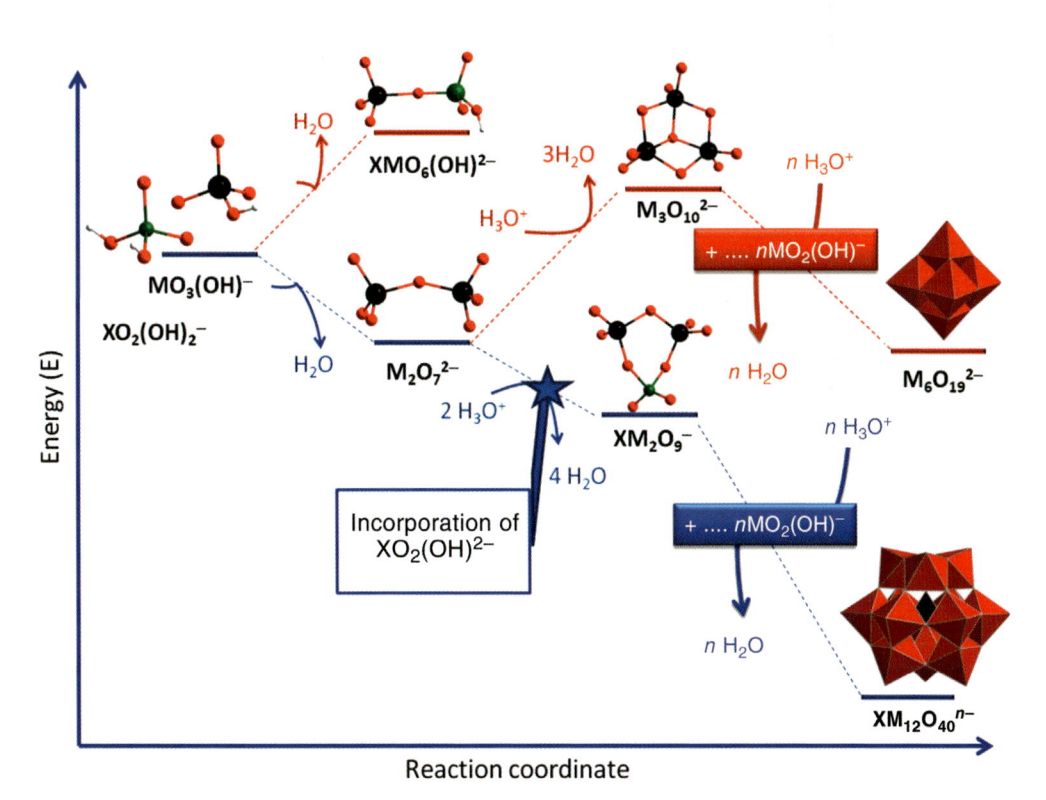

Figure 1.3 Schematic representation of the initial steps, dimer and trimer, of the formation mechanisms for the Lindqvist $[M_6O_{19}]^{2-}$ (M = W and Mo) and Keggin $[XM_{12}O_{40}]^{3-}$ (M = W, Mo and X = P and As) anions.

We have also investigated the first nucleation steps, dimer, trimer, and tetramer, in the formation of the Keggin-type anion. Stoichiometrically obtaining the Keggin anion $[XM_{12}O_{40}]^{n-}$ requires up to 10 protonations and 12 water condensations. As we have described for the Lindqvist anion, we postulate that once the dinuclear species have been formed, including the isodimer $[M_2O_6(OH)_2]^{2-}$ that also appears in the first formation step of the Lindqvist anion or the heterodimer $[MXO_5(OH)_3]^{2-}$, successive steps of protonation and water condensation with subsequent aggregation occurs to justify the clusters observed in the ESI-MS experiments. We propose that the heteroanion, $[PO_2(OH)_2]^-$ or $[AsO_2(OH)_2]^-$, is not incorporated into the polyanion in the first step of the nucleation, that is, forming a heterodimer, but in a later step [45]. Once the heterotrimer is formed, the heteroanion acts as a template for the formation of the Keggin anion. In both studies, Lindqvist and Keggin anions, we did not find any significant differences between tungstates and molybdates, see Figure 1.3. As a complementary work to our report, Lang *et al.* provided the whole thermodynamic analysis of the consecutive steps for the formation of $[PW_{12}O_{40}]^{3-}$ and the analysis of intermediate anions [46].

1.4
Isomerism in Polyoxometalates

Another interesting aspect of metal oxo clusters is that despite their structural robustness polyoxometalate structures present rotational isomerism. In fact, the POM clusters can be seen as a sort of "inorganic Rubik's Cube," see Figure 1.4. It has been known since two isomers of the Keggin silicomolybdic acid were described in the early 50s by Strickland [47]. Isomerism in POMs has been largely studied and their properties deeply analyzed experimentally and computationally, although some basic points are still not completely clear. The main interest in isomerism is related to the possibility of tuning some properties with

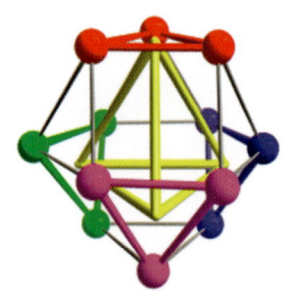

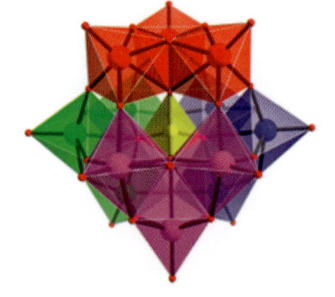

Figure 1.4 Rationalization of the 12 metal-centered Keggin heteropolyoxoanion structure, $[\alpha\text{-}XM_{12}O_{40}]^{n-}$. The anion is composed of four independent $[M_3O_{13}]$ triads (red, green, magenta, blue) interconnected by threefold nodes only. The structure exhibits idealized tetrahedral (T_d) symmetry as indicated by the inscribed tetrahedron (yellow) that is introduced by the central templating anion $[XO_4]^{n-}$.

controlled geometrical changes, such as the different location of a given atom (positional isomerism) or a rotation of a fragment of the molecule (rotational isomerism). Baker and Figgis in 1970 postulated the existence of five isomers for the Keggin anion $[XM_{12}O_{34}]^{3-}$ ($X = As^V, P^V$, etc.; $M = Mo^{VI}, W^{VI}$) and six isomers for the Wells–Dawson anion $[X_2M_{18}O_{62}]^{6-}$ ($X = As^V, P^V$, etc.; $M = Mo^{VI}, W^{VI}$) [48]. Among the five isomers of the $[PW_{12}O_{40}]^{3-}$, the α isomer is characterized by an assembly of four edge-sharing triads, W3 that share corners to each other in a tetrahedral fashion and has T_d symmetry. By successive 60° rotations of 1–4 triads (shaded octahedra), one gets the β (C_{3v}), γ (C_{2v}), δ (C_{3v}), and ε (T_d) isomers. The corresponding energy scale is commensurate with the experimental findings, showing that as the number of rotated triads increases, so does the energy of the metal oxide core, resulting in the order α < β < γ < δ < ε [48]. As per isomer definition, their molecular formula remains the same, but their electrochemistry varies from isomer to isomer. A combination of experimental and theoretical data prove that the redox properties vary from isomer to isomer; however, fully oxidized Keggin anions prefer to adopt the α arrangement, a result which has been further verified by theoretical evidence [49].

In the 1970, Baker and Figgis postulated six isomers for the Wells–Dawson anion: α, β, γ and α*, β*, γ*. The α- $[X_2M_{18}O_{62}]^{q-}$ anion is built up from two A-α-XM_9O_{34} half units joined by six common oxygen atoms, the structure belonging to D_{3h} point group. The β anion, see Figure 1.5, derives from α isomer by a formal rotation by π/3 of a polar (cap) M_3O_{13} group: the symmetry is lowered to C_{3v}. The formal rotation by π/3 of the second polar M_3O_{13} group restores the symmetry plane and the point group D_{3h} for the γ isomer. In all these anions, the hexagonal belts of both XM_9 moieties are symmetry related through the equatorial horizontal plane and their twelve tungsten atoms appear eclipsed along the direction of the C_3. If the two A-α-XM_9O_{34} subunits are related through an inversion center, as postulated by Wells [34] in 1945 for $[P_2W_{18}O_{62}]^{6-}$, the resulting anion named α* would belong to the D_{3d} point group. Rotation of one or both polar (cap) M_3O_{13} groups of this α* anion would generate the two remaining isomers, β* (C_{3v}) and γ* (D_{3d}), respectively. In that case, the hexagonal belts of both XM_9 moieties are symmetry related through the inversion center and their twelve tungsten atoms appear staggered along the direction of the C_3 axis. Zhang et al. carried out density functional theory calculations to investigate α, β, γ, α*, β*, and γ*-$[W_{18}O_{54}(PO_4)_2]^{6-}$ Wells–Dawson isomers, which exhibited stability in the order of α > β > γ > γ* > β* > α*, reproduced the experimental observations (α > β > γ), and confirmed the hypothesis of Contant and Thouvenot (γ* > β* > α*) [50].

Conventionally, redox-inactive anions, such as $[SO_4]^{2-}$ and $[PO_4]^{3-}$, are often used as templating anions in the formation of many POM clusters. A strategy to create new functional POMs involves the encapsulation of redox-active templates instead. By utilizing sulfite, selenite, tellurite, and periodate anions as templates, several new types of redox-active heteropolyoxometalates have been isolated. The POM cluster $[M_{18}O_{54}(SO_3)_2]^{3-}$ (M = W, Mo), which contains two embedded redox-active sulfite templates, can be activated by a metallic surface

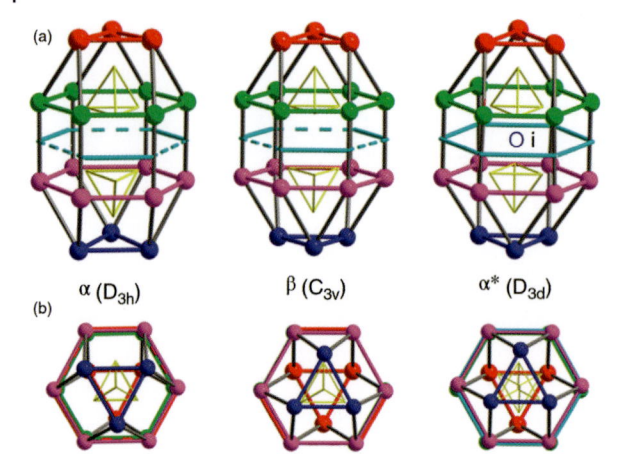

Figure 1.5 (a) Rationalization of the 18 metal-centered α, β, and α* Wells–Dawson structures. (b) Views along the C_3 axis. In the α isomer, both CAP (red and blue) triads appear eclipsed along the C_3. On the contrary, the CAP triads appear staggered in the β and α* isomers. This is due to a π/3 rotation of the blue CAP in the α isomer that leads to the β anion. Another rotation of the red CAP in the β isomer derives in the γ isomer, not depicted here. The short–long alternation in the distances between the oxygens that interconnect both BELT units, depicted in turquoise, is maintained in the α, β, and γ isomers. If both CAP-BELT and A-α-XM_9O_{34} subunits are related through an inversion center "i," depicted in the figure, we obtain the α* isomer with a D_{3d} symmetry. In this case, the distance alternation between central oxygens (turquoise) disappears. The same happens with β* and γ*, since the oxygens are related through the improper axis S_6. Like β and γ isomers, a π/3 rotation of the blue CAP in the α* isomer will lead to the β* and the rotation of both CAP (blue and red) to γ*. Note from part (b), vision along the C_3 axis, the XO_4^{n-} yellow octahedral appear eclipsed in α, β, consequently in γ isomers, whereas staggered in α* and subsequently in β*, γ.

and can reversibly interconvert between two electronic states [51]. Both templates can be replaced by a single template located in the center of the cluster to give a Dawson-like $\{W_{18}X\}$ POM [52]. The first member of this family to be discovered was actually an isopolyanion $\{W_{19}\}$ with a Dawson-type cage; the nineteenth tungsten is located at the center of the cluster instead of the two tetrahedral heteroatoms that are usually found inside conventional Dawson clusters [53]. Structural analysis of the cluster shows that the nineteenth tungsten could be replaced by other elements, such as Pt^{IV}, Sb^V, Te^{VI}, or I^{VII}. The POMs β*-$[H_3W_{18}O_{56}(IO_6)]^{6-}$ embedded with high-valent iodine [54] and γ*-$[H_3W_{18}O_{56}(TeO_6)]^{7-}$ that captures the tellurate anion $[TeO_6]^{6-}$ were discovered thereafter (Figure 1.6) [55].

ESI-MS and DFT studies have been carried out to analyze the relative stability for a series of nonclassical WD anions. In contrast, for the nonclassical WD anions the obtained DFT stability order is γ* > β* > α* > α > β > γ where the isomers γ*, β*, and α are the only anions of this type known to have been synthesized so far. The collision energy necessary to induce the total fragmentation of the $\{XW_{18}\}$ parent polyanion by ESI-MS was always found to be lower than the

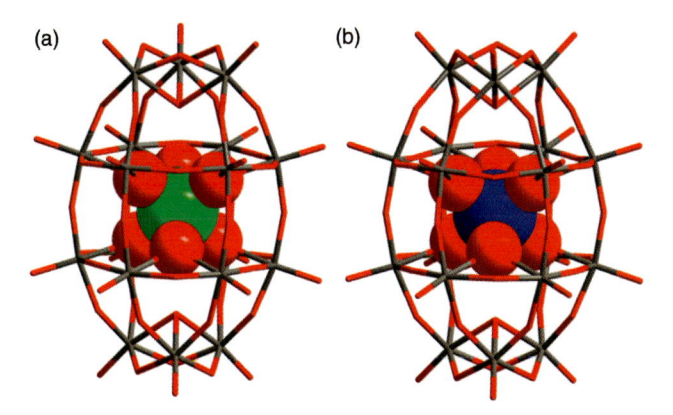

Figure 1.6 Structures of the new Dawson-like {$W_{18}X$} POM type; the {W_{18}} cages are shown as sticks and the central {XO_6} group is represented as space filling model. (a) γ^*-[$W_{18}O_{56}(XO_6)$]$^{10-}$ X = W^{VI} and Te^{VI}. (b) β^*-[$W_{18}O_{56}(IO_6)$]$^{9-}$ being the first example of β^* isomer.

homologous {W_{19}} species; binding energies predicted by DFT are in agreement with this general trend. We have been able to rationalize the isomerism in this new class of WD anions and explain their preference to adopt certain isomer structures [56].

We have also recently studied the transformation of the lacunary polyoxoanion [β_2-$SiW_{11}O_{39}$]$^{8-}$ into [γ-$SiW_{10}O_{36}$]$^{8-}$ using high-resolution electrospray mass spectrometry, density functional theory, and molecular dynamics. Using this approach we have demonstrated that the reaction mechanism proceeds through an unexpected {SiW_9} precursor that undertakes a direct $\beta \rightarrow \gamma$ isomerization via a rotational transformation. The remarkably low-energy transition state of this transformation could be identified through theoretical calculations. Moreover, we explore the significant role of the countercations for the first time in such studies. This combination of experimental and the theoretical studies can now be used to understand the complex chemical transformations of oxoanions, leading to the design of reactivity by structural control [57].

Recently, Kondinsky et al. investigated computationally the α-, γ-, and β-isomeric structures, relative stabilities, and the electronic and basicity properties of magnetic [$V^{IV}_{14}E_8O_{50}$]$^{12-}$ (hereafter referred to as {$V_{14}E_8$}) heteropolyoxovanadates (heteroPOVs) and their heavier chalcogenide-substituted [$V^{IV}_{14}E_8O_{42}X_8$]$^{12-}$ ({$V_{14}E_8X_8$}) derivatives for E = Si^{IV}, Ge^{IV}, and Sn^{IV} and X = S, Se, and Te. By using density functional theory (DFT) with scalar relativistic corrections in combination with the conductor-like screening model of solvation, they have accounted for the structure–property relations in heteroPOVs as well as to assist the synthesis and molecular deposition of these molecular vanadium-oxide spin clusters on surfaces. Their DFT calculations reveal stability trends $\alpha > \gamma > \beta$ for polyoxoanions {$V_{14}E_8$} and {$V_{14}E_8X_8$}, based on relative energies and HOMO–LUMO energy gaps. Among β and γ isomers, the hitherto unknown

$\gamma\text{-}[V_{14}Sn_8O_{50}]^{12-}$ and $\gamma\text{-}[V_{14}Sn_8O_{42}S_8]^{12-}$ seem to be the most viable targets for isolation. Furthermore, these Sn-substituted polyoxoanions are of high interest for electrochemical studies because of their capability to act as two-electron redox catalysts [58].

1.5
Building Blocks

One important factor that triggers the assembly of a set of building blocks into a particular POM species out of a vast number of possible candidates relates to the preferential stabilization of a specific building block library that can be used for the construction of larger aggregates. This point is brought into sharp focus when one realizes that, even though the POMs' structural features usually become the center of the researchers' attention, these POMs are still polyanions and cannot exist without the charge balancing cations that often define the network into which the anion is "complexed" and charge balance is achieved. In this way the cations themselves appear to be able to influence the existing equilibria in solution, provide stability of specific building blocks, and direct the assembly toward the formation and crystallization of a specific molecular candidate, see Figure 1.7.

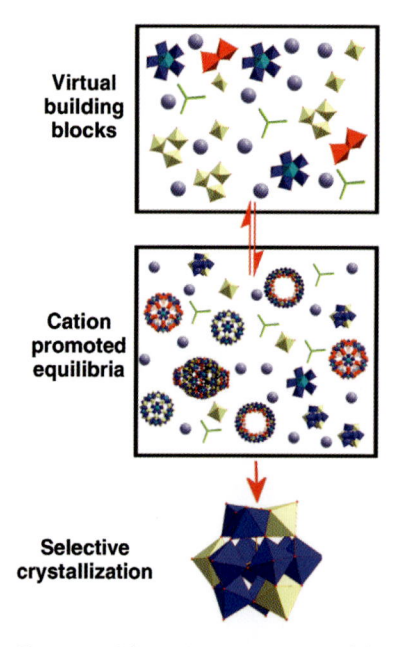

Figure 1.7 Schematic representation of the traditional "one-pot" synthesis of POM clusters leading to the formation of various structural archetypes in solution highlighting the role of counterions in selective stabilization, formation, and crystallization of a specific POM cluster.

Extensive research efforts over the last decades contributed towards our better understanding regarding the counterions' crucial effect on the self-assembly process that goes beyond simply maintaining the charge neutrality in the reaction mixture. Since the properties of the cations such as size, charge, coordination modes, symmetry, and solubility are found to modulate the reactivity as well as the stability of POM building blocks, these cations can clearly affect the nature of the product obtained from a POM synthesis [59]. Using a counterion directed self-assembly approach for the construction of novel POM species, there are two important points that need to be taken into consideration: (a) the generation of novel POM-based building block libraries and (b) promotion of their self-assembly in a controlled fashion to form novel architectures with potential useful functionality. The first option, in order to achieve these targets, is based on the use of bulky positively charged organic cations as counterions in the synthetic procedure [60–62]. The use of bulky cations such as hexamethylene tetramine (HMTA), triethanol amine (TEA), N,N-bis-(2-hydroxyethyl)-piperazine (BHEP), and morpholine prevents the rapid aggregation of POM-based synthons into clusters of stable and uniform spherical topology. Also, the use of cations in combination with transition metals as linker units are found to be capable of diversifying the population of the available constituents by stabilizing reactive secondary building units and directing their self-assembly into novel archetypes. The second option is based on the combined use of organic ligands and additional transition metals not only as counterions, but also as ligands, metal linkers, buffers, and even as redox reagents in some cases, in order to direct the self-assembly process toward a completely new direction [63,64]. Extensive use of the discussed approach gave researchers the opportunity to isolate a number of discrete iso- and heteropolyoxometalate clusters as well as many extended architectures using this simple but efficient concept.

1.6
Classic POM Synthesis

The incorporation of heteroatoms, heterometallic centers, lacunary building blocks, and cations and organic ligands have a profound effect on the self-assembly process and consequently the overall architecture. The architectural design principles used up until recently were based mainly on a fine balance of empirical observation and serendipity. In most cases, the utilized experimental procedures for producing POM-based clusters involves acidification of a solution of the chosen metallic salts, usually molybdates, tungstates, or vanadates [65] followed by a condensation process that involves the interaction of multiple building block libraries leading to the formation of a great variety of POM clusters. Traditionally, the synthesis of the POM cluster takes place in aqueous media and involves routine procedures requiring a small number or even just one step

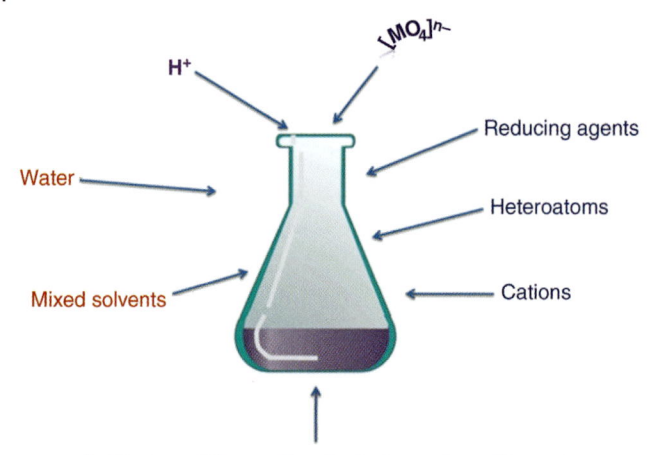

Figure 1.8 Parameters that are often adjusted in the synthesis/isolation of new POM clusters using the multiparameter one-pot method.

("one-pot" synthetic approach) with many variables as already discussed (Figure 1.8) [66]. Recently, chemists are making efforts to re-evaluate their synthetic methodologies and adopt novel approaches that will lead to the designed synthesis of unique architectures (trapping of functional metallic cores, molecular nanoparticles, site specific activity) and potentially the emergence of novel properties (dynamic molecular organization, controlled oscillatory nano-devices, autocatalytic features) that are discussed further in detail.

1.7
Novel Synthetic Approaches Using Flow Systems

Recently, we pioneered the use of a flow reactor system approach to both explore the mechanism and in the synthesis of complex polyoxometalate clusters. For example, by using the flow system we were able to generate a stationary kinetic state of the "intermediate" molybdenum-blue (MB) wheel, filled with a $\{Mo_{36}\}$ guest to give a host–guest complex [67]. The MB host–guest complex has the form $Na_{22}\{[Mo^{VI}_{36}O_{112}(H_2O)_{16}]\subset[Mo^{VI}_{130}Mo^{V}_{20}O_{442}(OH)_{10}(H_2O)_{61}]\}\cdot 180H_2O\{Mo_{36}\}\subset\{Mo_{150}\}$. Carrying out the reaction under controlled continuous flow conditions enabled selection for the generation of $\{Mo_{36}\}\subset\{Mo_{150}\}$ as the major product, and allowed the reproducible isolation of this host–guest complex in good yield, as opposed to the traditional "one-pot" batch synthesis that typically leads to crystallization of the $\{Mo_{154-x}\}$ species (Figure 1.9) [68]. Structural and spectroscopic studies identified the $\{Mo_{36}\}\subset\{Mo_{150}\}$ compound as the intermediate in the synthesis of MB wheels. It is interesting to note that,

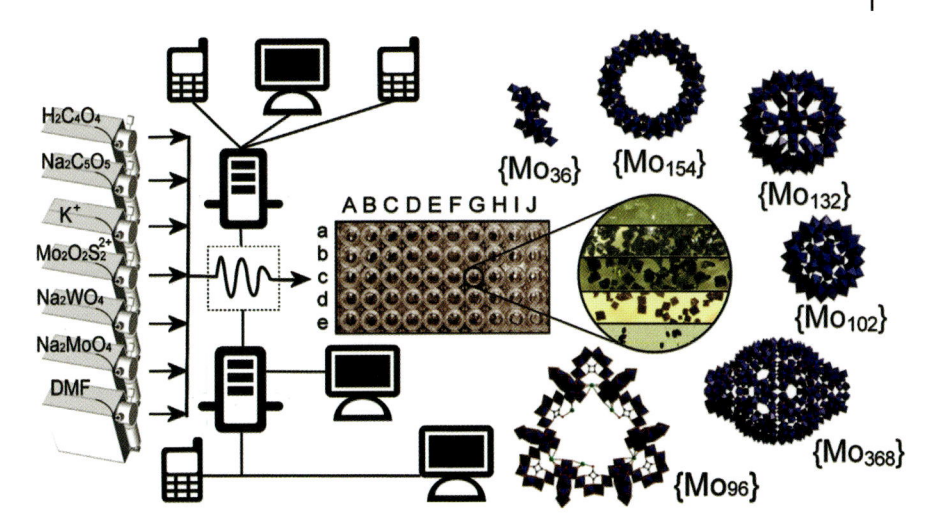

Figure 1.9 Scheme of an automated system controlled by a PC that allows the discovery of new species.

compared to the archetypal 28 electron reduced "empty" $\{Mo_{154}\}$ wheel, the $\{Mo_{150}\}$ is only 20 electron reduced. This is of crucial importance since further reduction of the wheel results in the expulsion of the $\{Mo_{36}\}$ guest indicating why this was not observed before reproducibly. Also, further experiments showed an increase in the yield and the formation rate of the $\{Mo_{154-x}\}$ wheels by deliberate addition of preformed $\{Mo_{36}\}$ to the reaction mixture. Dynamic light scattering (DLS) was also used to corroborate the mechanism of formation of the MB wheels through observation of the individual cluster species in solution. DLS measurement of the reaction mixtures, from which $\{Mo_{36}\}$ and $\{Mo_{150}\}$ crystallized, gave particle size distribution curves averaging 1.9 and 3.9 nm, respectively. The above approach allowed the use of size as a possible distinguishing feature of these key species in the reduced acidified molybdate solutions and direct observation of the molecular evolution of the available synthons to MB wheels [69].

Although the qualitative data obtained does not allow comprehensive kinetic studies at this stage, it brings us one step closer to understanding the formation of complicated systems like the MBs in solution. Using these techniques to follow the assembly of other self-assembled chemical systems in solution will open the door to further understanding and finally control of such complex self-assembly processes. The characterization of the "bottom-up" designed nanosized species in solution will also overcome the problems associated with product crystallization and isolation and will ultimately unveil the true potential of solution-processable nanosized metal oxides to be exploited in the manufacture of novel materials and molecular devices with engineered functionality.

1.8
Conclusions

One of the key aspects of the new developments at the frontiers of metal oxide cluster science is based on the finding that cluster structures are built on a hierarchy of template and templating subunits; however, this is not yet explored in detail. Also, it is just emerging that reaction networks based upon polyoxometalates can be increasingly treated as complex chemical systems containing interdependent networks of self-assembling, self-templating building blocks. Indeed, complex interacting "systems" defined using polyoxometalate building blocks may be used as the archetypal models to explore inorganic chemical networks.

The area of polyoxometalates is now entering into a new phase whereby it is possible to design and control both the structure and function of the systems. However, their dynamic nature with a seemingly endless structural diversity means that the assembly of functional nanomolecules and adaptive materials under nonequilibrium conditions will be developed. This approach will be used to access new building block libraries that will lead to the formation of novel nanomaterial structures and functions not accessible from near equilibrium processing techniques and will be focused on producing new materials, assemblies, and devices. Such processes may be driven using redox reactions, ion exchange, and metal unit substitution to drive, direct, and trap the self-assembly of molecular metal oxide-based building blocks, clusters, and materials in solution. By using such nonequilibrium-based processing, it will be a possible aim to engineer materials with unprecedented structures functionality and adaptive potential than possible with conventional, static, and near equilibrium self-assembly techniques. For example, in very recent work we have shown that it is possible to engineer cluster–guest compounds whereby a cluster-based oscillator is engineered, and the oscillation in the internal cluster template can be driven by the presence of a reducing amine in solution [70]. The fact that such dynamic behavior can be set up and observed in solution is exciting, and the coupling of such processes between the solution and solid-state has fantastic promise for the future design and discovery of polyoxometalate-based reaction systems and networks with emergent properties. We are confident that the structural explosion in the area of polyoxometalates will now lead to another explosion in functionality taking advantage of the transferable building blocks, and the ability to engineer nonequilibrium systems with unprecedented properties and emergent functionalities.

Acknowledgments

The authors would like to thank the University of Glasgow, WestCHEM, the EPSRC, Leverhulme Trust, the Royal Society/Wolfson Foundation, the Royal Society of Edinburgh, and Marie Curie actions for financial support as well as members of the Cronin laboratory, past and present.

References

1. West, M.L. (ed.); Hesiod. (1966) *Theogony: Edited with Prolegomena and Commentary*. Clarendon Press, Oxford.
2. Russell, B. (1970) *A History of Western Philosophy*, Harcourt Brace Jovanovich, New York.
3. Descartes, R., Miller, V.R., and Miller, R.P. (1983) *Principles of Philosophy/René Descartes: Translated, with Explanatory Notes by Valentine Rodger Miller and Reese P. Miller*, Reidel, Dordrecht, London.
4. Jones, R. (2008) *Soft Machines – Nanotechnology and Life*, Oxford University Press, Oxford.
5. Whitesides, G.M. and Grzybowski, B. (2002) *Science*, **295**, 2418–2421.
6. Altamura, L. (2016) Self-assembly: Latest research and news – Nature. Available at http://www.nature.com/subjects/self-assembly (accessed Nov. 2, 2016).
7. Bensaude-Vincent, B. (2006) Self-assembly, Self-organization: a philosophical perspective on converging technologies, paper prepared for France/Stanford Meeting Avignon, December 2006. Available at http://stanford.edu/dept/france-stanford/Conferences/Ethics/BensaudeVincent.pdf (accessed Nov. 2, 2016).
8. Prigogine, I., Stengers, I., and Toffler, A. (1984) *Order out of Chaos*, Bantam Books, New York.
9. Petrov, V., Gaspar, V., Maesere, J., and Showalter, K. (1993) *Nature*, **361**, 240–243.
10. England, J. (2015) *Nat. Nanotech.*, **10**, 919–923.
11. Macklem, P.T. (2008) *J. Appl. Physiol.*, **104**, 1844–1846.
12. Ozin, G.A., Hou, K., Lotsch, B.V., Cademartiri, L., Puzzo, D.P., Scotognella, F., Ghadimi, A., and Thomson, J. (2009) *Mater. Today*, **12**, 12–23.
13. Desiraju, G.R. (1989) *Crystal Engineering: The Design of Organic Solids*, Elsevier, New York.
14. Evans, D.F. and Wennerstrom, H. (1999) *The Colloidal Domain: Where Physics, Chemistry, Biology, and Technology Meet*, John Wiley & Sons, New York.
15. Jones, M.N. and Chapman, D. (1995) *Micelles, Monolayers, and Biomembranes*, Wiley–Liss, New York.
16. Jakubith, S., Rotermund, H.H., Engel, W., von Oertzen, A., and Ertl, G. (1990) *Phys. Rev. Lett.*, **65**, 3013–3016.
17. Hess, B. (2000) *Naturwissenschaften*, **87**, 199–211.
18. Whitesides, G.M. and Boncheva, M. (2002) *PNAS*, **99**, 4769–4774.
19. Cademartiri, L. and Ozin, G.A. (2009) *Concepts of Nanochemistry*, Wiley-VCH Verlag GmbH, Weinheim.
20. Drexler, E. (1986) *Engines of Creation: The Coming Era of Nanotechnology*, Anchor Books, New York.
21. Pope, M.T. and Müller, A. (1991) *Angew. Chem., Int. Ed. Engl.*, **30**, 34–48.
22. Berzelius, J. (1826) *Poggendorff's Ann. Phys.*, **6**, 369–380.
23. Keggin, J.F. (1934) *Proc. R. Soc. A.*, **144**, 75.
24. Hill, C.L. (1998) *Chem. Rev.*, **98**, 1–2.
25. Pope, M.T. and Müller, A. (1994) *Polyoxometalates: From Platonic Solids to Anti-Retroviral Activity*, Kluwer Academic Publishers, Dordrecht.
26. (a) Proust, A., Matt, B., Villanneau, R., Guillemot, G., Gouzerh, P., and Izzet, G. (2012) *Chem. Soc. Rev.*, **41**, 7605–7622; (b) Yu-Fei, S. and Ryo, T., (2012) *Chem. Soc. Rev.*, **41**, 7384–7402; (c) Miras, H.N., Yan., J., Long, D.-L., and Cronin, L., (2012) *Chem. Soc. Rev.*, **41**, 7403–7430; (d) Long, D.-L., Burkholder, E. and Cronin, L., (2007) *Chem. Soc. Rev.*, **36**, 105–121.
27. Pope, M.T. (1983) *Heteropoly and Isopoly Oxometalates*, Springer, New York.
28. Müller, A. and Roy, S. (2004) *The Chemistry of Nanomaterials: Synthesis, Properties and Applications*, Wiley-VCH Verlag GmbH, Weinheim.
29. (a) Miras, H.N., Wilson, E.F. and Cronin, L. (2009) *Chem. Commun.*, 1297–1311; (b) Wilson, E.F., Miras, H.N., Rosnes, M.H., and Cronin, L., (2011) *Angew. Chem., Int. Ed.*, **50**, 3720–3724; (c) Sartorel, A., Carraro, M., Scorrano, G., De Zorzi, R., Geremia, S., McDaniel, N.D., Bernhard, S., and Bonchio, M., (2008) *J. Am. Chem. Soc.*, **130**, 5006–5007; (d) Geletii, Y.V., Botar, B., Köegerler, P.,

Hillesheim, D.A., Musaev, D.G., and Hill, C.L., (2008) *Angew. Chem., Int. Ed.*, **47**, 3896–3899; (e) Sartorel, A., Miró, P., Salvadori, E., Romain, S., Carraro, M., Scorrano, G., Di Valentin, M., Llobet, A., Bo, C., and Bonchio, M., (2009) *J. Am. Chem. Soc.*, **131**, 16051–16053; (f) Zheng, Q., Vilà-Nadal, L., Busche, C., Mathieson, J.S., Long, D.-L., and Cronin, L. (2015) *Angew. Chem., Int. Ed.*, **54**, 7895–7899.

30 (a) Kamata, K., Yonehara, K., Sumida, Y., Yamaguchi, K., Hikichi, S., and Mizuno, N. (2003) *Science*, **300**, 964–966; (b) Kim, W.B., Voitl, T., Rodriguez-Rivera, G.J., and Dumesic, J.A. (2004) *Science*, **305**, 1280–1283; (c) Nyman, M., Bonhomme, F., Alam, T.M., Rodriguez, M.A., Cherry, B.R., Krumhansl, J.L., Nenoff, T.M., and Sattler, A.M. (2002) *Science*, **297**, 996–998; (d) Miras, H.N., Cooper, G.J.T., Long, D.-L., Bogge, H., Müller, A., Streb, C., and Cronin, L. (2010) *Science*, **327**, 72–74; (e) Rausch, B., Symes, M.D., Chisholm, G., and Cronin, L. *Science*, (2014) **345**, 1326–1330; (f) Douglas, T. and Young, M. (1998) *Nature*, **393**, 152–155; (g) Müller, A., Shah, S.Q.N., Bogge, H., and Schmidtmann, M. (1999) *Nature*, **397**, 48–50; (h) Neumann, R. and Dahan, M. (1997) *Nature*, **388**, 353–355; (i) Weinstock, I.A., Barbuzzi, E.M.G., Wemple, M.W., Cowan, J.J., Reiner, R.S., Sonnen, D.M., Heintz, R.A., Bond, J.S., and Hill, C.L. (2001) *Nature*, **414**, 191–195; (j) Busche, C., Vilà-Nadal, L., Yan, J., Miras, H.N., Long, D.-L., Georgiev, V.P., Asenov, A., Pedersen, R.H., Gadegaard, N., Mirza, M.M., Paul, D.J., Poblet, J.M., and Cronin, L. (2014) *Nature*, **515**, 545–549; (k) Shiddiq, M., Komijani, D., Duan, Y., Gaita-Ariño, A., Coronado, E., and Hill, S. (2016) *Nature*, **531**, 348–351; (l) Coronado, E., Galán-Mascarós, J.R., Gómez-García, C.J., and Laukhin, V. (2000) *Nature*, **408** 447–449.

31 Müller, A., Reuter, H., and Dillinger, S. (1995) *Angew. Chem., Int. Ed. Engl.*, **34**, 2328–2361.

32 Müller, A., Beckmann, E., Bögge, H., Schmidtmann, M., and Dress, A. (2002) *Angew. Chem., Int. Ed.*, **47**, 1162–1167.

33 Lindqvist, I. (1952) *Acta Crystallogr.*, **5**, 667.

34 (a) Keggin, F.J. (1933) *Nature*, **131**, 908; (b) Keggin, F.J. (1934) *Proc. R. Soc. Lond.*, **A144**, 75; (c) Wells, A.F. (1945) *Structural Inorganic Chemistry*, Oxford University, Oxford; (d) Dawson, B. (1953) *Acta Crystallogr.*, **6**, 113.

35 (a) Preyssler, C. (1970) *Bull. Soc. Chim. Fr.*, 30; (b) Alizadeh, M.H., Harmalker, S.P., Jeannin, Y., Martin-Frère, J., and Pope, M.T. (1985) *J. Am. Chem. Soc.*, **107**, 2662–2669.

36 (a) Müller, A., Bögge, H. and Diemann, E. (2003) *Inorg. Chem. Commun.*, **6**, 52–53; (b) Corrigendum: Müller, A., Bögge, H. and Diemann, E., (2003) *Inorg. Chem. Commun.*, **6**, 329; (c) Garcia-Ratés, M., Miró, P., Poblet, J.M., Bo, C., and Avalo, J.B., (2011) *J. Phys. Chem. B*, **115**, 5980–5992.

37 (a) Long, D.-L., Tsunashima, R., and Cronin, L. (2010) *Angew. Chem., Int. Ed.*, **49**, 1736–1758; (b) Long, D.-L. and Cronin, L. (2006) *Chem. Eur. J.*, **12**, 3698–3706.

38 (a) Cronin, L. (2004) *High Nuclearity Clusters: Iso and Heteropolyoxoanions and Relatives: Comprehensive Coordination Chemistry II*, vol. **7** (eds J.A. McCleverty and T.J. Meyer), Elsevier, Amsterdam, pp. 1–56; (b) Long, D-.L., Kögerler, P., Farrugia, L.J., and Cronin, L., (2003) *Angew. Chem., Int. Ed.*, **42**, 4180–4183; (c) Miras, H.N., Yan, J., Long, D.-L., and Cronin, L., (2008) *Angew. Chem., Int. Ed.*, **47**, 8420–8423.

39 Long, D.-L., Kögerler, P., and Cronin, L. (2004) *Angew. Chem., Int. Ed.*, **43**, 1817–1820.

40 Müller, A., Krickemeyer, E., Meyer, J., Bögge, H., Peters, F., Plass, W., Diemann, E., Nonnenbruch, F., Randerath, M., and Menke, C. (1995) *Angew. Chem., Int. Ed. Engl.*, **34**, 2122–2124.

41 Müller, A., Krickemeyer, E., Bögge, H., Schimidtmann, M., and Peters, F. (1998) *Angew. Chem., Int. Ed.*, **37**, 3359–3363.

42 (a) Symes, M.D., Kitson, P.J., Yan, J., Richmond, C.J., Cooper, G.J.T., Bowman, R.W., Vilbrandt, T., and Cronin, L. (2012) *Nat. Chem.*, **4**, 349–354; (b) Kitson, P.J., Marshall, R.J., Long, D.–L., Forgan, R.S., and Cronin, L., (2014) *Angew. Chem., Int. Ed.*, **53**, 12723–12728; (c) de la Oliva, A.R.,

Sans, V., Miras, H.N., Yan, J., Zang, H., Richmond, C.J., Long, D.-L., and Cronin, L., (2012) *Angew. Chem., Int. Ed.* **51**, 12759–12762.

43 (a) Vilà-Nadal, L., Rodríguez-Fortea, A., Yan, L.K., Wilson, E.F., Cronin, L., and Poblet, J.M. (2009) *Angew. Chem., Int. Ed.*, **48**, 5452–5456; (b) Vilà-Nadal, L., Rodríguez-Fortea, A., and Poblet, J.M. (2009) *Eur. J. Inorg. Chem.*, **2009**, 5125–5133; (c) Vilà-Nadal, L., Wilson, E.F., Miras, H.N., Rodríguez-Fortea, A., Cronin, L., and Poblet, J.M. (2011) *Inorg. Chem.*, **50**, 7811–7819; (d) Rodríguez-Fortea, A., Vilà-Nadal, L., and Poblet, J.M. (2008) *Inorg. Chem.*, **50**, 7745–7750.

44 Lang, Z.-L., Guan, W., Yan, L.-K., Wen, S.-Z., Su, Z.-M., and Haob, L.-Z. (2012) *Dalton Trans.*, **41**, 11361–11368.

45 Vilà-Nadal, L., Mitchell, S.G., Rodríguez-Fortea, A., Miras, H.N., Cronin, L., and Poblet, J.M. (2011) *Phys. Chem. Chem. Phys.*, **13**, 20136–20145.

46 Lang, Z.-L., Guan, W., Wub, Z.-J., Yan, L.-K., and Su, Z.-M. (2012) *Comput. Theor. Chem.*, **999**, 66–73.

47 (a) Strickland, J.D.H. (1952) *J. Am. Chem. Soc.*, **74**, 862–867; (b) Strickland, J.D.H. (1952) *J. Am. Chem. Soc.*, **74**, 872–876; (c) Strickland, J.D.H. (1952) *J. Am. Chem. Soc.*, **74**, 868–871.

48 (a) Baker, L.C.W. and Figgis, J.S. (1970) *J. Am. Chem. Soc.*, **92**, 3794–3797; (b) Hervé, G. and Tézé, A. (1977) *Inorg. Chem.*, **16**, 2115–2117.

49 López, X. and Poblet, J.M. (2004) *Inorg. Chem.*, **43**, 6863–6865.

50 Zhang, F.-Q., Guan, W., Yan, L.-K., Zhang, Y.-T., Xu, M.-T., Hayfron-Benjamin, E., and Su, Z.-M. (2011) *Inorg. Chem.*, **50**, 4967–4977.

51 (a) Fleming, C., Long, D.-L., McMillan, N., Johnston, J., Bovet, N., Dhanak, V., Gadegaard, N., Kögerler, P., Cronin, L., and Kadodwala, M. (2008) *Nat. Nanotechnol.*, **3**, 229–233; (b) Fay, N., Bond, A.M., Baffert, C., Boas, J.F., Pilbrow, J.R., Long, D.-L., and Cronin, L., (2007) *Inorg. Chem.*, **46**, 3502–3510.

52 Long, D.-L., Song, Y.F., Wilson, E.F., Kögerler, P., Guo, S.X., Bond, A.M., Hargreaves, J.S.J., and Cronin, L. (2008) *Angew. Chem., Int. Ed.*, **47**, 4384–4387.

53 Long, D.-L., Kögerler, P., Parenty, A.D.C., Fielden, J., and Cronin, L. (2006) *Angew. Chem., Int. Ed.*, **45**, 4796–4798.

54 Vilà-Nadal, L., Peuntinger, K., Busche, C., Yan, J., Lüders, D., Long, D.-L., Poblet, J.M., Guldi, D.M., and Cronin, L. (2013) *Angew. Chem., Int. Ed.*, **52**, 9695–9699.

55 Yan, J., Long, D.-L., Wilson, E.F., and Cronin, L. (2009) *Angew. Chem., Int. Ed.*, **48**, 4376–4380.

56 Vilà-Nadal, L., Mitchell, S.G., Long, D.-L., Rodríguez-Fortea, A., López, X., Poblet, J.M., and Cronin, L. (2012) *Dalton Trans.*, **41**, 2264–2271.

57 Cameron, J.M., Vilà-Nadal, L., Winter, R.S., Iijima, F., Murillo, J.C., Rodríguez-Fortea, A., Oshio, H., Poblet, J.M., and Cronin, L. (2016) *J. Am. Chem. Soc.*, **138**, 8765–8773.

58 Kondinski, A., Heine, T., and Monakhov, K.Y. (2016) *Inorg. Chem.*, **55**, 3777–3788.

59 (a) Contant, R. (1990) *Inorganic. Syntheses*, vol. **27**, John Wiley & Sons, New York, 109; (b) Contant, R. and Ciabrini, J.P., (1977) *J. Chem. Res. Miniprint*, 2601–2617; (c) Contant, R. and Ciabrini, J.P., (1977) *J. Chem. Res. Synop.*, 222; (d) Knoth, W.H. and Harlow, R.L., (1981) *J. Am. Chem. Soc.*, **103**, 1865–1867; (e) Canny, J., Tézé, A., Thouvenot, R., and Hervé, G., (1986) *Inorg. Chem.*, **25**, 2114–2119; (f) Kirby, J.F. and Baker, L.C.W., (1998) *Inorg. Chem.*, **37**, 5537–5543.

60 Yan, J., Long, D.-L., Miras, H.N., and Cronin, L. (2010) *Inorg. Chem.*, **49**, 1819–1825.

61 Miras, H.N., Stone, D.J., McInnes, E.J.L., Raptis, R.G., Baran, P., Chilas, G.I., Sigalas, M.P., Kabanos, T.A., and Cronin, L. (2008) *Chem. Commun.*, 4703–4705.

62 Miras, H.N., Ochoa, M.N.C., Long, D.-L., and Cronin, L. (2010) *Chem. Commun.*, **46**, 8148–8150.

63 Sartzi, C., Miras, H.N., Vilà-Nadal, L., Long, D.-L., and Cronin, L. (2015) *Angew. Chem., Int. Ed.*, **54**, 15708–15712.

64 Miras, H.N., Sorus, M., Hawkett, J., Sells, D.O., McInnes, E.J.L., and Cronin, L. (2012) *J. Am. Chem. Soc.*, **134**, 6980–6983.

65 Ritchie, C., Streb, C., Thiel, J., Mitchell, S.G., Miras, H.N., Long, D.-L., Peacock, R.D., McGlone, T., and Cronin, L. (2008) *Angew. Chem., Int. Ed.*, **47**, 6881–6884.

66 (a) Hill, C.L. and Prosser-McCartha, C.M. (1995) *Coord. Chem. Rev.*, **143**, 407–455; (b) Dolbecq, A., Dumas, E., Mayer, C.R., and Mialane, P., (2010) *Chem. Rev.*, **110**, 6009–6048.

67 Miras, H.N., Cooper, G.J.T., Long, D.-L., Bogge, H., Müller, A., Streb, C., and Cronin, L. (2010) *Science*, **327**, 72–74.

68 Miras, H.N., Richmond, C.J., Long, D.-L., and Cronin, L. (2012) *J. Am. Chem. Soc.*, **134**, 3816–3824.

69 Ibrahim, M., Mal, S.S., Bassil, B.S., Banerjee, A., and Kortz, U. (2011) *Inorg. Chem.*, **50**, 956–960.

70 Takashima, Y., Miras, H.N., Glatzel, S., and Cronin, L. (2016) *Chem. Commun.*, **52**, 7794–7797.

2
Inorganic Nanotubes and Fullerene–Like Nanoparticles from Layered (2D) Compounds

L. Yadgarov,[1] R. Popovitz-Biro,[2] and R. Tenne[1]

[1]Weizmann Institute, Department of Materials and Interfaces, Herzl St 234, Rehovot 76100, Israel
[2]Weizmann Institute, Department of Chemical Research Support, Herzl St 234, Rehovot 76100, Israel

2.1
Introduction

Inorganic nanotubes are investigated for more than 80 years now. Pauling [1] proposed that magnesium-silicates, like chrysotile (see Figure 2.1) and alumino-silicate, like kaolinite ($Al_2Si_2O_5(OH)_4$) with asymmetric structure along the c-axis, will bend and fold due to the tensile strain imposed on the outer silica tetrahedra layer and the compression strain imposed on the inner layer of the alumina octahedra. Chrysotile nanotubes (NTs) were first observed using transmission electron microscopy (TEM) by Bates *et al.* [2]. For the same reason, Pauling [1] suggested that spontaneous folding of layered compounds with symmetric structure along the c-axis, like molybdenite (MoS_2), cadmium chloride ($CdCl_2$), and others is not to be expected. Kroto *et al.* [3] found that C_{60} forms hollow closed nanostructures, that is, fullerenes, spontaneously. It was reasoned that flat graphite nanoclusters suffer from edge effect and consequently cannot tolerate the large chemical energy stored in the dangling bonds of the rim atoms of the small graphene sheet [4]. In the bulk graphite crystals, each carbon atom is bonded to three neighboring carbon atoms (sp^2 bond). However, the rim atoms of graphite are only twofold bonded. The number of surface atoms in the bulk is out-numbered by the threefold bonded "bulk" atoms and the relative energy stored in the rim atoms is therefore negligible. However, this picture does not hold in the nanosize regime where twofold bonded rim atoms are abundant. Consequently at high temperatures, graphite nanocrystals spontaneously reorganize into hollow-cage seamless structures (fullerene) made of 60 carbon atoms (C_{60}) with 20 hexagons and 12 pentagons. Here, all carbon atoms are threefold bonded via a distorted sp^2 bond. The elastic energy produced by the distorted sp^2 bonds is thus more compensated by the seaming of the dangling

---

Handbook of Solid State Chemistry, First Edition. Edited by Richard Dronskowski, Shinichi Kikkawa, and Andreas Stein.
© 2017 Wiley-VCH Verlag GmbH & Co. KGaA. Published 2017 by Wiley-VCH Verlag GmbH & Co. KGaA.

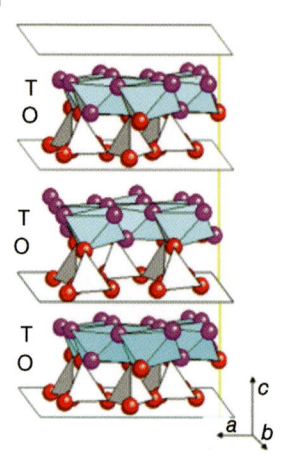

Figure 2.1 Chrysotile with asymmetric structure along the c-axis.

bonds in the closed-cage fullerenes. By introducing 12 disjoint pentagons to the otherwise hexagonal network, folding of the graphene nanocluster can occur along two axes forming thereby the fullerenes. Soon afterwards Iijima [5] reasoned that the same rationale holds also to carbon nanotubes (CNT), which are a stable form of graphene nanoribbons. Here though the graphitic sheets are folded along a single axis. The CNTs are closed from both ends by a hemispherical fullerene containing six pentagons, each.

Layered materials, such as WS_2 and MoS_2, were investigated intensively over the past 50 years due largely to their interesting anisotropic properties and numerous applications. Pauling was the first to realize that in analogy to graphite MoS_2 is a layered structure material with a 2H unit cell (2H stands for a unit cell consisting of two layers with hexagonal symmetry) [6]. Here each molybdenum layer is sandwiched between two layers of sulfur atoms in trigonal prismatic structure (see Figure 2.2). The layers are stacked together via weak van der Waals forces. While being sixfold bonded within the layer, Mo rim atoms are only fourfold bonded. Similarly, rim sulfur atoms are twofold bonded instead of being threefold bonded as in bulk MoS_2. This analogy led us to propose in 1992 [7] that the chemical energy stored in the dangling bonds of the rim atoms predominates in nanoplatelets of MoS_2 (WS_2). This extra stored energy leads to their folding and seaming into closed-cage (hollow) nanostructures that were designated as inorganic fullerene-like (IF) nanoparticles and inorganic nanotubes (INT).

The studies of the IF/INT led to discovery of various fascinating properties and suggested many potential applications in nanoelectronics, sensors, tribology, and so on. For instance, it was demonstrated that the tribological performance of IF-MoS_2 nanoparticles are superior compared to the corresponding bulk structures [8,9]. The bulk 2H-MoS_2 platelets are commonly used as a solid lubricant additive for oils. These layered structures are lubricious since the layers are able

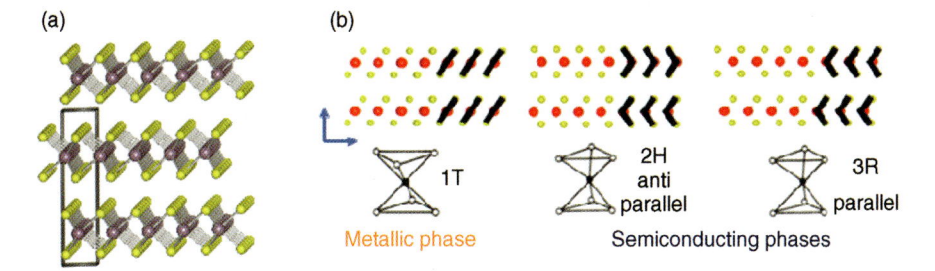

Figure 2.2 Schematic representation of (a) MoS$_2$ triatomic layers with hexagonal symmetry – 2H unit cell (b) 1T, 2H, and 3R stacking in c-direction. Red (or purple) circles represent Mo atoms and yellow represents S atoms. The layers are stacked together via weak van der Waals forces. While being sixfold bonded within the layer, Mo rim atoms are only fourfold bonded. Similarly, rim sulfur atoms are twofold bonded instead of being threefold-bonded as in bulk MoS$_2$.

to shear under load and form protective film on the surfaces in the contact [9,10]. The IF nanoparticle (NP) usually exhibits an outstanding tribological performance, particularly under the boundary-lubrication conditions. However, it should be noted that the tribological properties of these NP are highly influenced by their morphology structure, size, and test conditions [8,11–14].

Figure 2.3a shows a TEM micrograph of a multiwall IF-MoS$_2$ NP, which was synthesized by a high temperature reaction between MoO$_3$ and H$_2$S [15,16]. Each IF nanoparticle consists of a few MoS$_2$ closed concentric layers (walls) and have some structural defects. These NP are obtained in a pure phase under very narrow window of conditions. Typical nanoparticle of say 100 nm diameter

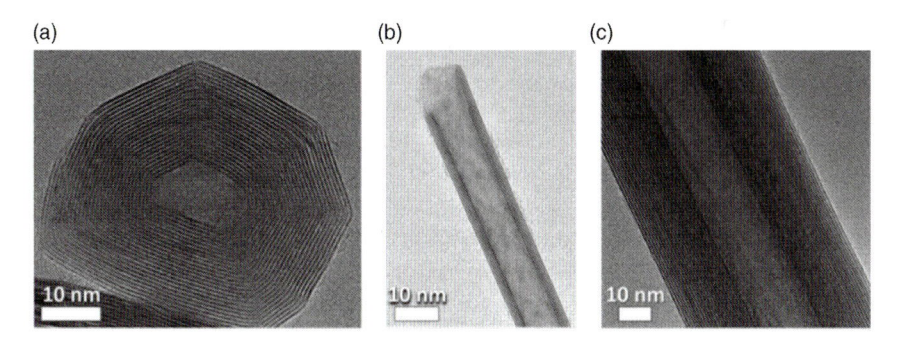

Figure 2.3 TEM micrographs of (a) IF-MoS$_2$ NP and (b and c) INT-WS$_2$. The INT presented in here were formed by two very different mechanisms: (b) Type I – Rosentsveig–Margolin mechanism [17] and (c) Type II – Rothschild–Zak mechanisms [18,19,20]. (Type I – formation of the tungsten suboxide NP, which is encapsulated by WS$_2$ followed by the breakthrough of the volatile phase through the encapsulating surface. Then the root of the nanotube is formed from the oxide protrusion on the top NP surface. During further growth of the INT, the volatile oxide that springs out of the nanotube's hollow core reacting with H$_2$S forms WS$_2$ NT at the tip of the tube. Type II – the growth of the tungsten suboxide nanowires followed by their rapid sulfidization).

contains more than a million atoms. The growth of such NP is self-limiting, that is, the IF does not grow indefinitely. Once the NP reaches the maximum size rather than growing further, a new one nucleates from the gas phase. The IF-MoS_2 seems to be stable almost indefinitely at room temperature.

The syntheses of multiwall INT-WS_2 and INT-MoS_2 were reported early in Refs [7,16] and they can be produced now as a pure phase in substantial amounts [18,21,22]. Large amounts of WS_2 nanotubes could be synthesized by sulfidization of WO_3 nanoparticles. Two kinds of growth modes for these INT were deciphered. Type I NTs are obtained spontaneously by the rapid solid–gas reaction of WO_{3-x} vapors and H_2S on the tip of the growing INT [23]. These INT-WS_2 are slender (5–10 walls and 20–40 nm in diameter) and are open-ended (Figure 2.3b). In an alternative mechanism, WO_{3-x} nanowhiskers are first obtained by the reaction of hydrogen with the oxide powder. The oxide nanowhiskers are subsequently converted into Type II INT-WS_2 via an outside-inwards reaction between the tungsten oxide nanowhisker and the H_2S gas [18,24]. Characteristically, these INT are thicker (20–30 walls; 50–150 nm in diameter), longer (up to 30 microns), and are often closed ended (Figure 2.3c). The elastic energy of folding of a single wall MoS_2 (WS_2) nanotube is about an order of magnitude larger than that of a single wall carbon nanotube of the same diameter [25]. Not surprisingly, therefore, inorganic layered compounds with bulky unit cells and interlayer separation of 5–7 Å prefer to form multiwall INT of appreciably larger diameter than their carbon analogues [26]. Here, the interlayer van der Waals interaction compensates for the excess bending energy of the MoS_2 layers. Single to triple wall NTs of compounds like WS_2 were nonetheless reported recently [27]. These INT were obtained under highly exergonic conditions, where the stabilizing van der Waals energy is of lesser importance. Under such circumstances the fast nanotube-forming reaction is carried out in a purely kinetic-controlled regime. Tight binding density functional theoretical calculations show that, in contrast to the bulk 2H-WS_2 polytype, single wall MoS_2 (WS_2) nanotubes with zigzag (n, 0) configuration possess a direct bandgap transition, suggesting that they can luminesce and offering them unique optical and electronic properties and potentially numerous applications [28]. An alternative strategy to produce single wall INT, of for example, MoO_3 [29] and MoS_2 [30] in a solution-based reaction was recently described. Single wall INT could reveal a host of interesting optoelectronic properties, which are entirely different from those of multiwall INTs and are briefly discussed in what follows.

Over the years, numerous strategies were developed for the synthesis of nanotubes from variety of inorganic layered compounds. These works were reviewed in great detail in the past, for example, Refs. [31–33], and will not be repeated here. Few recent developments are noteworthy, though. Ternary and quaternary misfit layered compounds (MLC) have been investigated quite intensively over the last few decades [34,35]. In particular, MLC of the type MX-TX_2, where M = Sn, Pb, Sb, Bi, Ln (Ln = Lanthanide atoms); T = Nb, Ta, Cr, V, and so on, and X = S, Se, Te were studied in quite detail. One way to visualize the MLC compounds is as a repeating sequence of (001) layer of the MX compound (O) with distorted rocksalt structure intercalated between two hexagonal TX_2 (*T*) layers. Thus, a superstructure with the sequence O–T or even more complex ones, like O–T–O–T–T are formed. The stability of such disparate structures is mostly attributed to van der Waals interactions. However, in other cases, like LnX-TX_2 MLC extra stability is gained by partial charge transfer from the Ln atom to the metal T atom. MLC NTs were recently synthesized and their structure was elucidated in some detail [36,37]. Large variety of NTs from chalcogenide MLC have been prepared in recent years. This important class of nanotubes are discussed in a greater detail in Section 2.2.1.

Another topic of great interest is the doping of the MS_2 (M = Mo, W) IF/INT and its effect on the physiochemical properties of such NPs. Careful control over the concentration of the dopant atoms and their position in the host lattice is mandatory in order to move the Fermi level up or down in the bandgap altering the physicochemical behavior of such nanostructures in a controllable fashion. Thus, a well-regulated doping of IF-MoS_2 with rhenium-atoms was recently demonstrated [38]. The Re atoms were shown to occupy a molybdenum substitutional site, that is, Re_{Mo} endowing the NPs extra negative charges at their surfaces. The Re concentration did not exceed 200 ppm, which provided $\approx 10^{16}$ cm^{-3} free carriers in the lattice (assuming only 1–2% of the Re atoms are ionized at room temperature). Further discussion of the effect of the doping on the properties of the NPs is reserved to Sections 2.2.2 and 2.3.2.

The optical, electrical, and mechanical properties of MS_2 (M = Mo, W) IF/INT have been investigated. The mechanical properties of such individual nanostructures have been studied for sometime [39,40] and are summarized in a recent review article [41]. As soon as the IF-WS_2 nanoparticles became available in substantial amounts, their tribological properties were investigated, demonstrating their usefulness as additives to variety of lubricating fluids [13]. More recently, the Re-doped IF-MoS_2 was shown to exhibit remarkable tribological characteristics [38], offering them numerous applications. The lubrication mechanism of the NPs was studied to a certain extent using variety of approaches. Most significantly, loading and shearing of individual IF/INT were carried out *in situ* within scanning electron microscopy (SEM), atomic force microscopy (AFM), and TEM; displaying the interplay of a few mechanisms in providing the superior lubrication behavior of such NPs [42]. This topic was discussed quite extensively before and will be discussed here very briefly.

The optical properties of WS_2 (MoS_2) nanotubes and fullerene-like NPs were investigated for some time now. First, the optical bandgap of the IF/INT was determined and was found, in accordance with the theoretical analysis [28], to be red-shifted compared to the bulk material [43]. Recently, great interest was directed toward light scattering by surface plasmon resonance (LSPR) of heavily doped semiconducting nanocrystals with electron carrier densities of $\sim 10^{19}$–10^{22} cm^{-3} [44,45]. IF-MoS_2 were recently been shown to exhibit a LSPR in the near-infrared region [46]. Remarkably, in addition to the plasmonic scattering, the IF NP also maintains some of the excitonic structure of the bulk counterparts. The recent progress in the understanding of the optical properties of the

IF will be presented in Section 2.3.1 of this review. High performance field effect transistors (FET) were fabricated using individual INT-WS_2 [42,47]. These together with other studies reporting the electrical properties of INT are discussed in Section 2.3.3.2.

Numerous recent publications in this field are concerned with potential applications of the IF/INT. Early on, the superior tribological properties of IF/INT attracted a great deal of attention. These studies were taken step forward by licensing the technology to "N.I.S., Inc." and its sister company "NanoMaterials, Ltd" (www.nisusacorp.com and www.apnano.com, respectively). Commercial exploitation of this technology required major efforts to scale-up the IF-WS_2 and to a lesser extent the INT NPs production and developing various formulated fluid lubricants. This effort paid-off and the two companies launched a large series of products based on these solid nanolubricants under the trademark name "NanoLub." N.I.S., Inc. and "NanoMaterials, Ltd" are targeting markets, like the automotive industry, trains and shipping, power generation, and mining. Among their recent achievements are a series of metal working fluids that provide improved finishing to various metal surfaces at saved manufacturing costs. Expanding markets for these products promises to make "NanoLub" based lubricants an industrial commodity in the foreseeable future. More recently, much attention was paid to the potential applications of such NPs for reinforcing polymer nanocomposites, which was the subject of several recent review articles [48,49].

Recent studies include reinforced biocomposites and bioceramic materials, reinforced Mg-alloys and concrete, and so on. The reinforcing mechanism and in particular the increase in the fracture toughness of different nanocomposites based on these nanoparticle is the focus of several studies. Perhaps, the most remarkable phenomenon in this connection is that, in contrast to CNT, IF/INT can be dispersed in the nanocomposite matrix relatively easily. Another interesting observation in this connection is the diminution in the fluid viscosity upon adding the IF/INT. Further discussion of the large body of potential applications is given in Section 2.4. Related to their biomedical applications is the issue of biotoxicity and biocompatibility of the IF/INT, which is also discussed in Section 2.4. Concluding remarks and future prospects of these, relatively new brand of nanomaterials is provided in Section 2.5.

2.2
Recent Developments in Synthetic Methods

2.2.1
Nanotubes from Misfit Layer Compounds (MLC)

Misfit layer chalcogenides of the formula $(MX)_{1+y}(TX_2)_m$ denoted hereafter MX-TX_2, with (M = Sn, Pb, Bi, Sb, rare earth; T = Sn, Ti, V, Cr, Nb, Ta; X = S, Se; $0.08 < y < 0.28$; $m = 1$–3 have been extensively studied and reviewed [34,50].

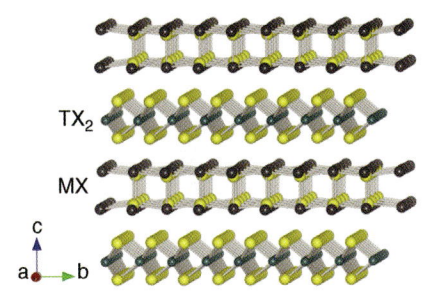

Figure 2.4 Schematic representation of the structure of the $(MX)_{1-y}(TX_2)$ misfit compound. Green, gray, and yellow circles are T, M, and X atoms respectively.

In general, the MLC have a planar composite structure, composed of two alternating layered subsystems, namely, MX and TX_2. The MX slab consists of a two-atom-thick slice of a distorted (structurally modulated) rock-salt structure. The TX_2 slab consists of a three-atom-thick sandwich, in which the transition metal is surrounded by chalcogen atoms in a trigonal prismatic or octahedral coordination. Layers of the two subsystems are stacked alternately or in a more complex sequence one atop the other, along a common c-axis that is perpendicular to the layers. The two subsystems have in general one common axis (b) and differ usually in one of their in-plane lattice (a) parameters and so they are incommensurate along that direction (see Figure 2.4). The MLC superstructure gains extra stability from the difference of the work function of the two components which leads to some degree of charge transfer between the MX and TX_2 slabs. Therefore, one can consider misfit compounds as being intercalation compounds of stage one ($n=1$), where an MX monolayer is intercalated into the galleries between each two TX_2 layers of a compound with hexagonal layered structure (O–T' superstructure). Generally speaking, the density of free electrons is higher in the MX layer, making it the natural donor and the TX_2 slab becomes the acceptor.

Nonplanar MLC (M = Pb, Bi; T = Nb) structures such as microcrystals of cylindrical and conical scrolls were obtained by vapor transport from stoichiometric mixtures of the elements [51]. Observation of the growth nuclei suggested a general mechanism, based on the tendency of curling-up of thin sheets of MLC, which stems from the difference in lattice parameters of the two constituents. Upon bending, this difference becomes smaller and thus the strain energy is reduced.

In the nanoregime, laser ablation of SnS_2 resulted in condensation of sulfur deficient nested IF nanoparticles of spherical and polyhedral shapes and short NTs [52]. HRTEM analysis of the nanostructures showed superstructures of periodically stacked SnS and SnS_2 layers, reminiscent of MLC. The misfit between the two sublattices is considered as a new stimulus for the folding of the layers, resulting in IF and INT structures. This driving force adds to the already established mechanism that is based on the annihilation of dangling bonds at the periphery of the 2D layers. Recently, unprecedented IF closed-cage nanostructures of PbS-SnS$_2$ MLC were synthesized by irradiating a mixture of Pb, SnS$_2$, and graphite with highly concentrated sunlight [53]. The reaction occurs in a kinetically controlled, inside-out layer-by-layer growth mechanism, where the large temperature gradient favors nucleation and growth from a small-condensed core or slab of PbS-SnS$_2$. Under these extremely high temperature conditions, thermal fluctuations induce enhanced bending, shearing, and folding of the molecular sheets, thus stimulating the formation of IF with rather small radii of curvatures. Surprisingly, calculation of the interface energies in these misfit superstructures revealed small amount of charge transfer from SnS$_2$ to PbS, which leads to extra stability due to the interlayer polarization forces.

In contrast, under thermodynamically controlled conditions, high temperature vapor transport of Sn–S species, in the presence of bismuth and antimony sulfide catalysts, yielded large amounts of SnS–SnS$_2$ nanoscrolls and NTs with various internal superstructures of the MLC type [36,54,55]. In this system, misfit occurs along two in-plane directions, which leads to different in-plane orientations between the two subsystems and so, different folding vectors and chiralities. Also, there is diversity in the stacking along the common c-axis, like O–T and O–T–O–T–T resulting in various stoichiometries. A recent theoretical study performed with density functional-based tight binding (DFTB) calculations clearly indicated a small charge transfer from the SnS$_2$ slab to the SnS layer, which leads to additional stabilization of the MLC nanostructures [56]. Under similar, two-step high-temperature annealing process, tubular and conical monocrystalline (PbS)$_{1.14}$NbS$_2$ structures, were also obtained. The NbS$_2$ and PbS layers were found to be stacked in an alternating fashion and have a single in-plane orientation with two folding vectors. Furthermore, a series of Ta-based nano- and microtubes of the MLCs type MS-TaS$_2$ (M = Pb, Sn, Sb, Bi) could be obtained, by chemical vapor transport growth technique, using chlorine as transport reagent [57]. The typically encountered outer diameter of these tubes, seem to be larger than their PbS–NbS$_2$ and SnS–SnS$_2$ counterparts, which may arise from higher "rigidity" of the misfit slab.

Interestingly, TX$_2$ compounds that are unstable in bulk phase can be stabilized upon stacking in MLC type MX-TX$_2$ structures, forming interface-modulated monolayers. For example, CrS$_2$ is a metastable structure which does not exist in the bulk form, but forms stable pseudohexagonal layers when stacked together with LaS, alternately, in (LaS)$_{1.2}$CrS$_2$ MLC [58]. This phenomenon is reminiscent of the stable Li intercalated CrS$_2$ compound [59]. A series of such lanthanide-based misfit layered NTs was recently prepared in high yields, via a simple chemical method. This synthesis comprises annealing a mixture of the corresponding oxides at high temperature, in the presence of H$_2$/H$_2$S gases [60]. The growth mechanism involves generation of strain during oxide to sulfide conversion, which is then released by growth of sulfide nanosheets perpendicular to the oxide substrate. Further scrolling of the growing nanosheets is induced by the misfit-strain between the two sulfide-sublattices. Whereas in MLC composed of two stable constituents, van der Waals forces play the major role in "sticking"

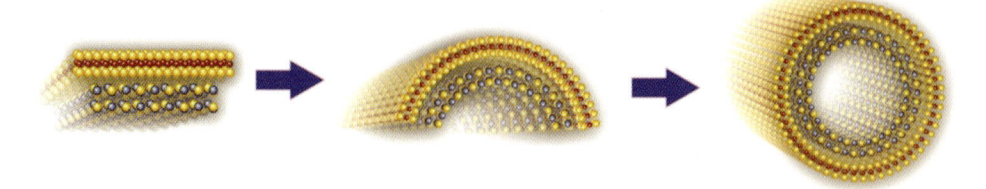

Figure 2.5 The formation mechanism of the MLC-NTs can be described through three main steps. (1) In the first stage nanoclusters of the MLC are self-assembled in the gas or in a condensed phase. (2) Partial relaxation of the misfit strain between the MX (O) layer with the distorted rocksalt structure and the hexagonal TX_2 (T) layer leads to folding of the MLC nanoclusters. (3) The dangling bonds at the rim of the MLC planes are healed by closure into seamless NTs.

the layers together, here the stability is gained mostly through charge transfer from the LaS slab to the CrS_2 layer. This charge transfer results in strong interaction as evident from the smaller interlayer spacing along the c-axis [37].

Nanotubular structures from a new family of misfit compounds LnS–TaS_2 (Ln = La, Ce, Nd, Ho, Er) and LaSe–$TaSe_2$ were recently synthesized by the chemical vapor transport method, using TaCl or TaBr as transport reagent [61]. In most of the NTs of this series, the classical behavior of one common b-axis coinciding with the tubule axis was observed, as shown schematically in Figure 2.4. The formation mechanism of the MLC-NTs can be described through three main steps (see Figure 2.5): (1) in the first stage nanoclusters of the MLC are self-assembled in the gas or in a condensed phase; (2) partial relaxation of the misfit strain between the MX (O) layer with the distorted rocksalt structure and the hexagonal TX_2 (T) layer leads to folding of the MLC nanoclusters; (3) the dangling bonds at the rim of the MLC planes are healed by closure into seamless NTs. In fact, this three-step mechanism resembles that of the asymmetric oxide NTs, like chrysotile and kaolinite, as discussed above [1,2].

Figure 2.6 displays a TEM and SEM image and selected area electron diffraction (SAED) of a LaS–TaS_2 nanotube. Six pairs of spots were assigned to the (11.0) and (10.0) planes of TaS_2, which is equal to the multiplicity factor of these planes and thus, indicating a single folding vector for the TaS_2 layers. Similarly, one folding vector was established for the LaS layers, since four pairs of spots were observed for the (110) and (220) reflections of LaS layer, which is equal to the multiplicity factor of these reflections. The chiral angles of the NTs could be determined as half of the azimuthal splitting of the $hk.0$ reflections of TaS_2 and $hk0$ reflections of LaS. Two pairs of LaS 020 spots match the 10.0 spots of TaS_2 and are parallel to the tubule axis. These coincident spots reveal the presence of a common commensurate in-plane b direction that coincides with the nanotube axis. The LaS and TaS_2 are stacked in an alternating O–T sequence along their common c-axis with a periodicity of 1.15 nm. The relatively small c-axis value in these MLC NTs is rationalized in terms of the greater stability of Ln(III) state compared to the respective II state. Thus, when intercalated between two

Figure 2.6 (a) TEM and (b) SEM images of LaS–TaS$_2$ tubular crystals with the LnS and TaS$_2$ layers stacked periodically. (a) Top: High magnification images, with medium and low magnification images shown as insets. Middle: SAED patterns acquired from the areas shown in the upper images. Spots corresponding to the same interplanar spacings are marked by large segmented ellipses or circles (red for TaS$_2$ and green for LnS) and the respective Miller indices are indicated. The tubule axes are marked by purple double arrows. Basal reflections are marked by small blue arrows. Chiral angles of 3.1° for the tubules were determined from the splitting of the spots, as discussed in the text. Bottom: Line profiles perpendicular to the tubules axes integrated along the rectangles marked in the upper images. (From Refs [37,61].)

hexagonal TaS_2 layers, a strong tendency for charge transfer from LnS to the TaS_2 layers is anticipated. Presumably, the charge transfer between the two sublattices is also responsible for the abundance of LnS-TaS_2 tubules having primarily a single folding vector. The strong interaction between the two subsystems is supported also by a high up-shift in the E_{2g} Raman mode of TaS_2 in these compounds [61]. Furthermore, density-functional tight-binding method was used to calculate the electronic structure of bulk LaS–TaS_2 revealing increased electron density on the Ta atoms and a decrease on the La atoms when compared to the isolated layers.

2.2.2
Doping of Inorganic Nanotubes and Fullerene-Like Nanoparticles

Substitutional doping of semiconductor nanocrystals and nanowires with a small amount of foreign atoms plays a major role in controlling their electrical, optical, and magnetic properties [62–64]. For instance, VO_x NTs were transformed from spin-frustrated semiconductors to ferrimagnets by doping with either holes/electrons [65]. Doped and alloyed NT have been reported for specific cases, such as Ti-doped INT-MoS_2; Mo-, C-, and Nb-doped INT-WS_2 [66–69] and INT-(W/Mo)S_2 [70].

Calculations indicated that n- and p-type doping of multiwall INT-MoS_2 could be accomplished by substituting the Mo lattice atoms with Re [66,71] or Nb [72], respectively. Also, Re was shown to replace the Mo atoms in the lattice of MoS_2 crystals (<0.1 at%), stabilizing the 3R polytype and serving as n-type donor [73,74]. Substituting sulfur by halogen atoms was also shown to produce n-type conductivity [75]. High concentrations of dopant (>1 at%), usually results in alloying and clustering of the dopant atoms leading to a poorly controlled electronic properties. [66,71,76] Notwithstanding this fact, many works refer to high concentrations of foreign atom (alloying) in the lattice as doping. Recently, Re-doped IF/INT were successfully synthesized using both *in situ* and a posteriori method [38]. The formal Re concentration in these nanostructures varied from 0.02 to 0.7 at%.

For the *in situ* synthesis of Re-doped IF-MoS_2 (Re:IF-MoS_2), a precursor $Re_xMo_{1-x}O_3$ ($x<0.01$) powder was prepared in an auxiliary reactor. Subsequently, evaporation of this powder was executed along a temperature gradient (770–820 °C) in a reducing atmosphere produced by a mixture of $N_2/H_2/H_2S$ gases, leading to MoO_{3-x}@MoS_2 core–shell NPs [38,77]. To complete this oxide to sulfide conversion a long (25–35 h) annealing process at 870 °C in the presence of H_2S and forming gas (H_2-10 wt%; N_2) was performed. At the end of this diffusion-controlled process a powder of Re:IF-MoS_2 NP with dopant concentration below 0.1 at% was obtained.

A posteriori doping of IF-WS_2 and INT-WS_2 was carried out by chemical vapor transport (CVT) in an evacuated quartz ampoule containing also ReO_3, or $ReCl_3$ and iodine [74]. In the case of $ReCl_3$, both the rhenium and the chlorine atoms (substitutional to sulfur atoms) served as n-type dopants. A temperature difference of 150 °C between the cold and the hot zone of the ampoule was used. The INT-WS$_2$ and IF-WS$_2$, which served as precursors for the doping reaction, were placed in the cold zone of the ampoule. They were prepared according to a procedure described in Refs. [24,78].

TEM and SEM analysis of the Re: IF-MoS$_2$, Re: IF-WS$_2$, and Re: INT-WS$_2$ revealed that their structure is very similar to their undoped counterparts that are shown in Figure 2.3. No impurity, like oxides, or platelets (2H) of MoS$_2$ could be found in the product powder of Re: IF-MoS$_2$. The line profile and the fast Fourier transform analysis show an interlayer spacing of 0.627 nm. Also, high resolution TEM (HRTEM) did not reveal any structural changes of the IF even for the samples with high Re concentration (0.71 at%). According to the TEM/SEM and X-ray diffraction (XRD) analyses there is no significant distinction between the posteriorly doped Re (Cl) INT/IF-WS$_2$ and undoped nanostructures. However, using HRTEM analysis it was discovered that the substitutional Re doping of the INT may induce partial 2H to 1T transformation [79]. This phase transition was known formerly only for WS$_2$ and MoS$_2$ that were intercalated by alkali metals [80–82]. To elucidate this phenomenon, DFTB calculations were applied for the related MoS$_2$ compound. It was found that the occurrence of a 1T-phase within the INT is stimulated by substitutional doping of Re atoms, which destabilizes the 2H-polytype. Explicitly, the Re impurity atoms within the lattice of MS$_2$ (M = Mo, W) serve as electron donors. Energy considerations suggest that, once Re atoms are placed within the MS$_2$ layer of the INT, there is no essential preference between 1T and 2H polytype. Obviously, the influence of the donor Re atoms is limited to their close vicinity and the larger is their concentration the larger become the 1T domains in the pristine 2H lattice.

Since the rhenium concentration in the IF/INT is less than 1 at%, many techniques, like XRD, X-ray photoelectron spectroscopy (XPS) and energy dispersive X-ray spectroscopy (EDS) are not suitable to determine its (Re) concentration quantitatively. However, the Re concentration in the IF/INT could be accurately determined by inductively coupled plasma mass spectrometry (ICP-MS). The Re level varied between approximately 0.01 and 0.5 at% (100–5000 ppm) depending on the loading of the dopant, temperature, and duration of the process. The doping level of the Re: IF-MoS$_2$, the Re level was found to be lower (by third) compared to the initial ReO$_3$ content of the oxide powder in the precursor. In a typical nanoparticle consisting of about $5-7 \times 10^5$ Mo atoms with Re concentration of ~0.03 at%, there are on the average about 200–300 Re atoms. Substitutional model for the Re-doped IF/INT was confirmed by use of X-ray absorption fine structure (XAFS) measurements [38]. Namely, using this technique, the dopant locations were determined by checking the local structure around the Re atoms. Furthermore, high-angle annular dark field in scanning TEM (HAADF – STEM) in aberration-corrected HRTEM was utilized to confirm the substitutional Re$_{Mo}$ site occupied by the rhenium atoms in the IF-MoS$_2$ lattice [83]. In conclusion, a reproducible and well-controlled process for Re-doping of IF-MoS$_2$ and INT-WS$_2$ was developed.

2.2.3
Core–Shell Inorganic Nanotube Superstructures

The long inner hollow cavity of NTs and their capillary forces were used to drive internal wetting and filling of CNT with liquids [84]. Salt encapsulation was shown to result in profound structural changes of the enclosed material relative to the bulk. Following these works, multiwall INT-WS$_2$ were used as host templates for the synthesis of core–shell nanotubular structures [85]. The relative large internal diameter (about 10 nm) of these NTs core and its wetting by the molten salt results in a conformal folding of the guest layers on the interior of the host template, thus leading to formation of rather defect-free core–shell inorganic nanotubular superstructures. Annealing of INT-WS$_2$ together with PbI$_2$ (or BiI$_3$), which is also a layered material with CdI$_2$ structure, at 500 °C for about one month, resulted in the formation of core–shell PbI$_2$@WS$_2$ (or BiI$_3$@WS$_2$) nanotubular superstructures. The encapsulated PbI$_2$ (or BiI$_3$) layers cover the inner walls of the template host NTs, as shown in the HRTEM image (Figure 2.7). An analogous result can be achieved while irradiating *in situ* in the TEM a powder of a layered compound with low melting temperature, such as SbI$_3$, in the presence of stable INT-WS$_2$. The growth mechanism proposed for the formation of such core–shell NTs is consistent with theoretical calculations [86]. Capillary imbibition of the molten halide into the hollow core of the host INT results in formation of a thin wetting layer, which may crystallize to form the inner tubular layers, upon cooling. The molten-salt was shown to have strong van der Waals interaction with the interior of the INT and thus permitting good surface wetting, which is also evident from its concave meniscus.

A different approach to achieve core–shell inorganic nanotubes is via a gas phase chemical synthesis, where the layered product employs the inner or outer surfaces of INT-WS$_2$ as a nucleation site. For example, conformal coating of MoS$_2$ layers atop template INT-WS$_2$ leads to formation of WS$_2$@MoS$_2$ core–shell NTs in a quasiepitaxial growth mode.

2.2.4
Single-to-Triple Wall Inorganic Nanotubes

The outer diameter of multiwall INT such as WS$_2$ (MoS$_2$) synthesized under high temperature conditions (>800 °C), varies from 20 to 40 nm and are made up of at least 5–10 layers. These NTs grow spontaneously from the vapor phase by sulfurizing WO$_3$ nanoparticles and are termed Type I. This size range agrees well with theoretical calculations that compared the stability range of NTs with that of nanostripes, having the same number of atoms [26]. These calculations indicate that single-to-triple layer NTs are metastable. Presumably, their synthesis would require highly exergonic conditions to drive the reactions far enough from equilibrium. Under these circumstances, the weaker van der Waals interaction (about 20 meV/atom) is less relevant and hence the energy difference between multiwall and few layers NTs diminishes. Indeed, 1–3 layer NTs with a

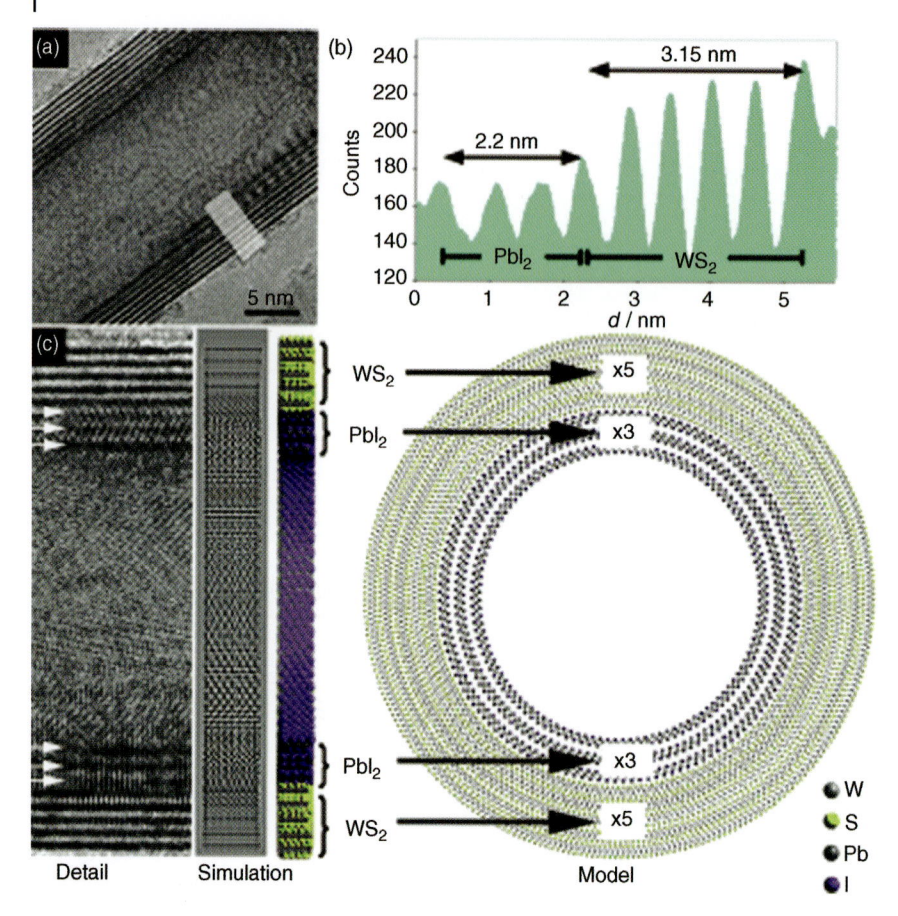

Figure 2.7 (a) HRTEM micrograph showing a core–shell $PbI_2@WS_2$ composite nanotube. (b) Line profile obtained from the indicated region in (a) showing two types of nanotube layers: five outer WS_2 layers with sharper contrast and an average spacing of 0.63 nm and three inner layers with more complex contrast and an average spacing of 0.73 nm, corresponding to three concentric PbI_2 nanotubes. (c) Detail from (a) showing the complex contrast of the inner PbI_2 layers (arrowed) relative to the outer WS_2 layers. To the right of the detail is a simulation and a cutaway space-filling model (left) and cross-sectional structure model (right) with both WS_2 (*aba* stacking) and PbI_2 layers (*abc* stacking) indicated. (From Ref. [85].)

diameter of 3–7 nm and a length of 20–100 nm were produced by irradiating multiwall INT-WS_2 with high-intensity inductively coupled radiofrequency plasma [27]. Under optimized conditions, that is, power of 600 W and 40 min treatment, about 80% of the parent NTs produced daughter (1–3 layers) NTs tethered to their surfaces as shown in Figure 2.8a.

It is hypothesized that the formation of the daughter NTs occurs through a strong interaction of the high energy plasma with point or line defects, which

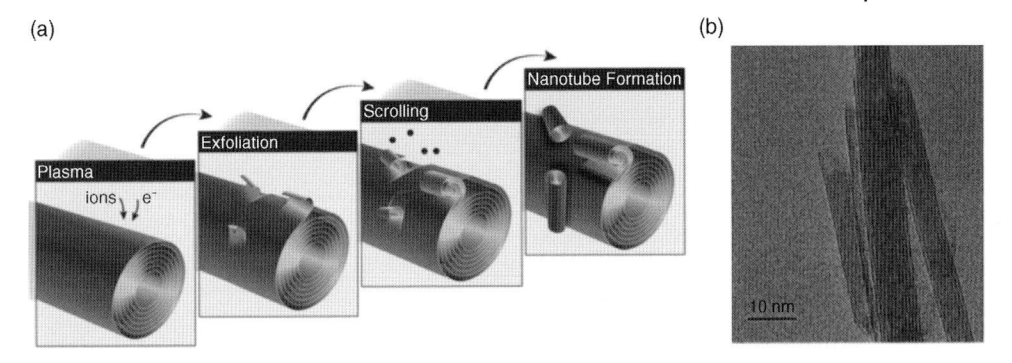

Figure 2.8 (a) Schematics of the proposed growth mechanism of the daughter nanotubes by plasma treatment of the multiwall mother nanotubes. (b) TEM images of daughter WS_2 nanotubes obtained by plasma ablation of multiwall INT at 600 W for 40 min. (From Ref. [27].)

could be produced by sulfur evaporation, on the outer surface of the parent (multiwall) nanotube. This interaction leads to unzipping and exfoliation of 1–3 WS_2 layer thick fragments, releasing thereby the large elastic strain, and followed by scrolling and closure into small daughter nanoscrolls or NTs as depicted in Figure 2.8b. Occasionally, WS_x nanoclusters are observed in the vicinity of daughter NTs, as a result of condensation of tungsten and sulfur atoms or WS_2 moieties from the vapor phase. The condensation of such clusters onto the tube edges could lead to further elongation or even the growth of an extra layer on the daughter nanotube.

In conclusion, it can be seen that new chemical strategies, which are being developed unceasingly lends itself to new nanotubes and fullerene-like nanoparticles of different layered compounds. Since, layered compounds encompass many of the elements in the periodic table new such nanostructures with different properties can be anticipated.

2.3
Properties

2.3.1
General Outlook

Recently, a number of excellent reviews that examined the different properties of the INT/IF were published [33,41,87]. In order to avoid repetition, only a brief review of most of the IF/INT properties will be presented here. The relationship between the synthetic methods and the structural or physical properties of the IF/INT were examined by Levi *et al.* [33]. It was shown that the properties of the INT and IF may be tuned by variation of the synthetic procedure. This point is also addressed in Section 2.3.3. Additionally, it was shown that the IF/INT properties may be also tuned using chemical modifications or surface functionalization. These findings are expected to lead to new strategies for contrast agents, drug delivery, and a variety of other medical application.

The optical and tribological properties including the lubrication mechanism of the IF/INT were reviewed by Visic et al. It was shown that adding small amounts of the IF/INT to lubricating fluids lead to major improvements in their performance with a large number of industrial applications developed over the last few years. One of the reasons to this beneficial effect is the IF/INT tendency to slowly deform and exfoliate under mechanical stress [8,12,88,89]. The IF/INT exhibit excellent tribological behavior even in humid atmosphere. In comparison, the bulk crystallites tend to deteriorate rapidly due to surface oxidation through the prismatic ($hk0$) edges, which contain many dangling bonds [88]. The reason for that phenomenon is the minimization of the dangling bonds in the IF/INT, compared to the crystalline material. Additionally, it was found that polymer nanocomposite films containing IF/INT have improved friction and wear behavior compared to the pure polymer [90,91–94]. The improved wear resistance and reduced friction of polymer nanocomposites with IF/INT can be attributed to the gradual release of the IF/INT to the polymer–metal interface during the tribological test. The heating of the polymer (epoxy resin) surface during the test leads to its softening that facilitates the release of the NPs to the polymer surface [95]. Moreover, the added NPs increase the fracture toughness and hardness of the resin matrix, which leads to an improved wear resistance of the matrix. This mechanism can be regarded as the polymer analogue of the Archard effect in metals [96].

The WS_2 and MoS_2 are semiconductors with an indirect bandgap of about 1.3 eV [97] and 1.2 eV [98], respectively. It was found that the electronic band structure of $2H-MoS_2$ ($2H-WS_2$) consists of several absorption thresholds that can be associated with maxima (minima) in the valence (conduction) bands. Here, the indirect gap between the Γ point (in the valence band) and the midpoint between the Γ and K (in the conduction band) corresponds to the lowest allowed transition. This indirect transition has minor feature at ~1040 nm (1.2 eV). Additionally, direct transition at the K point in the vicinity of 700 nm (1.8 eV) is assigned as A exciton. Here, K_4 and K_1 are the initial states and K_5 is the final state. The A and B excitons exist due to the spin–orbit splitting (~60 nm) and interlayer interaction. The third (Z) threshold is at ~500 nm (2.5 eV) and originates from a direct transition between the valence band and the conduction band at the M point.

The IF/INT preserve the semiconducting behavior of the bulk counterparts [38,99]. The optical properties of semiconducting IF/INT are of great interest, mainly due to the tunable characteristics of their physical behavior, such as the reduction of their bandgap with shrinking tube diameter. This effect attributed to the deterioration of the hybridization of the atomic orbitals due to the curvature. In contrast to nanoplatelets, the quantum size effects are not significant in these closed nanostructures.

The synthesis of the IF/INT-MS$_2$ (M = Mo, W) is executed under high-temperature conditions (ca. 800–950 °C) and thus might induce some intrinsic defects and sulfur vacancies in particular [19,24]. This assumption is supported by recent studies, which revealed that new energy states within the MS$_2$ bandgap are induced by the intrinsic defects in the layers [79,100]. Some of the defects arise due to the polytypes mixture in the IF/INT phase. While the 2H polytype is the stable form in ambient conditions, the IF/INT incorporate defects associated with patches of other polytypes, that is, 1T (metallic) and 3R (semiconducting) (Figure 2.2) [79,100,101]. These dislocations, defects, and metal "patches" in the lattice can induce an excess of charge carriers on the NPs surface, particularly in suspensions where different redox species may lead to extra charges residing at these surface sites. The charge carriers residing on the surface should induce charged-colloidal behavior of the MS$_2$ nanoparticles in solutions, which manifests itself, among others, through the negative zeta potential (ZP) of the IF suspensions and their improved stability [46].

Inspired by the newly discovered optical and electronic properties of the 2D MoS$_2$, the optical behavior of the IF counterparts were reconsidered [46]. It was found that, in addition to the established excitonic transitions, scattering peak appears at ~730 nm in the extinction spectra of the dispersed IF-MoS$_2$. Figure 2.9 provides a direct comparison between the extinction and the absolute absorbance spectrum of both IFs and platelets. Note that whereas the standard extinction measurements in a UV-vis spectrometer provide the combination of absorbance and scattering of the dispersed NP, integrating sphere allows retrieving the "pure" absorbance. This differentiation between the extinction and absolute absorbance is very important for the understanding of the optical properties of the semiconducting NP. And so, from the comparison of extinction and absorbance spectra it becomes clear that the extinction peak at 730 nm (Figure 2.9, blue curve) is primarily due to scattering rather than absorption.

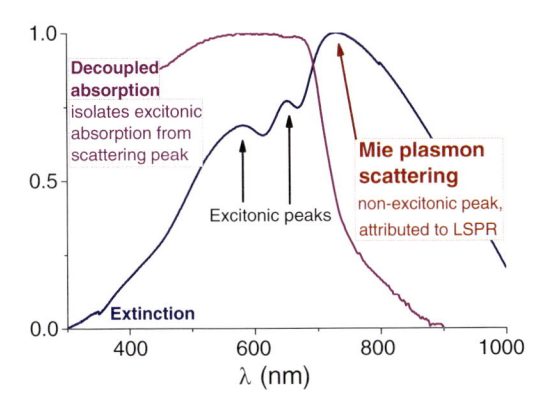

Figure 2.9 Extinction and absorbance spectra of IF-MoS$_2$ suspension. The purple curve marked represents the decoupled (true) absorption measured by the integrating sphere Quantaurus QY; the blue curve is the extinction measured by UV-vis spectrometer. (From Ref. [46].)

The dependence of this optical transition on the IF size, shape, and solvent refractive index, suggests that this transition arises from a surface plasmon resonance (SPR). This strong optical transition was attributed to the high density of free charge carriers ($\approx 10^{19}$ cm^{-3}), which are confined to the NPs' surface. Furthermore, the relatively large size (70–150 nm) of the NPs induce significant cross-section for light scattering. This assumption is further supported by ZP measurements, which confirmed that the IF NP have a distinct negative surface charge. Remarkably, the IF-MoS$_2$ supports the excitonic structure of the bulk while inducing strong plasmonic scattering (which is not displayed by the bulk material). The SPR occurs in the near-infrared spectral region, such that its higher energy part overlaps the A and B excitonic features of the IF. The strong light–matter interaction in the IF-MoS$_2$ suggests that they can operate as a plasmonic device at room temperature, which opens up new opportunities for their applications, such as saturable absorbers in a mode-locked laser, optical tracking during medical diagnostics (or drug delivery), and more.

2.3.2
The Effect of the Doping on the Properties of the Nanoparticles

2.3.2.1 Tribological Properties of the Re-Doped IF-MoS$_2$

In order to elucidate the influence of doping on the tribological properties of the IF, Re-doped IF-MoS$_2$ were added to poly-alpha-olephin synthetic oil and tested under variety of conditions (i.e., boundary, mixed lubrication, elastohydrodynamic regimes). The Re-doped IF were shown to significantly reduce friction and wear as compared to their undoped counterparts, IF-WS$_2$, and micron-sized platelets (Figure 2.10) [38,102]. Here, the friction coefficient was reduced by

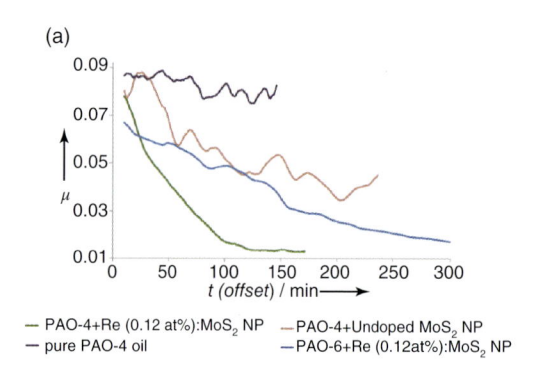

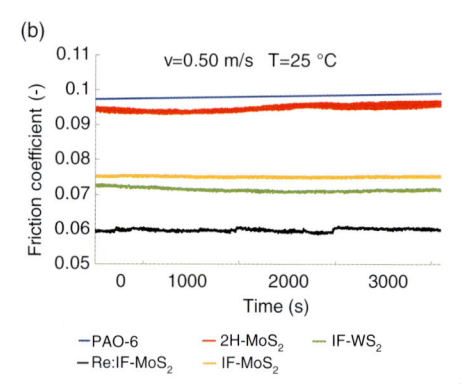

Figure 2.10 (a) Friction coefficient (μ) versus time measured with pin on disk set-up for different samples in PAO-4 oil (viscosity 18 cSt). Here pin (0.09 cm^2) made of AISI 1020 steel (hardness $H_V = 180$) was rubbed against a disk (AISI 4330 steel; $H_V = 550$). The applied loads are in the range of 500–600 N and the velocity is 0.24 m/s. (b) Evolution of friction coefficient: 1-h test in boundary lubrication steady condition. Here the tested samples were obtained by mixing the Re:IF-MoS$_2$, undoped IF-MoS$_2$, IF-WS$_2$, and platelets (bulk) of 2H-MoS$_2$ additives in a granular form with PAO-6 base oil for 30 min. (From Ref. [38,104].)

more than 40% compared to the base lubricant value. This effect was attributed to the increased conductivity of the doped NP. The substitutional Re atoms in the MoS_2 lattice, release free electrons that reside at the NP surface. The doping of the IF also endows more defects and strain to the NP lattice. Hence, the doped IF have higher tendency (compared to the undoped IF) to slowly deform and exfoliate under mechanical stress. It was shown before that the gradual exfoliation of the IF leads to a transfer of the MoS_2 nanosheets onto the asperities of the surfaces (third body) and provides effective lubrication [8,12,88,103]. Thus, easier exfoliation of the Re-doped NP layers is indeed expected to lead to significantly lower wear and friction coefficient.

The self-repelling nature of the Re-doped IF induces appreciably slower agglomeration and sedimentation producing more stable suspensions compared with the undoped NP [38,104]. Moreover, the Re-doped IF are more immune to the electrostatic charges that form during the tribological measurements (tribocharging). This finding further emphasizes the fact that the contribution of electronic effects to friction and wear is not negligible. The higher agglomeration of the undoped IF inhibits their free motion and thus the supply of the lubricant to the contact area. Furthermore, all the IF nanoparticles and in particular the doped ones, exhibit the "mending effect," that is, self-healing of the surface during the tribological test [38,104]. Here, the surface roughness of the wear trace was reduced from 1.5 to 0.15 mm. Apparently, the deep "valleys" on the surface of the disc are filled by the IF. This surface smoothing assembles an effective larger contact area, namely, lower apparent load, hence reduces the friction and wear.

More recently, the effect of the Re: $IF-MoS_2$ nanoparticles on lubricating soft matter was examined. Here, the force of insertion/traction of metallic and polyurethane tubes into a soft silicone (RTV) ring were measured. Small amounts of different NPs were mixed with a medical gel (Esracaine) and smeared on the tube surface. This study aimed to simulate the insertion/traction of an endoscope or catheter to the urethra [105,106]. The Re: $IF-MoS_2$ NPs outperformed all other IF reducing the friction between the metallic and polymer tubes by a factor of three compared to the pure gel. The load in this test is a few kPa that did not lead to any damage of the NPs. Consequently, the improved lubrication of the interface could be attributed to their rolling and sliding between the two surfaces. Furthermore, a distinct reduction of the gel viscosity could also help reducing the traction force of the tube from the silicone ring.

2.3.2.2 Optical Properties of the Re-Doped IF-MoS$_2$

The Burstein–Moss shift was initially discovered by Burstein and Moss during their research on the optical properties of InSb [107,108]. Here the apparent band gap of a semiconductor is increased since all states that are close to the conduction band are being populated and push the absorption edge to higher energies. This phenomenon occurs in heavily doped semiconductors when the electron carrier concentration exceeds the conduction band edge density of states. This effect arises from the Pauli "Exclusion Principle" and defines how adding electrons or holes to traditional semiconductors like ZnO, CdSe, and

InGaAs controls carrier concentration, modifies the bandgap, and enables control over the electronic structure [109–111].

Fundamental understanding of the optical response of these and other conventional semiconductors was thus partially responsible for the development of various electronic devices including photoelectrochemical cells, resonant tunneling devices, Bragg mirrors, and surface-emitting lasers [109–114]. Layered compounds like the 2H-MoS_2 (Figure 2.2) are also candidates for the investigation of Burstein–Moss effects, although early studies in which Re was used as the dopant did not display the classic signatures [74,115–117]. Re-doped IF-MoS_2 provide physical platform for the investigation of n-doping effects in a confined system. Tunability with strain and electric field suggest that such a systematic exploration will prove useful [118]. There is an especially pressing need to connect the actual dopant concentration to the number of carriers in Re: IF-MoS_2. This is an important issue for conventional bulk semiconductors [107,108] and quantum dots [119] since the Burstein–Moss model provides a framework under which a quantitative relationship can be established.

It was recently revealed that Re-doped IF-MoS_2 exhibit Burstein–Moss shift due to band filling [120]. Here, the optical properties of the Re-doped IF were compared to the properties of the undoped IF and the 2H-bulk material. It was shown that the small size of the IF NP induces confinement that softens the exciton positions and reduces spin–orbit coupling, whereas doping has the opposite effect. Moreover, the doping-induced blue shift of the A and B excitons, which compares well with predictions from the Burstein–Moss model [109–111] establishing a quantitative link between the actual (Re) dopant concentration and carrier density. Thus, additional (substitutional) Re doping induce Burstein–Moss effect that usually appear at heavily doped semiconductors. Figure 2.11a shows the transitions for excitons A and B in the undoped IF-MoS_2. For n-type (Re) doping, the extra carriers occupy the lower conduction band states, which pushes the excitonic transitions to higher energies inducing Burstein–Moss shift (Figure 2.11b).

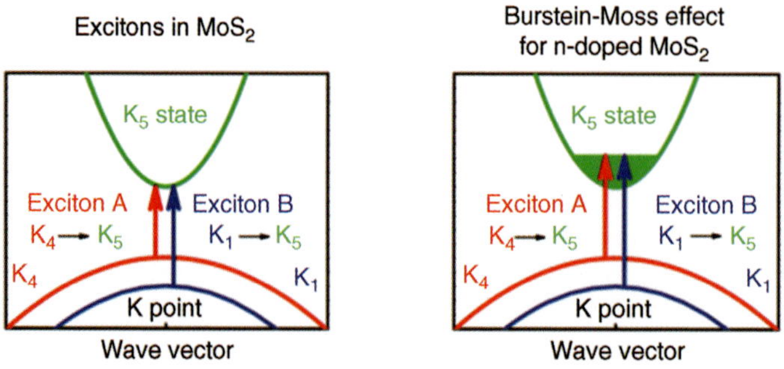

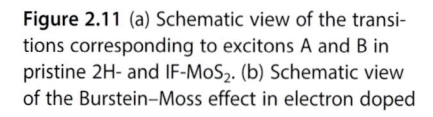

Figure 2.11 (a) Schematic view of the transitions corresponding to excitons A and B in pristine 2H- and IF-MoS_2. (b) Schematic view of the Burstein–Moss effect in electron doped MoS_2. In this work, we investigate Re doped IF-MoS_2. In both panels, excitons A and B are associated with optically allowed $K_4 \rightarrow K_5$ and $K_1 \rightarrow K_5$ transitions. (From Ref. [120].)

The doping-induced changes in spin–orbit coupling were extracted on the basis of exciton and oscillator strength trends. Using DFTB calculations, the activation barrier for the rhenium atom ionization in the MoS_2 lattice was estimated at 150–200 meV [71]. Using these findings it was demonstrated that the carriers are bound rather than free, that is, only 1–2% of the Re atoms are ionized in the MoS_2 lattice. These findings are important for understanding finite length scale and doping effects in transition metal dichalcogenides and the more complicated functional materials that emanate from this parent compound. It is important to note that the IF are intrinsically highly doped, that is, $10^{19}\,cm^{-3}$ (Section 2.3.1). The Re concentration in the examined IF is of the same order of magnitude (Section 2.2.2), however, as shown above only 1–2% of the Re atoms are ionized. At the same time, the local changes induced by the Re atoms, like the (local) 2H-1T polytypic transformation may push extra carriers to the conduction band of the IF nanoparticles. The use of the Burstein–Moss model allows to extend the quantitative connection between the Re concentration and the carrier density, and highlights the unique electronic character of Re-doped IF-MoS_2. It also illustrates the utility of reaching beyond traditional bulk semiconductors and quantum dots to explore layered systems like metal dichalcogenides.

2.3.3
Properties of Individual WS₂ Nanotubes

This part of the review will concentrate on the measurements and analysis of the properties of individual INT. Study of the properties of single INT are extremely challenging and far from being fully developed. Nevertheless, the understanding of the physiochemical behavior of a single INT provides a fundamental understanding of their electrical, optical, and the mechanical behaviors.

2.3.3.1 Field-Effect Transistors Based on INT-WS₂

Field-effect transistor (FET) based on individual INT-WS_2 was shown to exhibit mobility of up to $50\,cm^2/(V\,s)$ [47]. The density of free charge carrier of the examined INT was $\sim 10^{19}\,cm^{-3}$ and the current density was $\sim 2.4 \times 10^8\,A/cm$. Moreover, the current-carrying capacity of a single nanotube was shown to exceed 0.6 mA, which is surprisingly high, compared to other low-dimensional materials. Two types of devices were fabricated: side- and end-contacted INT-WS_2 FETs, which was useful in elucidating the current transport mechanism in the NTs. Here it was suggested that, for the INT-WS_2, several outer layers participate in the current transport. That finding is somewhat surprising, since it was shown that in multiwall CNTs the current is mostly carried by the outermost wall [121,122]. Temperature-dependent electrical measurements of INT-WS_2-based FETs confirmed their semiconducting characteristics. Thus, the significant mobility together with the high current-carrying capacity of the individual INT, suggest that INTs are either affected by a strong interlayer coupling or are heavily doped.

The role of the environment and the contact–nanotube interface in the determination of conductivity and mobility of the INT-WS$_2$ was further explored Levi et al. [47]. The devices were fabricated by dry depositing of the INT-WS$_2$ onto Si/SiO$_2$ wafers. Using the four-probe measurements, it was found that the resistance of the Au contacts was about two orders of magnitude larger than that of the nanotubes. Thus, it can be concluded that improvement of the contacts will considerably enhance the device performance. This conclusion was supported by DFTB simulations, in which Ti contacts were found to be superior to the frequently used Au [123]. Since solvent residues may results in surface trap states, its removal should improve the INT – contact electronic coupling. Indeed, FETs that were prepared by dry deposition [124] of the INT-WS$_2$ exhibited a positive effect on the FET features [47]. The measured differential conductivity values of the INT-WS$_2$ FETs in vacuum was higher by three orders of magnitude compared to the ambient environments, that is, $\sigma_{diff}(\text{vacuum}) = 1.2 \times 10^5 \, \Omega^{-1} \, \text{cm}^{-1}$ versus $\sigma_{diff}(\text{air}) = 1.6 \times 10^2 \, \Omega^{-1} \, \text{cm}^{-1}$ [125]. Since moieties in the ambient like carbon-containing residues, water, oxygen, and even nitrogen molecules may generate trap-states or act like scattering centers, the vacuum environment is indeed expected to have significant importance on the high current-carrying capacity of the INT-WS$_2$-FET.

The significant current-carrying capacity and the high mobility of individual INT makes them promising candidates for high-power nanoelectronic devices as well as for future sensorial applications.

2.3.3.2 Electromechanical Properties of INT-WS$_2$

Recently reported torsional electromechanical (EM) measurements presented complex and reproducible electrical response of the INT-WS$_2$ to the mechanical deformation (Figure 2.12) [126]. These measurements were combined with DFTB calculations in order to elucidate the coupling between the mechanical deformation (torsion and tension) and the electrical properties of the NTs. Here, it was shown that mechanical deformations induce electrical response that span over several orders of magnitude. Additionally, several modes of mechanical

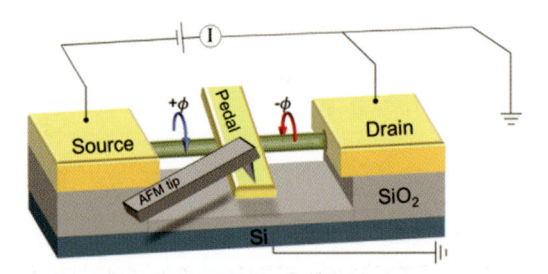

Figure 2.12 Schematic representation of the setup used for EM measurements of the INT-WS$_2$. The AFM tip presses the pedal down, while a bias is applied to the source electrodes. The blue and red arrows depict the direction of the torsion, resulting in intralayer shear of the outermost layer of the INT-WS$_2$. (From Ref. [126].)

deformation, like torsion and bending were detected simultaneously and their effect on the electrical response was analyzed.

Whereas the electrical properties of individual INT-WS_2 are dominated by their outermost layers, their mechanical properties are generally related to the layer ensemble [127]. In the earlier studies of the INT mechanical properties, the used NTs were produced by vapor solid (VS) growth [17]. Contrarily, the INT used in the current study were produced by outside-in sulfidization (OS), that is, Type II NTs [19]. These INT exhibited higher interlayer (mechanical) coupling for tension and torsion when compared to the INT produced by VS method (Type I NTs) [40,127,128]. The INT-WS_2 exhibited strong electrical response to the mechanical deformation. The decoupling between the electrical responses of the mechanical modes (torsion and tension) was accomplished mainly using detailed calculations. The bandgap–strain relations were studied using DFTB. It was shown that once the INT twisted around its own axis, the conduction band shifts. This shift is dominated by the unoccupied 3S sulfur states that are closer to the Fermi level and therefore the bandgap is reduced upon torsion [129]. In the course of the EM measurements, the INT-WS_2 exhibited high interlayer mechanical coupling. This finding indicates that the INT are suitable for use in nanoelectromechanical systems (NEMS), for example, as nanoactuators. Moreover, the EM response of the INT-WS_2 was highly sensitive to torsional strain and tension.

Currently, large varieties of new types of INT are synthesized. That, together with the controlled doping achieved recently is expected to result in many other EM effects. Thus, the large variety of tailorable properties positions the INTs as extraordinary building blocks for a range of NEMS.

2.4
Applications

Earlier in their research, IF/INT nanoparticles of WS_2 (MoS_2) were hypothesized to be superior solid lubricants. Bulk 2H-MoS_2 is known for its beneficial solid lubrication behavior and is used commercially since many years. The weak van-der-Walls forces between two adjacent MoS_2 layers provide facile shearing between each two adjacent layers when sliding across each other. Although the interlayer forces are not much different in the IF/INT NPs, shearing of the interlocked layers in the hollow cage structures is rather impossible. Notwithstanding this fact, a combination of several other factors contribute to their being superior solid lubricants, offering them a large number of applications. The first report on the superior solid lubrication properties of IF-WS_2 [13] as additive to fluid lubricants, led to a surge of interest in their tribological behavior. This topic was discussed extensively in previous reviews [89,130] and will not be rehearsed again. The superior solid lubrication behavior was attributed to a combination of rolling, sliding, and exfoliation of the layers. The exfoliated layers cover the asperities of the two mating surfaces providing thereby facile shearing of the two surfaces sliding past each other. The combination of these three mechanisms was confirmed by recent studies using AFM [131], TEM [132], and SEM [133].

Re:IF-MoS$_2$ nanoparticles have been synthesized and their properties are being studied for a number of years now [38,83, 104,105,120,134–136]. Exquisite control of the position of the Fermi level of the doped NPs was accomplished by introducing Re levels from say 20 to no larger than 500 ppm in the host IF/INT lattice. Their negative surface charge (self-repelling character) and atomically smooth surfaces was shown to bear on their improved tribological properties [38,102,104–106,123]. Indeed, their mutual repulsion, which prevents their agglomeration and their atomically smooth surface, was shown to produce outstandingly small friction [104,137]. Furthermore, their superior lubrication characteristics are probably responsible also for the reduced drag and dynamic viscosity when they are dispersed in fluids [98]. Thin film deposition of these nanoparticles onto soft silicone substrate from aqueous solutions was studied recently [124]. Self-organized tessellated films were thus obtained onto catheter surfaces. Substantially reduced encrustation of hydroxyapatite from artificial urine solutions onto Re:IF-MoS$_2$ coated catheter surfaces was observed [138]. These recent studies offer plethora of potential medical applications to these NPs.

The advantageous properties of the IF NPs are not limited to fluid lubricants. Advances reporting their reinforcing effect of polymer matrices; wear and friction modifiers of surfaces of different solids, like polymers and porous metallic matrices were published in the early 2000's [139–141]. Subsequently, various polymer matrices compounded with tiny amounts of these NPs were studied intensively. The main objective for this effort is for improving the mechanical properties of the nanocomposite, its thermal stability, as well as its tribological behavior.

Much of this work was recently reviewed and will not be deliberated here again [48,49]. The main advantages of IF/INT over such well-studied nanostructures as CNT and graphene are that they can be easily dispersed in variety of matrices and their safe use, which has been well documented in several recent reports (see for example, Refs [142,143]). On the other hand IF/INT are both heavier and physically weaker compared to the carbon-based nanostructures. From the practical point of view, these deficiencies are nevertheless of minor importance. First, the inorganic NPs are added in tiny amounts to the polymer matrix and the viscosity of the polymer fluid is sufficiently high to prevent their sedimentation during the blending and hardening of the matrix. Furthermore, the inorganic NPs are anyhow two to three orders of magnitude stronger than the polymer matrix, making them very suitable as reinforcing agents of the polymer matrix.

Also, recently a decent amount of work has been devoted for reinforcing of biocompatible and biodegradable polymers by incorporating these NPs in the matrix. This trend goes hand-in-hand with two contemporary paradigm shifts, that is, upsurge in research and applications of "green chemistry" methodologies and increasing usage of artificial parts in the body to replace ailing tissues.

Numerous biodegradable polymers and biocompatible polymers, for instance polylactic acid (PLA), polyvinyl alcohol (PVA) and thermoplastic polyurethane (t-PU) exhibit low strength and fracture toughness. Furthermore, biocompatible polymers, like polyether ether ketone (PEEK) that exhibits high strength and fracture toughness are used in hip replacement joints and knees. Nonetheless, slow release of wear debris leads to inflammation of the tissue surrounding the implant and shortens its lifetime. Compounding these polymers with small amounts of IF and more so INT-WS_2 (MoS_2) could slow the shedding of wear debris and elongate the lifetime of the artificial prosthesis.

Few examples will serve as showcase of the progress in this field. In one study [144], poly(propylene fumarate) (PPF) was loaded with INT-WS_2 (0.01–0.2 wt%). This biodegradable polymer is used as scaffold for bone tissue engineering. The NTs were found to disperse individually in the polymer matrix and enhance the cross-linking density of the polymer chains. Substantial improvements in the mechanical properties, such as compression and flexural strengths (up to 190%) were observed for the nanotube reinforced PPF. The main attributes of the NTs could be ascribed to bridging and pullout effects. Coupled with the fact that these NTs were found to be nontoxic, at least in low level loadings (up to 50 mg/ml), the findings of this study are very encouraging in terms of further development of INT-WS_2 based scaffolds for tissue engineering. In another publication, small amounts of INT-WS_2 were blended into poly(3-hydroxybutyrate) (PHB), which is a truly biodegradable polymer [145]. This polymer suffers from low crystallization rate that impair its mechanical properties and lead to its brittleness. Furthermore, long-term crystallization of the polymer produces crakes limiting its lifetime. The addition of small amounts (0.1 wt%) of INT-WS_2 to PHB promoted the crystallization of the polymer; its modulus (+17%) and hardness as well as its tribological characteristics. Also recently, the mechanical properties of cement (concrete) [146] and Mg-alloys [147] were shown to improve considerably by adding small amounts of INT-WS_2 to the matrix. While bridging and pullout were found to inhibit the propagation of cracks through the cement, grain refining and boundary dislocations produced the reinforcement effect in the magnesium alloys. Furthermore, the addition of the INT led to faster curing of the concrete (9 instead of 24 days), which could stimulate the commercialization of INT-reinforced concrete. These recent studies demonstrate the usefulness of the IF/INT inorganic nanophases as additives for improving the mechanical, thermal, and tribological properties of solid surfaces.

2.5 Conclusions

Hollow closed structures, that is, nanotubes (INT) and fullerene-like (IF) nanoparticles are shown to be a genuine metastable phase of layered compounds in the nanorange. The large extra chemical energy stored in the rim atoms of 2-D clusters forces them to bend and seam, primarily if heated to high temperatures. Variety of synthetic procedures have been pursued in order to obtain IF and INT from binary and more recently from ternary and quaternary layered compounds. Most strikingly, nanotubes of numerous ternary and quaternary "misfit" compounds were synthesized and their structure was elucidated, recently. Stable dispersions and exceedingly low friction was achieved by *in situ* doping of the IF and INT of WS_2 (MoS_2) with Re which impart negative charge on the nanoparticle surface. Optical extinction in the visible and near IR ranges was attributed to light scattering by surface plasmon resonance. The mechanical, electrical, and electromechanical properties of individual WS_2 nanotubes were investigated, using both experimental techniques and theoretical calculations. Once available in sufficient amounts, their properties were studied and distinct differences from the response of bulk phases were observed. IF-WS_2 nanoparticles are produced now in appreciable amounts and are used commercially for improved solid lubrication; metal finishing and in the future also to reinforce and improve the thermal stability of variety of nanocomposites.

References

1 Pauling, L. (1930) The structure of the chlorites. *Proc. Natl. Acad. Sci. USA*, **16**, 578.

2 Bates, T.F., Sand, L.B., and Mink, J.F. (1950) Tubular crystals of chrysotile asbestos. *Science*, **111**, 512–513.

3 Kroto, H.W., Heath, J.R., O'Brien, S.C., Curl, R.F., and Smalley, R.E. (1985) C60: buckminsterfullerene. *Nature*, **318**, 162–163.

4 Kroto, H. (1987) The stability of the fullerenes C_n, with $n = 24, 28, 32, 36, 50, 60$ and 70. *Nature*, **329**, 529–531.

5 Iijima, S. (1991) Helical microtubules of graphitic carbon. *Nature*, **354**, 56–58.

6 Dickinson, R.G. and Pauling, L. (1923) The crystal structure of molybdenite. *J. Am. Chem. Soc.*, **45**, 1466–1471.

7 Tenne, R., Margulis, L., Genut, M., and Hodes, G. (1992) Polyhedral and cylindrical structures of tungsten disulphide. *Nature*, **360**, 444–446.

8 Cizaire, L. *et al.* (2002) Mechanisms of ultra-low friction by hollow inorganic fullerene-like MoS_2 nanoparticles. *Surf. Coat. Technol.*, **160**, 282–287.

9 Rosentsveig, R. *et al.* (2009) Fullerene-like MoS_2 nanoparticles and their tribological behavior. *Tribol. Lett.*, **36**, 175–182.

10 Martin, J., Donnet, C., Le Mogne, T., and Epicier, T. (1993) Superlubricity of molybdenum disulphide. *Phys. Rev. B*, **48**, 10583.

11 Paskvale, S., Remškar, M., and Čekada, M. (2016) Tribological performance of TiN, TiAlN and CrN hard coatings lubricated by MoS_2 nanotubes in Polyalphaolefin oil. *Wear*, doi: 10.1016/j.wear.2016.01.020.

12 Joly-Pottuz, L. *et al.* (2005) Ultralow-friction and wear properties of IF-WS_2 under boundary lubrication. *Tribol. Lett.*, **18**, 477–485.

13 Rapoport, L. *et al.* (1997) Hollow nanoparticles of WS_2 as potential solid-state lubricants. *Nature*, **387**, 791–793.

14 Lahouij, I., Vacher, B., Martin, J.-M., and Dassenoy, F. (2012) IF-MoS_2 based lubricants: influence of size, shape and crystal structure. *Wear*, **296**, 558–567.

15 Margulis, L., Salitra, G., Tenne, R., and Talianker, M. (1993) Nested fullerene-like structures. *Nature*, **365**, 113–114.

16 Feldman, Y., Wasserman, E., Srolovitz, D., and Tenne, R. (1995) High-rate, gas-phase growth of MoS_2 nested inorganic

fullerenes and nanotubes. *Science*, **267**, 222.
17 Margolin, A., Rosentsveig, R., Albu-Yaron, A., Popovitz-Biro, R., and Tenne, R. (2004) Study of the growth mechanism of WS_2 nanotubes produced by a fluidized bed reactor. *J. Mater. Chem.*, **14**, 617–624.
18 Rothschild, A., Sloan, J., and Tenne, R. (2000) Growth of WS_2 nanotubes phases. *J. Am. Chem. Soc.*, **122**, 5169–5179.
19 Zak, A. et al. (2011) Large-scale synthesis of WS_2 multiwall nanotubes and their dispersion, an update. *Sens. Transducers J.*, **12**, 1–10.
20 Rothschild, A., Frey, G., Homyonfer, M., Tenne, R., and Rappaport, M. (1999) Synthesis of bulk WS_2 nanotube phases. *Mater. Res. Innov.*, **3**, 145–149.
21 Remškar, M., Škraba, Z., Cleton, F., Sanjines, R., and Levy, F. (1998) MoS_2 microtubes: an electron microscopy study. *Surf. Rev. Lett.*, **5**, 423–426.
22 Zak, A. et al. (2010) Scaling up of the WS_2 nanotubes synthesis. *Fullerenes Nanotubes Carbon Nanostruct.*, **19**, 18–26.
23 Rosentsveig, R., Margolin, A., Feldman, Y., Popovitz-Biro, R., and Tenne, R. (2002) WS_2 nanotube bundles and foils. *Chem. Mater.*, **14**, 471–473.
24 Zak, A., Sallacan-Ecker, L., Margolin, A., Genut, M., and Tenne, R. (2009) Insight into the growth mechanism of WS_2 nanotubes in the scaled-up fluidized-bed reactor. *Nano: Brief Rep. Rev.*, **4**, 91–98.
25 Enyashin, A.N., Gemming, S., and Seifert, G. (2007) *Materials for Tomorrow*, Springer, pp. 33–57.
26 Seifert, G., Köhler, T., and Tenne, R. (2002) Stability of metal chalcogenide nanotubes. *J. Phys. Chem. B*, **106**, 2497–2501.
27 Brüser, V. et al. (2014) Single-to triple-wall WS_2 nanotubes obtained by high-power plasma ablation of WS_2 multiwall nanotubes. *Inorganics*, **2**, 177–190.
28 Seifert, G., Terrones, H., Terrones, M., Jungnickel, G., and Frauenheim, T. (2000) Structure and electronic properties of MoS_2 nanotubes. *Phys. Rev. Lett.*, **85**, 146–149.
29 Hu, S. and Wang, X. (2008) Single-walled MoO_3 nanotubes. *J. Am. Chem. Soc.*, **130**, 8126–8127.
30 Ni, B., Liu, H., Wang, P.-p., He, J., and Wang, X. (2015) General synthesis of inorganic single-walled nanotubes. *Nat. Commun.*, **6**, 8756.
31 Evarestov, R. (2015) Symmetry and ab initio calculations of nanolayers, nanotubes and nanowires, *Theoretical Modeling of Inorganic Nanostructures*, Springer, Berlin.
32 Rao, C. and Govindaraj, A. (2009) Synthesis of inorganic nanotubes. *Adv. Mater.*, **21**, 4208–4233.
33 Levi, R., Bar-Sadan, M., and Tenne, R. (2013) *Springer Handbook of Nanomaterials* (ed. Robert Vajtai), Springer, Berlin, pp. 605–638.
34 Wiegers, G. (1996) Misfit layer compounds: structures and physical properties. *Prog. Solid State Chem.*, **24**, 1–139.
35 Makovicky, E. and Hyde, B. (1981) *Inorganic Chemistry*, vol. **46**, Springer Verlag, pp. 101–170.
36 Radovsky, G., Popovitz-Biro, R., Stroppa, D.G., Houben, L., and Tenne, R. (2013) Nanotubes from chalcogenide misfit compounds: Sn–S and Nb–Pb–S. *Acc. Chem. Res.*, **47**, 406–416.
37 Panchakarla, L.S. et al. (2014) Nanotubes from misfit layered compounds: a new family of materials with low dimensionality. *J. Phys. Chem. Lett.*, **5**, 3724–3736.
38 Yadgarov, L. et al. (2012) Controlled doping of MS_2 (M = W, Mo) nanotubes and fullerene-like nanoparticles. *Angew. Chem., Int. Ed.*, **51**, 1148–1151.
39 Kis, A. et al. (2003) Shear and Young's moduli of MoS_2 nanotube ropes. *Adv. Mater.*, **15**, 733–736.
40 Kaplan-Ashiri, I. et al. (2006) On the mechanical behavior of WS_2 nanotubes under axial tension and compression. *Proc. Natl. Acad. Sci. USA*, **103**, 523–528.
41 Kaplan-Ashiri, I. and Tenne, R. (2015) On the mechanical properties of WS_2 and MoS_2 nanotubes and fullerene-like nanoparticles: *in situ* electron microscopy measurements. *JOM*, 1–17. doi: 10.1007/s11837-015-1659-2

42 Maharaj, D. and Bhushan, B. (2015) Friction, wear and mechanical behavior of nano-objects on the nanoscale. *Mater. Sci. Eng.: R*, **95**, 1–43.

43 Frey, G.L., Elani, S., Homyonfer, M., Feldman, Y., and Tenne, R. (1998) Optical-absorption spectra of inorganic fullerenelike MS_2 (M = Mo, W). *Phys. Rev. B*, **57**, 6666–6671.

44 Faucheaux, J.A., Stanton, A.L., and Jain, P.K. (2014) Plasmon resonances of semiconductor nanocrystals: physical principles and new opportunities. *J. Phys. Chem. Lett.*, **5**, 976–985.

45 Luther, M.J., Jain, K.P., Ewers, T., and Alivisatos, A.P. (2011) Localized surface plasmon resonances arising from free carriers in doped quantum dots. *Nat. Mater.*, **10**, 361–366.

46 Yadgarov, L. *et al.* (2014) Dependence of the absorption and optical surface plasmon scattering of MoS_2 nanoparticles on aspect ratio, size, and media. *ACS Nano*, **8**, 3575–3583.

47 Levi, R., Bitton, O., Leitus, G., Tenne, R., and Joselevich, E. (2013) Field-effect transistors based on WS_2 nanotubes with high current-carrying capacity. *Nano. Lett.*, **13**, 3736–3741.

48 Naffakh, M., Díez-Pascual, A.M., Marco, C., Ellis, G.J., and Gómez-Fatou, M.A. (2013) Opportunities and challenges in the use of inorganic fullerene-like nanoparticles to produce advanced polymer nanocomposites. *Prog. Polym. Sci.*, **38**, 1163–1231.

49 Otorgust, G., Sedova, A., Dodiuk, H., Kenig, S., and Tenne, R. (2015) Carbon and tungsten disulfide nanotubes and fullerene-like nanostructures in thermoset adhesives: a critical review. *Rev. Adhes. Adhes.*, **3**, 311–363.

50 Rouxel, J., Meerschaut, A., and Wiegers, G. (1995) Chalcogenide misfit layer compounds. *J. Alloy. Compd.*, **229**, 144–157.

51 Bernaerts, D., Amelinckx, S., Van Tendeloo, G., and Van Landuyt, J. (1997) Microstructure and formation mechanism of cylindrical and conical scrolls of the misfit layer compounds $PbNb_nS_{2n+1}$. *J. Cryst. Growth*, **172**, 433–439.

52 Hong, S.Y., Popovitz-Biro, R., Prior, Y., and Tenne, R. (2003) Synthesis of SnS_2/SnS fullerene-like nanoparticles: a superlattice with polyhedral shape. *J. Am. Chem. Soc.*, **125**, 10470–10474.

53 Brontvein, O. *et al.* (2015) Solar synthesis of $PbS-SnS_2$ superstructure nanoparticles. *ACS Nano*, **9**, 7831–7839.

54 Radovsky, G. *et al.* (2011) Synthesis of copious amounts of SnS_2 and SnS_2/SnS nanotubes with ordered superstructures. *Angew. Chem., Int. Ed.*, **50**, 12316–12320.

55 Radovsky, G., Popovitz-Biro, R., and Tenne, R. (2012) Study of tubular structures of the misfit layered compound SnS_2/SnS. *Chem. Mater.*, **24**, 3004–3015.

56 Lorenz, T., Joswig, J.-O., and Seifert, G. (2014) Combined $SnS@SnS_2$ double layers: charge transfer and electronic structure. *Semicond. Sci. Tech.*, **29**, 064006.

57 Radovsky, G., Popovitz-Biro, R., and Tenne, R. (2014) Nanotubes from the misfit layered compounds $MS-TaS_2$, where M = Pb, Sn, Sb, or Bi: synthesis and study of their structure. *Chem. Mater.*, **26**, 3757–3770.

58 Kato, K., Kawada, I., and Takahashi, T. (1977) Die Kristallstruktur von $LaCrS_3$. *Acta Crystallogr. B:*, **33**, 3437–3443.

59 Marseglia, E. (1983) Transition metal dichalcogenides and their intercalates. *Int. Rev. Phys. Chem.*, **3**, 177–216.

60 Panchakarla, L.S., Popovitz-Biro, R., Houben, L., Dunin-Borkowski, R.E., and Tenne, R. (2014) Lanthanide-based functional misfit-layered nanotubes. *Angew. Chem.*, **126**, 7040–7044.

61 Radovsky, G. *et al.* (2015) Tubular structures from the $LnS-TaS_2$ (Ln = La, Ce, Nd, Ho, Er) and $LaSe-TaSe_2$ misfit layered compounds. *J. Mater. Chem. C*, **4**, 89–98.

62 Derycke, V., Martel, R., Appenzeller, J., and Avouris, P. (2002) Controlling doping and carrier injection in carbon nanotube transistors. *Appl. Phys. Lett*, **80**, 2773.

63 Kong, J., Zhou, C., Yenilmez, E., and Dai, H. (2000) Alkaline metal-doped n-type semiconducting nanotubes as quantum dots. *Appl. Phys. Lett*, **77**, 3977.

64 Zhou, C., Kong, J., Yenilmez, E., and Dai, H. (2000) Modulated chemical doping of individual carbon nanotubes. *Science*, **290**, 1552.

65 Krusin-Elbaum, L. et al. (2004) Room-temperature ferromagnetic nanotubes controlled by electron or hole doping. *Nature*, **431**, 672–676.

66 Zhu, Y. et al. (2001) Nb-doped WS_2 nanotubes. *Chem. Phys. Lett.*, **342**, 15–21.

67 Zhu, Y.Q. et al. (2001) Tungsten–niobium–sulfur composite nanotubes. *Chem. Commun.*, 121–122.

68 Hsu, W.K. et al. (2001) Titanium-doped molybdenum disulfide nanostructures. *Adv. Funct. Mater.*, **11**, 69–74.

69 Hsu, W. et al. (2000) Mixed-phase $W_xMo_yC_zS_2$ nanotubes. *Chem. Mater.*, **12**, 3541–3546.

70 Nath, M., Mukhopadhyay, K., and Rao, C. (2002) $Mo_{1-x}W_xS_2$ nanotubes and related structures. *Chem. Phys. Lett.*, **352**, 163–168.

71 Deepak, F.L. et al. (2008) Fullerene-like $Mo(W)_{1-x}Re_xS_2$ nanoparticles. *Chem. Asian J.*, **3**, 1568–1574.

72 Ivanovskaya, V.V., Heine, T., Gemming, S., and Seifert, G. (2006) Structure, stability and electronic properties of composite $Mo_{1-x}Nb_xS_2$ nanotubes. *Phys. Status Solidi B*, **243**, 1757–1764.

73 Wildervanck, J. and Jellinek, F. (1971) The dichalcogenides of technetium and rhenium. *J. Less Common Met.*, **24**, 73–81.

74 Tiong, K., Liao, P., Ho, C., and Huang, Y. (1999) Growth and characterization of rhenium-doped MoS_2 single crystals. *J. Cryst. Growth*, **205**, 543–547.

75 Späh, R., Elrod, U., Lux-Steiner, M., Bucher, E., and Wagner, S. (1983) pn Junctions in tungsten diselenide. *Appl. Phys. Lett*, **43**, 79.

76 Tahir, M.N. et al. (2010) Synthesis and functionalization of chalcogenide nanotubes. *Phys. Status Solidi B*, **247**, 2338–2363.

77 Zak, A., Feldman, Y., Alperovich, V., Rosentsveig, R., and Tenne, R. (2000) Growth mechanism of MoS_2 fullerene-like nanoparticles by gas-phase synthesis. *J. Am. Chem. Soc.*, **122**, 11108–11116.

78 Feldman, Y., Zak, A., Popovitz-Biro, R., and Tenne, R. (2000) New reactor for production of tungsten disulfide hollow onion-like (inorganic fullerene-like) nanoparticles. *Solid State Sci.*, **2**, 663–672.

79 Enyashin, A.N., Bar-Sadan, M., Houben, L., and Seifert, G. (2013) Line defects in molybdenum disulfide layers. *J. Phys. Chem. C*, **117**, 10842–10848.

80 Py, M. and Haering, R. (1983) Structural destabilization induced by lithium intercalation in MoS_2 and related compounds. *Can. J. Phys.*, **61**, 76–84.

81 Imanishi, N., Toyoda, M., Takeda, Y., and Yamamoto, O. (1992) Study on lithium intercalation into MoS_2. *Solid State Ion.*, **58**, 333–338.

82 Wypych, F. and Schöllhorn, R. (1992) 1T-MoS_2, a new metallic modification of molybdenum disulfide. *J. Chem. Soc., Chem. Commun.*, 1386–1388.

83 Yadgarov, L. et al. (2012) Investigation of rhenium-doped MoS_2 nanoparticles with fullerene-like structure. *Z. Anorg. Allg. Chem.*, **638**, 2610–2616.

84 Sloan, J., Kirkland, A.I., Hutchison, J.L., and Green, M.L. (2002) Structural characterization of atomically regulated nanocrystals formed within single-walled carbon nanotubes using electron microscopy. *Acc. Chem. Res.*, **35**, 1054–1062.

85 Kreizman, R. et al. (2009) Core–shell $PbI_2@WS_2$ inorganic nanotubes from capillary wetting. *Angew. Chem., Int. Ed.*, **48**, 1230–1233.

86 Enyashin, A.N., Kreizman, R., and Seifert, G. (2009) Capillary imbition of PbI_2 melt by inorganic and carbon nanotubes. *J. Phys. Chem. C*, **113**, 13664–13669.

87 Visic, B. and Tenne, R. (2015) Inorganic fullerene-like nanoparticles and nanotubes: tribological, mechanical and optical properties, in Layered Nanomaterials, Graphene, Chalchogenides, Metal Oxides and More (eds H. Terrones and M. Terrones), Wiley-VCH Verlag GmbH.

88 Chhowalla, M. and Amaratunga, G.A.J. (2000) Thin films of fullerene-like MoS_2

88 nanoparticles with ultra-low friction and wear. *Nature*, **407**, 164–167.
89 Rapoport, L., Fleischer, N., and Tenne, R. (2005) Applications of WS_2 (MoS_2) inorganic nanotubes and fullerene-like nanoparticles for solid lubrication and for structural nanocomposites. *J. Mater. Chem.*, **15**, 1782–1788.
90 Naffakh, M., Díez-Pascual, A.M., Remškar, M., and Marco, C. (2012) New inorganic nanotube polymer nanocomposites: improved thermal, mechanical and tribological properties in isotactic polypropylene incorporating INT-MoS_2. *J. Mater. Chem.*, **22**, 17002–17010.
91 Hou, X., Shan, C., and Choy, K.L. (2008) Microstructures and tribological properties of PEEK-based nanocomposite coatings incorporating inorganic fullerene-like nanoparticles. *Surf. Coat. Technol.*, **202**, 2287–2291.
92 Remškar, M. et al. (2013) Friction properties of polyvinylidene fluoride with added MoS_2 nanotubes. *Phys. Status Solidi A*, **210**, 2314–2319.
93 Naffakh, M. et al. (2007) Influence of inorganic fullerene-like WS_2 nanoparticles on the thermal behavior of isotactic polypropylene. *J. Polym. Sci. B: Polym. Phys.*, **45**, 2309–2321.
94 Shneider, M., Dodiuk, H., Kenig, S., and Tenne, R. (2010) The effect of tungsten sulfide fullerene-like nanoparticles on the toughness of epoxy adhesives. *J. Adhes. Sci. Technol.*, **24**, 1083–1095.
95 Shneider, M. et al. (2013) Tribological performance of the epoxy-based composite reinforced by WS_2 fullerene-like nanoparticles and nanotubes. *Phys. Status Solidi A*, **210**, 2298–2306.
96 Zhang, Z., Zhang, L., and Mai, Y.-W. (1995) Particle effects on friction and wear of aluminium matrix composites. *J. Mater. Sci.*, **30**, 5999–6004.
97 Ballif, C. et al. (1996) Preparation and characterization of highly oriented, photoconducting WS_2 thin films. *Appl. Phys. A*, **62**, 543–546.
98 Coehoorn, R. et al. (1987) Electronic structure of $MoSe_2$, MoS_2, and WSe_2. I. Band-structure calculations and photoelectron spectroscopy. *Phys. Rev. B*, **35**, 6195–6202.
99 Frey, G.L., Tenne, R., Matthews, M.J., Dresselhaus, M.S., and Dresselhaus, G. (1998) Optical properties of MS_2 (M = Mo, W) inorganic fullerene-like and nanotube material. *J. Mater. Res.*, **13**, 2412–2417.
100 Enyashin, A.N. et al. (2011) New route for stabilization of $1T$-WS_2 and MoS_2 phases. *J. Phys. Chem. C*, **115**, 24586–24591.
101 Houben, L. et al. (2012) Diffraction from disordered stacking sequences in MoS_2 and WS_2 fullerenes and nanotubes. *J. Phys. Chem. C*, **116**, 24350–24357.
102 Rapoport, L. et al. (2012) High lubricity of Re-doped fullerene-like MoS_2 nanoparticles. *Tribol. Lett.*, **45**, 257–264.
103 Rapoport, L. et al. (2005) Friction and wear of fullerene-like WS_2 under severe contact conditions: friction of ceramic materials. *Tribol. Lett.*, **19**, 143–149.
104 Yadgarov, L. et al. (2013) Tribological studies of rhenium doped fullerene-like MoS_2 nanoparticles in boundary, mixed and elasto-hydrodynamic lubrication conditions. *Wear*, **297**, 1103–1110.
105 Sedova, A. et al. (2014) Re-doped fullerene-like MoS_2 nanoparticles in relationship with soft lubrication, *Nanomater. Energy*, **4**, 30–38.
106 Goldbart, O. et al. (2014) Lubricating medical devices with fullerene-like nanoparticles. *Tribol. Lett.*, **55**, 103–109.
107 Moss, T. (1954) The interpretation of the properties of indium antimonide. *Proc. Phys. Soc. London, Sect. B*, **67**, 775.
108 Burstein, E. (1954) Anomalous optical absorption limit in InSb. *Phys. Rev.*, **93**, 632.
109 Tenne, R. et al. (1990) Transport and optical properties of low-resistivity CdSe. *Phys. Rev. B*, **42**, 1763.
110 Sarkar, A., Ghosh, S., Chaudhuri, S., and Pal, A. (1991) Studies on electron transport properties and the Burstein–Moss shift in indium-doped ZnO films. *Thin Solid Films*, **204**, 255–264.
111 Muñoz, M. et al. (2001) Burstein–Moss shift of n-doped $In_{0.53}Ga_{0.47}As/InP$. *Phys. Rev. B*, **63**, 233302.

References

112 Tenne, R., Flaisher, H., and Triboulet, R. (1984) Photoelectrochemical etching of ZnSe and nonuniform charge flow in Schottky barriers. *Phys. Rev. B*, **29**, 5799.

113 Lee, J.J., Yang, C.S., Park, Y.S., Kim, K.H., and Kim, W.T. (1999) The Burstein–Moss effect in Cu_2GeSe_3: Co^{2+} single crystals. *J. Appl. Phys.*, **86**, 2914–2916.

114 Wu, J. *et al.* (2002) Effects of the narrow band gap on the properties of InN. *Phys. Rev. B*, **66**, 201403.

115 Evans, B. and Young, P. (1965) Optical absorption and dispersion in molybdenum disulphide. *Proc. R. Soc. Lond. A Mater*, **284**, 402–422.

116 Connell, G., Wilson, J., and Yoffe, A. (1969) Effects of pressure and temperature on exciton absorption and band structure of layer crystals: molybdenum disulphide. *J. Phys. Chem. Solids*, **30**, 287–296.

117 Yen, P., Hsu, H., Liu, Y., Huang, Y., and Tiong, K. (2004) Temperature dependences of energies and broadening parameters of the band-edge excitons of Re-doped WS_2 and 2H-WS_2 single crystals. *J. Phys. Condens. Matter*, **16**, 6995.

118 Kou, L. *et al.* (2012) Tuning magnetism and electronic phase transitions by strain and electric field in zigzag MoS_2 nanoribbons. *J. Phys. Chem. Lett.*, **3**, 2934–2941.

119 Amasha, S. *et al.* (2013) Pseudospin-resolved transport spectroscopy of the Kondo effect in a double quantum dot. *Phys. Rev. Lett.*, **110**, 046604.

120 Sun, Q.C. *et al.* (2013) Observation of a Burstein–Moss shift in Rhenium-doped MoS_2 nanoparticles. *ACS Nano*, **7**, 3506–3511.

121 Frank, S., Poncharal, P., Wang, Z., and de Heer, W.A. (1998) Carbon nanotube quantum resistors. *Science*, **280**, 1744–1746.

122 Delaney, P., Di Ventra, M., and Pantelides, S. (1999) Quantized conductance of multiwalled carbon nanotubes. *Appl. Phys. Lett.*, **75**, 3787–3789.

123 Popov, I., Seifert, G., and Tománek, D. (2012) Designing electrical contacts to MoS_2 monolayers: a computational study. *Phys. Rev. Lett.*, **108**, 156802.

124 Tevet, O. *et al.* (2010) Nanocompression of individual multilayered polyhedral nanoparticles. *Nanotechnology*, **21**, 365705.

125 Zhang, C. *et al.* (2012) Electrical transport properties of individual WS_2 nanotubes and their dependence on water and oxygen absorption. *Appl. Phys. Lett.*, **101**, 113112.

126 Levi, R. *et al.* (2015) Nanotube electromechanics beyond carbon: the case of WS_2. *ACS Nano*, **9**, 12224–12232.

127 Tang, D.-M. *et al.* (2013) Revealing the anomalous tensile properties of WS_2 nanotubes by *in situ* transmission electron microscopy. *Nano. Lett.*, **13**, 1034–1040.

128 Nagapriya, K. *et al.* (2008) Torsional stick-slip behavior in WS_2 nanotubes. *Phys. Rev. Lett.*, **101**, 195501.

129 Seifert, G. (2000) *Electronic Properties of Novel Materials—Molecular Nanostructures: XIV International Winterschool/Euroconference*, AIP Publishing, pp. 415–418.

130 Tannous, J. *et al.* (2011) Understanding the tribochemical mechanisms of IF-MoS_2 nanoparticles under boundary lubrication. *Tribol. Lett.*, **41**, 55–64.

131 Maharaj, D. and Bhushan, B. (2013) Effect of MoS_2 and WS_2 nanotubes on nanofriction and wear reduction in dry and liquid environments. *Tribol. Lett.*, **49**, 323–339.

132 Lahouij, I., Dassenoy, F., Vacher, B., and Martin, J.M. (2012) Real time TEM imaging of compression and shear of single fullerene-like MoS_2 nanoparticle. *Tribol. Lett.*, **45**, 131–141.

133 Tevet, O. *et al.* (2011) Friction mechanism of individual multilayered nanoparticles. *Proc. Natl. Acad. Sci. USA*, **108**, 19901–19906.

134 Sun, Q.-C. *et al.* (2013) Spectroscopic determination of phonon lifetimes in rhenium-doped MoS2 nanoparticles. *Nano. Lett.*, **13**, 2803–2808.

135 Chhetri, M. *et al.* (2015) Beneficial effect of Re-doping on the electrochemical HER activity of MoS 2 fullerenes. *Dalton Trans*, **44**, 16399–16404.

136 Woo, S.H. et al. (2015) Fullerene-like re-doped MoS$_2$ nanoparticles as an intercalation host with fast kinetics for sodium ion batteries. *Isr. J. Chem.*, **55**, 599–603.

137 Tomala, A. et al. (2015) Interaction between selected MoS$_2$ nanoparticles and ZDDP tribofilms. *Tribol. Lett.*, **59**, 1–18.

138 Ron, R. et al. (2014) Attenuation of encrustation by self-assembled inorganic fullerene-like nanoparticles. *Nanoscale*, **6**, 5251–5259.

139 Leshchinsky, V. et al. (2002) Inorganic nanoparticle impregnation of self lubricated materials. *Int. J. Powder Metall.*, **38**, 50–57.

140 Zhang, W. et al. (2003) Use of functionalized WS$_2$ nanotubes to produce new polystyrene/polymethylmethacrylate nanocomposites. *Polymer*, **44**, 2109–2115.

141 Leshchinsky, V. et al. (2002) Self lubricating bearing materials impregnated with WS$_2$ fullerene-like nanoparticles. *Int. J. Powder Metall.*, **38**, 50–57.

142 Pardo, M., Shuster-Meiseles, T., Levin-Zaidman, S., Rudich, A., and Rudich, Y. (2014) Low cytotoxicity of inorganic nanotubes and fullerene-like nanostructures in human bronchial epithelial cells: relation to inflammatory gene induction and antioxidant response. *Environ. Sci. Technol.*, **48**, 3457–3466.

143 Rashkow, J.T., Talukdar, Y., Lalwani, G., and Sitharaman, B. (2015) Interactions of 1D-and 2D-layered inorganic nanoparticles with fibroblasts and human mesenchymal stem cells. *Nanomedicine*, **10**, 1693–1706.

144 Lalwani, G. et al. (2013) Tungsten disulfide nanotubes reinforced biodegradable polymers for bone tissue engineering. *Acta Biomater.*, **9**, 8365–8373.

145 Naffakh, M. et al. (2014) Novel poly(3-hydroxybutyrate) nanocomposites containing WS$_2$ inorganic nanotubes with improved thermal, mechanical and tribological properties. *Mater. Chem. Phys.*, **147**, 273–284.

146 Nadiv, R., Shtein, M., Peled, A., and Regev, O. (2015) WS$_2$ nanotube–reinforced cement: dispersion matters. *Constr. Build. Mater.*, **98**, 112–118.

147 Huang, S.-J., Ho, C.-H., Feldman, Y., and Tenne, R. (2016) Advanced AZ31 Mg alloy composites reinforced by WS$_2$ nanotubes. *J. Alloy. Compd.*, **654**, 15–22.

3
Layered Materials: Oxides and Hydroxides

Ida Shintaro

Kyushu University, Department of Applied Chemistry, Faculty of Engineering, 744 Motooka, Nishi-ku, Fukuoka 819-0395, Japan

3.1
Layered Perovskite Oxides

Layered perovskite oxides have two-dimensional (2D) perovskite layers, such as $CaTiO_3$ and $BaTiO_3$ in their structure, and they are mainly classified into three types according to the different 2D perovskite layers parallel with the (100), (110), and (111) faces; the three types of layered perovskites are denoted here as (100)-layered perovskite, (110)-layered perovskite, and (111)-layered perovskite [1]. The thickness of the perovskite layer is represented by the number of BO_6 octahedra layers. The (100)-layered perovskite structure can be classified into three homologous series, the Ruddlesden–Popper [2], Dion–Jacobson [3–5], and Aurivillius [6–8] phases. Layered perovskite oxides exhibit electrical and magnetic properties such as high-T_c superconductivity, high thermoelectric properties, and photocatalytic properties. In this section, the fundamental structure of layered perovskite oxides is described.

3.1.1
The (100)-Layered Perovskite Oxides

3.1.1.1 Ruddlesden–Popper Phase ($A_{n+1}B_nO_{3n+1}$)

$Sr_3Ti_2O_7$ is a representative Ruddlesden–Popper (RP) phase [2]. S. N. Ruddlesden and P. Popper first determined the crystal structures of $Sr_3Ti_2O_7$ [2]. In the case of $Sr_3Ti_2O_7$, the chemical composition is not an integral multiple of the simple perovskite $SrTiO_3$, but the amounts of Sr and O are in excess. The excess amount of Sr–O forms rock-salt (RS) layers between the perovskite layers. Thus, the chemical composition of $Sr_3Ti_2O_7$ can be presented by $SrO\text{-}2(SrTiO_3)$, which reflects the layer structure. In this case, the number 2 in front of $SrTiO_3$ indicates the number of BO_6 octahedra in the 2D perovskite layer. Layered perovskites that have (100)-perovskite layers with a thickness of n BO_6 octahedra

Handbook of Solid State Chemistry, First Edition. Edited by Richard Dronskowski, Shinichi Kikkawa, and Andreas Stein.
© 2017 Wiley-VCH Verlag GmbH & Co. KGaA. Published 2017 by Wiley-VCH Verlag GmbH & Co. KGaA.

layers in the thickness direction linked with the RS-type interlayer structure (chemical composition: AO-n(ABO$_3$) or A$_{n+1}$B$_n$O$_{3n+1}$) are called RP phases. Many layered perovskites (A$_{n+1}$B$_n$O$_{3n+1}$) with RP phase have been reported. Figure 3.1a–c shows a structural model of Sr$_{n+1}$Ti$_n$O$_{3n+1}$, ($n = 1$–3) that has periodic 2D perovskite layers with a thickness corresponding to n layers of octahedra [9]. The perovskite layers along the ab-plane are formed by corner-shared BO$_6$ octahedra. The adjacent perovskite layers are arranged so that they are mutually shifted in the ab direction. The thinnest layered perovskites ($n = 1$, Sr$_2$TiO$_4$) in the RP phases are referred to as K$_2$NiF$_4$ type. Although the RP phase can be presented with the general formula A$_{n+1}$B$_n$O$_{3n+1}$ or AO-n(ABO$_3$), depending on the components, the chemical formula is also written as A$'_2$A$_{n-1}$B$_n$O$_{3n+1}$, where the A$'$ cation in the interlayer is distinguished from the A cation in the perovskite layer. The [A$_{n-1}$B$_n$O$_{3n+1}$] represents a perovskite layer containing oxygen at its terminal surface. Therefore, an RP phase can be described as that where A$'$ cations reside between negatively charged [A$_{n-1}$B$_n$O$_{3n+1}$] perovskite layers. The A$'$ cation in the A$'_2$A$_{n-1}$B$_n$O$_{3n+1}$ can be ion-exchanged with other cations such as alkali metal ions or protons. The coordination environment of the A$'$ atom is generally different from the 12-coordination A atom in the perovskite layer, and the A$'$ atom can be ion-exchanged relatively easily with different chemical species. If a part of the A and/or B cations is replaced with other atoms having different valences, then the interlayer

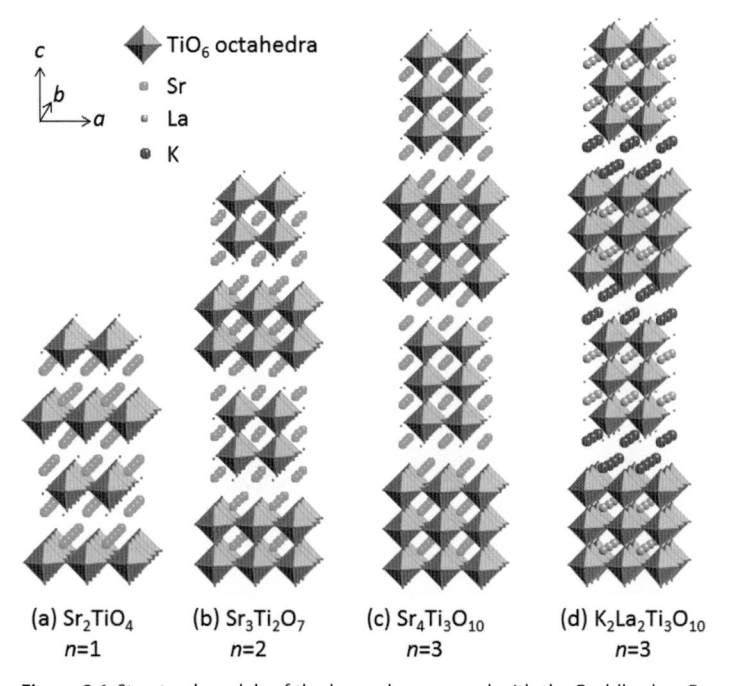

(a) Sr$_2$TiO$_4$ $n=1$

(b) Sr$_3$Ti$_2$O$_7$ $n=2$

(c) Sr$_4$Ti$_3$O$_{10}$ $n=3$

(d) K$_2$La$_2$Ti$_3$O$_{10}$ $n=3$

Figure 3.1 Structural models of the layered compound with the Ruddlesden–Popper phase: (a) Sr$_2$TiO$_4$, (b) Sr$_3$Ti$_2$O$_7$, (c) Sr$_4$Ti$_3$O$_{10}$, and (d) K$_2$La$_2$Ti$_3$O$_{10}$.

charge density is changed, which results in a change in the amount of A' cations in the interlayer. Figure 3.1d shows a structural model of $K_2La_2Ti_3O_{10}$, which is one of the materials in the series of $A'_2A_{n-1}B_nO_{3n+1}$.

3.1.1.2 Dion–Jacobson Phase ($A'A_{n-1}B_nO_{3n+1}$)

The layered perovskite Dion–Jacobson (DJ) phase can be represented by the general formula $A'A_{n-1}B_nO_{3n+1}$. The monovalent A' cations reside between negatively charged $[A_{n-1}B_nO_{3n+1}]$, two-dimensional (100)-perovskite layers composed of corner-shared BO_6 octahedra. The DJ phase has three different structure types that differ with respect to the displacement direction of adjacent perovskite layers [10]. Figure 3.2 shows structural models of the three different types of DJ phases using $RbCa_2Nb_3O_{10}$, $KCa_2Nb_3O_{10}$, and $LiCa_2Nb_3O_{10}$ as examples [10]. Figure 3.2a shows a structural model of $RbCa_2Nb_3O_{10}$. The layers of NbO_6 octahedra in $RbCa_2Nb_3O_{10}$ are not displaced. There is a height difference between the NbO_6 octahedra and the A cations perpendicular to the observed plane of approximately 0.2 nm, which consists of the Nb—O bond length and half of the octahedron body diagonal. This structure is realized for very large A cations such as Rb^+ or Cs^+. Figure 3.2b shows a structural model of $KCa_2Nb_3O_{10}$, where the adjacent perovskite layers are mutually displaced by a displacement vector of $a/2$. This structure is realized for large A cations such as K^+. Figure 3.2c shows a structural model of $LiCa_2Nb_3O_{10}$. The adjacent perovskite layers in $LiCa_2Nb_3O_{10}$ are mutually displaced with the displacement vector of $(a + b)/4$, where the a- and b-axes are involved in the perovskite plane because this compound belongs to the tetragonal crystal system. This structure

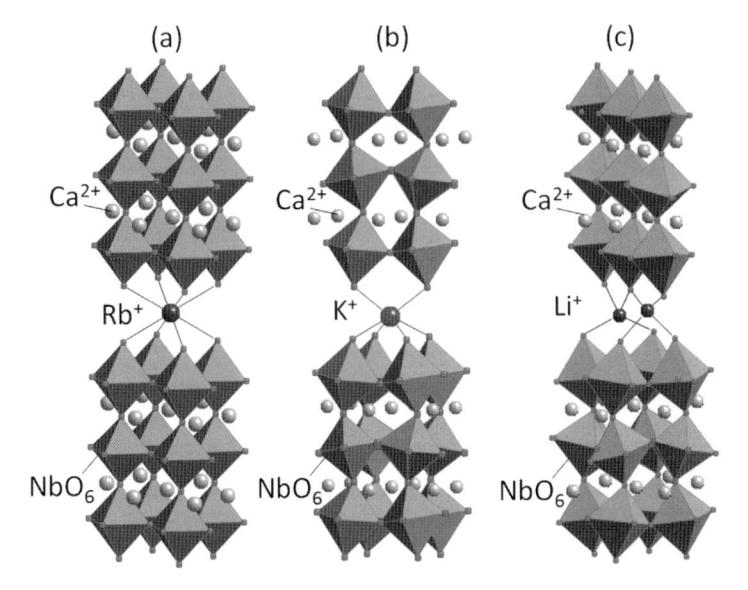

Figure 3.2 Structural models of the layered compound with the DJ phase: (a) $RbCa_2Nb_3O_{10}$, (b) $KCa_2Nb_3O_{10}$, and (c) $LiCa_2Nb_3O_{10}$.

is realized for large A cations such as Na^+ or Li^{2+}. The potassium ion is coordinated by six oxygen atoms, and the cesium ion is coordinated by eight oxygen atoms. The average bond lengths between the alkali metal cations and the oxygen atoms are 2.78 and 3.17 Å as for the K and Cs compounds, respectively. The Cs compound has a larger coordination number and larger bond lengths than the K compound. The stacking manner of the $Ca_2Nb_3O_{10}$ layers depends on the size of the interlayer cations. These materials have been studied in the field of photocatalyst [11,12].

3.1.1.3 Aurivillius phase (Bi_2O_2-$A_{n-1}B_nO_{3n+1}$)

The layered perovskite Aurivillius phase can be represented by the general formula Bi_2O_2-$A_{n-1}B_nO_{3n+1}$, where a fluorite [Bi_2O_2] layer resides between negatively charged [$A_{n-1}B_nO_{3n+1}$] two-dimensional (100)-perovskite layers composed of corner-shared BO_6 octahedra [6–8]. The number of octahedra layers in the perovskite layer is represented by n. The A site can be occupied by large cations such as Ca^{2+}, Sr^{2+}, Ba^{2+}, Na^+, K^+, Pb^{2+}, Bi^{3+}, and Ln^{3+} (Ln = rare earth element), and the B site by cations such as Ti^{4+}, Nb^{5+}, W^{6+}, Fe^{3+}, and Cr^{3+}. However, the [Bi_2O_2] layer is considered much less flexible for replacement with other chemical species. Figure 3.3 shows structural models of two representative layered perovskites with the Aurivillius phase; $n=2$ ($SrBi_2Ta_2O_9$) and $n=3$ ($Bi_4Ti_3O_{12}$). This crystal structure was first described in 1949 by B. Aurivillius. The Aurivillius phases were originally of interest for their ferroelectric

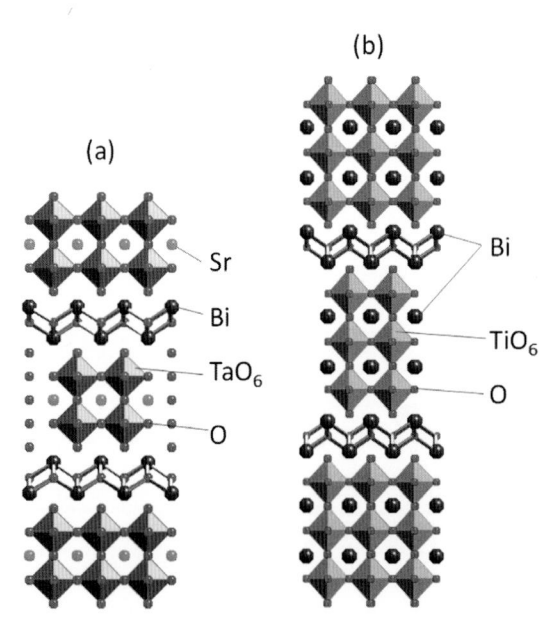

Figure 3.3 Structural models of layered perovskite with the Aurivillius phase: (a) $SrBi_2Ta_2O_9$ and (b) $Bi_4Ti_3O_{12}$.

3.1.2
The (110)-Layered Perovskite Oxides

The (110)-layered perovskite oxides can be represented by the general formula $A_nB_nO_{3n+2}$, and are derived from the ABO_3 perovskite structure by the separation of layers of corner-sharing BO_6 octahedra along the (110) plane [1,17]. Figure 3.4 shows structural models of several types of $A_nB_nO_{3n+2}$ projected along the a-axis and b-axis. Within the layers, the corner-shared BO_6 octahedra extend zigzag-like along the b direction and chain-like along the a-axis. $La_4Ti_4O_{14}$ is a typical (110)-layered perovskite with $n=4$. $La_4Ti_4O_{14}$ ($A_4B_4O_{14}$) can be written as $La_2Ti_2O_7$ ($A_2B_2O_7$), which is often used in the literature. It should be noted that some layered materials with the formula $A_2B_2O_7$ are pyrochlores. For example, $Sr_2Ta_2O_7$, $Nd_2Ti_2O_7$, $Sr_2Nb_2O_7$, and Ca_2Nb_2O are (110)-layered perovskites, whereas $Sm_2Ti_2O_7$, $Eu_2Ti_2O_7$, $Lu_2Ti_2O_7$, and $Y_2Ti_2O_7$ are pyrochlores. For smaller rare earth titanates such as $Eu_2Ti_2O_7$, the pyrochlore structure is dominant [17]. There is a relationship between the layer thickness and the lattice for $n=4$ and 5; when the perovskite layer has a theoretical perovskite structure without distortion, the layered material with $n=4$ takes a base-centered (C) lattice, while the layered material with $n=5$ takes a body-centered lattice. The space group for materials with odd-numbered n is $Cmcm$, while that for even-numbered n is $Immm$. In contrast, the space group for the theoretical

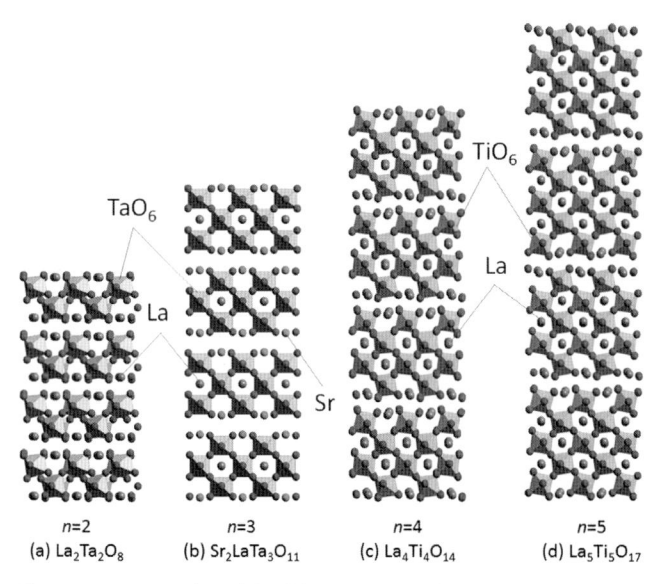

Figure 3.4 Structural models of the (110)-layered perovskite oxides: (a) $La_2Ta_2O_8$, (b) $Sr_2LaTa_3O_{11}$, (c) $La_4Ti_4O_{14}$, and (d) $La_5Ti_5O_{17}$.

RP phase is $I4/mmm$, regardless of the value of n. With regard to the charge balance of the structure, the formula for the 2D perovskite layer in $Sr_4Ta_4O_{14}$ is $[SrTaO_3]^+$, which does not satisfy electrical neutrality. Therefore, the constituent atoms, especially metal atoms, are largely displaced along the b-axis from the center of the perovskite layer toward the terminal surface, which results in large electric dipole in each layer of the BO_6 octahedra.

3.1.3
Intercalation Properties of the Layered Perovskite Oxides

The layered oxides $A'_2A_{n-1}B_nO_{3n+1}$ (RP phase) and $A'A_{n-1}B_nO_{3n+1}$ (DJ phase) exhibit intercalation properties [18,19]. The A cations in the interlayer between the 2D perovskite layers are easily exchanged with protons by acid treatment using diluted HCl and HNO, to form the protonated layered oxides $H_2A_{n-1}B_nO_{3n+1}$ and $HA_{n-1}B_nO_{3n+1}$. $K_2Ln_2Ti_3O_{10}$ (RP phase with $n=3$, Ln: lanthanide ions), $NaLnTiO_3$ (RP phase with $n=1$, Ln: lanthanide ions), and $A'A_2B_3O_{10}$ (DJ phase with $n=3$, A': Li, Na, K, Rb, Cs; A: Ca, Sr, Ba; B: Nb, Ta) can be converted to the respective protonated forms $H_2Ln_2Ti_3O_{10}$, $HLnTiO_3$, and $HA_2B_3O_{10}$ by acid treatment in aqueous solution [20]. Some layered perovskites with the Aurivillius phase also exhibit ion-exchange properties [21–23]. For materials with the Aurivillius phase, the $[Bi_2O_2]^{2+}$ layer between the perovskite layer $[A_{n-1}B_nO_{3n+1}]$ is dissolved in acid solutions such as HCl, and protons are intercalated into the interlayer to compensate the charge balance so that a protonated layered oxide $H_2A_{n-1}B_nO_{3n+1}$ could be formed. For example, the $Bi_2ANaB_3O_{12}$ (A: Sr, Ca; B: Nb, Mn, Ta), $Bi_2SrB_2O_9$ (B: Nb, Ta), and $Bi_2W_2O_9$ structures are converted into the respective $H_2ANaB_3O_{10}$, $H_2SrB_2O_7$, and $H_2W_2O_7$ protonated layered oxides. According to the chemical formula Bi_2O_2-$A_{n-1}B_nO_{3n+1}$ with the Aurivillius phase, the chemical formula for the protonated form derived from the Aurivillius phase should theoretically be the same as that derived from the $A'_2A_{n-1}B_nO_{3n+1}$ RP phase. However, the amount of protons in the protonated form derived from the Aurivillius phase is less than that derived from the RP phase. The main reason for this result could be that a part of the A site cations in the perovskite layer are exchanged with Bi^{3+} in the Bi_2O_2 layer when the parent layered compound is prepared:

$$[Bi_2A_{0.2}O_2]^{1.8+}[Bi_{0.2}A_{0.8}NaB_3O_{10}]^{1.8-}$$
$$\rightarrow H_{1.8}Bi_{0.2}A_{0.8}NaB_3O_{10} \text{ (A : Sr, Ca; B : Nb, Ta)}.$$

3.1.4
Conversion from 2D Perovskite to 3D Perovskite

When protonated layered compounds such as $H_2A_{n-1}B_nO_{3n+1}$ and $HA_{n-1}B_nO_{3n+1}$ are heated, a 3D perovskite is typically formed because the 2D perovskite layers are linked through dehydration. In this reaction, A-site defect perovskites are obtained [20,24]. $H_2La_2Ti_3O_{10}$ is converted to $La_{2/3}TiO_3$ by the heat treatment. $H_2SrTa_{2-x}Nb_xO_7$ and $H_2Sr_{1.5}Nb_3O_{10}$ are converted to $A_{0.5}BO_3$ (A: Sr; B: Nb, Ta) as a quasi-stable phase. The dehydration of layered protonated oxides results in the formation of A-site defect perovskites. As an approach to prepare 3D perovskite without defects, a method that simultaneously reduces the metal ions and removes oxygen from the material by hydrogen treatment has been reported. For example, $MEu_2Ti_2NbO_{10}$ (M: Na, Li) is converted $MEu_2Ti_2NbO_9$. In this conversion, a two-step reaction is used, where monovalent A cations in the interlayer are exchanged with divalent cations to decrease the amount of A cations in the interlayer, followed by the removal of some oxygen by heat treatment with hydrogen gas, as shown in Figure 3.5. $K_2Eu_2Ti_3O_{10}$ is converted to $MEu_2Ti_3O_{10}$ (M: Ca, Sr, Ni, Cu, Zn) by ion-exchange reaction, and is subsequently converted to $MEu_2Ti_3O_9$ by heat treatment with hydrogen gas [25].

3.1.5
Reaction with Other Chemical Reagents

The interlayer A cations can be ion exchange using molten salts such as nitrates and chlorides. $Na_2La_2Ti_3O_{10}$ and $NaLaTiO_4$ are converted to $MLa_2Ti_3O_{10}$ and $M_{0.5}LaTiO_4$ (M: Co, Cu, Zn) in the molten salt, respectively [26,27]. This reaction is a conversion reaction from the DJ phase to the RP phase. Conversion from the RP phase to the Aurivillius phase is also possible using BiOCl. $K_2La_2Ti_3O_{10}$, $K_2SrNb_2O_7$, and $NaLaTiO_4$ are converted to $[Bi_2O_2]^{2+}$ $[La_2Ti_3O_{10}]^{2-}$, $[Bi_2O_2]^{2+}$ $[SrNb_2O_7]^{2-}$, and $[BiO]^+$ $[LaTiO_4]^-$, respectively, by reaction

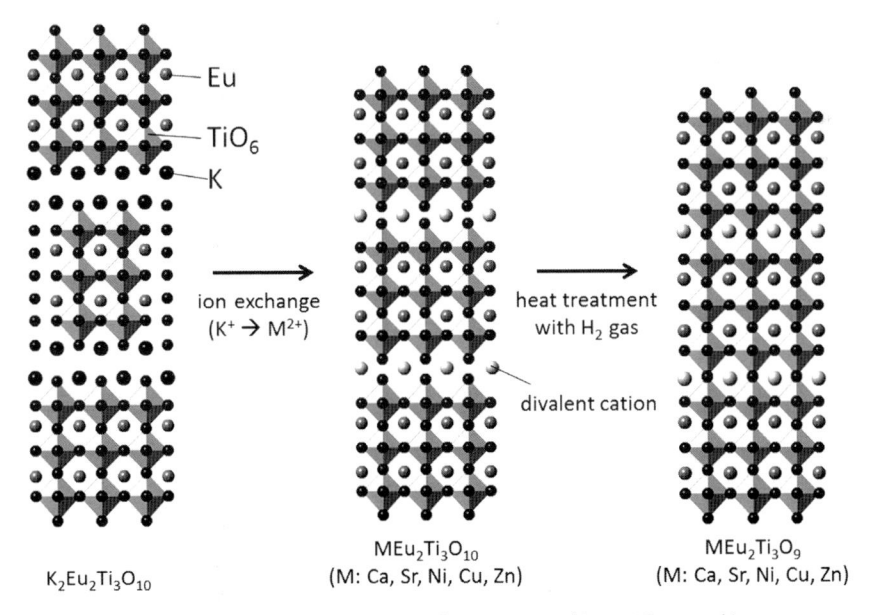

Figure 3.5 Schematic illustration of conversion from 2D perovskite to 3D perovskite.

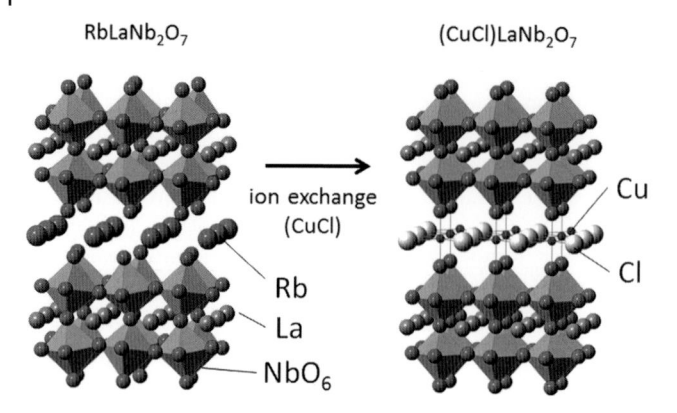

Figure 3.6 Schematic illustration of conversion from $RbLaNb_2O_7$ to $(CuCl)LaNb_2O_7$.

with BiOCl [28,29]. When the DJ phase reacts with metal iodides such as $NiBr_2$ and $FeBr_2$, the density of A cations in the interlayer is decreased. $RbLaNb_2O_7$ is converted to $M_{0.5}LaNb_2O_7$ (M: Ni, Fe) by reaction with metal iodide [30]. During this reaction, the intercalation of metal ions and chloride/iodide ions also occurs. $ALaNb_2O_7$ (M: Li, Na, K, Rb) is converted to $(FeCl)LaNb_2O_7$ by reaction with $FeCl_2$ [31]. $RbNdNb_2O_7$, $RbLaTa_2O_7$, and $RbCa_2Nb_3O_{10}$ are converted to $(CuCl) NdNb_2O_7$, $(CuCl) LaTa_2O_7$, and $(CuCl) Ca_2Nb_3O_{10}$, respectively, by reaction with CuCl, as shown in Figure 3.6 [32–35]. In this reaction, a halogen-metal network is formed in the interlayer. As an ion-exchange reaction in aqueous solution, $K_2La_2Ti_3O_{10}$ is converted to $[VO]^{2+} [La_2Ti_3O_{10}]^{2-}$ in $VOSO_4$ aqueous solution [28].

3.2
Layered Metal Oxides

The layered metal oxides are represented by the general formula $A_xM_yO_z$ (A: alkaline or alkaline earth ions, M: transition metal), in which the monovalent or divalent A ions are sandwiched between negatively charged 2D $[M_yO_z]^{m-}$ layers. In Section 3.1, the layered oxides with perovskite layers are described; however, there are many layered oxides without the 2D perovskite layer, such as $K_4Nb_6O_{17}$, $KMnO_2$, and $LiCoO_2$. These layered oxides exhibit various electronic and optical properties according to their compositions and structures. Therefore, control of the layered oxide composition is an important factor in the design of their properties. The intercalation method is often used for the purpose. In this section, the intercalation properties and structures of the layered transition oxides are introduced with several typical layered transition oxides as examples. Most of the layered compounds have alkaline ions in the interlayer sandwiched between transition metal oxide layers with layers of edge- or corner-shared BO_6 octahedra.

The charge density of the 2D transition layer is generally large, so that the electrostatic interaction between the 2D layer and guest ions in the interlayer is relatively strong. Therefore, ion-exchange reactions are slow in the layered transition metal oxides, and the intercalation of organic ions does not readily occur. In most of the layered transition metal oxides, the chemical species that can be directly intercalated in the interlayer is limited to small inorganic ions. When the layered oxide is treated with acid, the guest metal ions are easily exchanged with protons to form a proton-intercalated layered oxide. The protonated form is capable of intercalating basic organic molecules such as organic amines, in which the interlayer protons act as Brønsted acid sites and pull the organic amines into the interlayer. The intercalated organic amine becomes an organic ammonium ion through bonding with a proton. The type of organic amine intercalated is dependent on the acid strength of the protonated layered transition metal oxide (transition metal oxyacid). For example, in the case of $HTiNbO_5$, the lower pK_a limit of the guest species that can be intercalated in the protonated layered oxide is 5.7 (although there are cases in which molecules with weaker basicity are intercalated due to the effect of hydrogen bonding) [36]. In the case of $H_2Ti_2O_9$, the pK_a is around 9.0. The orientation of intercalated organic amines can be estimated from the length of the alkyl chain and the interlayer distance. The reactivity of the layered oxide with organic amines is given as an index of the ability for the formation of a layered compound. However, there are cases in which the organic amines destroy the layer structure due to the basicity of the organic amine, such as alkyl amines, and mesostructures are formed via nontopotactic processes such as dissolution and redeposition.

When the host layer of a layered transition metal oxide is partially reduced, the negative charge of the host layer is increased and cations are intercalated into the interlayer to compensate the charge:

$$xA^+ + xe^- + [H] \rightarrow (A^+)_x[H]^{x-} \quad (A : \text{cation}, [H] : \text{host layer}).$$

There are two types of redox intercalation that are the result of chemical methods using reducing agents and that by electrochemical reduction. $NaBH_4$, LiC_4H_9, and H_2 are generally used as reducing agents, whereby Na^+, Li^+, and H^+ are intercalated. Li^+ or Na^+ ions can be reversibly intercalated by controlling the electrode potential with the electrochemical reduction method. ABO_2 (A: Li, Na; B: Ti, V, Cr, Mn, Fe, Co, Ni, Nb) with the α-$NaFeO_2$ related structure has been studied as a material that exhibits redox intercalation properties [37].

3.3
Layered Co Oxides

Most of the layered Co oxides have 2D CoO_2 layers that consist of edge-shared CoO_6 octahedra, where the cobalt atoms form a triangular lattice. Alkaline ions are sandwiched between the CoO_2 layers and various layered structures are formed according to their stacking forms. The layered Co oxides exhibit various

3 Layered Materials: Oxides and Hydroxides

interesting properties based on the layered structure. For example, $\gamma\text{-Na}_x\text{CoO}_2$ exhibits interesting thermoelectric properties, the hydrated compound of $\text{Na}_{0.35}\text{CoO}_2\cdot 1.3\,\text{H}_2\text{O}$ exhibits superconductivity, and LiCoO_2 is widely used as a positive electrode material in Li batteries [38]. In this section, the typical structures and properties of layered Co oxide such as NaCoO_2, LiCoO_2, and layered oxides with a RS layer are described.

3.3.1
Li$_x$CoO$_2$

The structures of Li_xCoO_2 are classified into several types of layered structures such as O1, O2, O3, O6, T$^\#$2, and P3, based on the shape of the coordination polyhedron having an alkaline (Li$^+$) ion in the center, and the number of CoO$_2$ layers, where P, T, or O represents the type of alkali polyhedron (prismatic, tetrahedral, or octahedral, respectively), and the number 1, 2, or 3 indicates the number of CoO$_2$ layers within the cell [38–43]. Figure 3.7 shows structural models of various types of layered Li_xCoO_2. LiCoO_2 is used as a cathode material for Li batteries. The insertion and desorption of Li$^+$ in LiCoO_2 occurs at around 4.0 V versus Li, so that when Li$^+$ is deintercalated from LiCoO_2 by electrochemical reaction, the valence of the Co ion is changed from +3 to +4. The size of the Co^{4+} ion is larger than that of Co^{3+}; therefore, the size of CoO$_6$ is reduced by the

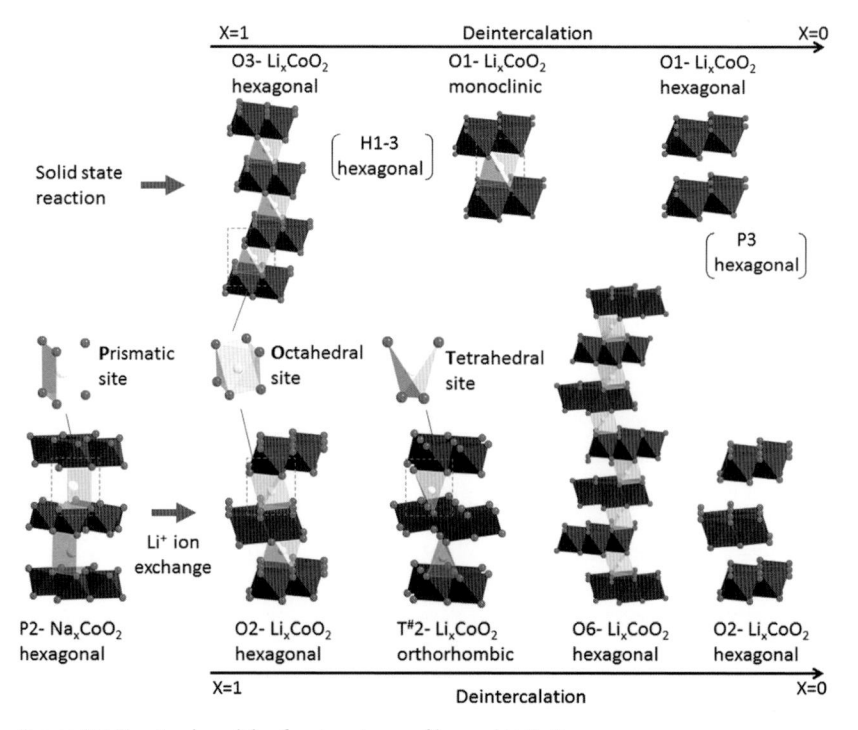

Figure 3.7 Structural models of various types of layered Li$_x$CoO$_2$.

deintercalation of Li^+. The electrostatic interactions between CoO_2 layers are weakened because the negative charge of the CoO_2 layer and the amount of Li^+ ions (cations) in the interlayer is decreased, which results in an increase of the interlayer distance. Thus, the crystal lattice extends along the stacking direction, while the CoO_2 layer is contracted.

O3-type $LiCoO_2$ is the most thermodynamically stable phase that can be prepared by conventional solid-state reaction. The hexagonal O3-$LiCoO_2$ is converted to monoclinic O1-type Li_xCoO_2 at $x=0.45$–0.5 by the electrochemical deintercalation of Li ions (whereas during deintercalation, a hybrid of the O1 and O3 structures known as the H1–3 structure is formed.). By further deintercalation, the monoclinic O1-type Li_xCoO_2 is converted to hexagonal O1-type Li_xCoO_2. This hexagonal O1-type Li_xCoO_2 structure is often mixed with the metastable P3-type Li_xCoO_2.

Although O3-type $LiCoO_2$ is stable, O2-type Li_xCoO_2 is thermodynamically unstable, and cannot therefore be prepared by conventional one-step solid-state reaction. However, it can be prepared by ion exchange of Na^+ in P2-type $Na_{0.7}CoO_2$ with Li^+ ions. When Na^+ ions in $NaCoO_2$ are exchanged with Li^+, the Na^+ ions with a large ionic radius in the interlayer are exchanged by Li^+ ions with a small ionic radius, which results in displacement of the CoO_2 layer, and the coordination of alkaline ions in the interlayer is then changed from triangular prismatic to octahedral, which is O2-type $LiCoO_2$. O2-type Li_xCoO_2 also undergoes electrochemical Li^+ ion insertion and desorption. During the deintercalation of Li ions, the structure of Li_xCoO_2 is changed, and $T^\#2$-type Li_xCoO_2 ($0.52<x<0.72$) and O6-type Li_xCoO_2 ($0.33<x<0.42$) are observed (Figure 3.7). The deintercalation of Li^+ from O3-type $LiCoO_2$ also occurs by the chemical oxidation method. $Li_{0.31}CoO_2$, $Li_{0.47}CoO_2$, and $Li_{0.91}CoO_2$ are formed by oxidation using Cl_2, Br_2, and I_2, respectively. The deintercalation reaction also proceeds with NO_2BF_4, until $x=0$ (Li_xCoO_2, $x=0$).

3.3.2
Na_xCoO_2

Layered sodium cobalt oxides have been examined for use as a positive electrode material in sodium ion batteries since the 1980s. Although these batteries are not being vigorously developed compared to lithium ion batteries, positive electrodes that are prepared using inexpensive raw materials such as sodium with little restriction on resources have attracted attention recently due to the limited worldwide reserves of lithium [43–46]. In addition, recent work has revealed that Na_xCoO_2 exhibits interesting thermoelectric properties and superconductivity [46–49]. The structures of Na_xCoO_2 are classified into P2, P3, and O3 based on the shape of coordination polyhedron with the alkaline (Na^+) ion in the center, and the number of CoO_2 layers, where P and O represent the type of alkali polyhedron (prismatic or octahedral, respectively) and the number 1, 2, or 3 indicates the number of layers within the hexagonal cell. Figure 3.8 shows structural models of various types of Na_xCoO_2. In the case of Na_xCoO_2, for

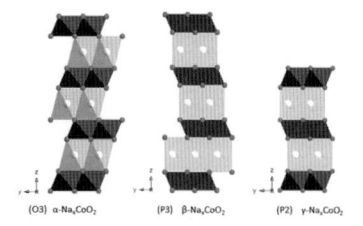

Figure 3.8 Structural models of various types of Na_xCoO_2.

example, there are P3-type Na_xCoO_2 ($0.55 < x < 0.63$), P2-type Nb_xCoO_2 ($0.64 < x < 0.74$), and O3-type Nb_xCoO_2 ($x = 1$) structures. The change in the crystal structure is caused by the displacement (translation) of the CoO_2 layer. These structures are changed by intercalation reactions; the O3-type Na_xCoO_2 is converted to the P3-type Na_xCoO_2 by the deintercalation of Na^+, whereas the P3-type Na_xCoO_2 is converted to the O3-type Na_xCoO_2 by the intercalation of Na^+. The sodium ions occupy two different sites in Na_xCoO_2; the Na-1 position shares faces with the CoO_6 octahedra in the layers above and below, while the Na-2 position shares edges only. The γ-Na_xCoO_2 (P2) also shows relatively large thermopower ($x = 0.5$–0.6) and low resistivity, and a layered Co oxide with superconductivity can be prepared from γ-$Na_{0.7}CoO_2$. $Na_{0.7}CoO_2$ is formed when γ-$Na_{0.7}CoO_2$ is oxidized with Br_2. When the layered compound is immersed in water, intercalation of water molecules occurs with the deintercalation of Na^+ and the interlayer distance is then increased. In the obtained layered oxide ($Na_{0.35}CoO_2 \cdot 1.3\,H_2O$), Na ions and water molecules are sandwiched between CoO_2 layers, as shown in Figure 3.9. This compound exhibits superconductivity at around 5 K. The observed superconductivity is considered to be induced by the large separation of the CoO_2 layers by the introduction of H_2O molecules [46,48].

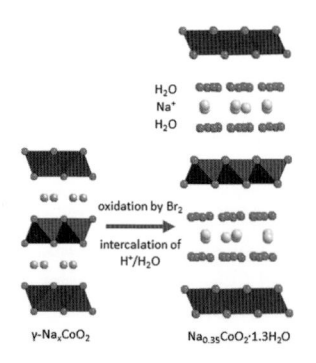

Figure 3.9 Schematic illustration of conversion from γ-Na_xCoO_2 to $Na_{0.35}CoO_2 \cdot 1.3\,H_2O$.

3.3.3
Other Layered Co Oxides

The physical properties of layered Co oxides is related to the interlayer distance, which is changed according to the intercalated ions. The interlayer distances of layered compounds intercalated with protons, Li^+, Na^+, and K^+ are 0.45, 0.5, 5.5, and 0.6 nm, respectively, where the distance is increased with the ionic radius. When the layered oxide is hydrated, the distance expands up to 0.98 nm, and superconductivity is observed. When the RS layer is formed in the interlayer, the distance expands up to 1.1 nm, and a large thermoelectric effect is observed. $[Ca_2CoO_3]_{0.6}CoO_2$ is a high-temperature thermoelectric p-type oxide with a dimensionless figure of merit (ZT) of 0.83 (at 800 °C) for a single-crystalline sample. Figure 3.10 shows a structural model of the layered Co oxide, in which a square Ca_2CoO_3 RS-type layer is sandwiched between hexagonal CoO_2 layers. The subsystems of the Ca_2CoO_3 RS-type and hexagonal CoO_2 layers have incommensurate b-axis lengths, that is, $[Ca_2CoO_3]_{0.6}CoO_2$ possesses a misfit-layered structure in the interface between the $[Ca_2CoO_3]$ and $[CoO_2]$ layers, while the RS layer and hexagonal CoO_2 layer are connected through van der Waals bonding. In the Co-layered oxide, with the Ca-based RS layer, strong ionic bonding tightly binds the RS layer with the hexagonal CoO_2 layer, which causes significant interfacial stress. Electronic conduction occurs mainly within the CoO_2 layer with the Ca_2CoO_3 layer serving as a charge reservoir, and while it is difficult to determine in which layer the lattice thermal conductivity is higher, the misfit between layers is expected to hinder the cross-plane phonon transport. Therefore, anisotropy in the thermal and electrical transport properties is expected due to the 2D character of this layered structure. $Bi_2Sr_2Co_2O_y$ is also a layered Co oxide containing a Bi-based RS layer, in which a RS-type $Bi_2Sr_2O_4$ layer is sandwiched between hexagonal CoO_2 layers, which also has a large ZT value. $Bi_2Sr_2Co_2O_y$ is another composite crystal with b-axis lattice misfit, where the CdI_2-type CoO_2 block and the NaCl-type $Bi_2Sr_2O_4$ block are alternately stacked along the c-axis [47].

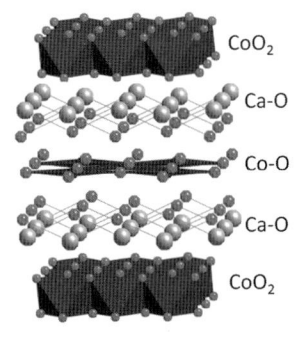

Figure 3.10 Structural model of $[Ca_2CoO_3]_{0.6}CoO_2$.

3.4
Layered Manganese Oxides

The layered manganese oxide consists of 2D manganese oxide layers of edge-shared or corner-shared MnO$_6$ octahedra with Mn^{3+} or Mn^{4+}, and metal ions sandwiched between them. Most of these layered compounds exhibit intercalation properties, and the valences of the Mn ions can change between Mn^{3+} and Mn^{4+} during the intercalation reaction [50]. Therefore, metal ions intercalated between 2D manganese oxide layers can deintercalated or intercalated by electrochemical and/or oxidation/reduction reactions. These properties have attracted attention in battery and electrochemical capacitor research. In this section, the typical crystal structure of the layered manganese oxides and their intercalation properties are introduced.

Various types of layered and tunnel manganese oxide structures have been reported. Figure 3.11 shows several types of layered manganese oxides. The reason why layered manganese oxide has many types of layered structure could be that defects are easily generated in the lattice because the valences of Mn ions are easily changed between Mn^{3+} and Mn^{4+}. Turner and Buseck classified layered and tunnel manganese oxide structures into three groups ((1×n), (2×n), and (3×n)) according to the number of layers of MnO$_6$ octahedra in the vertical

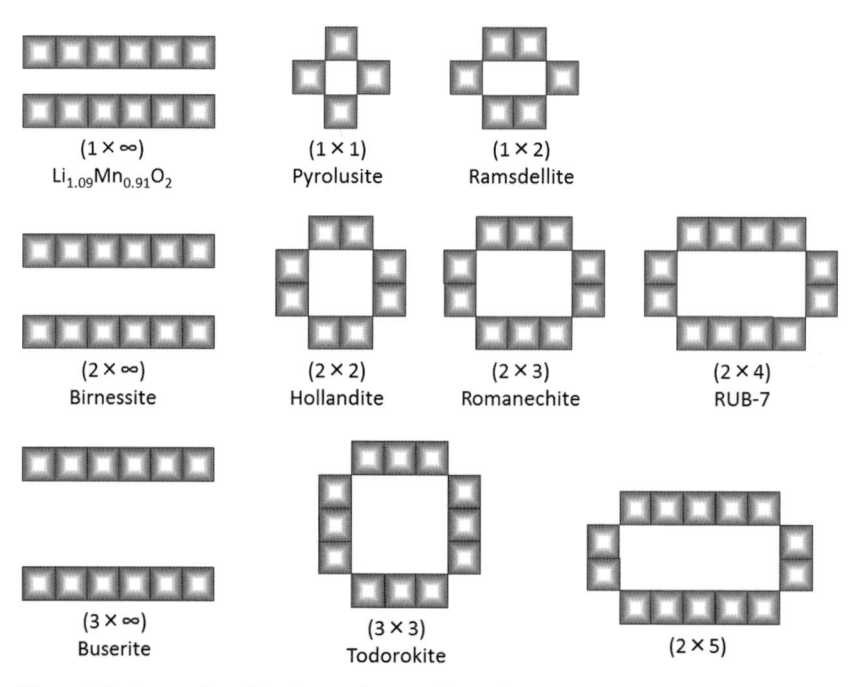

Figure 3.11 Structural models of several types of layered manganese oxides and tunnel manganese oxides.

(first number) and horizontal (second number) directions [51]. When the number of MnO_6 octahedra layers in the horizontal direction is infinite, there are three types of layered oxides: $(1\times n)$, $(2\times n)$, and $(3\times n)$, which correspond to $Li_{1.09}Mn_{0.91}O_2$, layered birnessite and layered buserite, respectively. The order of the layer distance is $Li_{1.09}Mn_{0.91}O_2$ (0.47 nm) < layered birnessite (0.7 nm) < layered buserite (1.0 nm), and the charge density of the manganese oxide layers is decreased in the same order. The intercalation properties of the birnessite and buserite are well known. One-layer water of crystallization and alkaline ions are intercalated in the interlayer of birnessite, while two-layer water of crystallization is intercalated in the interlayer of buserite. When the one-layer water of crystallization is deintercalated from buserite, the layered structure is converted to birnessite.

One feature of the MnO_6 octahedra in birnessite and buserite is that vacancies or defect sites are present in the manganese oxide layers. There are vacancy sites with or without periodicity (although sites with periodicity correspond to defect-type sites), and defect sites with periodicity can be classified into several crystal groups. There are two types of crystal structure, α-$NaMnO_2$ and β-$NaMnO_2$, in layered $NaMnO_2$ consisting of Mn^{3+}. The structure of α-$NaMnO_2$ is similar to birnessite, and Na^+ ions are sandwiched between manganese oxide layers composed of edge-shared octahedra. However, there are no defect sites with periodicity in the MnO_6 octahedral structure. In birnessite, the molar ratio of Na^+/Mn can be changed in the range from 0.2 to 0.7, while the structure is converted to α-$NaMnO_2$ by the dehydration of interlayer water beyond a Na^+/Mn ratio of 0.7. β-$NaMnO_2$ has a zigzag manganese oxide layer with edge-shared MnO_6 octahedra, and Na^+ ions are intercalated between the interlayer [52].

3.4.1
Intercalation Reaction of Layered Manganese Oxides

Layered manganese oxides such as birnessite and buserite have relatively low charge densities, so that the layer distance is easily expanded by the intercalation reaction. In the ion-exchange reaction of birnessite, the basal space is not significantly changed by the intercalation of monovalent metal ions; however, the basal space is expanded by the intercalation of di- or trivalent metal ions because the multivalent metal ions are intercalated into the interlayer as hydrated ions. Monovalent cations in the interlayer are exchanged with protons by acid treatment. Furthermore, disproportionation reaction of Mn^{3+} ($2Mn^{3+} \rightarrow Mn^{4+} + Mn^{2+}$) occurs with acid treatment, so that MnO_6 octahedra with Mn^{4+} are formed in the manganese oxide layer, while some Mn^{2+} ions are dissolved in the solution and some remain in the interlayer. The protons in the protonated form of the layered oxide can be exchanged with organic amines by the intercalation reaction, which results in a large expansion of the interlayer distance.

3.4.2
Tunnel Structure Manganese Oxides

Tunnel structure manganese oxide can be prepared by hydrothermal reaction using birnessite as a starting material. The Na^+ ions in birnessite can be easily exchanged with Li^+, K^+, Rb^+, Mg^{2+}, and Ba^{2+} by the intercalation reaction, although the birnessite structure is retained. When the ion-exchanged layered oxide is treated under hydrothermal conditions, the layered structures are converted to tunnel structures. Figure 3.12 shows a schematic illustration of the typical reaction process for the formation of tunnel structure manganese oxides from layered manganese oxide. Various types of tunnel structures can be prepared by the ion-exchange reaction of birnessite with Li^+, K^+, Ba^{2+}, Rb^+, and Mg^{2+} [53–56].

3.5
Layered Copper Oxides

Layered copper oxides are important materials in the field of superconductivity. $(La,Ba)_2CuO_4$ with a layered structure was the first high-T_c superconductor. Since its discovery, a variety of superconductive copper-oxide phases with a number of different layers have been synthesized. Although some layered copper oxides, such as $La_{2-x}Sr_xCuO_4$ (LSCO), exhibit a relatively large thermoelectric

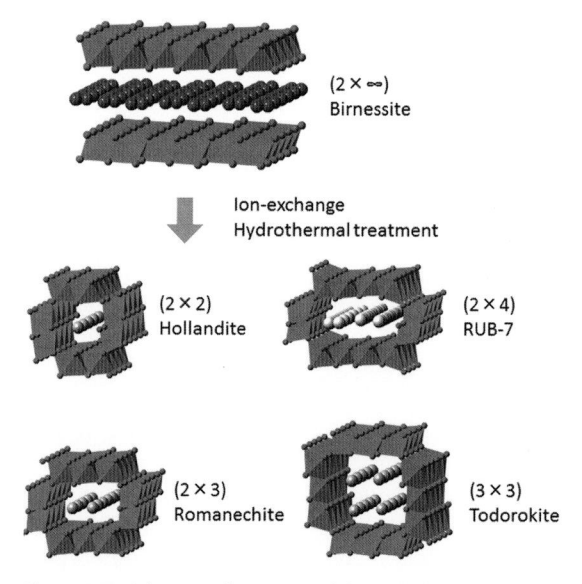

Figure 3.12 Schematic illustration of the typical reaction process for the formation of tunnel structure manganese oxides from layered manganese oxide.

effect [57], most of the research on layered copper oxides has been performed in the field of superconducting materials. In this section, only some important layered materials are explained due to space limitations. For more information regarding the superconducting materials, please refer to the literature references provided [58–62].

3.5.1
Crystal Structure of Layered Copper Oxides

The structures of superconducting layered copper oxides consist of three different types of layers: the perovskite blocks including the so-called infinite-layer block, the RS layer, and the fluorite (FL) layer. The RS layer with a positive charge is considered to act as a hole donor for the copper oxide layer, while the FL layer with negative charge functions as an electron donor. Depending on the layer structure, the layered copper oxides are categorized into two main categories: Category-A, which contains only copper oxide and RS layers, and Category-B, which has all three layers [60]. The structures of category-A contain alternating $M_mO_{m\pm\delta}$ charge reservoir-blocking blocks with m units of $MO1_{\pm\delta/m}$ layers (M: Cu, Bi, Pb, Tl, Hg, Al, Ga, C, B, etc.) and superconductive $Q_{n-1}Cu_nO_{2n}$ blocks with $nCuO_2$ planes and $n-1$ Q metal layers [Q: Ca, rare earth element]. The $M_mO_{m\pm\delta}$ and $Q_{n-1}Cu_nO_{2n}$ blocks are separated from each other by a single AO layer (A: Ba, Sr, La, etc.). The of Category-B structures contain FL-structured layer blocks of B-$[O_2-B]_{s-1}$ [B: (Ce,R)] (R; rare earth element), a blocking layer, and superconductive blocks with CuO_2 planes. An interesting point to note is that whereas each Category-A phase has only one type of (nonsuperconductive) blocking block besides the superconductive block that contains the CuO_2 planes, Category-B phases have two distinct blocking blocks. One of these blocks is the common block with RS- and/or perovskite-structured layers. This block is attached (through AO layers) to the adjacent CuO_2 planes by sharing of the apical oxygen atoms of the CuO_5 pyramids that constitute the CuO_2 planes. The other blocking block of $(Ce,R)-[O_2-(Ce,R)]_{s-1}$ (valence states, Ce(IV) and R(III), are assumed) is then inserted between the basal planes of the CuO_2-plane pyramids, and is of the fluorite structure. In both categories, blocking blocks are considered to not only provide appropriate spacing between the superconductive CuO_2 planes, but also to control the hole-doping level of these planes.

3.5.2
$La_{2x}Ba_xCuO_4$

$La_{2x}Ba_xCuO_4$ has a K_2NiF_4-type structure, which is one of the (110)-layered perovskite oxides [59]. When the barium content (x) is 0.2, superconductivity was observed at 35 K. Although the copper atoms form CuO_6 octahedra with oxygen, the Cu—O bond length in the direction of the c-axis is longer than that of the a,b-axis due to the Jahn–Teller effect. It is considered that the CuO_2 plane has an important role for the superconductivity of this structure. In La_2CuO_4, the valence of the copper ions is +2. This material is an insulator and antiferromagnetic material with the lone pair electrons arranged antiparallel to the CuO_2 plane. When La^{3+} ions are replaced with Ba^{2+}, some Cu^{2+} ions become Cu^{3+} to maintain charge neutrality in the material. When the average valence of the Cu ions becomes +2.2, superconductivity is observed.

3.5.3
$YBa_2Cu_3O_{7-\delta}$

The most thoroughly characterized superconducting layered oxide is $YBa_2Cu_3O_{7-\delta}$ [63], the Y-123 phase, which has a layered structure with the layer sequence, BaO-$CuO_{1-\delta}$-BaO-CuO_2-Y-CuO_2, and thus a single $CuO_{1-\delta}$ chain with an oxygen-deficient perovskite-type arrangement of the Cu and O atoms as the charge-reservoir constituent. This is the first superconducting material with T_c (93 K) higher than the boiling temperature (77 K) of liquid nitrogen. The basic structure is perovskite, which is similar to the layered structure with three perovskite-type unit cells stacked along the c-axis. Figure 3.13 shows a structural model of the Y-123 phase. In the top and bottom units, Ba^{2+} is located in the center, and Y is located in the center of the middle unit. Each unit is $ACuO_3$ (A: Ba or Y), so that the three layers correspond to $(Ba_2Y)Cu_3O_9$. A comparison of the $(Ba_2Y)Cu_3O_9$ composition with the actual composition $(YBa_2Cu_3O_7)$ reveals two more oxygen atoms than the actual composition. Thus, two oxygen atoms

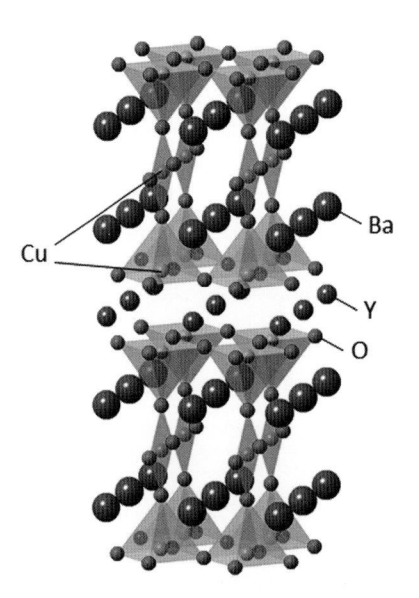

Figure 3.13 Structural model of $YBa_2Cu_3O_{7-\delta}$.

are extracted from the center unit in the actual structure, and the composition of the top and bottom units is represented by the formula $BaCuO_{2.5}$, so the coordination number of Ba becomes 10 (the coordination number of Ba in a perovskite structure is typically 12). The composition of the center part is $YCuO_2$; therefore, the coordination number of Y becomes 8. In this structure, there are two different Cu positions; the five-coordinated pyramidal layer, and the four-coordinated planar layer. The bottom plane of the pyramidal layer is the c plane, which forms a 2D plane, while the CuO_4 plane of the planar layer forms a Cu-O chain in the c-axis direction.

3.6
Layered Titanium Oxide and Niobium Oxide

There are several types of layered titanium and niobium oxides. The layered titanium and niobium oxides exhibit photocatalytic properties for hydrogen production from water using sunlight [64,65]. Figure 3.14 shows typical structural models of layered titanium and niobium oxides: $K_2Ti_4O_9$ [66], $KTiNbO_5$ [67], and $K_4Nb_6O_{17}$ [68]. In the layered structure, alkaline ions reside between 2D oxide layers composed of edge-shared and/or corner-shared TiO_6 and NbO_6 octahedra, as shown in Figure 3.14. Although the structure of the host layers is slightly different, these layered oxides structure and intercalation, optical and electrical properties in common. In most cases, the ion-exchange capacity of the layered oxide is not very high, and the only ion that can be directly and quantitatively exchanged with interlayer ions is basically limited to H^+. However, the protonated layered oxides exhibit interesting intercalation properties for various ions. Some intercalation properties of the layered oxide are introduced here, with $K_4Nb_6O_{17}$ as an example. $K_4Nb_6O_{17}$ has 2D niobium oxide layers composed of corner-shared and edge-shared NbO_6 octahedra, and has two-types of interlayers (interlayer I and interlayer II), which are arranged alternately. Figure 3.14c shows a structural model of $K_4Nb_6O_{17}$. Under ambient conditions, only interlayer I is hydrated, so that the chemical formula can be written as $K_4Nb_6O_{17} \cdot 3 H_2O$.

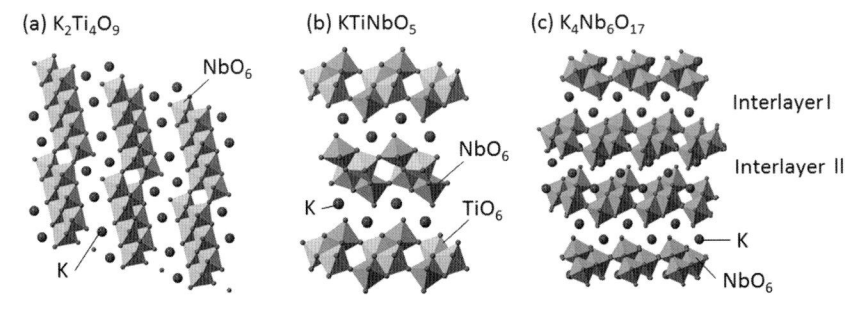

Figure 3.14 Structural models of layered titanium and niobium oxides: (a)$K_2Ti_4O_9$, (b)$KTiNbO_5$, and (c)$K_4Nb_6O_{17}$.

$K_4Nb_6O_{17}$ has relatively high ion-exchange capacity compared with other layered titanium or niobium oxides, and therefore various types of ions such as inorganic and organic ions can be directly intercalated into the interlayer of $K_4Nb_6O_{17}$ [68–70]. The ion-exchange capacity of interlayer I is higher than that of interlayer II. For example, monovalent alkaline ions are intercalated into both interlayers I and II, while divalent ions such as alkaline earth ions are intercalated only into interlayer I. In the case of intercalation of alkyl ammonium ions, the reaction proceeds sequentially from interlayer I to interlayer II. Therefore, if intercalation is stopped at the appropriate time, then a layered oxide where alkyl ammonium ions are intercalated only into interlayer I can be obtained.

3.7
Layered Double Hydroxides

Layered double hydroxides (LDHs), which are one of a series of clay materials, have anion exchange ability, and are therefore known as an anion-exchangeable clay or anionic clay [71,72]. Some LDHs occur naturally and a typical LDH is hydrotalcite (CO_3-intercalated Mg-Al LDH). The LDHs consist of positively charged brucite-like host layers and hydrated exchangeable anions intercalated into the interlayer to maintain the charge balance, as shown in Figure 3.15. The charge of the brucite-like layers arises due to the substitution of some divalent ions with trivalent ions. The chemical composition of LDHs can be expressed by the general formula $[M_{1-x}^{2+}M_x^{3+}(OH)_2][A_{x/n}^{n-} \cdot mH_2O]$, where M^{2+} and M^{3+} represent di- and trivalent metal ions within the brucite-like layers, and A^{n-} is an interlayer anion, where the range of x is typically 0.2–0.33. Combinations of Mg^{2+}–Al^{3+}, Zn^{2+}–Al^{3+}, Fe^{2+}–Al^{3+}, Co^{2+}–Al^{3+}, Co^{2+}–Fe^{3+}, Co^{2+}–Co^{3+}, and Ni^{2+}–Gd^{3+} (M^{2+}–M^{3+}) have been reported, with anions (X) such as Cl^-, Br^-, NO_3^-, CO_3^{2-}, and SO_4^{2-} located in the interlayer [71,73–79]. The anions located in the interlayer can be easily exchanged with various ions such as

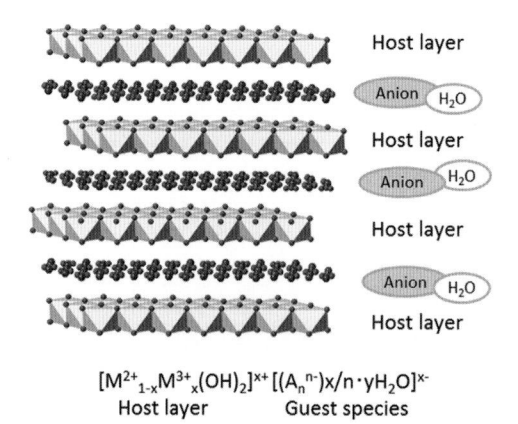

$[M^{2+}_{1-x}M^{3+}_x(OH)_2]^{x+} [(A_n^{n-})x/n \cdot yH_2O]^{x-}$
Host layer   Guest species

Figure 3.15 Structural model of layered double hydroxide.

inorganic anions, organic anions, and complex biomolecules. The applications of LDHs are broad and diverse, ranging from CO_2 adsorbents [80], catalyst [81], ion exchangers [82], fire-retardant additive [83], polymers composites [84], delivery agents [85], and cement additive [86].

3.8
Exfoliation of Layered Structures

Nanosheets prepared by the exfoliation of layered compounds have attracted attention as a method for the controlled fabrication of layered structures because they correspond to a mono-host layer in the layered compounds. There are many types nanosheets, including oxide [21,22,65,87–96] and hydroxide [97–102] nanosheets, which can be prepared by exfoliation of various types of layered oxides and hydroxides. Figure 3.16 shows the exfoliation process for layered cesium titanate [92]. Cesium titanate has a layered structure composed of 2D TiO host layers with guest cesium ions. The guest cesium ions are intercalated with protons in acid solution to form the protonated layered oxide. In this case, exfoliation occurs by the intercalation of tetrabutylammonium ions. One of the important points for exfoliation is the ability of the protonated layered oxide to intercalate organic bases. Therefore, the acidity of protons in the host layer and the basicity of organic species are important considerations for the exfoliation reaction. Figure 3.17 shows a typical atomic force microscopy (AFM) image of a titania nanosheet. Each nanosheet is a single crystal with 2D structure, a thickness of approximately one nanometer, and a lateral size from several hundreds of nanometers to several micrometers. The exfoliated nanosheets can be stacked in layers using layer-by-layer (LBL) [103–105] and Langmuir–Blodgett (LB) [106] techniques. Therefore, nanosheets prepared by the exfoliation process are promising building blocks for the precise fabrication of layered structures by a bottom-up process. Many materials that exhibit superior physical properties, including superconductive, thermoelectric, and photocatalytic properties, are layered compounds. The origin of these properties is based on the interaction

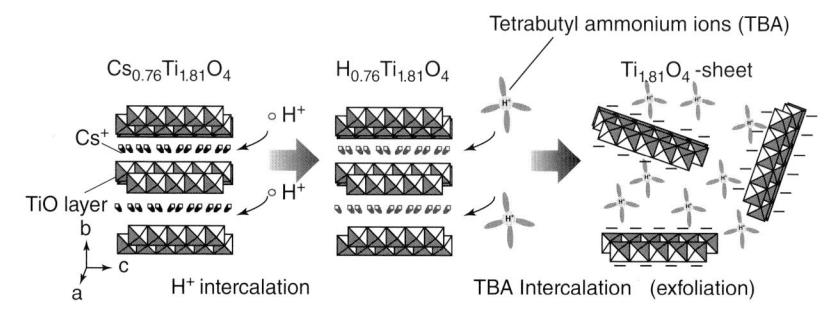

Figure 3.16 Schematic illustration of exfoliation process of layered cesium titanate.

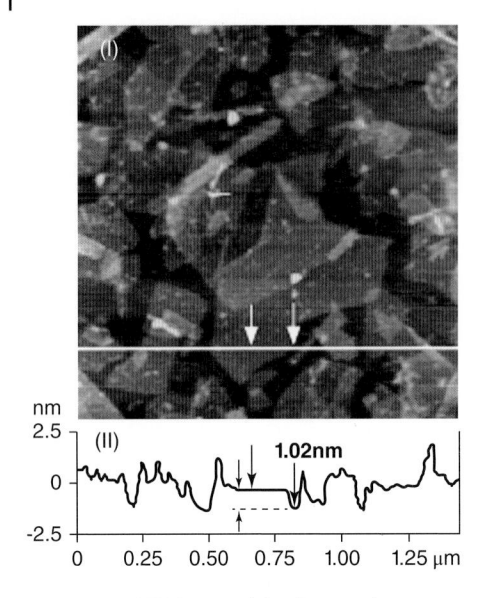

Figure 3.17 AFM image of titania nanosheet

between the individual layers. It is expected that multilayer films of nanosheets will exhibit significantly different properties from the individual nanosheets themselves and lead to potentially new functionalities [107,108].

References

1. Lichtenberg, F., Herrnberger, A., and Wiedenmann, K. (2008) *Prog. Solid State Chem.*, **36**, 253–387.
2. Ruddlesden, S.N. and Popper, P. (1958) *Acta Crystallogr.*, **11**, 54–55.
3. Dion, M., Ganne, M., and Tournoux, M. (1981) *Mater. Res. Bull.*, **16**, 1429–1435.
4. Ebina, Y., Sasaki, T., Harada, M., and Watanabe, M. (2002) *Chem. Mater.*, **14**, 4390–4395.
5. Jacobson, A.J., Lewandowski, J.T., and Johnson, J.W. (1986) *J. Less Common Met.*, **116**, 137–146.
6. Aurivillius, B. (1949) *Ark. Kemi*, **1**, 463–480.
7. Aurivillius, B. (1949) *Ark. Kemi*, **1**, 499–512.
8. Aurivillius, B. (1950) *Ark. Kemi*, **2**, 519–527.
9. Haeni, J.H., Theis, C.D., Schlom, D.G., Tian, W., Pan, X.Q., Chang, H., Takeuchi, I., and Xiang, X.D. (2001) *Appl. Phys. Lett.*, **78**, 3292–3294.
10. Fukuoka, H., Isami, T., and Yamanaka, S. (2000) *J. Solid State Chem.*, **151**, 40–45.
11. Domen, K., Ebina, Y., Ikeda, S., Tanaka, A., Kondo, J.N., and Maruya, K. (1996) *Catal. Today*, **28**, 167–174.
12. Domen, K., Ebina, Y., Sekine, T., Tanaka, A., Kondo, J., and Hirose, C. (1993) *Catal. Today*, **16**, 479–486.
13. A-Paz De Araujo, C., Cuchlaro, J.D., McMillan, L.D., Scott, M.C., and Scott, J.F. (1995) *Nature*, **374**, 627–629.
14. Cummins, S.E. and Cross, L.E. (1968) *J. Appl. Phys.*, **39**, 2268–2274.
15. Al-Areqi, N.A.S., Al-Alas, A., and Beg, S. (2010) *Russ. J. Phys. Chem. A*, **84**, 2334–2344.
16. Kendall, K.R., Navas, C., Thomas, J.K., and Zur Loye, H.C. (1996) *Chem. Mater.*, **8**, 642–649.

References

17 Henderson, N.L., Baek, J., Halasyamani, P.S., and Schaak, R.E. (2007) *Chem. Mater.*, **19**, 1883–1885.

18 Kudo, A. and Kaneko, E. (1998) *Micropor. Mesopor. Mat.*, **21**, 615–620.

19 Toda, K., Watanabe, J., and Sato, M. (1996) *Mater. Res. Bull.*, **31**, 1427–1435.

20 Schaak, R.E. and Mallouk, T.E. (2002) *Chem. Mater.*, **14**, 1455–1471.

21 Ida, S., Ogata, C., Unal, U., Izawa, K., Inoue, T., Altuntasoglu, O., and Matsumoto, Y. (2007) *J. Am. Chem. Soc.*, **129**, 8956–8957.

22 Schaak, R.E. and Mallouk, T.E. (2002) *Chem. Commun.*, (7), 706–707.

23 Tsunoda, Y., Shirata, M., Sugimoto, W., Liu, Z., Terasaki, O., Kuroda, K., and Sugahara, Y. (2001) *Inorg. Chem.*, **40**, 5768–5771.

24 Schaak, R.E. and Mallouk, T.E. (2000) *J. Solid State Chem.*, **155**, 46–54.

25 Schaak, R.E. and Mallouk, T.E. (2000) *J. Am. Chem. Soc.*, **122**, 2798–2803.

26 Hyeon, K.A. and Byeon, S.H. (1999) *Chem. Mater.*, **11**, 352–357.

27 Kim, S.Y., Oh, J.M., Park, J.C., and Byeon, S.H. (2002) *Chem. Mater.*, **14**, 1643–1648.

28 Gopalakrishnan, J., Sivakumar, T., Ramesha, K., Thangadurai, V., and Subbanna, G.N. (2000) *J. Am. Chem. Soc.*, **122**, 6237–6241.

29 Sivakumar, T., Seshadri, R., and Gopalakrishnan, J. (2001) *J. Am. Chem. Soc.*, **123**, 11496–11497.

30 Viciu, L., Liziard, N., Golub, V., Kodenkandath, T.A., and Wiley, J.B. (2004) *Mater. Res. Bull.*, **39**, 2147–2154.

31 Viciu, L., Koenig, J., Spinu, L., Zhou, W.L., and Wiley, J.B. (2003) *Chem. Mater.*, **15**, 1480–1485.

32 Caruntu, G., Kodenkandath, T.A., and Wiley, J.B. (2002) *Mater. Res. Bull.*, **37**, 593–598.

33 Kodenkandath, T.A., Kumbhar, A.S., Zhou, W.L., and Wiley, J.B. (2001) *Inorg. Chem.*, **40**, 710–714.

34 Kodenkandath, T.A., Lalena, J.N., Zhou, W.L., Carpenter, E.E., Sangregorio, C., Falster, A.U., Simmons J Jr., W.B., O'Connor, C.J., and Wiley, J.B. (1999) *J. Am. Chem. Soc.*, **121**, 10743–10746.

35 Viciu, L., Golub, V.O., and Wiley, J.B. (2003) *J. Solid State Chem.*, **175**, 88–93.

36 Sasaki, T., Izumi, F., and Watanabe, M. (1996) *Chem. Mater.*, **8**, 777–782.

37 Schoellhorn, R. (1980) *Angew. Chem., Int. Ed. Engl.*, **19**, 983–1004.

38 Mizushima, K., Jones, P.C., Wiseman, P.J., and Goodenough, J.B. (1980) *Mater. Res. Bull.*, **15**, 783–789.

39 Basch, A., Campo, L.De., Albering, J.H., and White, J.W. (2014) *J. Solid State Chem.*, **220**, 102–110.

40 Carlier, D., Saadoune, I., Ménétrier, M., and Delmas, C. (2002) *J. Electrochem. Soc.*, **149**, A1310–A1320.

41 Carlier, D., Van der Ven, A., Delmas, C., and Ceder, G. (2003) *Chem. Mater.*, **15**, 2651–2660.

42 Okumura, T., Yamaguchi, Y., Shikano, M., and Kobayashi, H. (2012) *J. Mater. Chem.*, **22**, 17340–17348.

43 Whittingham, M.S. (2004) *Chem. Rev.*, **104**, 4271–4301.

44 Berthelot, R., Carlier, D., and Delmas, C. (2011) *Nature Mater.*, **10**, 74–80.

45 Motohashi, T., Katsumata, Y., Ono, T., Kanno, R., Karppinen, M., and Yamauchi, H. (2007) *Chem. Mater.*, **19**, 5063–5066.

46 Takada, K., Osada, M., Izumi, F., Sakurai, H., Takayama-Muromachi, E., and Sasaki, T. (2005) *Chem. Mater.*, **17**, 2034–2040.

47 Ohtaki, M. (2011) *J. Ceram. Soc. Jpn.*, **119**, 770–775.

48 Takada, K., Sakurai, H., Takayama-Muromachi, E., Izumi, F., Dilanian, R.A., and Sasaki, T. (2003) *Nature*, **422**, 53–55.

49 Van Nong, N., Pryds, N., Linderoth, S., and Ohtaki, M. (2011) *Adv. Mater.*, **23**, 2484–2490.

50 Yabuuchi, N. and Komaba, S. (2014) *Sci. Technol. Adv. Mater.*, **15**, 043501.

51 Turner, S. and Buseck, P.R. (1981) *Science*, **212**, 1024–1027.

52 Feng, Q., Kanoh, H., and Ooi, K. (1999) *J. Mater. Chem.*, **9**, 319–333.

53 Feng, Q., Kanoh, H., Miyai, Y., and Ooi, K. (1995) *Chem. Mater.*, **7**, 1722–1727.

54 Feng, Q., Yanagisawa, K., and Yamasaki, N. (1998) *J. Porous Mat.*, 5, 153–161.

55 Feng, Q., Yanagisawa, K., and Yamasaki, N. (1996) *Chem. Commun.*, 14, 1607–1608.

56 Shen, Y.F., Zerger, R.P., DeGuzman, R.N., Suib, S.L., McCurdy, L., Potter, D.I., and O'Young, C.L. (1993) *Science*, **260**, 511–515.

57 Komoto, H. and Takeuchi, T. (2009) *J. Electron. Mater.*, **38**, 1365–1370.

58 Anderson, P.W. (1987) *Science*, **235**, 1196–1198.

59 Bednorz, J.G. and Müller, K.A. (1986) *Zeitschrift für Physik B Condens.Matter*, **64**, 189–193.

60 Karppinen, M. and Yamauchi, H. (1999) *Mater. Sci. Eng. R: Reports*, **26**, 51–96.

61 Yamauchi, H. and Karppinen, M. (1997) *Superlattice. Microst.*, **21**, 127–152.

62 Yamauchi, H., Karppinen, M., and Tanaka, S. (1996) *Physica C*, **263**, 146–150.

63 Wu, M.K., Ashburn, J.R., Torng, C.J., Hor, P.H., Meng, R.L., Gao, L., Huang, Z.J., Wang, Y.Q., and Chu, C.W. (1987) *Phys. Rev. Lett.*, **58**, 908–910.

64 Kudo, A., Sayama, K., Tanaka, A., Asakura, K., Domen, K., Maruya, K., and Onishi, T. (1989) *J. Catal.*, **120**, 337–352.

65 Kudo, A., Tanaka, A., Domen, K., Maruya, K.i., Aika, K.i., and Onishi, T. (1988) *J. Catal.*, **111**, 67–76.

66 Izawa, H., Kikkawa, S., and Koizumi, M. (1982) *J. Phys. Chem.*, **86**, 5023–5026.

67 Wadsley, A.D. (1964) *Acta Crystallogr.*, **17**, 1545.

68 Gasperin, M. and Bihan, M.T.Le. (1982) *J. Solid State Chem.*, **43**, 346–353.

69 Kinomura, N., Kumada, N., and Muto, F. (1985) *J. Chem. Soc. Dalton.*, 2349–2351.

70 Nakato, T., Kuroda, K., and Kato, C. (1992) *Chem. Mater.*, **4**, 128–132.

71 Khan, A.I. and O'Hare, D. (2002) *J. Mater. Chem.*, **12**, 3191–3198.

72 Wang, Q. and Ohare, D. (2012) *Chem. Rev.*, **112**, 4124–4155.

73 Benito, P., Herrero, M., Labajos, F.M., and Rives, V. (2010) *Appl. Clay Sci.*, **48**, 218–227.

74 Cavani, F., Trifirò, F., and Vaccari, A. (1991) *Catal. Today*, **11**, 173–301.

75 Evans, D.G. and Slade, R.C.T. (2006) *Struct. Bonding (Berlin)*, **119**, 1–87.

76 He, J., Wei, M., Li, B., Kang, Y., Evans, D.G., and Duan, X. (2006) In *Layered Double Hydroxides*, vol. **119**, pp. 89–119.

77 Khan, A.I., Ragavan, A., Fong, B., Markland, C., O'Brien, M., Dunbar, T.G., Williams, G.R., and O'Hare, D. (2009) *Ind. Eng. Chem. Res.*, **48**, 10196–10205.

78 Manzi-Nshuti, C., Wang, D., Hossenlopp, J.M., and Wilkie, C.A. (2008) *J. Mater. Chem.*, **18**, 3091–3102.

79 Williams, G.R., Khan, A.I., and O'Hare, D. (2006) *Struct. Bonding (Berlin)*, **119**, 161–192.

80 Wang, Q., Luo, J., Zhong, Z., and Borgna, A. (2011) *Energy Env. Sci.*, **4**, 42–55.

81 Xu, X., Lu, R., Zhao, X., Xu, S., Lei, X., Zhang, F., and Evans, D.G. (2011) *Appl. Catal. B: Environ.*, **102**, 147–156.

82 Millange, F., Walton, R.I., Lei, L., and O'Hare, D. (2000) *Chem. Mater.*, **12**, 1990–1994.

83 Nyambo, C., Songtipya, P., Manias, E., Jimenez-Gasco, M.M., and Wilkie, C.A. (2008) *J. Mater. Chem.*, **18**, 4827–4838.

84 Leroux, F. and Besse, J. (2001) *Chem. Mater.*, **13**, 3507–3515.

85 Alcântara, A.C.S., Aranda, P., Darder, M., and Ruiz-Hitzky, E. (2010) *J. Mater. Chem.*, **20**, 9495–9504.

86 Plank, J., Zhimin, D., Keller, H., Hössle, F. v., and Seidl, W. (2010) *Cement Concrete Res.*, **40**, 45–57.

87 Abe, R., Shinohara, K., Tanaka, A., Hara, M., Kondo, J.N., and Domen, K. (1998) *J. Mater. Res.*, **13**, 861–865.

88 Han, Y.S., Park, I., and Choy, J.H. (2001) *J. Mater. Chem.*, **11**, 1277–1282.

89 Ida, S., Ogata, C., Eguchi, M., Youngblood, W.J., Mallouk, T.E., and Matsumoto, Y. (2008) *J. Am. Chem. Soc.*, **130**, 7052–7059.

90 Omomo, Y., Sasaki, T., Wang, L., and Watanabe, M. (2003) *J. Am. Chem. Soc.*, **125**, 3568–3575.

91 Ozawa, T.C., Fukuda, K., Akatsuka, K., Ebina, Y., and Sasaki, T. (2007) *Chem. Mater.*, **19**, 6575–6580.

92 Sasaki, T., Watanabe, M., Hashizume, H., Yamada, H., and Nakazawa, H. (1996) *J. Am. Chem. Soc.*, **118**, 8329–8335.

93 Schaak, R.E. and Mallouk, T.E. (2000) *Chem. Mater.*, **12**, 3427–3434.

94 Schaak, R.E. and Mallouk, T.E. (2000) *Chem. Mater.*, **12**, 2513–2516.

95 Sugimoto, W., Terabayashi, O., Murakami, Y., and Takasu, Y. (2002) *J. Mater. Chem.*, **12**, 3814–3818.

96 Takagaki, A., Lu, D., Kondo, J.N., Hara, M., Hayashi, S., and Domen, K. (2005) *Chem. Mater.*, **17**, 2487–2489.

97 Adachi-Pagano, M., Forano, C., and Besse, J.P. (2000) *Chem. Commun.*, 91–92.

98 Ida, S., Shiga, D., Koinuma, M., and Matsumoto, Y. (2008) *J. Am. Chem. Soc.*, **130**, 14038–14039.

99 Leroux, F., Adachi-Pagano, M., Intissar, M., Chauvière, S., Forano, C., and Besse, J.P. (2001) *J. Mater. Chem.*, **11**, 105–112.

100 Ma, R., Liu, Z., Takada, K., Iyi, N., Bando, Y., and Sasaki, T. (2007) *J. Am. Chem. Soc.*, **129**, 5257–5263.

101 Ma, R., Takada, K., Fukuda, K., Iyi, N., Bando, Y., and Sasaki, T. (2008) *Angew. Chem,. Int. Ed.*, **47**, 86–89.

102 Nadeau, P.H., Wilson, M.J., McHardy, W.J., and Tait, J.M. (1984) *Science*, **225**, 923–925.

103 Fang, M. (1999) *Chem. Mater.*, **11**, 1526–1532.

104 Li, L., Ma, R., Ebina, Y., Fukuda, K., Takada, K., and Sasaki, T. (2007) *J. Am. Chem. Soc.*, **129**, 8000–8007.

105 Osada, M. and Sasaki, T. (2012) *Adv. Mater.*, **24**, 210–228.

106 Muramatsu, M., Akatsuka, K., Ebina, Y., Wang, K., Sasaki, T., Ishida, T., Miyake, K., and Haga, M.A. (2005) *Langmuir*, **21**, 6590–6595.

107 Ida, S., Sonoda, Y., Ikeue, K., and Matsumoto, Y. (2010) *Chem. Commun.*, **46**, 877–879.

108 Li, B.W., Osada, M., Ozawa, T.C., Ebina, Y., Akatsuka, K., Ma, R., Funakubo, H., and Sasaki, T. (2010) *ACS Nano*, **4**, 6673–6680.

4
Organoclays and Polymer-Clay Nanocomposites

M.A. Vicente[1] and A. Gil[2]

[1]*Universidad de Salamanca, GIR–QUESCAT, Departamento de Química Inorgánica, Plaza de la Merced, S/N, 37008 Salamanca, Spain*
[2]*Universidad Pública de Navarra, Departamento de Química Aplicada, Campus de Arrosadia, s/n, 31006 Pamplona, Spain*

4.1
Introduction

Technological advances have generated a demand for functionalized materials, where metals, ceramics, or plastics do not meet all the requirements for some new applications [1]. Hybrid organic–inorganic materials have become one of the main trends in research in several scientific areas. These composites represent an interesting approach to the preparation of novel materials for academic and industrial uses. Moreover, they could display improved characteristics for applications demanding advanced functional materials [2,3].

The properties of a hybrid composite can be tailored by changing its composition or managing the generated interface. The versatile applications of hybrid organic–inorganic materials stem from the combination of the characteristics of the organic component (plasticity, flexibility, conductivity, absorptivity, and reactivity, among others) with the functionality of the inorganic constituents (inertia, thermal, chemical, and mechanical resistance, and morphological structure, among others). Indeed, the interface developed between the organic and inorganic components generates distinct multifunctional properties. Moreover, the synergism between the inorganic and organic domains overcomes their individual features [1,4].

Hybrid materials can be prepared by several methodologies, and they can be applied in sectors such as motor vehicle catalysts, cosmetics, dyes, fibers, anticorrosive products, materials for medical imaging, and anticancer drugs. The applications of hybrid materials have been widely reviewed [2–11].

Natural and/or synthetic clays can be used in the synthesis of hybrid materials. Natural clays are versatile supports for the insertion of a wide array of organic and/or inorganic hosts, using various methods as intercalation, functionalization,

---

Handbook of Solid State Chemistry, First Edition. Edited by Richard Dronskowski, Shinichi Kikkawa, and Andreas Stein.
© 2017 Wiley-VCH Verlag GmbH & Co. KGaA. Published 2017 by Wiley-VCH Verlag GmbH & Co. KGaA.

and/or pillaring, and organometallic complexes [12,13]. The bidimensional character and facile swelling of the interlayer space of some clays make them promising materials for the insertion of organic components via functionalization of the surface, intercalation, or adsorption. Layered solids can undergo exfoliation and/or delamination as well as intercalation reactions with organic molecules via ion exchange and surface functionalization, among other processes [14].

The clay-polymer nanocomposites (CPNs) area has received significant attention in the recent years. Work on layered clays has gone on for quite some time as it is of great importance to researchers studying soils, adsorbents, catalysts, in general, functionalized materials. A novel idea of generating materials with a superstructure similar to olefins was published by Ballard and Rideal [15]. The concept presented was to synthesize materials with lamellar structures connected with the molecules as found in polyethylene, polypropylene, and so on. This was accomplished with a completely unrelated molecularly structured material, an inorganic clay. The authors used vermiculite, a layered clay mineral, and *tied* the lamellae together with organic cations. The resulting film showed flexibility in a stress–strain experiment superior to a poly(ethylene terephthalate) (PET) film, higher modulus, and energy to break; however, the new material lacked the recoverability that PET can display at low extensions. This is related to the nature of the clay–cation interaction being ionic versus the covalent bonding between the molecules and lamellae in semicrystalline polymers. Researchers from Toyota developed nylon 6-montmorillonite clay hybrids [16,17]. These new materials have found application in the automotive industry due to their temperature stability. The nylon 6-clay hybrids display physical properties that are much improved over the neat polymer. The clay hybrid displayed increased heat distortion temperature, higher modulus and tensile strength, greater impact strength, less water adsorption, lower permeability, and thermal expansion coefficient than nylon 6. The research in the clay-polymer nanocomposite area has expanded to include such polymers as polypropylene, poly(ethylene oxide), polystyrene, among others.

Before describing organoclay structure and chemistry, a rudimentary understanding of the polymer nanocomposite itself is required. A traditional composite containing micron or larger particles/fibers/reinforcement can best be thought of as containing two major components, the bulk polymer and the filler/reinforcement, and a third, very minor component, or interfacial polymer (see Figure 4.1). Poor interfacial bonding between the bulk polymer and filler can result in an undesirable balance of properties, or at worst, material failure under mechanical, thermal, or electrical load. In a polymer nanocomposite, since the reinforcing particle is at the nanometer scale, it is actually a minor component in terms of total weight or volume percent in the final material. If the nanoparticle is fully dispersed in the polymer matrix, the bulk polymer also becomes a minor, and in some cases, a nonexistent part of the final material. With the nanofiller homogenously dispersed in the polymer matrix, the entire polymer becomes an interfacial polymer, and the properties of the material begin to change. Changes in properties of the interfacial polymer become magnified in

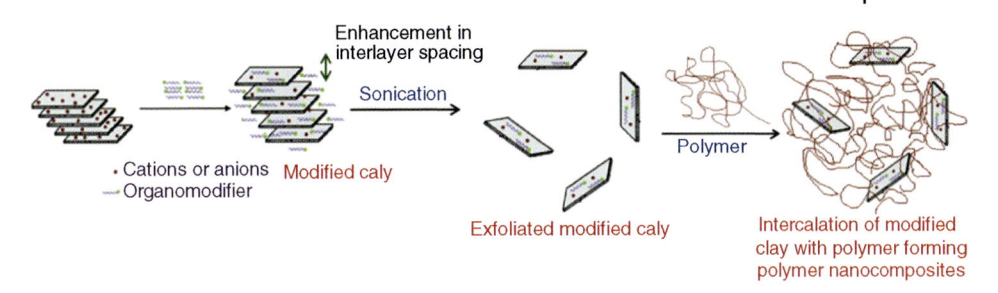

Figure 4.1 Schematic diagram showing clay modification and intercalation of polymer to form clay-polymer nanocomposites (CPNs). (Reproduced from Ref. [18] with permission.)

the final material, and great improvements in properties are seen. Therefore, a polymer nanocomposite is a composite where filler and bulk polymer are minor components, and the interfacial polymer is the component that dictates material properties. With this in mind, the design of the nanoparticle is critical to nanocomposite structure, and careful understanding of nanoparticle chemistry and structure is needed.

4.2 Organophilization of Clay Minerals

Clay minerals are used in their organophilic form when employed in the formulation of CPNs. The reason is that pristine clays are strongly polar and hydrophilic, thus being incompatible with the hydrophobic polymers [19]. In fact, to convert the originally hydrophilic clays into organophilic ones is important also for other applications, in general, for those implying their interaction with an organic molecule, as the adsorption of pollutants, rheological control agents, paints, cosmetics, personal care products, oil-well drilling fluids, and so on [20]. Besides, the intercalation of organic species in the interlayer region of the clay minerals causes the attractive forces between contiguous platelets to decrease, making their exfoliation easier and favouring their interaction with the polymers and, in general, with other organic species. Industrially, organophilic bentonites are usually prepared in large amounts by reaction of the bentonites with quaternary alkylammonium ions, without removing the excess cationic surfactants [21]. However, for specific applications, a large variety of organic species are used for the organophilization of clay minerals, not only bentonites but also other clay minerals, particularly important is the case of kaolinite that cannot be organophilized by direct treatment with organic cations. Due to the importance of having good organophilic clays for the preparation of CPNs, this process is now briefly summarized. Longer reviews of this process can be found elsewhere [20,22].

Some general applications of organophilic clays have already been mentioned, but new applications have been developed in the last years. A complete list would be very long, but some examples should be cited to illustrate the importance of this process: biomedical applications (controlled drug delivery, adjuvants in vaccines, regenerative medicine), agriculture (herbicide formulations), enzyme immobilization, sensors, antibacterial agents, luminescent or catalytic materials (themselves, or incorporating a metallic cation), or removal of organic pollutants, among others [10,23–34].

4.2.1
Organophilization of Smectites

The smectites, or 2:1 expandable phyllosilicates, are characterized by their layered structure and their negative charge in the layers, compensated by exchangeable cations located in the interlayer region (cation exchange capacity, CEC). The most used clay mineral in this group is montmorillonite, saponite, and hectorite being other representative minerals. Accordingly, smectites can be organophilized by treatment with organic cations or with neutral molecules. In the first case, the process involves the changing of the natural cations of the clay (Na^+, K^+, Ca^{2+}, etc.) by the organic cations, which are intercalated into the interlayer space of the clay, usually with expansion of this region, while in the second case, the neutral molecules are similarly intercalated into the interlayer region, but without exchanging the natural cations.

Tetra-substituted ammoniums are by far the most used cations for organophilization, and montmorillonite modified with quaternary ammonium salts is commercially available with the name of Cloisite*. This includes tetraalkylammonium cations (the most common), but also tetraaryl substituents, and mixtures between them, that is, a lot of organic alkyl or aryl groups have been considered as substituents in these cations. The use of tetraalkylammonium cations has been largely studied and, as indicated before, is a routine method for industrial organophilization. The studies carried out have involved the variation of the nature of the clay mineral and of the organic cations, the length of the organic chains, the CEC of the clay, among other factors. The logical tendency is that the polar part of the ammonium cation is in contact with the clay layers, while the organic chains are repelled by them. However, the amount of organic cations needed for compensating the CEC (it may be considered that the CEC can largely vary from one clay mineral sample to other), the size of the organic chains, and so on, influence the orientation and self-assembly of the organic cations, from a monolayer, bilayer, pseudotrilayer, and even paraffinic arrangements (see Figure 4.2).

Besides tetrasubstituted ammoniums, tetrasubstituted phosphoniums have also been used as organophilization agents. The process of organophilization is similar, although an advantage in the behavior of the phosphonium clays, due to their enhanced thermal stability, has been claimed in the melt processing of CPNs [35]. Also, better adsorbent properties toward organic pollutants have been reported, being attributed to the lower degree of hydration of the phosphonium cations compared to the ammonium ions [36]. Recent examples of

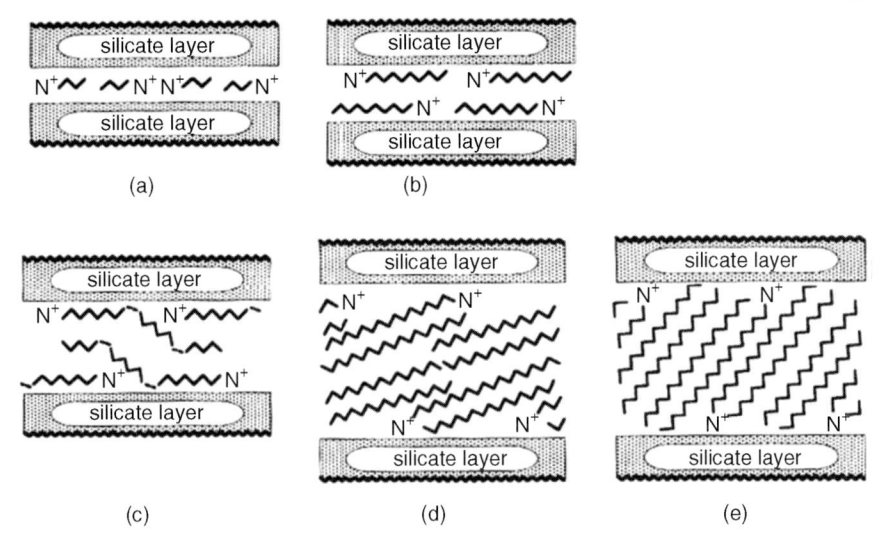

Figure 4.2 Arrangements of alkylammonium cations in the interlayer space of smectites: (a) monolayers, (b) bilayers, (c) pseudotrilayers, (d and e) paraffin-type arrangements. (Reproduced from Ref. [22] with permission.)

phosphonium-organophilized smectites have been reported by Bouzid et al. [37] and Sáenz Ezquerro et al. [38].

Dyes and metallic complexes have also been frequently intercalated into smectite clay minerals. Some of them have cationic forms, for example, the dyes stibazolium and methylene blue or the complexes $[Cu(en)_2]^{2+}$, $[Ru(bpy)_3]^{2+}$, and $[Co(en)_3]^{3+}$, and their interaction with the clay minerals causes their organophilization. Logically, there are a very large number of cationic metallic complexes, but it is necessary that the ligands are organic in nature to affect the organophilization of the clay minerals.

As indicated, organophilization can also be carried out with neutral molecules, which enter into the interlayer space, usually with expansion of this region, but without removing the original exchange cations of the clay mineral. The nature of the molecules that can be used can vary widely. One of the most common includes alcohols, both containing one or more alcohol groups. In fact, ethylene glycol (1,2-ethanediol) has been widely used for evaluating the expansion capacity of clay minerals. Other molecules used comprise amines, ethers, crown ethers, anhydrides, carboxylic acids, cryptands, aminoacids, neutral dyes, or neutral metallic complexes, among others. Special mention can be made of the so-called biomolecules, including peptides, enzymes, or proteins. Also, the use of alkoxysilanes (octyltriethoxysilane, (3-aminopropyl)triethoxysilane, 3-mercaptopropyltrimethoxysilane, and much others) merits special mention. The alkoxy groups react with the OH groups present on the surface of the clay minerals, producing an organoclay material after hydrolysis and condensation, all the process being usually called silanization, and allowing to incorporate Si to the initial clay [39,40].

4.2.2
Organophilization of Kaolinite

Kaolinite is a very abundant and versatile clay mineral belonging to the 1 : 1 group of phyllosilicates (other clay minerals in this group are nacrite, dickite, and halloysite), and thus its organophilization is attractive. However, a priori kaolinite cannot be intercalated and in fact it was considered nonexpansible for a long time. First, it does not have exchangeable cations, because its layers are neutral, and consequently no compensating cations are needed in its structure. In addition, the interaction between the layers is very strong, involving hydrogen bonds, dipole–dipole interactions, and van der Waals forces, thus preventing the easy insertion of molecules in the interlayer region.

Fortunately, this can be solved by a host–guest displacement methodology, involving two or more steps (see Figure 4.3). It has been observed that kaolinite can be intercalated by small and highly polar species, such as dimethylsulfoxide (DMSO), by compounds forming hydrogen bonds, such as hydrazine, urea, and formamides, and by alkaline salts of small carboxylates (acetates, propionates, butyrates, or isovalerates of potassium, rubidium, cesium, or ammonium), and this intercalation is the first step of the mentioned host–guest displacement methodology. The intercalation of these species separates the kaolinite layers by several angstroms, allowing that in the second step these species can be substituted by new ones. Amines, alcohols, aminoacids, metallic alkoxides, metallic complexes, or porphyrins, among other compounds, have been incorporated into kaolinite in this way [22].

4.2.3
Organophilization of Fibrous Clays

Sepiolite and palygorskite are the most representative clay minerals in this group. Their structure makes these minerals not expansible, because of a periodical inversion of the SiO_4 tetrahedra in their sheets, which generates the fibers and also the tunnels between them, along one direction. Consequently, the

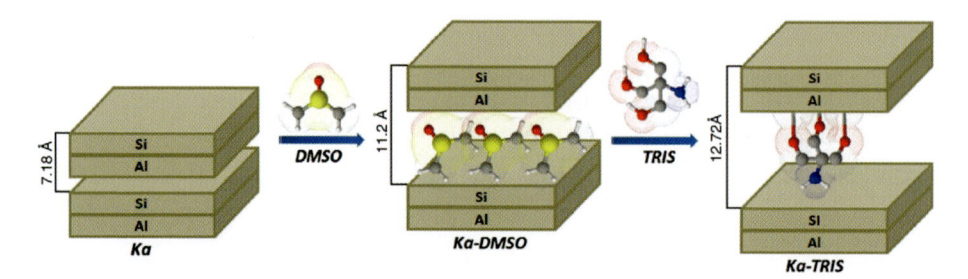

Figure 4.3 Schematic representation of the functionalization of kaolinite with tris(hydroxymethyl)aminomethane by the host–guest displacement methodology. (Reproduced from Ref. [39] with permission.)

intercalation of these minerals is not possible. Although the organic species should penetrate into the channels of the structure, the interaction mainly takes place in the external surface of the clays, where a lot of silanol groups are located (in fact, in bigger amount than in smectite clays, due to the discontinuity of the layers). Thus, the organophilization of the surface of these clay minerals by grafting is favored, mainly by organosilanes, whose active groups react with the silanol clay groups giving rise to stable siloxane bridges [41].

4.2.4
Organophilization of Other Clays

Although the main groups of organophilic clay minerals have been reported in the previous points, some clay minerals have certain peculiarities that merit brief comments.

Laponite is a commercially available synthetic hectorite. Laponite exfoliates very easily in aqueous dispersion, and it does not show significant long-range order in the solid state. Although its organophilization has not been widely studied, grafting appears to be the most adequate procedure [42].

Vermiculite is a layered, expandable mineral, very similar to smectites, but with higher layer charge, which in fact restricts its expansion. Its organophilization has been carried out mainly by using tetraalkyl-substituted ammonium salts [43–45].

The development of new hydrothermal methods for preparing synthetic clays has allowed the direct synthesis of organophilized clays by adding the organic species, mainly tetraalkyl-substituted ammonium salts, to the initial mixture, before submitting it to the hydrothermal treatment. This method has been mainly applied for the preparation of organophilic hectorites [46,47].

4.3
Synthesis, Structures, and Physicochemical Characterizations

Three common methods have been developed for the synthesis of clay-polymer nanocomposites [48]. First, an organically modified clay can be combined with a polymer by a solution process. It is also possible to combine a molten polymer with a modified clay without solvent. The last procedure entails blending monomer with the clay, which is in general organically modified, followed by subsequent polymerization of the monomer within the clay galleries.

CPNs generally have two subgroupings, intercalated or delaminated/exfoliated (see Figure 4.4). The intercalated type of clay-polymer hybrid has highly extended single chains confined between the clay sheets, within the gallery regions. The clay sheets retain a well-ordered, periodic, stacked structure. The intercalation process can be monitored by tracking the increasing basal spacing from X-ray scattering, since the galleries must expand to accommodate larger molecules. The delaminated or exfoliated structure ideally has well dispersed

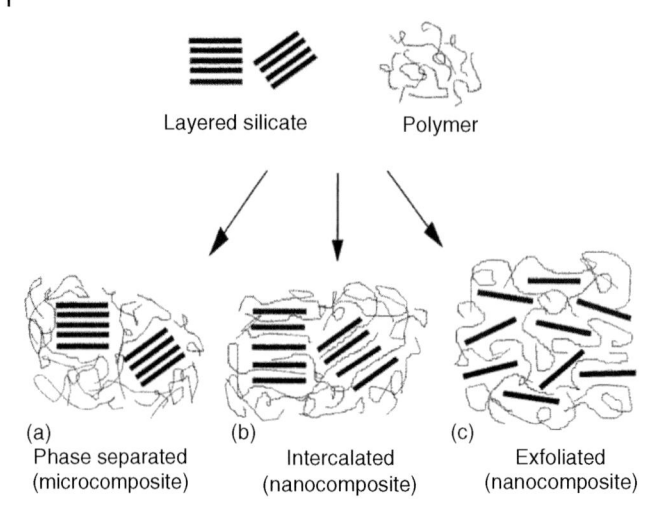

Figure 4.4 Differences between the structures of microcomposites and intercalated or exfoliated nanocomposites. (Reproduced from Ref. [48] with permission.)

and randomized clay sheets within a matrix of the polymer chains. In this case, the sheets have lost their stacked orientation, and if the structure is truly random then no distinct interlayer reflections should be observable by X-ray scattering. In general, it is difficult to distinguish between intercalated and delaminated nanocomposites, and there is a lack of accord in the literature.

CPNs are usually characterized by various techniques, X-ray diffraction (XRD), transmission electron microscopy (TEM), thermal analyses, and infrared spectroscopy (FTIR) being the most used. ^{1}H, ^{13}C, ^{27}Al, ^{29}Si, and ^{19}F, nuclear magnetic resonance (NMR) has also been applied [49,50]. It should be noted that the characterization of the clay mineral component is difficult if its amount in the final CPNs is very low.

The position, form and intensity of the (001) reflection in XRD diffractograms give important and immediate information, the position of this reflection informs if the clay mineral has been intercalated (shift to lower angles, higher basal spacings) or if it has been delaminated/exfoliated (the c-ordering decreases or completely disappears), and its form and intensity is also related to the crystallinity of the clay, its periodicity in the c-dimension (see Figure 4.5). This technique also allows obtaining a similar information on the transformation from the pristine clay to the organophilic one, if such treatment is carried out. Obviously, this information can be obtained from smectite or kaolinite clay minerals, not from fibrous clays (not expandable neither exfoliating) or from those clays already exfoliated before the treatments.

The information given by XRD can be complemented by TEM that allows observing the dispersion of the clay mineral particles into the polymeric matrix at a nanometric scale. This technique can be combined with a focused ion beam (FIB), that is, bombarding the solid with ions instead of electrons.

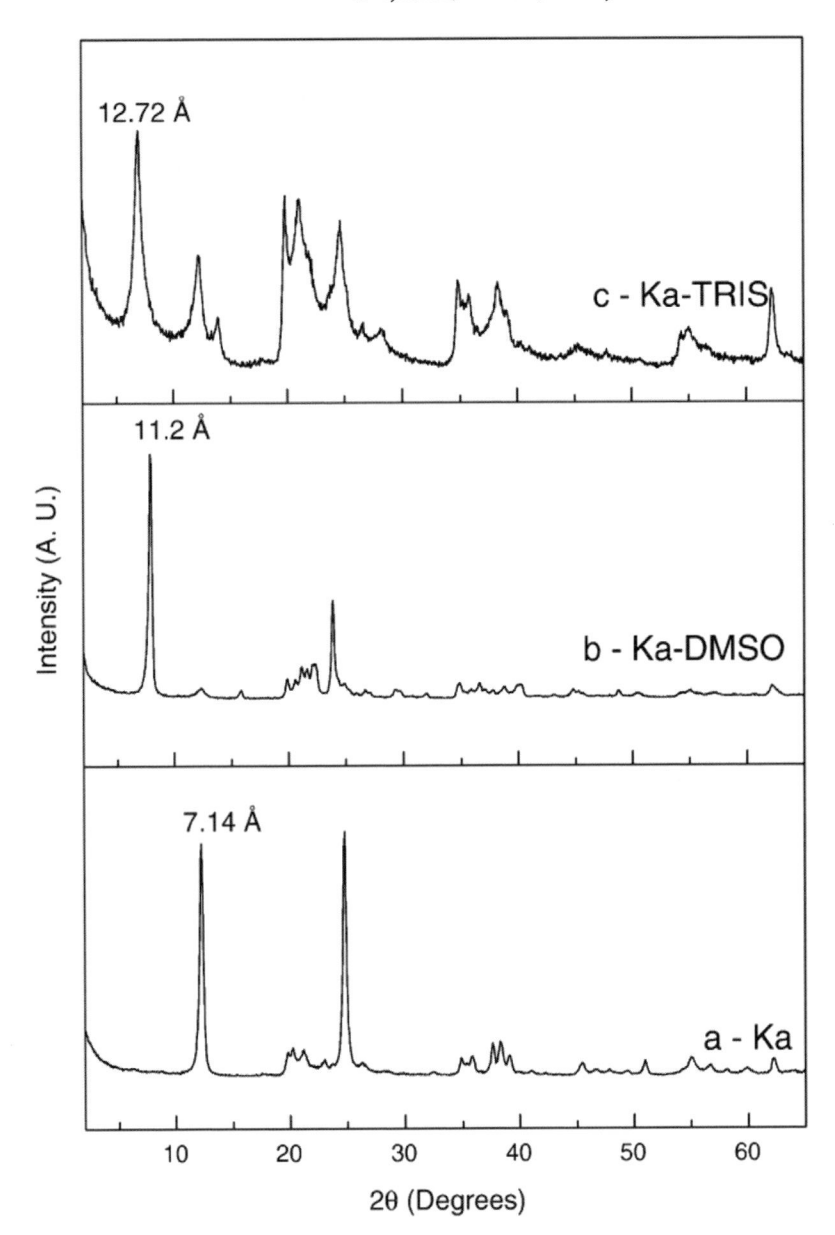

Figure 4.5 X-ray diffractograms showing the successive intercalation of kaolinite with DMSO and tris(hydroxymethyl)aminomethane (process schematized in Figure 4.3). (Reproduced from Ref. [39] with permission.)

Infrared spectroscopy allows observing the polymerization of the monomers, as the vibrations of the groups participating in this reaction are affected. Information is obtained also on the short distance ordering of the polymer, by analysis of the hydrogen bonds signals. In some cases, particular information can be obtained depending on the nature of the polymer, for example, on the trans conformation of polyamide from the vibrations of its methylene groups. Bands from the clay minerals are difficult to observe if its amount in the final nanocomposite is very low; changes in the hydroxyl vibrations should be observed, although the polymer moieties do not always interact with the clay minerals.

Thermal analyses are applied to characterize the CPNs and also for studying their important thermal properties (*vide infra*). Differential thermal analysis (DTA) is used to determine the temperatures of melting, crystallization, glass transition, or decomposition of the polymer, which depends on the amount of the clay minerals, while thermogravimetry gives complementary information on the mass loss in each step.

4.4
Clay-Polymer Nanocomposites: Properties and Applications

This section will focus on where clay nanocomposites have been used to improve property performance, especially to yield improvement in more than one area, and also where the improvements have led to commercial use. The use of clay nanoparticles as additives to enhance polymer performance has been reported in recent years. CPNs are of particular interest because of their enhancements, relative to an unmodified polymer, in a large number of physical properties including mechanical properties, thermal resistance, electrical properties, and gas permeation, among other characteristics [18,49,51,52].

4.4.1
Mechanical Properties

The most common use of CPNs has been in mechanical reinforcement of thermoplastics, especially polyamide-6 (PA) and polypropylene (PP), using the elastic modulus as the most important mechanical property. Other properties could also be evaluated as impact resistance, tensile strength, and elongation at break. The modulus can estimate the stresses and strains exerted by external forces. The restricted mobility of the polymers at the filler interface has been reported to be the main cause of the high modulus of CPNs and its effect is more pronounced above the glass transition temperature, T_g (see Figure 4.6) [53].

Clay-polyamide-6 and -polypropylene nanocomposites were used for automotive applications to replace metal components near the engine block and to increase the flexural/tensile modulus maintaining the impact performance but yielding some weight savings [54].

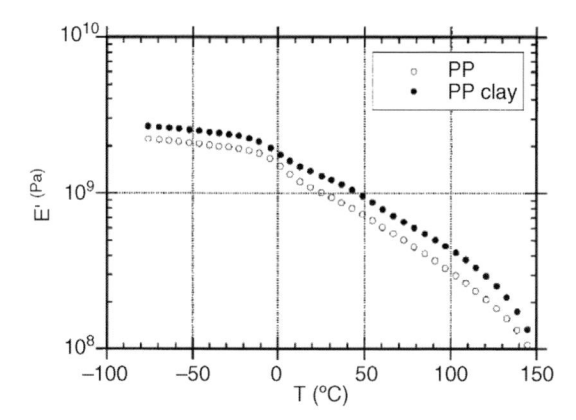

Figure 4.6 Dynamic elastic modulus of a clay-polypropylene nanocomposite as a function of temperature. (Reproduced from Ref. [53] with permission.)

4.4.2
Thermal Properties and Fire Retardance

The thermal stability of a material is usually assessed by thermogravimetric analysis (TGA). Generally, the incorporation of a low clay amount, about 2.5–5 wt%, into the polymer matrix was found to enhance thermal stability by acting as a superior insulator and mass transport barrier to the volatile products generated during decomposition. In the early stages of the thermal decomposition, the clay would shift the decomposition to higher temperature. After that, this heat barrier effect would result in a reverse thermal stability, accelerating the decomposition process, in addition to the heat flow supplied by the outside heat source. The first work on the improved thermal stability of a CPN, polymethylmethacrylate and montmorillonite, was reported by Blumstein [55]. The nanocomposites were prepared by free radical polymerization of the monomer methyl methacrylate intercalated in the clay, showing by XRD analysis an increase of 0.76 nm in the basal spacing. The author showed that the polymer intercalated between the clay layers degraded at a temperature 40–50 °C higher than the degradation of the pure unfilled polymer matrix. The thermal stability could be related to a decrease in the relative amount of polymer end-capped by the carbon–carbon double bond. The author proposed that the enhanced thermal stability of the CPN was not only due to the difference in chemical structure but also to restricted thermal motion of the monomer in the silicate layer. Since this first work and in the 1990s, various groups have drawn attention to the thermal stabilization brought by the CPNs [56,57] proposing several possible origins for the observed thermal stability improvement such as inactivation of the centers active in the polymers by interaction with the filler, the chemical nature of the studied polymeric material and its degradation mechanism.

Another possibility to the evaluation of the thermal stability of synthesized nanocomposites is considering the ISO 305:1990 standard [58], where two methods are specified to determine the thermal stability of chlorinated polymers. The samples are exposed, in sheet form, at elevated temperatures. The first method evaluates the color development at several intervals of time, for a given temperature (e.g., 180 °C). In the second method, the temperature is reduced (e.g., 70 °C) and the time is increased up to several days [59] (see Figure 4.7).

The decreased flammability of CPNs is one of the most important effects due to the presence of clay minerals. The clay minerals reduce the maximum heat release rate (HRR) and hence minimize the flame propagation. The presence of the clay produces the formation of a durable carbon–silicate char observed in the burn tests of thermoplastic polymers as polypropylene (PP). Clay minerals are used in combination with traditional flame retardant fillers, such as magnesium oxide, brominated polycarbonate/Sb_2O_3, or decabromodiphenyl oxide/Sb_2O_3.

4.4.3
Electrical and Electrochemical Properties

The first application of CPNs in this topic was as cable insulators related to their low electric conductivity [50]. Moreover, CPNs based on poly(ethylene oxide) have been reported to be anisotropic solid electrolyte materials presenting ionic conductivity values various orders of magnitude higher than the clay, when the clay is exchanged by small cations such as Li^+ or H^+ [51]. Comparing this type of CPNs with electrolytes in solution, the silicate layer acts as the anion. Therefore, ionic conductivity should exclusively involve a cationic transport mechanism. These nanocomposites have interesting potential applications as components in electrochemical devices, such as rechargeable solid-state batteries, membranes, and electrochemical sensors.

4.4.4
Gas Permeation

Another common application of CPNs is for gas-barrier materials. Clay nanoparticles create a complex network or "tortuous path" in the polymer matrix, such that various gases either diffuse very slowly or not at all through polymer chains in thin films or thicker polymer parts (see Figure 4.8). The tortuosity factor is defined as the ratio of the actual distance d' that the gas must travel to the shortest distance d that it would travel in the absence of barriers. This relation could also be expressed in terms of the length L, the width W, and the volume fraction of the sheets ϕ_s as

$$\tau = \frac{d'}{d} = 1 + \frac{L}{2W}\phi_s.$$

From this expression, it is possible to see that a sheet-like morphology is efficient at maximizing the path length due to the large length-to-width ratio. According to the model proposed by Nielsen [60], the effect of tortuosity on the

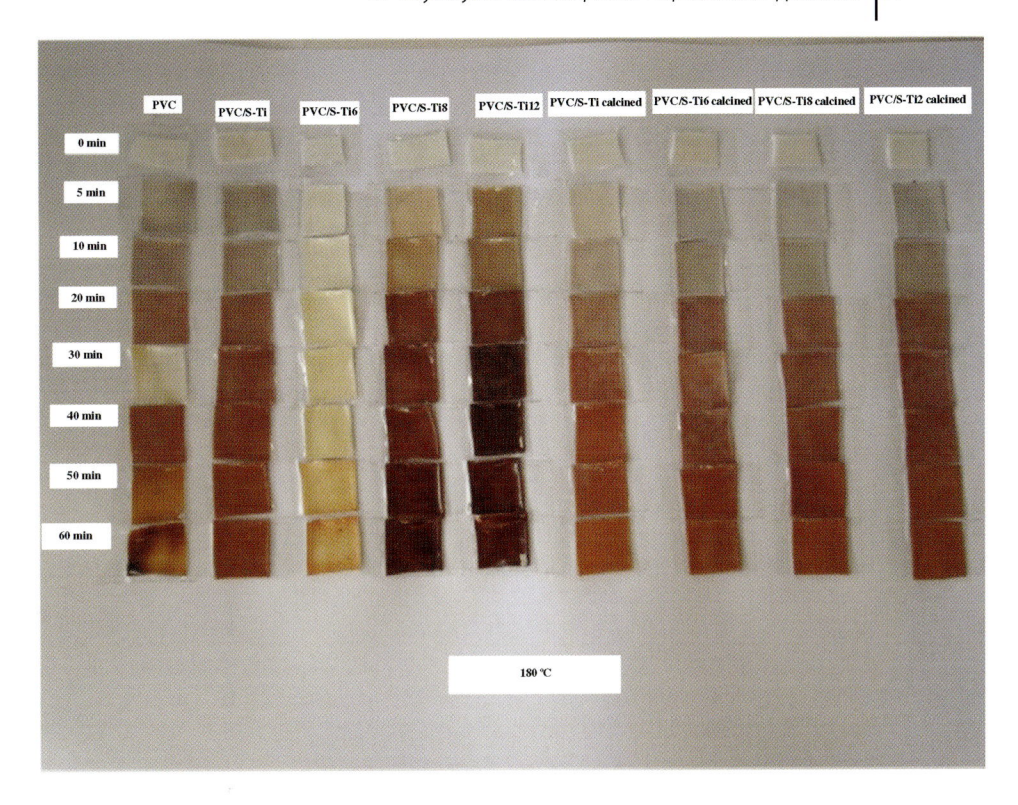

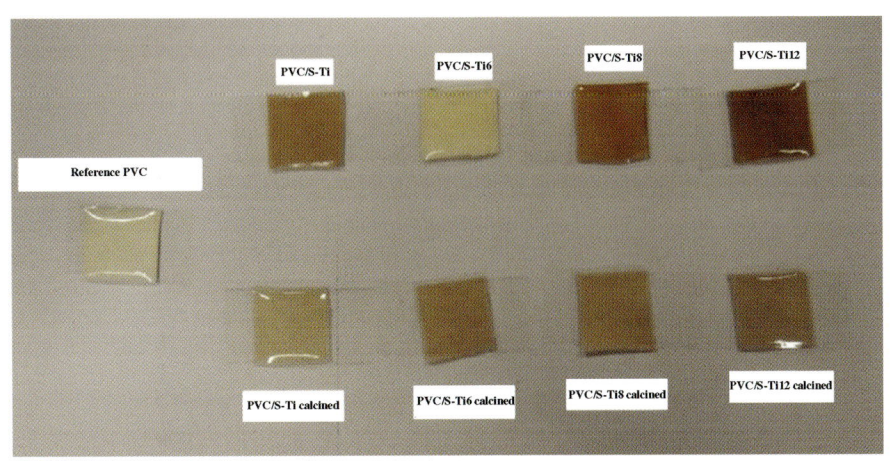

Figure 4.7 Evolution with time of the color and thermal resistance of titanium organoclay PVC nanocomposites. (Reproduced from Ref. [59] with permission.)

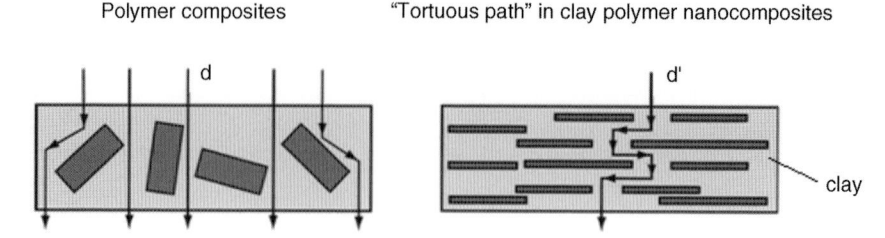

Figure 4.8 Formation of tortuous path in CPNs.

permeability can be expressed as

$$P_{CPN}/P_P = (1 - \phi_s)/\tau,$$

where P_{CPN} and P_P represent the permeability of the nanocomposite and the pure polymer.

The direct benefit of the formation of such path is observed in CPNs by improved barrier properties, with a decrease in the permeability of small gases as O_2, H_2O, CO_2, among others [61–64]. It has been reported that gas permeability can be reduced to 20% of its initial value using 4.8% of nanoclay dispersion [65]. This is also the case of polymers that are used as packaging material as polyethylene terephthalate and nylon.

4.5
Future Applications

Clay polymer nanocomposites are used in many applications to enhance existing properties of the polymers. CPNs will continue to be used for enhanced mechanical, flammability, and gas barrier properties, but fundamental limits in clay chemistry prevent them from being used easily in applications requiring electrical/thermal conductivity or optical applications. Along those lines, combinations of organoclays with other nanofillers to obtain a true multifunctional material will likely arise in the future. Combining an organoclay with carbon nanotubes, or quantum dots, could yield a very interesting nanocomposite with enhanced mechanical, flammability, thermal, and electrical properties, allowing it to be a drop-in replacement for many materials in a complex part.

Acknowledgments

The authors thank the partial support from the Spanish Ministry of Economy and Competitiveness (MINECO) and the European Regional Development Fund (FEDER) (Grant MAT2013-47811-C2-R).

References

1. Kickelbick, G. (2007) Introduction to hybrid materials, in *Hybrid Materials Synthesis, Characterization and Application* (ed. G. Kickelbick), Wiley-VCH Verlag GmbH, Weinheim, pp. 1–48.
2. Sanchez, C., Shea, K.J., and Kitagawa, S. (2011) *Chem. Soc. Rev.*, **40**, 471–472.
3. Sanchez, C., Belleville, P., Popall, M., and Nicole, L. (2011) *Chem. Soc. Rev.*, **40**, 696–753.
4. Sanchez, C., Julia, B., Belleville, P., and Popall, M. (2005) *J. Mater. Chem.*, **15**, 3559–3592.
5. Gómez–Romero, P. and Sanchez, C. (2004) Hybrid materials, functional applications. An introduction, in *Functional Hybrid Materials* (eds P. Gómez–Romero and C. Sanchez), Wiley-VCH Verlag GmbH, Weinheim, pp. 1–14.
6. Clemente–León, M., Coronado, E., Martí–Gastaldo, C., and Romero, F.M. (2011) *Chem. Soc. Rev.*, **40**, 473–497.
7. Orilall, M.C. and Wiesner, U. (2011) *Chem. Soc. Rev.*, **40**, 520–535.
8. Férey, G., Serre, C., Devic, T., Maurin, G., Jobic, H., Llewellyn, P.L., Weireld, G.D., Vimont, A., Daturi, M., and Chang, J. (2011) *Chem. Soc. Rev.*, **40**, 550–562.
9. Zhao, L. and Lin, Z. (2012) *Adv. Mater.*, **24**, 4353–4368.
10. Ruiz-Hitzky, E., Aranda, P., Darder, M., and Rytwo, G. (2010) *J. Mater. Chem.*, **20**, 9306–9321.
11. Zamboulis, A., Moitra, N., Moreau, J.J.E., Cattoën, X., and Man, M.W.C. (2010) *J. Mater. Chem.*, **20**, 9322–9338.
12. Takagi, S., Eguchi, M., Tryk, D.A., and Haruo, I. (2006) *J. Photochem. Photobiol. C*, **7**, 104–126.
13. Nagendrappa, G. (2011) *Appl. Clay Sci.*, **53**, 106–138.
14. Nagendrappa, G. (2002) *Resonance*, **7**, 64–77.
15. Ballard, D.G.H. and Rideal, G.R. (1983) *J. Mater. Sci.*, **18**, 545–561.
16. Usuki, A., Kojima, Y., Kawasumi, M., Okada, A., Fukushima, Y., Kurauchi, T., and Kamigaito, O. (1993) *J. Mater. Res.*, **8**, 1174–1178.
17. Usuki, A., Kojima, Y., Kawasumi, M., Okada, A., Kurauchi, T., and Kamigaito, O. (1993) *J. Mater. Res.*, **8**, 1179–1183.
18. Kotal, M. and Bhowmick, A.K. (2015) *Prog. Polym. Sci.*, **51**, 127–187.
19. Christidis, G.E. (2013) Assessment of industrial clays, in *Handbook of Clay Science*, 2nd edn, Developments in Clay Science, Vol **5B** (eds F. Bergaya and G. Lagaly), Elsevier, pp. 425–449.
20. De Paiva, L.B., Morales, A.R., and Valenzuela Díaz, F.R. (2008) *Appl. Clay Sci.*, **42**, 8–24.
21. Lagaly, G. and Dékány, I. (2013) Colloid clay science, in *Handbook of Clay Science*, 2nd edn, Developments in Clay Science, Vol **5A** (eds F. Bergaya and G. Lagaly), Elsevier, pp. 243–345.
22. Lagaly, G., Ogawa, M., and Dékány, I. (2013) Clay mineral–organic interactions, in *Handbook of Clay Science*, 2nd edn, Developments in Clay Science, Vol **5A** (eds F. Bergaya and G. Lagaly), Elsevier, pp. 435–505.
23. Sedaghat, M.E., Ghiaci, M., Aghaei, H., and Soleimanian–Zad, S. (2009) *Appl. Clay Sci.*, **46**, 131–135.
24. Jarraya, I., Fourmentin, S., Benzina, M., and Bouaziz, S. (2010) *Chem. Geol.*, **275**, 1–8.
25. de Faria, E.H., Nassar, E.J., Ciuffi, K.J., Vicente, M.A., Trujillano, R., Rives, V., and Calefi, P.S. (2011) *ACS Appl. Mater. Inter.*, **3**, 1311–1318.
26. Bouwe, R.G.B., Tonle, I.K., Letaief, S., Ngameni, E., and Detellier, C. (2011) *Appl. Clay Sci.*, **52**, 258–265.
27. de Faria, E.H., Ricci, G.P., Marçal, L., Nassar, E.J., Vicente, M.A., Trujillano, R., Gil, A., Korili, S.A., Ciuffi, K.J., and Calefi, P.S. (2012) *Catal. Today*, **187**, 135–149.
28. Dedzo, G.K., Letaief, S., and Detellier, C. (2012) *J. Mater. Chem.*, **22**, 20593–20601.
29. Olivero, F., Carniato, F., Bisio, C., and Marchese, L. (2012) *J. Mater. Chem.*, **22**, 25254–25261.
30. Holešová, S., Samlíková, M., Pazdziora, E., and Valášková, M. (2013) *Appl. Clay Sci.*, **83–84**, 17–23.

31 Lovo de Carvalho, A., Ferreira, B.F., Gomes Martins, C.H., Nassar, E.J., Nakagaki, S., Sippel Machado, G., Rives, V., Trujillano, R., Vicente, M.A., Gil, A., Korili, S.A., de Faria, E.H., and Ciuffi, K.J. (2014) *J. Phys. Chem. C*, **118**, 24562–24574.

32 Malek, N.A.N.N. and Ramli, N.I. (2015) *Appl. Clay Sci.*, **109–110**, 8–14.

33 Calabrese, I., Cavallaro, G., Lazzara, G., Merli, M., Sciascia, L., and Liveri, M.L.T. (2016) *Adsorption*, **22**, 105–116.

34 Čeklovský, A., Boháč, P., and Czímerová, A. (2016) *Appl. Clay Sci.*, **126**, 68–71.

35 Mittal, V. (2012) *Appl. Clay Sci.*, **56**, 103–109.

36 Kukkadapu, R.K. and Boyd, S.A. (1995) *Clays Clay Miner.*, **43**, 318–323.

37 Bouzid, S., Khenifi, A., Bennabou, K.A., Trujillano, R., Vicente, M.A., and Derriche, Z. (2015) *Chem. Eng. Commun.*, **202**, 520–533.

38 Sáenz Ezquerro, C., Ibarz Ric, G., Crespo Miñana, C., and Sacristán Bermejo, J. (2015) *Appl. Clay Sci.*, **111**, 1–9.

39 de Faria, E.H., Ciuffi, K.J., Nassar, E.J., Vicente, M.A., Trujillano, R., and Calefi, P.S. (2010) *Appl. Clay Sci.*, **48**, 516–521.

40 Avila, L.R., de Faria, E.H., Ciuffi, K.J., Nassar, E.J., Calefi, P.S., Vicente, M.A., and Trujillano, R. (2010) *J. Colloid Interface Sci.*, **341**, 186–193.

41 Ruiz–Hitzky, E., Aranda, P., Darder, M., and Fernandes, F.M. (2013) Fibrous clay mineral–polymer nanocomposites, in *Handbook of Clay Science*, 2nd edn, Developments in Clay Science, Vol. **5A** (eds F. Bergaya and G. Lagaly), Elsevier, pp. 721–741.

42 Borsacchi, S., Geppi, M., Ricci, L., Ruggeri, G., and Veracini, C.A. (2007) *Langmuir*, **23**, 3953–3960.

43 Brigatti, M.F., Galán, E., and Theng, B.K.G. (2013) Structure and mineralogy of clay minerals, in *Handbook of Clay Science*, 2nd edn, Developments in Clay Science, Vol **5A** (eds F. Bergaya and G. Lagaly), Elsevier, pp. 21–81.

44 Slade, P.G. and Gates, W.P. (2007) *Clays Clay Miner.*, **55**, 131–139.

45 Holešová, S., Valášková, M., Plevová, E., Pazdziora, E., and Matějová, K. (2010) *J. Colloid Interface. Sci.*, **342**, 593–597.

46 Carrado, K.A. (2000) *Appl. Clay Sci.*, **17**, 1–23.

47 Zhou, C.H., Tong, D., and Li, X. (2010) Synthetic hectorite: preparation, pillaring and applications in catalysis, in *Pillared Clays and Related Catalysts* (eds A. Gil, S.A. Korili, R. Trujillano, and M.A. Vicente), Springer, pp. 67–97.

48 Alexandres, M. and Dubois, Ph. (2000) *Mater. Sci. Eng.*, **28**, 1–63.

49 Pavlidou, S. and Papaspyrides, C.D. (2008) *Prog. Polym. Sci.*, **33**, 1119–1198.

50 Lambert, J.-F. and Bergaya, F. (2013) Smectite–polymer nanocomposites, in *Handbook of Clay Science*, 2nd edn, Developments in Clay Science, Vol **5A** (eds F. Bergaya and G. Lagaly), Elsevier, pp. 679–706.

51 Pinnavaia, T.J. and Beall, G.W. (eds) (2000) *Polymer-Clay Nanocomposites*, John Wiley & Sons, Ltd, Chichester.

52 Sinha Ray, S. and Okamoto, M. (2003) *Prog. Polym. Sci.*, **28**, 1539–1641.

53 Gloaguen, J.M. and Lefebvre, J.M. (2000) *Polymer*, **42**, 5841–5847.

54 Hasegawa, N., Kawasumi, M., Kato, M., Usuki, A., and Okada, A. (1998) *J. Appl. Polym. Sci.*, **67**, 87–92.

55 Blumstein, A. (1965) *J. Polym. Sci. A*, **3**, 2665–2673.

56 Burnside, S.D. and Giannelis, E.P. (1995) *Chem. Mater.*, **7**, 1597–1600.

57 Wang, S.J., Long, C.F., Wang, X.Y., Li, Q., and Qi, Z.N. (1998) *J. Appl. Polym. Sci.*, **69**, 1557–1561.

58 ISO/DIS 305 (2001 *Plastics–Determination of Thermal Stability of Poly (vinyl Chloride), Related Chlorine–Containing Homopolymers and Copolymers and Their Compounds–Discoloration Method*, European Committee for Standardization, Brussels.

59 Albeniz, S., Vicente, M.A., Trujillano, R., Korili, S.A., and Gil, A. (2014) *Appl. Clay Sci.*, **99**, 72–82.

60 Nielsen, L.E. (1967) *J. Macromol. Sci. A*, **1**, 929–942.

61 Yano, K., Usuki, A., Okada, A., Kurauchi, T., and Kamigaito, O. (1993) *J. Polym. Sci. Pol. Chem.*, **31**, 2493–2498.

62 Ogasawara, T., Ishida, Y., Ishikawa, T., Aoki, T., and Ogura, T. (2006) *Compos. Part A*, **37**, 2236–2240.

63 Herrera–Alonso, J.M., Marand, E., Little, J., and Cox, S.S. (2009) *Polymer*, **50**, 5744–5748.

64 Lewis, E.L.V., Duckett, R.A., Ward, I.M., Fairclough, J.P.A., and Ryan, A.J. (2003) *Polymer*, **44**, 1631–1640.

65 Messersmith, P.B. and Giannelis, E.P. (1995) *J. Polym. Sci. Polym. Chem.*, **33**, 1047–1057.

5
Zeolite and Zeolite-Like Materials

Watcharop Chaikittisilp and Tatsuya Okubo

The University of Tokyo, Department of Chemical System Engineering, 7-3-1 Hongo, Bunkyo-ku, 113–8656 Tokyo, Japan

5.1
Introduction

Porous solids have enormous impacts on our human lives. As one of remarkable porous solids, zeolites have found several applications across a wide range of industries, for example, as detergent builders and catalysts that can selectively crack heavy components in crude oil into smaller, useful chemicals. The annual markets worldwide for natural and synthetic zeolites are estimated to be about 3×10^6 and 2×10^6 tons, respectively [1,2]. In 1756, the first zeolite, stilbite, was discovered as a mineral in nature by the Swedish mineralogist Axel Fredrik Cronstedt [3]. He found that this mineral intumesced when heated in a blowpipe flame due to a large amount of water present in the mineral interiors. He therefore named this class of minerals "*zeolite*," derived from Greek words "*zeo*" (to boil) and "*lithos*" (stone).

In modern chemistry, zeolites are referred to as crystalline microporous aluminosilicates that are constructed from corner-sharing, tetrahedrally coordinated $TO_{4/2}$ primary units (where T is a tetrahedral atom, that is, silicon or aluminum) [4]. During the past decades, however, the research and development on zeolites has advanced beyond the aluminosilicate compositions; therefore, the usage of the term "zeolite" has evolved to include nonaluminosilicate counterparts. To be strict with the definition of "zeolite" (exclusively for aluminosilicates), the terms "*zeolite-like*" and "*zeotype*" materials have been introduced to describe similar materials but with different compositions, that is, pure silica, other metallosilicates (where a substituent can be boron, gallium, germanium, titanium, zinc, etc.), and nonsilicates such as aluminophosphates [5]. In this chapter, the term "zeolite" is used in its broadened meaning, covering pure silica and other metallosilicate compositions.

This chapter addresses the fundamental aspects of zeolites and zeolite-like materials, starting with a brief description of zeolite framework structures. The following contents will be focused on the synthesis of zeolites. In addition, it is our intention of including recent topics in the field such as zeolites with hierarchical porous structures, new synthetic methods toward scalable production of zeolites, and some recent applications such as catalytic conversions of biomass into monomers for a 100% renewable commodity plastic.

5.2
Structure and Classification

In general, the framework structure and the chemical composition are the most fundamental determinations reflecting the intrinsic properties of zeolites. Structurally, the primary units of zeolite frameworks are $SiO_{4/2}$ and $AlO_{4/2}$ tetrahedra. Each tetrahedron is linked to four neighboring tetrahedral, yielding structural building units such as four-ring (*4r*), six-ring (*6r*), and double four-ring (*d4r*). Infinite connection of $SiO_{4/2}$ and $AlO_{4/2}$ tetrahedra then results in the three-dimensional, four-connected zeolite frameworks (see Figure 5.1) [4–6].

It is noteworthy that Al–O–Al linkages have not been observed in aluminosilicate zeolites because the cluster of negative charges arising from the adjoining aluminate tetrahedra is energetically unstable. Löwenstein first rationalized the absence of Al–O–Al linkages in aluminosilicates due to the unfavorable interactions [7]. He suggested that whenever two aluminate tetrahedra are adjacent, at least one of them must have a coordination number larger than four (i.e., five or six). This Löwenstein's Al–O–Al avoidance rule explains the maximum aluminum substitution of 50% (i.e., Si/Al = 1) in zeolites. This rule is also applicable to Al–O–Al linkages in aluminophosphates and Ga–O–Ga linkages in gallosilicate zeolites.

To date, there are 232 framework structures of zeolite and zeolite-like materials approved by the Structure Commission of the International Zeolite

$TO_{4/2}$ Tetrahedra   Structural building unit   Extended, infinite zeolite framework

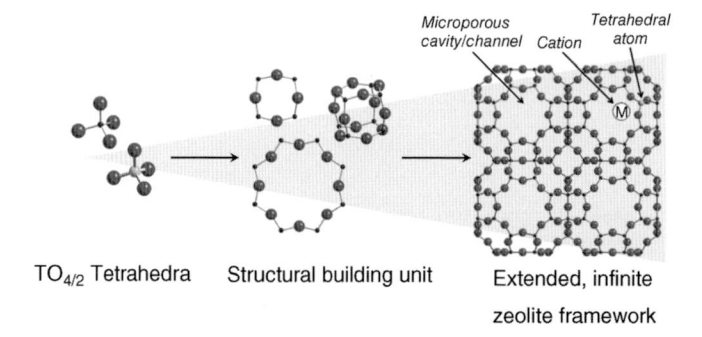

Figure 5.1 An extended, infinite zeolite framework, constructed from $TO_{4/2}$ primary units, consisting of different functional features, that is, substituting tetrahedral atoms, counter-balanced cations, and microporous cavities/channels.

FAU   *BEA   MOR

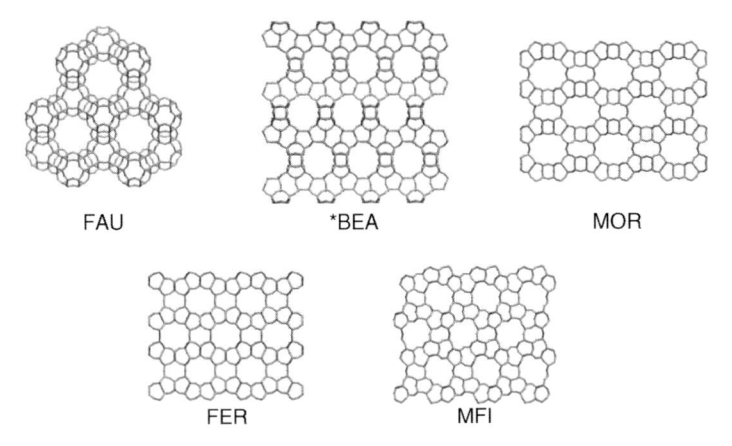

FER   MFI

Figure 5.2 Framework structures of the "big five" zeolites. These zeolites dominate most of the commercial production for use as a catalyst. Other zeolites that are practically used as catalysts in industrial processes include AEL, CHA, EUO, LTL, MTW, MWW, and RHO.

Association (IZA) that assigns a three-letter code (Framework Type Code, FTC) to each framework structure[1] [5]. Framework structures of the "big five" zeolites for catalysis (FAU, *BEA, MFI, MOR, and FER) are depicted in Figure 5.2.

Classification of zeolites can be made in several ways such as by the size of largest pores, the connectivity of channel system, and the chemical composition. Extra-large-, large-, medium-, and small-pore zeolites have the largest pores formed from more than 12, 12, 10, and 8 tetrahedral atoms, respectively. Rings linked to form the zeolite frameworks are defined as a number of tetrahedral atoms in each ring, for example, 6-ring (6r) comprises six tetrahedral atoms (and also six oxygen atoms). The theoretical maximum pore apertures of each ring are listed in Table 5.1. In addition to pore sizes, an amount of substituting

Table 5.1 Apertures formed from different numbers of tetrahedral atoms [4].

No. of T-atoms in ring	Maximum free aperture/Å
4	1.6
5	1.5
6	2.8
8	4.3
10	6.3
12	8.0
18	15

1) Database of Zeolite Structures, http://www.iza-structure.org/databases/.

aluminum or a Si/Al molar ratio is used to categorize zeolites into low-silica (Si/Al ≤ 2), intermediate silica (2 < Si/Al ≤ 5), and high-silica (Si/Al > 5) zeolites, according to Flanigen's notation [8].

Basically, there are three different levels to functionalize each zeolite: (i) the type and amount of tetrahedral atoms substituting in the frameworks, (ii) the type and structure of extra-framework, cationic species that balance the negative charges of zeolites, and (iii) the guest species encapsulated in the microporous cavities/channels of zeolites (see Figure 5.1). In aluminosilicate zeolites, isomorphous substitution of trivalent aluminum for tetravalent silicon in the tetrahedral site of the zeolite frameworks generates negative charges that constitute Brönsted acid sites when they are counter-balanced by protons. In addition to protons, the counterions can be metal ion species, having various structures and coordination, being able to exhibit extraordinary redox behaviors. Typical counterions found in natural zeolites are the cationic form of group IA (e.g., sodium and potassium) and IIA (such as calcium) elements. In general, the empirical formula of zeolites can be chemically expressed as

$$M_{2/n}O \cdot Al_2O_3 \cdot xSiO_2 \cdot yH_2O$$

where M is the counterion, n is the cation valence, x is equal to or greater than 2, and y is the water content in hydrated crystals.

Alternating the types and amounts of substituting tetrahedral atoms can significantly affect the properties of zeolites. In the case of aluminosilicate zeolites, for example, hydrophobicity can be controlled by altering a content of aluminum in the zeolite frameworks. Zeolites with high Si/Al ratios are hydrophobic, whereas low-silica zeolites are hydrophilic. Therefore, aluminosilicate zeolites with low Si/Al ratios can be used as drying agents for removal of water from organic solvents and as pervaporation membranes for dehydration of alcohols [9–11]. In addition, such aluminosilicate zeolites with high Al substitutions have been utilized as detergent builders and adsorbents because of their large ion-exchange capacities [1,2]. Incorporation of other heteroatoms other than aluminum, such as gallium, germanium, tin, and titanium, can result in zeolites with new framework structures and/or catalytic properties that are hardly to achieve from conventional aluminosilicate compositions [12–18].

In addition to the framework tetrahedral atoms, the properties of zeolites can be altered depending on the extra-framework cationic species that can be exchanged, making ion-exchanged zeolites capable of utilizing as gas adsorbents [19–21], redox catalysts [22–25], and optoelectronic devices [26,27]. In an air separation process by pressure swing adsorption (PSA), for instance, lithium-exchanged low-silica X (Li-LSX) zeolite shows better separation performance than the sodium-exchanged counterpart, thereby reducing the separation cost [19,28]. In catalysis, Cu(II)- and Fe(II)-exchanged zeolites have proven to be quite promising catalysts for NO_x treatment [22–24]. Furthermore, if there is mobility of charge-compensating cations, zeolites can exhibit ion-conductive properties [29–31]. Tuning the effective pore size of zeolites can be realized via ion exchange. For example, the sodium form of zeolite A ("zeolite 4A") has the effective pore size of about 4 Å. Exchanging the sodium cations with potassium can reduce the pore size to 3 Å ("zeolite 3A") because potassium is larger than sodium. By contrast, the pore size is increased to about 5 Å ("zeolite 5A") by exchanging with calcium. This is because calcium is divalent and thus only half number of calcium is required; as a result, the location of cations shifts to not shield the pore opening.

Thanks to the well-ordered, multidimensional spatial arrangements on the nanometer length scale of zeolites, functionalization of zeolites can also be achieved by introduction of guest species such as organic molecules [32,33] and metal clusters [34,35] into their microporous cavities/channels, which can broaden the application scope of zeolites beyond ion exchange, adsorption, and catalysis.

5.3 Zeolite Synthesis

5.3.1 Historical and Fundamental Views

Zeolites are typically synthesized under hydrothermal conditions; however, nonaqueous synthesis of giant crystals was also reported [36]. In general, the components of zeolite synthesis include silica, alumina (and sources of other T atoms), mineralizing agents (hydroxide or fluoride), and water. The first synthetic zeolite, levynite, was reported in 1862 by St. Claire Deville [37]. At the beginning stage, zeolites were synthesized at mimicking geologic conditions, that is, high pressure (over 100 bar) and temperature (over 200 °C) [8]. The first synthetic zeolite without a natural counterpart, much later identified as KFI zeolite, was successfully synthesized by Barrer [38]. Pioneering works of Barrer have provided us the fundamental knowledge on the hydrothermal chemistry of zeolites [39].

In the 1950s, the landmark discovery of synthetic zeolites at the Linde Division of Union Carbide Corporation by Milton, Breck, and their colleagues has tremendously influenced the gas and petrochemical industries [40–44]. They developed the hydrothermal synthesis of zeolites at lower temperature ($\sim$100 °C) and pressure (autogenous) using "reactive" alkali-metal aluminosilicate gels prepared from sodium aluminate and sodium silicate solutions, leading to the discovery of two new synthetic, commercially important zeolites, Linde A (Si/Al = 1) [40,41] and Linde X (Si/Al $\leq$ 1.5) [42,43], having LTA and FAU framework type codes, respectively. Both have shown to have excellent ion-exchange and gas separation properties. Later, the FAU zeolite with higher silica compositions, named Linde Y (Si/Al = 1.5 to ca. 2.8), was synthesized [44]. After post-modifications, this zeolite Y, designated as ultrastable Y (USY), is an excellent catalyst that crack the distilled crude oil to gasoline-range hydrocarbons. This type of synthesis has involved solely the inorganic components, and the crystallization of zeolites is directed by hydrated alkali-metal cations. All the synthetic zeolites from this system have a low-silica content until organic cations, the so-called organic structure-directing agents (OSDAs) were introduced into the synthesis reagents.

In 1961, one of the milestones in zeolite science, Barrer and Denny reported the synthesis of zeolite using quaternary ammonium cations, tetramethylammonium (TMA^+) cations [45]. By adding TMA^+ into sodium aluminosilicate gels, they were able to synthesize zeolite A with a higher silica composition, designated as zeolite N-A (nitrogenous-type of zeolite A). Independently, at about the same time, Kerr and researchers at Mobil Oil Corporation used TMA^+ for synthesizing ZK-4, an analog of zeolite A [46,47]. In both cases, the addition of TMA^+ cations to sodium aluminosilicate gels increased the Si/Al ratios of zeolite A products.

The researchers at Mobil became the first research group to synthesize zeolites with new framework structures by using organic cations. They succeeded in making two of the "big five" zeolites, beta and ZSM-5 (whose structures are *BEA and MFI, respectively), from aluminosilicate gels with higher silica contents by using tetraethylammonium (TEA^+) and tetrapropylammonium (TPA^+) cations, respectively. Both zeolite beta and ZSM-5 were the first high-silica synthetic zeolites obtained with the Si/Al ratios greater than 5. The utilization of OSDAs together with high-silica aluminosilicate gels for zeolite synthesis has led to the discovery of many new zeolite structures [48].

One of essential components in zeolite synthesis is a mineralizing agent, typically hydroxide ion. It assists the mineralization of silicate, aluminate, and other sources of T atoms by providing the right equilibrium of reversible Si—O—T bond formation and hydrolysis. If a concentration of mineralizing agents is too high, the equilibrium is shifted toward the dissolution of silicate species and thereby avoiding the solidification. On the contrary, a very low concentration of mineralizing agents cannot provide enough soluble silicate species, resulting in an amorphous solid.

Besides hydroxide, fluoride can be used as the mineralizing agent, offering the synthetic mixtures at near neutral pH. In the pioneering work of Flanigen and Patton, pure-silica MFI zeolite, silicalite-1, was synthesized by this fluoride route [49]. A group at Mulhouse led by Guth and Kessler has extended fluoride chemistry to synthesize various pure-silica zeolites [50]. They showed that crystals grown in the fluoride media are larger and contain fewer defects than those obtained in the conventional hydroxide chemistry. The great advancement in this field has been done by Corma's group at Valencia, resulting in the discoveries of numerously novel zeolite structures including those with low framework densities [51]. It is worth noting that the concentration of the synthetic mixtures was remarkably higher than the typical synthesis using hydroxide as the mineralizing agent. Interestingly, zeolites with less dense framework structures can be prepared from the reactants with low H_2O/SiO_2 ratios, while the more dilute synthetic mixtures lead to zeolites with higher framework densities [12,52,53].

As mentioned above, the formation of zeolite is directed by organic cations (OSDAs) and hydrated alkali-metal cations. In addition to cations, T atoms that

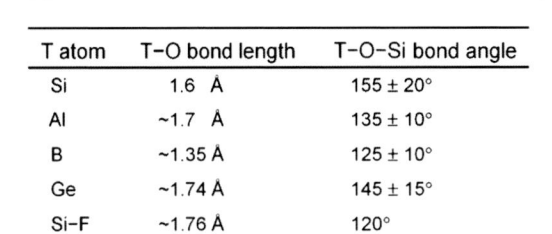

(a)

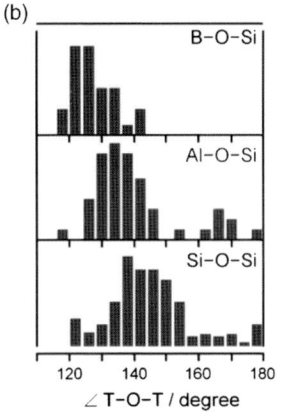

T atom	T–O bond length	T–O–Si bond angle
Si	1.6 Å	155 ± 20°
Al	~1.7 Å	135 ± 10°
B	~1.35 Å	125 ± 10°
Ge	~1.74 Å	145 ± 15°
Si-F	~1.76 Å	120°

(b)

Figure 5.3 (a) Changes in T—O bond lengths and average T—O—Si bond angles with different substituting T atoms (data taken from ref. [54]). (b) Histograms of T—O—Si bond angles experimentally observed in various solids (replotted from ref. [55]).

substitute for silicon in the zeolite frameworks can intrinsically stabilize some structural building units, thereby directing the formation of zeolites because such substituting T atoms possess different T—O bond lengths, T—O—T bond angles, and/or charges (Figure 5.3). Several T atoms other than aluminum are able to substitute for silicon at the tetrahedral positions in the zeolite frameworks, for example, divalent beryllium and zinc, trivalent boron and gallium, and tetravalent germanium, tin, and titanium. Substitution of beryllium and zinc favors the formation of three-ring (3r) silicates, while germanium can stabilize the double four-ring (d4r) unit [12]. It should be noted that substitution at low levels may not affect the formation of structural units and zeolites.

In typical tetrahedral frameworks (not limit to zeolites), Si—O and Al—O bond lengths are 1.61 and 1.73 Å, respectively, with O—T—O bond angles close to the tetrahedral angle. In the frameworks without constraints (e.g., quartz and cristobalite), Si—O—Si and Al—O—Si bond angles were found to be in a narrow range close to 145° [56]. Analysis of the crystallographic data on pure-silica zeolites revealed that the mean value of the Si—O bond length and the O–Si–O angle are close to the typical values, being 1.594 Å and 109.5°, respectively but the Si–O–Si angle ranges from 133.6 to 180°, with a mode value of 148° and a mean value of 154 ± 9°, reflecting the variation and flexibility of zeolite frameworks [57].

At the same coordination number (i.e., tetrahedral framework), T—O bond distances depend on the atomic radii, for example, 1.35 Å for B—O (note that some studies suggested that B—O in tetrahedral frameworks can be as long as ~1.5 Å), 1.73 Å for Al—O, and 1.81 Å for Ga—O [13]. Interestingly, the B—O—Si and Ga—O—Si (~135° [58]) angles are, on average, smaller than the Al—O—Si (and also Si—O—Si) angle, although their T—O bond distance is either shorter or longer, respectively. It is noteworthy that the substitution of beryllium and zinc into the tetrahedral frameworks also leads to smaller T—O—Si angles [12].

On a basis of their differences in structural chemistry, introduction of boron and gallium into zeolite synthesis can result in the new zeolite frameworks that have not yet been observed in the conventional aluminosilicate frameworks, for example, ITQ-52 (IFW-type) [59,60] and SSZ-82 (SEW-type) [61,62] for borosilicates, and ECR-34 (ETR-type) [63] for gallosilicates.

One of the very basic and interesting structural units found in the zeolite frameworks is the double four-ring (*d4r*) unit. At its full symmetry (O_h) with the O—T—O angles of regular tetrahedra, the T—O—T angle becomes 148.4° [64], close to the ideal value for the unstrained Si—O—Si (145°). This is also observed in the *d4r* unit built from 4 silicate and 4 aluminate tetrahedra linked in an alternate sequence, as found in zeolite A. Therefore, the *d4r*-containing zeolite frameworks can be synthesized in either pure-silica or aluminosilicate (with the Si/Al of 1) compositions. However, the *d4r* can exist at lower symmetry by narrowing the T—O—T angles. For example, if the symmetry is reduced to T_h, the T—O—T angle is decreased to 129.6° while maintaining regular tetrahedral [56]. As the *d4r* units in the zeolite frameworks are not always present at the highest symmetry, the introduction of germanium as the substituting T atom is an effective way for the synthesis of *d4r*-containing zeolites due to the smaller Ge—O—Ge angle (~130°). However, in the *d4r*-containing zeolites germanium is selectively incorporated into the *d4r* units by avoiding the formation of Ge—O—Ge linkages, akin to the Löwenstein's rule for aluminosilicates [65,66]. Using this concept, intensively studied by Corma's group and others, several (alumino)germanosilicate zeolites with novel framework structures were discovered; many of them contain multidimensional, large (or extra-large) pores [51,67–71]. It should be mentioned that the *d4r* units can also be stabilized if zeolites are synthesized from the fluoride-containing synthetic mixture as the fluoride ions are encapsulated inside the *d4r* units in the as-synthesized zeolites [72].

In addition to the conventional zeolite synthesis in which the reaction system consists of an aqueous solution of dissolved components and solid species present as precipitated (hydro)gels and/or particulates suspended in solution, zeolites can be synthesized by conversion of a dry, solid aluminosilicate gel in contact with vapors of volatile OSDAs and/or water, referred to as the dry gel conversion (DGC) method [73]. In this method, the solid gels are kept separated from the liquid solution and never come in direct contact with liquid water. In the first demonstration reported in 1990, a dry sodium aluminosilicate gel was converted to MFI zeolite by contacting the gel with vapors of ethylenediamine, triethylamine, and water [74]. The DGC method can be classified into vapor-phase transport (VPT) and steam-assisted crystallization (SAC), according to sources of vapors. In the VPT method, a solid gel prepared by drying all reactants excepting OSDAs is brought into contact with vapors of amine OSDAs and water [74,75]. In the latter method, on the contrary, a solid gel contains OSDAs and vapors are water alone (i.e., steam). The SAC method is more useful as many OSDAs are quaternary ammonium having low vapor pressure (nonvolatile under zeolite synthesis conditions). Miller at Chevron Corporation separately developed a similar dry-gel approach for the synthesis of powder and shaped zeolites and has recently brought into a commercial scale, *ZeolitePlus* technology, for the production of zeolites such as beta, ZSM-5, and L[2)] [76].

5.3.2
Recent Developments

5.3.2.1 OSDAs: From Design of Efficient OSDAs to OSDA-Free Synthesis

As briefly discussed in the abovementioned section, the use of OSDAs is one of the important key strategies to discovery new zeolite frameworks, starting with the introduction of simple quaternary ammonium cations such as TMA^+, TEA^+, TPA^+, and TBA^+ (tetrabutylammonium). Many fundamental bases for the design of efficient OSDAs have come from the solid works of Davis at Caltech and Zones at Chevron Corporation [48,77–81]. The molecular structures and properties of OSDAs, such as size, shape, rigidity/flexibility, and hydrophobicity/hydrophilicity, are of significance in the determination of the resulting zeolites. For example, small OSDAs with a spherical shape tend to direct the formation of small cages, thereby resulting in nonporous frameworks, namely, clathrasil [82,83], while larger ones can lead to porous zeolites. Linear OSDAs typically yield one-dimensional medium-pore zeolites due to their flexibility of alkyl chains [48,77]. In contrast, large and rigid OSDAs tend to give multidimensional large-pore zeolites such as CIT-1 (CON-type) [84], MCM-68 (MSE-type) [85], and ITQ-38 (ITG-type) [86].

In the synthesis of high-silica zeolites, the formation of OSDA–silicate composite species, driven by the intermolecular interactions between silicate species and OSDAs (enthalpic effects) as well as by the release of water molecules from the hydrophobic hydration shells surrounding OSDAs into bulk water (entropic effects), is the primary step toward the structure direction [87,88]. Therefore, there should exist an optimal degree of hydrophobicity of efficient OSDAs as the hydrophobic hydration shells have to be interrupted by silicate species to form the OSDA–silicate composites. The optimal hydrophobic/hydrophilic character of OSDAs for the synthesis of high-silica zeolites has been determined by the water-to-chloroform phase-transfer experiments [80,81]. The suitable OSDAs are those having C/N^+ ratios of 11–15 as they partition well between both aqueous and organic phases because of their good balance between hydrophobicity and hydrophilicity.

In the search for new zeolites, the scope of OSDAs has been extended to quaternary phosphonium cations that unlike their ammonium analogues do not easily proceed the Hofmann degradation, becoming more thermally stable than conventional quaternary ammonium OSDAs and thus allowing more severe synthesis conditions [89]. Although monoquaternary phosphonium OSDAs are known to give conventional zeolites such as *BEA, MEL, and MFI [90], combining with the fluoride and germanium chemistry some new multidimensional

---

2) http://www.chevrontechnologymarketing.com/ZeolitePlus/.

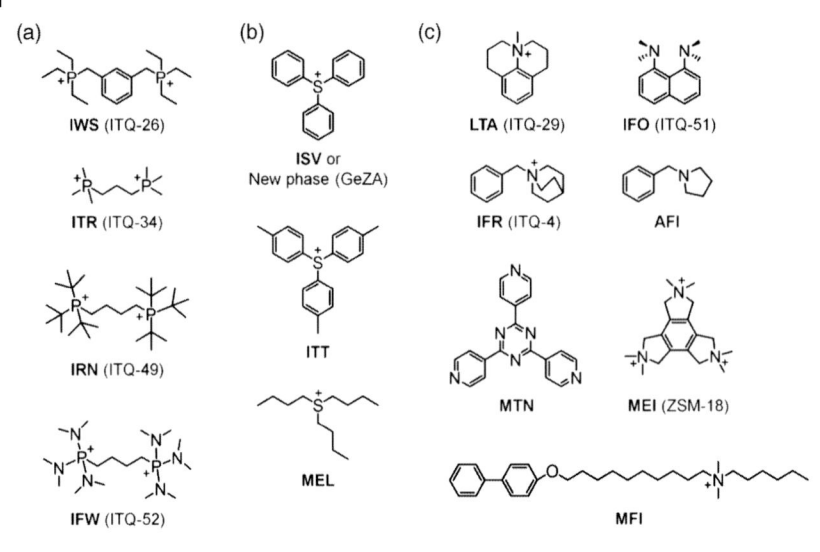

Figure 5.4 Examples of OSDAs having (a) diquaternary phosphonium, (b) tertiary sulfonium, and (c) aromatic functional groups used for the synthesis of different zeolites, denoted below the OSDA chemical structures.

large-pore zeolites, such as ITQ-27 (IWV-type) [89] and ITQ-53 (its structure has not yet been approved by IZA) [91] were discovered. Exploratory investigations of phosphonium OSDAs have been further made on diquaternary cations, some of them shown in Figure 5.4a, again resulting in several new, multidimensional pore zeolites [92–95].

Unlike ammonium OSDAs, the phosphonium OSDAs cannot be removed completely from the as-made zeolites by conventional thermal calcination in the presence of oxygen. As a result, a certain amount of P_2O_5-like phosphorus debris may remain inside the zeolite pores. A new method for removal of the phosphonium OSDAs has been developed by heating the as-made zeolites under a flow of hydrogen-containing gas mixtures [95]. Under this reductive condition, phosphorus species do not undergo oxidation, but rather are decomposed into low-molecular-weight phosphines. Note that caution has to be taken for properly treating the outlet gas streams due to the high toxicity of phosphines. Very recently, the scope of applicable OSDAs for zeolite synthesis has been broadened to tertiary sulfonium compounds (Figure 5.4b) [96]. Learning from the demonstrations of phosphonium OSDAs, ones can expect that more new zeolites will be discovered from this new sulfonium chemistry.

As more complex and bulker OSDAs have been designed and used to discover new zeolites, such organic compounds sometimes become too hydrophobic and therefore are not suitable for the *hydro*thermal synthesis of zeolites. However, the addition of aromatic rings into the bulky organic cations has been found to be an efficient way to make them function as OSDAs because their π–π interactions can increase the intermolecular interaction between OSDAs through supramolecular assembly [97]. Some effective OSDAs with the aromatic moieties are shown in Figure 5.4c. Of particular interest is the use of the supramolecular self-assembled, π-stacked dimers of quinolinium-derived molecules for the synthesis of pure- and high-silica LTA zeolites (ITQ-29) [98]. Furthermore, the aromatic compounds tethering cationic-ended alkyl chains can direct the formation of single-crystalline mesostructured MFI nanosheets [99,100].

Since the first use of organic cations in zeolite synthesis [45], the design of OSDAs has significantly advanced over decades. However, the design of a specific OSDA to synthesize a targeted zeolite is still a long-standing challenge in zeolite synthesis [101]. Molecular mechanics (MM) modeling has been employed to *a priori* predict the zeolite frameworks (as pure silica) to be obtained from a particular synthetic system by optimization of the interactions between OSDAs and inorganic zeolite frameworks [102]. MM calculations have been used for the *de novo* design of efficient OSDAs for a given zeolite structure. Although these calculations allow the *de novo* prediction of OSDAs, one of the paramount shortcomings limiting the practical, designed synthesis of zeolites is that the designed OSDAs are difficult or impossible to be synthesized. As succeeded in a pharmaceutical industry for drug design, a genetic algorithm applying chemical transformation (by known reactions) to a library of commercially available reagents has been employed to design "chemically synthesizable" OSDAs [103].

However, the prediction of efficient OSDAs has been successfully demonstrated mostly for pure-silica zeolites [102,104,105] and aluminophosphate zeolite-like materials [106,107], with rare successes for aluminosilicate zeolites [108]. This is partly because the accurate force fields for MM calculations are only available for describing the complex of OSDAs and zeolites with neutrally charged frameworks (i.e., pure silica and aluminophosphate) where van der Waals interactions are predominant. To account for the anionic aluminosilicate frameworks, density functional theory-based method with energy corrections (e.g., periodic DFT-D) has been used to explain the structure direction of OSDAs [109].

Thus far, OSDAs have held a promise for the discovery of new zeolites. However, such new OSDAs possess more and more complex structures, thereby becoming more expensive. To make the production of new zeolites economically feasible, one of the issues is to reduce the cost of OSDAs, which can be achieved by using the mixture of expensive OSDAs and cheaper molecules functioning as a pore-filling agent, by using recyclable OSDAs, and by synthesizing zeolites without using OSDAs [54]. Seed-directed synthesis of zeolites without using OSDAs has been developed to be a low-cost, eco-friendly route, potentially for the commercial manufacture of zeolites [110]. It is worth mentioned that the OSDA-free synthesis was a general approach for the mass production of "big five" zeolites, namely, FAU, FER, MFI, and MOR [111–113].

This OSDA-free approach was revisited and the seed-assisted, OSDA-free synthesis of *BEA zeolite, the last remaining "big five" zeolite that needs an OSDA for its synthesis, was demonstrated by Xiao's group in 2008 [114]. To broaden the range of zeolites that can be synthesized by this seed-assisted methodology systematically, Okubo and Itabashi proposed the working hypothesis, namely, the composite building unit (CBU) hypothesis [115]. They suggested that the target zeolite to be synthesized by the OSDA-free method should be added as seeds to a gel that yields a zeolite containing the common CBUs when the gel is heated without seeds. At least four requirements for the successful synthesis of zeolites by this method were suggested as (i) the zeolite seeds should not be completely dissolved prior to the onset of crystal growth during the hydrothermal treatment, (ii) the spontaneous nucleation of other crystalline phases should be avoided before completion of the crystal growth of the target zeolite, (iii) the external surfaces of seed crystals should be exposed to the liquid phase and the precursor must be able to access to the clean surfaces, and (iv) the chemical composition of the reactant gel and the seeds should be optimized [110].

5.3.2.2 Synthesis of Zeolites from Layered Silicates

Although the approach toward synthesis of zeolites by design becomes more and more rational and strategic (*vide supra*), the trial-and-error method is still a basic practice for discovering new structures. One of efforts to *a priori* synthesize zeolites is based on a bottom-up assembly of two-dimensional, crystalline precursors, namely, layered silicates [116]. Topotactic conversion of layered silicates into three-dimensionally connected zeolite frameworks proceeds upon the thermal treatment (calcination) of the layered precursors during which the dehydration–condensation of two facing silanol groups on the surfaces of neighboring layers ($\equiv$Si—OH + HO—Si$\equiv$ → $\equiv$Si—O—Si$\equiv$ + H_2O) occurs simultaneously with the removal of any interlayer organic compounds, if present, resulting in the three-dimensionally connected zeolites. The framework structures of the resultant zeolites resemble those of the parent layered silicate precursors, thereby allowing us to predict the resulting structures. MCM-22 (MWW-type) was the first zeolite obtained via topotactic conversion of the MCM-22(P) layered silicate precursor [117]. Other zeolites include MWW, FER, CDO, HEU, NSI, RWR, RRO, and SOD. There are several chemical and structural prerequisites for the successful topotactic conversion such as the positions of the silanol groups and the interlayer distances [116,118].

As illustrated in Figure 5.5, following successes in the topotactic conversion, several new (multi-dimensional pore) zeolites were discovered by a 3D-to-2D-to-3D transformation, starting with the existing zeolites [119–123]. This approach was described as the inverse sigma transformation [119] and the ADOR (assembly–disassembly–organization–reassembly) strategy [120]. One of the important prerequisites of this new approach is that the starting zeolites must have chemically selective weakness as they have to be selectively disassembled into layered silicates (selective 3D-to-2D transformation) [122]. Germanium-containing UTL zeolite is an ideal parent zeolite because germanium atoms are selectively located in the *d4r* units present in the UTL framework. This germanium-containing UTL can be selectively transformed into layered silicates simply by acid treatment. Once layered silicates are achieved, topotactic conversion can be applied to prepare new zeolites. The first zeolite discovered was COK-14 (OKO-type) [119].

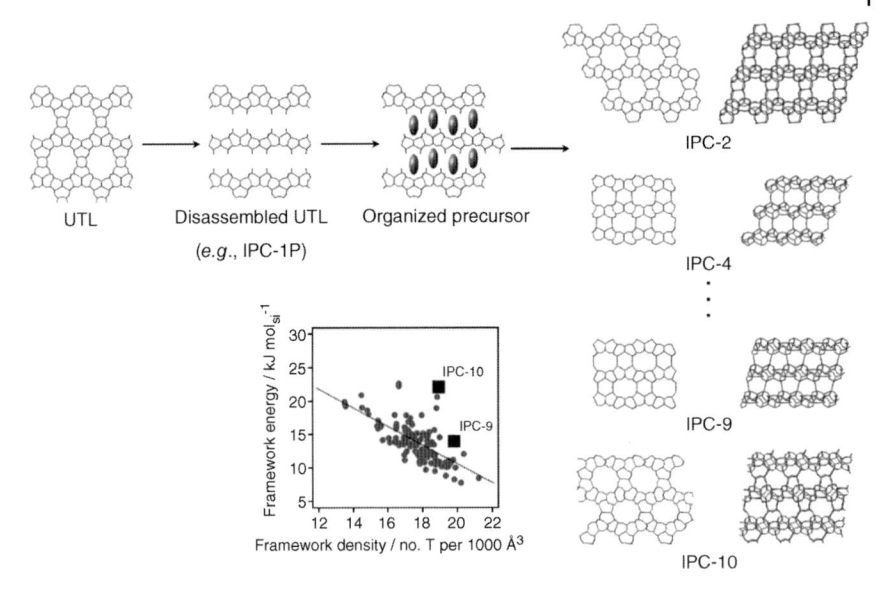

Figure 5.5 ADOR (assembly–disassembly–organization–reassembly) strategy for the synthesis of zeolites, starting with the *assembly* of a parent zeolite (UTL in this particular case) from the starting reagents, followed by the controlled *disassembly* of the parent zeolite typically by acid treatments, the *organization* of the disassembled zeolite, e.g., by reintercalation of some organic compounds, and the *reassembly* of the organized layered precursors, mainly by thermal calcination (for the dehydration–condensation of surface silanol groups and removal of organic compounds), respectively. A graph shown at the bottom center is the framework energy/density plot of existing conventional zeolites with the values of IPC-9 and 10.

The organization of the disassembled zeolites is a very important step that determines the framework structures of the final zeolites. The easiest way to alter the organization of the disassembled layers is the adjustment of pH, which affects the amounts of protons, waters, and dissolved silicate/germanate species in the interlayer space [119,120]. Hydrolysis of UTL zeolite at very high acidity (e.g., using 12 M HCl) eliminates a single 4-ring from the *d4r* unit, the so-called inverse sigma transformation [119]. The eliminated single 4-rings are shifted into the pores of the hydrolyzed UTL zeolite. After thoroughly washing with water to remove the eliminated single 4-rings from the solids, COK-14 is achieved by calcination. At lower acidity, on the contrary, a single 4-ring cannot selectively be eliminated from the *d4r* unit. Hydrolysis at low acidity (i.e., using 0.1 M HCl) gives the layered silicate, IPC-1P. Intercalation of silicate species into IPC-1P, followed by calcination gives IPC-2 zeolite. Note that IPC-2 is also OKO-type zeolite but its symmetry is different from COK-14. By adjusting pH, zeolites with continuously tunable porosity can be prepared [121].

As is well documented, intercalation of organic species into the interlayer space of the layered silicates can alter the interlayer distance and shift individual layers along directions parallel to the layers [118]. Several organic compounds were intercalated into the hydrolyzed UTL zeolites, resulting in the organized layered precursors with different stacking structures [120,123]. Very recently, two new zeolites synthesized by the ADOR mechanism, IPC-9 and IPC-10, possess unusual high framework energy (see the framework energy/density plot in Figure 5.5) [123]. In general, syntheses of both zeolites are considered to be unfeasible. This suggests that several energetically unfeasible zeolites may be discovered by the 3D-to-2D-to-3D transformation strategy [124].

5.3.2.3 Hierarchically Porous Zeolites

When utilized in catalysis and other applications relevant to bulky molecules, the only presence of micropores in zeolite frameworks can become problematic as it limits molecular diffusion, and therefore, restricts the ability of bulky molecules to transport through the zeolite micropores. Shortening the effective diffusion path lengths of zeolite crystals, for example, by synthesis of nanosized zeolites, synthesis or exfoliation of layered zeolites (e.g., MCM-22(P)), and introduction of mesopores into zeolite crystals, can solve this problem. Note that according to the International Union of Pure and Applied Chemistry (IUPAC), pores with widths below about 2 nm are denoted micropores, whereas those in the range of 2–50 nm are mesopores [125].

Among the abovementioned possible solutions, the introduction of mesopores into zeolites has gained particular attention. Synthesis of new zeolites having multidimensional micro- and mesopores is a direct, but most difficult, way. This is partly because zeolites with "intrinsic" mesopores tend to be thermodynamically less stable than their counterparts having smaller micropores [126]. By combining the germanium and fluoride chemistry with exotic OSDAs, only two mesoporous zeolites, ITQ-37 (-ITV-type) [127] and ITQ-43 [128], were discovered thus far. It is noteworthy that both zeolites possess interrupted frameworks.

Alternatively, the fabrication of zeolites with "secondary" mesopores resulting in the hierarchical zeolites with micro- and mesoporosity is more promising [129]. The molecular transport within the zeolite bodies can be enhanced by taking the advantages of the unique combination of the intrinsic micropores originated from zeolite frameworks and the bypass-interconnected mesopores [130]. The methods for the fabrication of hierarchically micro- and mesoporous zeolites can be divided to two main categories: top–down (e.g., desilication and dealumination) and bottom-up (e.g., hard- and soft-templating) approaches.

Postsynthetic treatments of zeolites in alkali or acidic conditions, resulting in desilication and dealumination, respectively, can creat mesopores in the zeolite particles. As is well known, the commercial USY zeolite (e.g., CBV780 manufactured by Zeolyst International) contains some mesopores generated during its dealumination process by steam treatments followed by acid leaching [131]. Alternatively, by treating zeolite in dilute alkali solutions under mild conditions, Ogura *et al.* were the first to report that the desilication of zeolites can generated

MFI zeolites with uniform-sized mesopores while preserving the framework crystallinity [132,133]. Later, this desilication technique has been investigated intensively by other groups, confirming their applicability to several zeolites with a wide range of Si/Al ratios [134]. By adding the alkyltrimethylammonium surfactants during the desilication process, zeolites with highly uniform, tunable mesopores have been realized [135].

Using periodically arranged silica nanoparticles as a hard template, three-dimensionally ordered mesoporous-imprinted (3DOm-i) zeolites (MFI, *BEA, FAU, LTA, and LTL) were synthesized by the direct hydrothermal synthesis or the SAC method (Figure 5.6) [136–138]. Unlike the hard-templating method requiring multistep procedures, organic soft templates functioning as mesopore-generating agents (mesoporogens) can be used by simply introducing into the typical zeolite synthesis mixtures, offering a single-step method for the fabrication of hierarchical zeolites [129]. Several amphiphilic, bifunctional mesoporogens containing both hydrophobic alkyl chains and zeolite structure-directing agents have been designed by Ryoo's group [139–141], with the most innovative molecules being a series of Gemini-type multiammonium surfactants as they can direct the formation the hexagonally ordered mesoporous zeolites [141].

Recently, the mesoporogen-free, nontemplating approach using small OSDAs has been demonstrated for the fabrication of hierarchical zeolites by enhancement of zeolite intergrowth [142,143]. Self-pillared zeolite nanosheets possessing

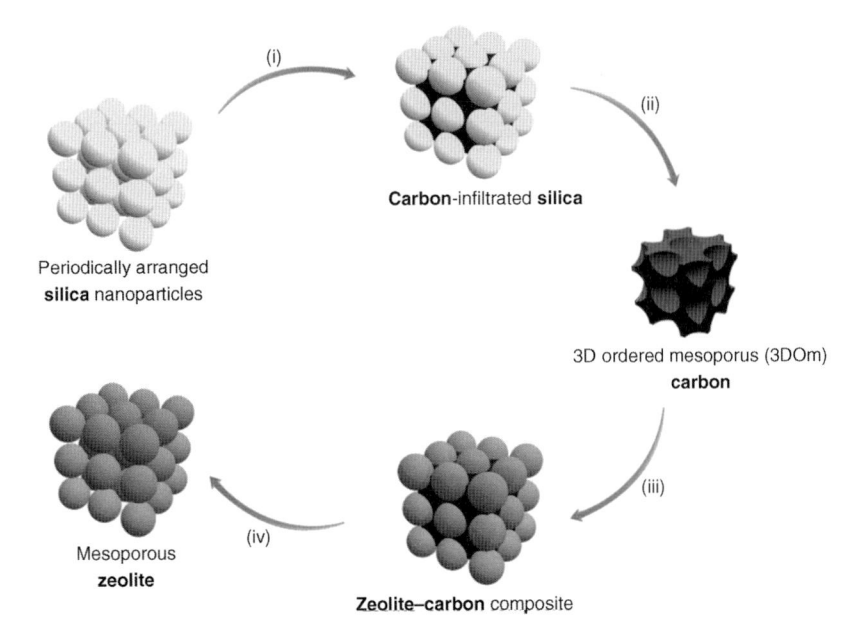

Figure 5.6 The fabrication of 3DOm-i zeolites: (i) infiltration of carbon precursors, followed by carbonization, (ii) dissolution of silica nanoparticles by basic or HF treatment, (iii) synthesis of zeolites in confined spaces by either hydrothermal or SAC methods, and (iv) removal of 3DOm carbons by thermal calcination.

hierarchical micro- and mesoporosity were synthesized by using a simple TBA^+ or tetrabutylphosphonium cation as an OSDA. Under the optimal conditions, repetitive branching (intergrowth) during the formation of MFI zeolite occurs, leading to hierarchical zeolites built from orthogonally connected nanosheets with a house-of-cards arrangement [142]. Through a different intergrowth mechanism, as depicted in Figure 5.7, hierarchically organized MFI zeolites having three classes of porosity in one body (i.e., intrinsic micropores of zeolite, mesopores created within the zeolite sheets, and macropores originated from the complex intergrown structure) have been achieved by sequential intergrowth using a diquat-C5 OSDA [143].

5.4
Summary and Outlook

As described in this chapter, the synthesis of zeolites has greatly advanced as a result of many masterpieces of researches contributed from scientists and engineers in both academia and industry worldwide. Launched by the pioneering works of Barrer, Breck, and Milton, the successes of Union Carbide, Mobil Oil, and other companies in the commercialization of zeolites and their utilization in industrial adsorption and catalysis processes have accelerated the development in the field. Following these, the creative works of Davis, Zones, Corma, and others have opened the new pages of zeolites synthesis using the efficient OSDAs as well as the new synthetic chemistry based on fluoride and other T atoms.

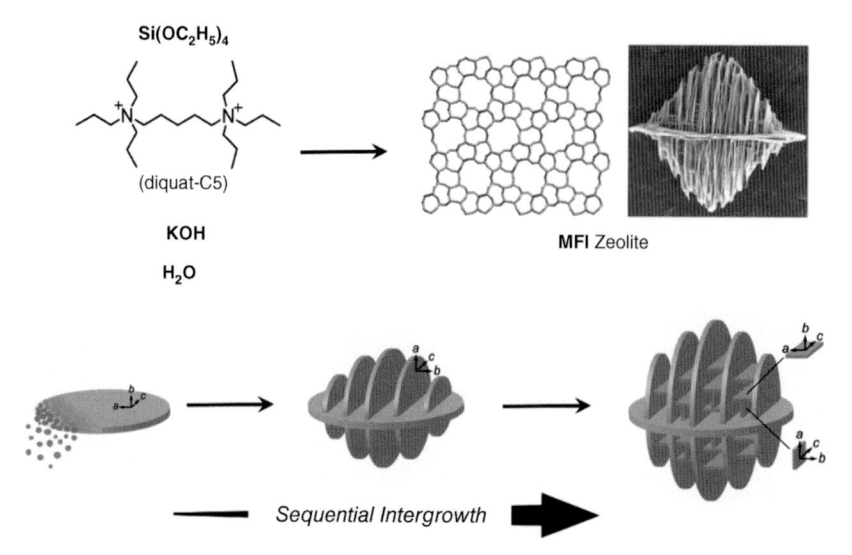

Figure 5.7 Schematic of the formation of hierarchically organized MFI zeolite by sequential intergrowth.

5.4 Summary and Outlook

Zeolites are one of the most important industrial materials and have been expected to play more and more important roles for the sustainable society. Currently, synthetic zeolites have been used in the detergent application to replace the environmentally unfriendly phosphate component (LTA and GIS), in the catalytic processes being fluid catalytic cracking (FCC) of crude oil, hydrocracking and chemical/petrochemical synthesis (mainly dominated by the "Big Five" zeolites), and in the adsorption applications such as drying and purification of natural gas, separation of paraffins and aromatics, and air separation (e.g., LTA, FAU, and MFI), while the natural zeolites have been used mainly to enhance the mechanical strength of cement.

Zeolites are finding their places in the future global development for which the sustainability with the least environment burdens while maintaining the economic growth becomes a core issue. As the major components of zeolites (oxygen, silicon, and aluminum) are the three most abundant elements in the Earth's crust, zeolites are abundant and cheap in general, and are thermally/hydrothermally stable due to their strong oxide frameworks. In the sense of sustainability, zeolites have shown recently several promising applications, including removal of radionucleotides (e.g., ^{137}Cs and ^{129}I) in radioactive wastes [144], treatment of automotive exhaust emissions such NO_x and hydrocarbons [22–24,145], and energy-efficient separations by zeolite membranes [10,11,146,147].

Ever-increasing demand for clean and cheap energy resources has stimulated the exploration of new energy sources to replace/substitute crude oil. The recent shale gas revolution has changed the energy scenario. Although the abundant natural gas can be used directly as the energy source, it cannot replace crude oil as the direct conversion of methane, a major component in the natural gas, into easily transportable, liquid fuels and other useful chemicals is difficult. Several metal (ion or oxo-complex)-modified zeolites have shown great promise to overcome the inertness of methane C—H bond, although the substantial improvement in catalytic activity is required [148–150].

Biomass is an alternative platform for renewable energy and resources. The discovery of tin-containing zeolites [151] has shed new light on the productions of renewable commodity chemicals derived from biomass by many Lewis acid-catalyzed reactions [16–18]. Isolated Lewis acid centers in hybrophobic environments of tin-zeolites are capable of metalloenzyme-like catalyzing the isomerization of sugar, particularly, the conversion of glucose into fructose, in water [17,18]. By using conventional aluminosilicate zeolites (as a Brönsted acid) and tin-zeolites, a new pathway for the 100% renewable production of poly(ethylene terephthalate) (PET) has been suggested (Figure 5.8) [152]. PET is synthesized by copolymerization of ethylene glycol and terephthalic acid (or dimethylterephtalate). Both monomers can be synthesized from biomass in which several reactions involving can be catalyzed by aluminosilicate or tin-containing zeolites. The ethylene glycol is typically prepared by oxidation of ethylene. Ethylene can be obtained from bioethanol by acid-catalyzed dehydration, in which aluminosilicate zeolites can be used. To produce another monomer, terephthalic acid or dimethylterephtalate, from biomass, five-step reactions are

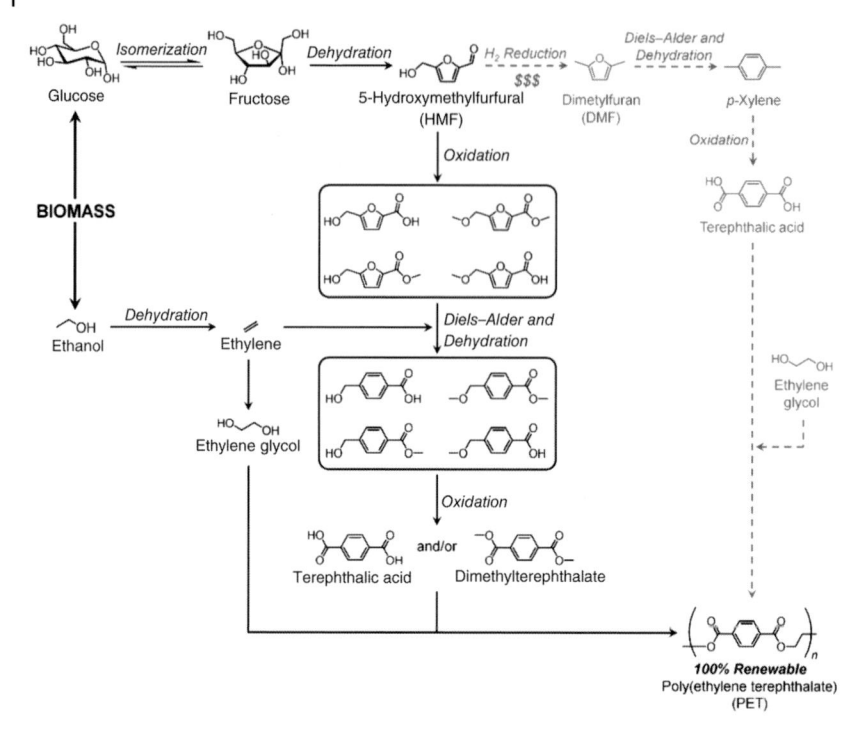

Figure 5.8 Reaction pathways for a 100% renewable production of PET from biomass using Diels–Alder reactions (adopt from ref. [152]). The route in black represent the 100% renewable route whereas that in light gray requires hydrogen and expensive metal-based catalysts for the hydrogenation of HMF.

required starting from glucose. In addition to the isomerization of glucose, Lewis acidic zeolites can be applied for the Diels–Alder dehydration reaction between ethylene and various oxidized derivatives of 5-hydroxymethylfurfural [153].

In addition to PET, several biomass-derived commodity chemicals can be synthesized from sugars via lactic acid, including acrylic acid and acrylonitrile. Again, Lewis acidic zeolites can be used in catalytic conversion of fructose to lactic acid, involving fragmentation of fructose by retro-aldol reaction and 1,2-intramolecular hydride shift reaction to produce alkyl lactate [154,155]. Very recently, the efficient route to produce L,L-lactide, an effective monomer for poly (lactic acid), by dehydration of lactic acid over aluminosilicate zeolites has been reported, providing an economically reasonable route to produce the 100% renewable, biomass-derived poly(lactic acid) [156].

As briefly described above, zeolites have found their solid positions as key materials in detergent, adsorption, and catalysis applications and will continue to be the dominant industrial materials in future. Several new applications toward the sustainable development have been suggested. To bring the proposed applications to the commercial scales, several improvements including the stability under real operation conditions and the optimization of catalyst and reaction would be required. In addition, in order to supply enough zeolites for increasing demands arising from such new applications, the reasonable mass production of zeolites at a larger scale has to be developed. One of the possible choices is the development of continuous-flow synthesis of zeolites. To realize this, zeolites must be obtained in short synthesis time, within residence time for the practical flow reactors. Very recently, using a tubular reactor with rapid heating continuous-flow syntheses of CHA zeolite and AFI aluminophosphate were demonstrated at a laboratory scale [157,158].

References

1 Vermeiren, W. and Gilson, J.-P. (2009) *Top. Catal.*, **52**, 1131–1161.
2 Kulprathipanja, S. (2010) *Zeolites in Industrial Separation and Catalysis*, Wiley-VCH, Weinheim.
3 Cronstedt, A.F. (1756) *Kongl Svenska Vetenskaps Academiens Handlingar Stockholm*, vol. **17**, pp. 120–123.
4 Breck, D.W. (1974) *Zeolite Molecular Sieves: Structure, Chemistry, and Use*, John Wiley & Sons, Inc., New York.
5 Baerlocher, C., McCusker, L.B., and Olson, D.H. (2007) *Atlas of Zeolite Framework Types*, 6th Revised edn, Elsevier, Amsterdam.
6 Li, Y. and Yu, J. (2014) *Chem. Rev.*, **114**, 7268–7316.
7 Lowenstein, W. (1954) *Am. Mineral.*, **39**, 92–96.
8 Davis, M.E. and Lobo, R.F. (1992) *Chem. Mater.*, **4**, 756–768.
9 Williams, D.B.G. and Lawton, M. (2010) *J. Org. Chem.*, **75**, 8351–8354.
10 Bowen, T.C., Noble, R.D., and Falconer, J.L. (2004) *J. Membr. Sci.*, **245**, 1–33.
11 Sato, K., Aoki, K., Sugimoto, K., Izumi, K., Inoue, S., Saito, J., Ikeda, S., and Nakane, T. (2008) *Microporous Mesoporous Mater.*, **115**, 184–188.
12 Corma, A. and Davis, M.E. (2004) *ChemPhysChem*, **5**, 304–313.
13 Fricke, R., Kosslick, H., Lischke, G., and Richter, M. (2000) *Chem. Rev.*, **100**, 2303–2405.
14 Millini, R., Perego, G., and Bellussi, G. (1999) *Top. Catal.*, **9**, 13–34.
15 Notari, B. (1996) *Adv. Catal.*, **41**, 253–334.
16 Corma, A. and García, H. (2002) *Chem. Rev.*, **102**, 3837–3892.
17 Román-Leshkov, Y. and Davis, M.E. (2011) *ACS Catal.*, **1**, 1566–1580.
18 Van de Vyver, S. and Román-Leshkov, Y. (2015) *Angew. Chem., Int. Ed.*, **54**, 12554–12561.
19 Baksh, M.S.A., Kikkinides, E.S., and Yang, R.T. (1992) *Sep. Sci. Technol.*, **27**, 277–294.
20 Torres, F.J., Vitillo, J.G., Civalleri, B., Ricchiardi, G., and Zecchina, A. (2007) *J. Phys. Chem. C*, **111**, 2505–2513.
21 Pham, T.D., Liu, Q., and Lobo, R.F. (2013) *Langmuir*, **29**, 832–839.
22 Vanelderen, P., Vancauwenbergh, J., Sels, B.F., and Schoonheydt, R.A. (2013) *Coord. Chem. Rev.*, **257**, 483–494.
23 Brandenberger, S., Kröcher, O., Tissler, A., and Althoff, R. (2008) *Catal. Rev.*, **50**, 492–531.
24 Ogura, M., Itabashi, K., Dedecek, J., Onkawa, T., Shimada, Y., Kawakami, K., Onodera, K., Nakamura, S., and Okubo, T. (2014) *J. Catal.*, **315**, 1–5.
25 Deimund, M.A., Labinger, J., and Davis, M.E. (2014) *ACS Catal.*, **4**, 4189–4195.
26 Wada, Y., Okubo, T., Ryo, M., Nakazawa, T., Hasegawa, Y., and Yanagida, S. (2000) *J. Am. Chem. Soc.*, **122**, 8583–8584.
27 Wada, Y., Sato, M., and Tsukahara, Y. (2006) *Angew. Chem., Int. Ed.*, **45**, 1925–1928.
28 Castle, W.F. (2002) *Int. J. Refrig.*, **25**, 158–172.
29 Simon, U. and Franke, M.E. (2003) *Host-Guest Systems Based on Nanoporous Crystals* (eds F. Laeri, F. Schuth, U.

Simon, and M. Wark), Wiley-VCH, Weinheim, pp. 364–378.
30 Yamamoto, N. and Okubo, T. (2000) *Microporous Mesoporous Mater.*, **40**, 283–288.
31 McKeen, J.C. and Davis, M.E. (2009) *J. Phys. Chem. C*, **113**, 9870–9877.
32 Calzaferri, G. (2012) *Langmuir*, **28**, 6216–6231.
33 Kim, H.S., Lee, S.M., Ha, K., Jung, C., Lee, Y.-J., Chun, Y.S., Kim, D., Rhee, B.K., and Yoon, K.B. (2004) *J. Am. Chem. Soc.*, **126**, 673–682.
34 Goldbach, A. and Saboungi, M.-L. (2005) *Acc. Chem. Res.*, **38**, 705–712.
35 De Cremer, G., Sels, B.F., Hotta, J.-i., Roeffaers, M.B.J., Bartholomeeusen, E., Coutiño-Gonzalez, E., Valtchev, V., De Vos, D.E., Vosch, T., and Hofkens, J. (2010) *Adv. Mater.*, **22**, 957–960.
36 Kuperman, A., Nadimi, S., Oliver, S., Ozin, G.A., Garces, J.M., and Olken, M.M. (1993) *Nature*, **365**, 239–242.
37 de St. Claire Deville, H. (1862) *C. R. Acad. Sci.*, **54**, 324–327.
38 Barrer, R.M. (1948) *J. Chem. Soc.*, 127–132.
39 Barrer, R.M. (1982) *Hydrothermal Chemistry of Zeolites*, Academic Press, London.
40 Breck, D.W., Eversole, W.G., Milton, R.M., Reed, T.B., and Thomas, T.L. (1956) *J. Am. Chem. Soc.*, **78**, 5963–5972.
41 Reed, T.B. and Breck, D.W. (1956) *J. Am. Chem. Soc.*, **78**, 5972–5977.
42 Milton, R.M. (1959) US Patent 2,882,243.
43 Milton, R.M. (1959) US Patent 2,882,244.
44 Breck, D.W. (1974) US Patent 3,130,007.
45 Barrer, R.M. and Denny, P.J. (1961) *J. Chem. Soc.*, 971–982.
46 Kerr, G.I. and Kokotailo, G.T. (1961) *J. Am. Chem. Soc.*, **83**, 4675.
47 Kerr, G.I. (1966) *Inorg. Chem.*, **5**, 1537–1539.
48 Lobo, R.F. and Davis, M.E. (1995) *J. Incl. Phenom. Mol. Recognit. Chem.*, **21**, 47–78.
49 Flanigen, E.M. and Patton, R.L. (1978) US Patent 4,073,865.
50 Guth, J.L., Kessler, H., and Wey, R. (1986) *Stud. Surf. Sci. Catal.*, **28**, 121–128.
51 Corma, A., Diaz-Cabanas, M.J., Martinez-Triguero, J., Rey, F., and Rius, J. (2002) *Nature*, **418**, 514–517.
52 Camblor, M.A., Villaescusa, L.A., and Diaz-Cabanas, M.J. (1999) *Top. Catal.*, **9**, 59–76.
53 Zones, S.I., Darton, R.J., Morris, R., and Hwang, S.-J. (2005) *J. Phys. Chem. B*, **109**, 652–661.
54 Zones, S.I. (2011) *Microporous Mesoporous Mater.*, **144**, 1–8.
55 Geisinger, K.L., Gibbs, G.V., and Navrotsky, A. (1985) *Phys. Chem. Miner.*, **11**, 266–283.
56 O'Keeffe, M. and Yaghi, O.M. (1999) *Chem. Eur. J.*, **5**, 2796–2801.
57 Wragg, D.S., Morris, R.E., and Burton, A.W. (2008) *Chem. Mater.*, **20**, 1561–1570.
58 McCusker, L.B., Meier, W.M., Suzuki, K., and Shin, S. (1986) *Zeolites*, **6**, 388–391.
59 Simancas, R., Jordá, J.L., Rey, F., Corma, A., Cantín, A., Peral, I., and Popescu, C. (2014) *J. Am. Chem. Soc.*, **136**, 3342–3345.
60 Smeets, S., McCusker, L.B., Baerlocher, C., Xie, D., Chen, C.-Y., and Zones, S.I. (2015) *J. Am. Chem. Soc.*, **137**, 2015–2020.
61 Burton, A.W. (2010) US Patent 7,820,141.
62 Xie, D., McCusker, L.B., and Baerlocher, C. (2011) *J. Am. Chem. Soc.*, **133**, 20604–20610.
63 Strohmaier, K.G. and Vaughan, D.E.W. (2003) *J. Am. Chem. Soc.*, **125**, 16035–16039.
64 O'Keeffe, M. and Hyde, B.G. (1996) *Crystal Structures I: Patterns and Symmetry*, Mineralogical Society of America, Washington DC.
65 Blasco, T., Corma, A., Díaz-Cabañas, M.J., Rey, F., Vidal-Moya, J.A., and Zicovich-Wilson, C.M. (2002) *J. Phys. Chem. B*, **106**, 2634–2642.
66 Sastre, G., Vidal-Moya, J.A., Blasco, T., Rius, J., Jordá, J.L., Navarro, M.T., Rey, F., and Corma, A. (2002) *Angew. Chem., Int. Ed.*, **41**, 4722–4726.
67 Corma, A., Navarro, M.T., Rey, F., Rius, J., and Valencia, S. (2001) *Angew. Chem., Int. Ed.*, **40**, 2277–2280.
68 Corma, A., Rey, F., Valencia, S., Jordá, J.L., and Rius, J. (2003) *Nat. Mater.*, **2**, 493–497.
69 Castañeda, R., Corma, A., Fornés, V., Rey, F., and Rius, J. (2003) *J. Am. Chem. Soc.*, **125**, 7820–7821.

70 Tang, L., Shi, L., Bonneau, C., Sun, J., Yue, H., Ojuva, A., Lee, B.-L., Kritikos, M., Bell, R.G., Bacsik, Z., Mink, J., and Zou, X. (2008) *Nat. Mater.*, **7**, 381–385.

71 Mathieu, Y., Paillaud, J.-L., Caullet, P., and Bats, N. (2004) *Microporous Mesoporous Mater.*, **75**, 13–22.

72 Pulido, A., Corma, A., and Sastre, G. (2006) *J. Phys. Chem. B*, **110**, 23951–23961.

73 Matsukata, M., Ogura, M., Osaki, T., Rao, P.R.H.P., Nomura, M., and Kikuchi, E. (1999) *Top. Catal.*, **9**, 77–92.

74 Xu, W., Dong, J., Li, J., Li, J., and Wu, F. (1990) *J. Chem. Soc., Chem. Commun.*, 755–756.

75 Kim, M.-H., Li, H.-X., and Davis, M.E. (1993) *Microporous Mater.*, **1**, 191–200.

76 Miler, S.J. (1996) US Patent 5,558,851.

77 Zones, S.I., Nakagawa, Y., Lee, G.S., Chen, C.Y., and Yuen, L.T. (1998) *Microporous Mesoporous Mater.*, **21**, 199–211.

78 Zones, S.I., Nakagawa, Y., Yuen, L.T., and Harris, T.V. (1996) *J. Am. Chem. Soc.*, **118**, 7558–7567.

79 Wagner, P., Nakagawa, Y., Lee, G.S., Davis, M.E., Elomari, S., Medrud, R.C., and Zones, S.I. (2000) *J. Am. Chem. Soc.*, **122**, 263–273.

80 Kubota, Y., Helmkamp, M.M., Zones, S.I., and Davis, M.E. (1996) *Microporous Mater.*, **6**, 213–229.

81 Goretsky, A.V., Beck, L.W., Zones, S.I., and Davis, M.E. (1999) *Microporous Mesoporous Mater.*, **28**, 387–393.

82 Liebau, F. (1983) *Zeolites*, **3**, 191–193.

83 Gies, H. and Marler, B. (1992) *Zeolites*, **12**, 42–49.

84 Lobo, R.F. and Davis, M.E. (1995) *J. Am. Chem. Soc.*, **117**, 3766–3779.

85 Dorset, D.L., Weston, S.C., and Dhingra, S.S. (2006) *J. Phys. Chem. B*, **110**, 2045–2050.

86 Moliner, M., Willhammar, T., Wan, W., González, J., Rey, F., Jorda, J.L., Zou, X., and Corma, A. (2012) *J. Am. Chem. Soc.*, **134**, 6473–6478.

87 Burkett, S.L. and Davis, M.E. (1994) *J. Phys. Chem.*, **98**, 4647–4653.

88 Burkett, S.L. and Davis, M.E. (1995) *Chem. Mater.*, **7**, 920–928.

89 Dorset, D.L., Kennedy, G.J., Strohmaier, K.G., Diaz-Cabañas, M.J., Rey, F., and Corma, A. (2006) *J. Am. Chem. Soc.*, **128**, 8862–8867.

90 Tuel, A. and Taarit, Y.B. (1994) *Microporous Mater.*, **2**, 515–524.

91 Yun, Y., Hernández, M., Wan, W., Zou, X., Jordá, J.L., Cantín, A., Rey, F., and Corma, A. (2015) *Chem. Commun.*, **51**, 7602–7605.

92 Dorset, D.L., Strohmaier, K.G., Kliewer, C.E., Corma, A., Díaz-Cabañas, M.J., Rey, F., and Gilmore, C.J. (2008) *Chem. Mater.*, **20**, 5325–5331.

93 Corma, A., Diaz-Cabanas, M.J., Jorda, J.L., Rey, F., Sastre, G., and Strohmaier, K.G. (2008) *J. Am. Chem. Soc.*, **130**, 16482–16483.

94 Hernández-Rodríguez, M., Jordá, J.L., Rey, F., and Corma, A. (2012) *J. Am. Chem. Soc.*, **134**, 13232–13235.

95 Simancas, R., Jordá, J.L., Rey, F., Corma, A., Cantín, A., Peral, I., and Popescu, C. (2014) *J. Am. Chem. Soc.*, **136**, 3342–3345.

96 Jo, C., Lee, S., Cho, S.J., and Ryoo, R. (2015) *Angew. Chem., Int. Ed.*, **54**, 12805–12808.

97 Moliner, M. (2015) *Top. Catal.*, **58**, 502–512.

98 Corma, A., Rey, F., Rius, J., Sabater, M.J., and Valencia, S. (2004) *Nature*, **431**, 287–290.

99 Xu, D., Ma, Y., Jing, Z., Han, L., Singh, B., Feng, J., Shen, X., Cao, F., Oleynikov, P., Sun, H., Terasaki, O., and Che, S. (2014) *Nat. Commun.*, **5**, 4262.

100 Singh, B.K., Xu, D., Han, L., Ding, J., Wang, Y., and Che, S. (2014) *Chem. Mater.*, **26**, 7183–7188.

101 Moliner, M., Rey, F., and Corma, A. (2013) *Angew. Chem., Int. Ed.*, **52**, 13880–13889.

102 Burton, A.W. (2007) *J. Am. Chem. Soc.*, **129**, 7627–7637.

103 Pophale, R., Daeyaert, F., and Deem, M.W. (2013) *J. Mater. Chem. A*, **1**, 6750–6760.

104 Sastre, G., Cantin, A., Diaz-Cabañas, M.J., and Corma, A. (2005) *Chem. Mater.*, **17**, 545–552.

105 Schmidt, J.E., Deem, M.W., and Davis, M.E. (2014) *Angew. Chem., Int. Ed.*, **53**, 8372–8374.

106 Lewis, D.W., Willock, D.J., Catlow, C.R.A., Thomas, J.M., and Hutchings, G.J. (1996) *Nature*, **382**, 304–606.

107 Lewis, D.W., Sankar, G., Wyles, J.K., Thomas, J.M., Catlow, C.R.A., and Willock, D.J. (1997) *Angew. Chem., Int. Ed. Engl.*, **36**, 2675–2677.

108 Boal, B.W., Schmidt, J.E., Deimund, M.A., Deem, M.W., Henling, L.M., Brand, S.K., Zones, S.I., and Davis, M.E. (2015) *Chem. Mater.*, **27**, 7774–7779.

109 Pulido, A., Moliner, M., and Corma, A. (2015) *J. Phy. Chem. C*, **119**, 7711–7720.

110 Iyoki, K., Itabashi, K., and Okubo, T. (2014) *Microporous Mesoporous Mater.*, **189**, 22–30.

111 Arika, J., Miyazaki, H., Igawa, K., and Itabashi, K. (1985) US Patent 4,562,055.

112 Arika, J., Miyazaki, H., Igawa, K., and Itabashi, K. (1987) US Patent 4,650,654.

113 Arika, J., Miyazaki, H., Itabashi, K., and Aimoto, M. (1987) US Patent 4,664,898.

114 Xie, B., Song, J., Ren, L., Ji, Y., Li, J., and Xiao, F.-S. (2008) *Chem. Mater.*, **20**, 4533–4553.

115 Itabashi, K., Kamimura, Y., Iyoki, K., Shimojima, A., and Okubo, T. (2012) *J. Am. Chem. Soc.*, **134**, 11542–11549.

116 Roth, W.J., Nachtigall, P., Morris, R.E., and Čejka, J. (2014) *Chem. Rev.*, **114**, 4807–4837.

117 Leonowicz, M.E., Lawton, J.A., Lawton, S.L., and Rubin, M.K. (1994) *Science*, **264**, 1910–1913.

118 Moteki, T., Chaikittisilp, W., Sakamoto, Y., Shimojima, A., and Okubo, T. (2011) *Chem. Mater.*, **23**, 3564–3570.

119 Verheyen1, E., Joos, L., Van Havenbergh, K., Breynaert, E., Kasian, N., Gobechiya, E., Houthoofd, K., Martineau, C., Hinterstein, M., Taulelle, F., Van Speybroeck, V., Waroquier, M., Bals, S., Van Tendeloo, G., Kirschhock, C.E.A., and Martens, J.A. (2012) *Nat. Mater.*, **11**, 1059–1064.

120 Roth, W.J., Nachtigall, P., Morris, R.E., Wheatley, P.S., Seymour, V.R., Ashbrook, S.E., Chlubná, P., Grajciar, L., Položij, M., Zukal, A., Shvets, O., and Čejka, J. (2013) *Nat. Chem.*, **5**, 628–633.

121 Wheatley, P.S., Chlubná-Eliášová, P., Greer, H., Zhou, W., Seymour, V.R., Dawson, D.M., Ashbrook, S.E., Pinar, A.B., McCusker, L.B., Opanasenko, M., Čejka, J., and Morris, R.E. (2014) *Angew. Chem., Int. Ed.*, **53**, 13210–13214.

122 Morris, R.E. and Čejka, J. (2015) *Nat. Chem.*, **7**, 381–388.

123 Mazur, M., Wheatley, P.S., Navarro, M., Roth, W.J., Položij, M., Mayoral, A., Eliášová, P., Nachtigall, P., Čejka, J., and Morris, R.E. (2016) *Nat. Chem.*, **8**, 58–62.

124 Trachta, M., Bludský, O., Čejka, J., Morris, R.E., and Nachtigall, P. (2014) *ChemPhysChem*, **15**, 2972–2976.

125 Thommes, M., Kaneko, K., Neimark, A.V., Olivier, J.P., Rodriguez-Reinoso, F., Rouquerol, J., and Sing, K.S.W. (2015) *Pure Appl. Chem.*, **87**, 1051–1069.

126 Jiang, J., Yu, J., and Corma, A. (2010) *Angew. Chem., Int. Ed.*, **49**, 3120–3145.

127 Sun, J., Bonneau, C., Cantin, A., Corma, A., Diaz-Cabañas, M.J., Moliner, M., Zhang, D., Li, M., and Zou, X. (2009) *Nature*, **458**, 1154–1157.

128 Jiang, J., Jorda, J.L., Yu, J., Baumes, L.A., Mugnaioli, E., Diaz-Cabanas, M.J., Kolb, U., and Corma, A. (2011) *Science*, **333**, 1131–1134.

129 Na, K., Choi, M., and Ryoo, R. (2013) *Microporous Mesoporous Mater.*, **166**, 3–19.

130 Pérez-Ramírez, J. (2012) *Nat. Chem.*, **4**, 250–251.

131 Janssen, A.H., Koster, A.J., and K. P., deJong. (2001) *Angew. Chem., Int. Ed.*, **4**, 1102–1104.

132 Ogura, M., Shinomiya, S., Tateno, J., Nara, Y., Kikuchi, E., and Matsukata, M. (2000) *Chem. Lett.*, 882–883.

133 Ogura, M. (2008) *Catal. Surv. Asia*, **12**, 16–27.

134 Verboekend, D. and Pérez-Ramírez, J. (2011) *Catal. Sci. Technol.*, **1**, 879–890.

135 Garcia-Martinez, J., Xiao, C., Cychosz, K.A., Li, K., Wan, W., Zou, X., and Thommes, M. (2014) *ChemCatChem*, **6**, 3110–3115.

136 Fan, W., Snyder, M.A., Kumar, S., Lee, P.-S., Yoo, W.C., McCormick, A.V., Penn, R.L., Stein, A., and Tsapatsis, M. (2008) *Nat. Mater.*, **7**, 984–991.

137 Lee, P.-S., Zhang, X., Stoeger, J.A., Malek, A., Fan, W., Kumar, S., Yoo, W.C., Hashimi, S.A., Penn, R.L., Stein, A., and

Tsapatsis, M. (2011) *J. Am. Chem. Soc.*, **133**, 493–502.

138 Chen, H., Wydra, J., Zhang, X., Lee, P.-S., Wang, Z., Fan, W., and Tsapatsis, M. (2011) *J. Am. Chem. Soc.*, **133**, 12390–12393.

139 Choi, M., Cho, H.S., Srivastava, R., Venkatesan, C., Choi, D.-H., and Ryoo, R. (2006) *Nat. Mater.*, **5**, 718–723.

140 Choi., M., Na, K., Kim, J., Sakamoto, Y., Terasaki, O., and Ryoo, R. (2009) *Nature*, **461**, 246–249.

141 Na, K., Jo, C., Kim, J., Cho, K., Jung, J., Seo, Y., Messinger, R.J., Chmelka, B.F., and Ryoo, R. (2011) *Science*, **333**, 328–332.

142 Zhang, X., Liu, D., Xu, D., Asahina, S., Cychosz, K.A., Agrawal, K.V., Wahedi, Y.A., Bhan, A., Hashimi, S.A., Terasaki, O., Thommes, M., and Tsapatsis, M. (2012) *Science*, **336**, 1684–1687.

143 Chaikittisilp, W., Suzuki, Y., Mukti, R.R., Suzuki, T., Sugita, K., Itabashi, K., Shimojima, A., and Okubo, T. (2013) *Angew. Chem., Int. Ed.*, **52**, 3355–3359.

144 Shimada, A., Sakatani, K., Kameo, Y., and Takahashi, K. (2015) *J. Radioanal. Nucl. Chem.*, **303**, 1137–1140.

145 Elangovan, S.P., Ogura, M., Davis, M.E., and Okubo, T. (2004) *J. Phys. Chem. B*, **108**, 13059–13061.

146 Lai, Z., Bonilla, G., Diaz, I., Nery, J.G., Sujaoti, K., Amat, M.A., Kokkoli, E., Terasaki, O., Thompson, R.W., Tsapatsis, M., and Vlachos, D.G. (2003) *Science*, **300**, 456–460.

147 Pham, T.C.T., Nguyen, T.H., and Yoon, K.B. (2013) *Angew. Chem., Int. Ed.*, **52**, 8693–8698.

148 Choudhary, V.R., Kinage, A.K., and Choudhary, T.V. (1997) *Science*, **275**, 1286–1288.

149 Gao, J., Zheng, Y., Jehng, J.-M., Tang, Y., Wachs, I.E., and Podkolzin, S.G. (2015) *Science*, **348**, 686–690.

150 Grundner, S., Markovits, M.A.C., Li, G., Tromp, M., Pidko, E.A., Hensen, E.J.M., Jentys, A., Sanchez-Sanchez, M., and Lercher, J.A. (2015) *Nat. Commun.*, **6**, 7546.

151 Corma, A., Nemeth, L.T., Renz, M., and Valencia, S. (2001) *Nature*, **412**, 423–425.

152 Davis, M.E. (2015) *Top. Catal.*, **58**, 405–409.

153 Pacheco, J.J. and Davis, M.E. (2014) *Proc. Natl. Acad. Sci. USA*, **111**, 8363–8367.

154 Holm, M.S., Saravanamurugan, S., and Taarning, E. (2010) *Science*, **328**, 602–605.

155 Orazov, M. and Davis, M.E. (2015) *Proc. Natl. Acad. Sci. USA*, **112**, 11777–11782.

156 Dusselier, M., Van Wouwe, P., Dewaele, A., Jacobs, P.A., and Sels, B.F. (2015) *Science*, **349**, 78–80.

157 Liu, Z., Wakihara, T., Nishioka, D., Oshima, K., Takewaki, T., and Okubo, T. (2014) *Chem. Mater.*, **26**, 2327–2331.

158 Liu, Z., Wakihara, T., Oshima, K., Nishioka, D., Hotta, Y., Elangovan, S.P., Yanaba, Y., Yoshikawa, T., Chaikittisilp, W., Matsuo, T., Takewaki, T., and Okubo, T. (2015) *Angew. Chem., Int. Ed.*, **54**, 5683–5687.

6
Ordered Mesoporous Materials

Michal Kruk[1,2]

[1]*College of Staten Island, City University of New York, Department of Chemistry, 2800 Victory Boulevard, Staten Island 10314, NY, USA*
[2]*Ph.D. Program in Chemistry, Graduate Center of City University of New York, 365 Fifth Avenue, New York, NY 10016, USA*

6.1
Mesoporous Materials

According to the well-established classification of the International Union of Pure and Applied Chemistry (IUPAC) [1], pores are divided into three categories based on their width: micropores of width below 2 nm, mesopores of width between 2 and 50 nm, and macropores of width above 50 nm. This classification is rooted in a gas adsorption behavior of the pores: micropores are sufficiently small to adsorb gases at low pressures through the micropore filling, mesopores exhibit monolayer–multilayer adsorption of molecules followed by the complete filling of pores via capillary condensation, while macropores exhibit virtually unrestricted monolayer–multilayer adsorption. Since the time when the above classification has been established, the knowledge of gas adsorption behavior in pores has evolved, and even for most commonly used nitrogen gas at −196 °C, pores of sizes on the borderlines between the aforementioned ranges may exhibit adsorption mechanisms different from those noted above [2]. For instance, pores of diameter slightly above 50 nm may exhibit capillary condensation. Still, the above classification of pores by size is useful and is commonly used. It should be noted that the term "micropore" is sometimes also used to describe pores that are microns in size, but since the pore size difference between this definition and the one discussed above is by three orders of magnitude, the misunderstanding is usually avoided.

Mesoporous materials were known for a long time [3]. Many silica gels belong to this category and some of them have been used for decades in liquid chromatography [4]. It should be noted that silica is hydroxylated silicon dioxide ($SiO_{2-x/2}(OH)_x$), which is a network solid that is amorphous on atomic scale, and in which silicon atoms are connected with one another via oxygen atoms, resulting in ≡Si—O—Si≡ linkages called siloxane bridges. The hydroxyl groups bonded to

Handbook of Solid State Chemistry, First Edition. Edited by Richard Dronskowski, Shinichi Kikkawa, and Andreas Stein.
© 2017 Wiley-VCH Verlag GmbH & Co. KGaA. Published 2017 by Wiley-VCH Verlag GmbH & Co. KGaA.

silicon atoms (i.e., ≡Si—O—H moieties) are called silanols. Aluminas, which are very useful as catalysts and catalyst supports [5], are often mesoporous, too. Carbon blacks often exhibit well-developed mesoporous structure and even activated carbons, which are primarily known as microporous adsorbents, may feature some mesopores [6]. All the above materials are disordered, that is, they do not exhibit any periodic arrays of mesopores. Also, their pore size distributions are typically broad. However, many mesoporous materials can currently be synthesized with periodic (or sometimes disordered) structures with pores of well-defined size and shape thanks to advances in the synthesis methodologies, including the soft templating [7] and hard templating [6]. The soft templating involves the formation of the mesoporous structure around surfactant micelles or other templates that have limited rigidity and often form via self-assembly. The hard templating involves the development of the mesoporous structure around rigid structures, which can be colloidal crystals or materials formed via soft templating, for instance. Even though ordered mesoporous materials (OMMs), some of which were mentioned above, were not known to the scientific community before 1992, the number of their currently known families is appreciable. In particular, metal–organic frameworks (MOFs) and covalent organic frameworks (COFs), which are typically microporous, can reach pore sizes in the mesopore range [8,9] and thus constitute novel atomic-scale ordered materials with well-defined mesopores. Some anodic aluminas and colloidal crystals, which are typically macroporous materials, may also exhibit voids in the mesopore range [10,11]. Moreover, phase-separated block copolymers can serve as precursors for OMMs, if one of the blocks can be selectively removed while the domains formed by the other block are preserved throughout the process [12]. Hereafter, the discussion will be primarily restricted to a vast family of surfactant-micelle-templated OMMs.

Due to their size, mesopores can accommodate a variety of molecules, both small and large (such as proteins) [13]. Mesoporous materials tend to have high specific surface areas, which are often on the order of several hundred square meters per gram, but can reach or even exceed $1200\,m^2/g$ [14]. They also tend to exhibit large pore volumes, which can reach or exceed $1\,cm^3/g$ [14]. Because of these and other remarkable characteristics, mesoporous materials are useful in many applications, including gas- and liquid-phase adsorption, heterogeneous catalysis (including enzymatic catalysis), electronics, templated synthesis of nanostructures, and so forth [13,15,16].

6.2
Surfactants

As noted above, surfactants are an important family of templates for the synthesis of well-defined mesoporous materials. Surfactants consist of molecules featuring hydrophilic parts (charged or neutral, but polar) and hydrophobic parts (often large, such as long alkyl chains, referred to as tails, or nonpolar/weakly polar oligomeric or polymeric moieties) and are known from their ability to self-assemble in aqueous solutions into well-defined aggregates, called micelles [17]. The hydrophobic parts of the surfactant molecules form a core of the micelle, thus reducing their contact with water, while the hydrophilic parts (called headgroups, if they are relatively small in size) reside on the micelle periphery and interact with water. Micelles may adopt a variety of shapes, including spherical and cylindrical [17], or may be branched, for instance in the case of their gyroidal structure [18]. Spherical micelles may form liquid crystalline (ordered) phases, including body-centered cubic or face-centered cubic structures, while cylindrical micelles may assemble into two-dimensional (2-D) hexagonal structures with micelles being parallel to one another [19].

6.3
Micelle-Templated Ordered Mesoporous Materials

In 1992, a groundbreaking method for the synthesis of high-surface-area mesoporous materials was reported, in which surfactant micelles serve as templates for periodically arranged mesopores of uniform size and well-defined shape [14]. Under basic conditions, alkyltrimethylammonium surfactants were shown to assemble with silicates to form ordered arrays of surfactant micelles embedded in the silicate framework. These products were named an M41S family of materials [14]. Two years earlier, the use of surfactants to induce rearrangements in layered silicates and obtain silicas with disordered pore systems of controlled pore size was reported [20], while a similar strategy later rendered also ordered mesoporous silicas [21]. The mesopores, that is the voids in the framework, of these materials are occupied by surfactant molecules, which need to be removed (typically by calcination, which is a heating at high temperature, for example, 550 °C, in the atmosphere of air or inert gas, or by solvent extraction) to make the pores fully accessible [14,22]. The best known member of the M41S family is MCM-41 silica with a two-dimensional (2-D) hexagonal structure of cylindrical pores [14] (see Figure 6.1) of diameter ranging from 2 to 7 nm [14,23,24] that are templated by

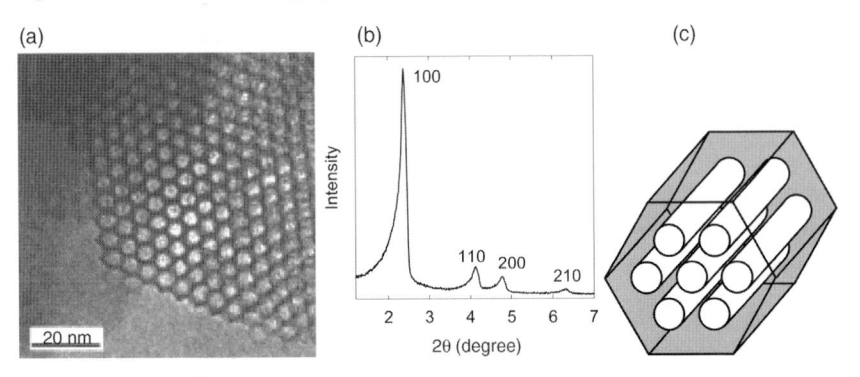

Figure 6.1 (a) Transmission electron microscopy image of MCM-41 silica. (Reproduced with permission from Ref. [72]. Copyright 2000 American Chemical Society.) (b) X-ray diffraction pattern of MCM-41. (Data taken from Ref. [26]) (c) Scheme of the structure of MCM-41.

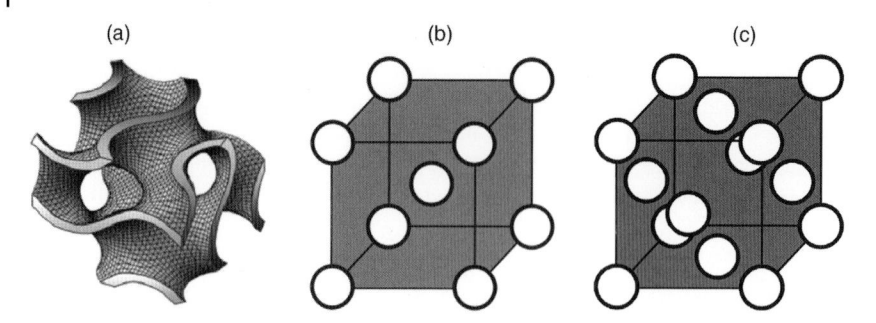

Scheme 6.1 Arrangements of mesopores in most common cubic structures of ordered mesoporous materials: (a) gyroidal structure of *Ia3d* symmetry. (b) body-centered cubic structure of *Im3m* symmetry, and (c) face-centered cubic structures(cubic close-packed) of *Fm3m* symmetry. Note that in the case of body-centered and face-centered cubic structures, the connections between the mesopores are not shown.

cylindrical micelles. Another well-known M41S material is MCM-48 silica with cubic *Ia3d* symmetry [14] (see Scheme 6.1) templated by a gyroidal micellar structure [18]. A lamellar structure in which layers of silica are separated by layers of surfactant molecules (the latter being attached to the silicate surface via ionic bonding), dubbed MCM-50, was also obtained [14]. However, this structure normally collapses as the surfactant is removed [25], highlighting the importance of 3-D framework cross-linking in obtaining structures sustainable upon the surfactant template removal.

The liquid crystal templating mechanism of the formation of M41S materials was initially postulated, which presumes that liquid crystalline micellar arrays template the inorganic framework that forms around the surfactant micelles [14]. While this mechanism was not found to be operative in most syntheses of OMMs [18,23,27,28], it was highly inspirational in focusing the attention on surfactant micelles in the quest for novel porous materials with unprecedented properties. Notably, the use of appropriate surfactants allowed one to obtain ordered mesoporous silicas of cubic Pm3n [28] and 3-D hexagonal (P6$_3$/mmc) [29] structures templated by spherical micelles. The formation of OMMs typically involves cases where there are no liquid crystalline surfactant phases in the solution prior to the addition of the framework precursor, although individual micelles are present, either of the shape similar to or different from that observed in the OMM eventually recovered [14]. Later studies of the formation of OMMs through evaporation-induced self-assembly [30] showed that surfactant micelles do not have to be present even at the moment of the addition of the framework precursor.

The hypothesis of a cooperative assembly mechanism based on the charge density matching between the surfactant and the framework precursor (with possible mediation by small ions) [18,23,28,29] was another major step forward, being important in the synthesis of various framework compositions templated by surfactants [31,32]. According to this mechanism, silicates (or other species that would form the framework) interact with surfactant molecules and form units consisting of framework precursor units (for instance oligosilicates) ionically bonded with one or more surfactant molecules. The composition of these units is dictated by charge density matching. The units assemble into ordered framework-surfactant (for instance, silicate-surfactant) composite materials. Therefore, the shape of the micelles before the addition of the framework precursor (for instance silicate) may or may not be the same as that in the final surfactant-templated material.

Silica is by far the most common composition of surfactant-templated OMMs, especially as it can be doped with heteroatoms or functionalized by organic groups in a direct synthesis. In particular, cationic surfactant micelles may template cocondensation of a silica precursor with other precursors containing heteroatoms, such as aluminum, titanium, vanadium, and many others, which is useful in the synthesis of heterogeneous catalyst materials [14,15]. However, because of the difference in rates of condensation of different framework precursors, a careful selection of the synthesis conditions may be needed to ensure incorporation of heteroatoms in the silica framework. Moreover, the loading of heteroatoms may be limited. A successful synthesis of silicas with pendant organic groups through cocondensation of tetraalkoxysilanes with organotrialkoxysilanes was another demonstration of the robustness of silica-based frameworks [33,34]. In this case, it is believed that organic groups are present on the surface of the silica framework, which may disrupt the interactions of silica with surfactant. Consequently, the loading of the organic groups is often limited to one group (or less) per four silicon atoms, even though in some cases more than half [35] or even all silicon atoms [36] may carry pendent organic groups. Organic groups can also be incorporated into the silica framework, if precursors with two or more alkoxysilyl groups attached to an organic group (such as bis(trialkoxysilyl)organic precursors) are used [37–39]. Such hybrid organic–inorganic frameworks in which a fraction of siloxane linkages is replaced by Si–R–Si linkages (where R is an organic group) are often referred to as periodic mesoporous organosilicas (PMOs). The bridging organic group, R, is typically either small ($-CH_2-$, $-CH_2-CH_2-$, $-CH=CH-$) or rigid (such as phenyl, thiophene, and biphenyl). Some of these materials exhibit periodic arrangements of groups in their frameworks in addition to the ordering of the mesopores [40], which is different from many other OMMs that exhibit amorphous frameworks.

6.4
Nonionic Poly(Ethylene Oxide)-Based Surfactants as Templates for OMMs

The use of nonionic surfactants was another major development in the surfactant-templated OMM synthesis. Neutral amines were recognized first [41], and were extensively used to template a family of silicas dubbed hexagonal mesoporous silicas (HMSs), even if their degree of structural ordering was low. Soon the focus largely shifted to surfactants with hydrophilic poly(ethylene oxide), PEO, block

(s) [42–44] and a hydrophobic part of many different compositions (typically long-chain alkyl groups or polymer blocks, mostly poly(propylene oxide), PPO). Pluronic poly(ethylene oxide)–poly(propylene oxide)–poly(ethylene oxide) (PEO–PPO–PEO) triblock copolymer surfactants [44] have gained particular importance, as they are readily available in a range of molecular weights and PEO/PPO ratios. The best-known material templated by Pluronics is SBA-15 silica of two-dimensional (2-D) hexagonal (honeycomb) structure of cylindrical mesopores, which is templated typically by Pluronic P123 ($EO_{20}PO_{70}EO_{20}$) triblock copolymer surfactant [44]. Although SBA-15 features a 2-D hexagonal structure akin to that of MCM-41 [14], its framework occludes the PEO chains of the Pluronic surfactant [45], presumably as a result of the framework formation around the PEO chains [42]. After their removal, PEO chains leave behind micropores in the silica wall, which constitute interconnections between the mesopores in SBA-15 [46]. Other block copolymer-templated mesoporous silicas include materials with body-centered cubic structure (SBA-16) of *Im3m* symmetry (see Scheme 6.1 and Figure 6.2) [44] and cubic close-packed (face-centered cubic) structure of *Fm3m* symmetry (known as FDU-12 [47] or KIT-5 [48] or FDU-1 [49,50]) (see Scheme 6.1), both with spherical mesopores and typically templated by Pluronic F127 ($EO_{106}PO_{70}EO_{106}$). Another structure is of *Ia3d* symmetry (known as FDU-5 [51] or KIT-6 [52]) (see Scheme 6.1), and is often synthesized using Pluronic P123 with appropriate additives, such as butanol [52] or sodium dodecyl sulfate (SDS) [53]. Many PMOs with diverse organic groups homogeneously distributed in the silica-based frameworks were also synthesized using block copolymer

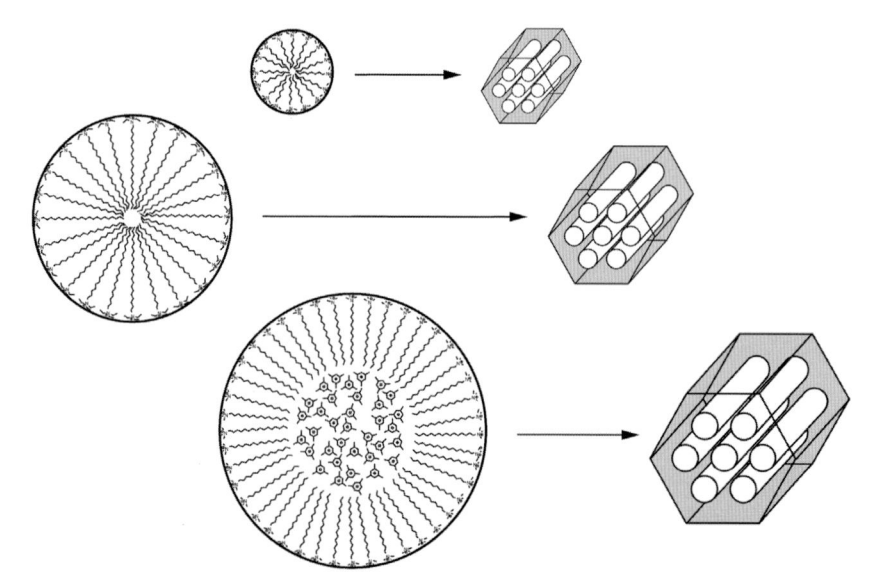

Scheme 6.2 Pore size control in micelle-templated materials via the selection of the surfactant size or addition of the micelle-swelling agent. (Reprinted with permission from Ref. [73]. Copyright 2012 American Chemical Society.)

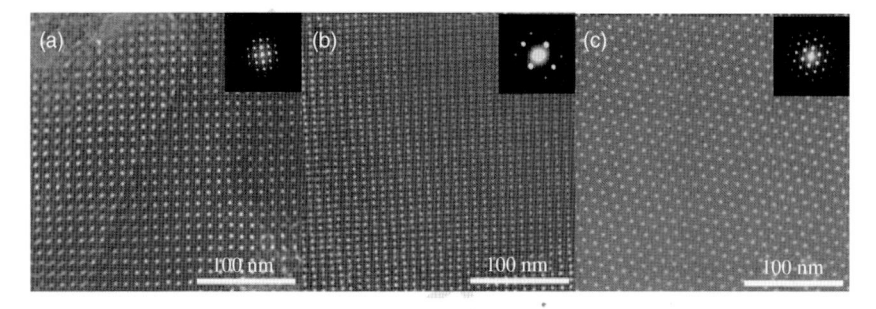

Figure 6.2 Transmission electron microscopy images showing different projections of SBA-16 silica. Insets show the corresponding electron diffraction patterns. Reprinted with permission from Ref. [63]. Copyright 2004 American Chemical Society.)

surfactant templates [54–58]. PEO-based surfactants were found to be particularly useful in the synthesis of OMMs with nonsilica frameworks, such as alumina, titania, zirconia, and so forth under nonaqueous synthesis conditions [7,31,59–61]. In fact, even porous polymers and carbons were obtained [59,62].

SBA-15 has become extremely popular [7,60], because it is easy to synthesize, highly hydrothermally stable, and its pore size (typically around 9 nm) is larger than that of MCM-41 and other materials templated by smaller surfactants, which is desirable in many applications [7,60]. Unique pore connectivity in the framework of SBA-15 opens opportunities in the templated synthesis of inverse replicas, including 2-D hexagonal arrangements of nanowires made of carbon and many other compositions [64,65] as well as of carbon nanotubules [66].

The study of the mechanism of formation of SBA-15 at 25–60 °C showed that spherical micelles of surfactant (Pluronic P123 or P103, $EO_{17}PO_{59}EO_{17}$) exist in solution before the addition of the silica precursor, and subsequently the silica-decorated micelles aggregate and form SBA-15 structure, which involves the rearrangement of the templating micelles to cylindrical shape [67]. On the other hand, the formation of SBA-16 silica with body-centered cubic structure templated by Pluronic F108 ($EO_{132}PO_{50}EO_{132}$) at 35 °C involves the silica-decorated spherical micelles that aggregate and form the ordered structure without change in the micelle shape [68] Moreover, KIT-6 silica of cubic $Ia3d$ symmetry appears to form through the liquid crystal templating mechanism, that is, with a presence of preformed cubic template before the silica precursor is added. These findings underscore the diversity of pathways through which silicas templated by Pluronics may form in dilute surfactant solutions.

6.5
Structure Control

Surfactants in aqueous solutions form micelles of different shapes, and the micelles can form a variety of liquid-crystalline structures. The tendency of a small-molecule surfactant to adopt a particular micellar geometry can be described using the surfactant packing parameter, $g = V/(a_0 l)$ [23]. This relation involves the effective hydrophilic head-group area at the water-micelle interface, a_0, the volume of hydrophobic surfactant tail(s) (possibly with cosolvent between the tails), V, and the kinetic surfactant tail length, l. When the surfactant in solution has a low value of the packing parameter ($g < 1/3$), which arises when the head-group area is large relative to the surfactant volume/length ratio, the surfactant molecules tend to form spherical micelles (they pack next to each other like cone-shaped objects would). For somewhat larger value of the packing parameter ($1/3 < g < 1/2$), the surfactant would form cylindrical micelles, so its molecules tend to pack like wedges would. Notably, the surface curvature of the micelles decreases as the surfactant packing parameter increases. For a higher value of the packing parameter, that is, when the surfactant molecule has a small head group and/or two tails, it tends to pack into low-curvature phases, such as gyroid (for $1/2 < g < 2/3$) or lamella (for $g = \sim 1$). Based on the above considerations, one can design surfactants that would favor particular phases. For instance, common cetyltrimethylammonium surfactants tend to form cylindrical micelles when they template materials, which in the case of silicas often results in the formation of MCM-41 with 2-D hexagonally ordered structure of cylindrical mesopores. On the other hand, the increase in the surfactant head group size (replacement of methyl by ethyl groups in trimethylammonium head group) renders materials with spherical mesopores (templated by spherical micelles) instead of cylindrical ones [23]. However, it needs to be understood that the predictive power of the surfactant packing parameter has its limitations, as the same surfactant may form MCM-41, gyroidal MCM-48, and lamellar phase under somewhat different conditions [14], and different phases (MCM-41, MCM-48, and lamellar) may transform from one to another during the synthesis [69].

Similar considerations can be made for nonionic surfactants with PEO-blocks, considering PEO as a counterpart of the head-group [70]. Moreover, if the hydrophobic part of the surfactant is polymeric (for instance PPO), it can be considered as a counterpart of the alkyl chain. However, one needs to consider that PEO (and hydrophobic polymeric blocks like PPO) can coil to an appreciable extent and moreover PEO becomes embedded in the framework of the templated material, so it is more beneficial to consider the PEO "head-group" volume rather than the head-group area [70]. Practically, one can vary the length of PEO block, while keeping the hydrophobic part approximately constant. For $C_nH_{2n+1}(OCH_2CH_2)_xOH$ surfactants such as Brij, several PEO block lengths may be available for a particular alkyl chain length, while for Pluronics (PEO–PPO–PEO), different PEO block lengths are available for surfactants with similar PPO block size [70]. Fine-tuning can be achieved through mixing of two surfactant with similar hydrophobic block and different PEO length [70], which allows one to overcome the limitations in the number of available surfactants. Silicas with spherical mesopores are obtained if PEO/hydrophobic-block ratio is high, that is, the volume of PEO domain is large in comparison to the PPO domain. As PEO/hydrophobic-block ratio is decreased, 2-D hexagonal silicas with cylindrical mesopores form, and a further decrease may afford lamellar structures [70]. While the above considerations can serve as a useful approximate guideline, it needs to be kept in mind that depending on conditions, the same surfactant may template different structures [71].

6.6 Pore Size Control

The surfactant micelle templating offers a variety of pore size control mechanisms, some of which stem from properties of micelles, while others are related to the features of frameworks of the materials (Scheme 6.2). First, the size of the micelles can be adjusted through the use of surfactants of different molecular size [14]. Cationic surfactants interact with templated frameworks at well-defined interface, so the size of the micelles dictates the diameter of voids in the framework. For the most commonly used alkyltrimethylammonium ($C_nH_{2n+1}(CH_3)_3N^+$) surfactants involved in MCM-41 and MCM-48 syntheses, the micelle radius can be estimated on the basis of the extended length of the surfactant alkyl tail that is equal to $n \times 0.125$ nm. For instance, when cetyltrimethylammonium surfactant ($C_{16}H_{33}(CH_3)_3N^+$) is used as a template, the pore radius of MCM-41 after the surfactant removal is close to 2.0 nm [72], the latter being the extended length of cetyl (hexadecyl) chain. In the above considerations, the contribution of trimethylammonium head group to the surfactant size is neglected, but also the shrinkage of the framework during the surfactant removal is not considered. These two assumptions lead to some inaccuracies, but the resulting errors are expected to cancel one another to an appreciable extent.

In the case of commonly used PEO-based surfactants, one can assume that the diameter of hydrophobic domains of the micelles determines the pore diameter, because PEO is expected to be occluded in the silica framework [74]. However, for block copolymer-templated surfactants, the extent of coiling of the hydrophobic block can be significant, so the pore size tends to be much smaller than the length of fully extended hydrophobic block, and is significantly dependent on the synthesis conditions. For instance, the pore diameter of SBA-15 silica templated by Pluronic P123 ($EO_{20}PO_{70}EO_{20}$) triblock copolymer is typically ~9 nm [46], even though the extended length of the PPO block is ~28 nm. Still, Pluronic P123 renders a larger pore size than its smaller Pluronic P103 counterpart [75], so the pore size is clearly controlled by the surfactant size, even though a relation is less clear than in the case of alkylammonium surfactants discussed above.

The use of alkylammonium surfactants provided access to silicas with cylindrical mesopores of diameter ranging from 2 to 5 nm [14,72], which was further extended through the hydrothermal restructuring processes to 7 nm [23,24]. The application of Pluronic block copolymers provided an easy access to silicas with cylindrical mesopores of diameters ranging from 5 to ~12 nm [76,77]. In this case, a typical way to increase the pore size is to increase the temperature and/or duration of the second step of the synthesis [44], which is referred to as a hydrothermal treatment, and is performed without stirring. While the structure type and even the unit-cell parameter (for as-synthesized samples) are often determined by the initial step of the synthesis, at which the surfactant-templated structure forms under stirring, the hydrothermal treatment leads to reduced wall thickness, and decreased shrinkage during calcination.

Major advances in the pore size enlargement of ordered mesoporous silicas were achieved through the use of custom-made surfactants. Their use afforded 2-D hexagonal (SBA-15) structures with (100) interplanar spacing, d_{100}, up to ~17 nm (d_{100} multiplied by 1.155 provides the unit-cell parameter, which is the distance between the pore centers) and pore diameter up to ~17 nm using PEO–poly(methyl acrylate)-based surfactants [78]. Moreover, body-centered cubic (SBA-16) structures (using PEO–poly(butylene oxide)-based surfactants) with pore diameters up to ~10 nm [63,79] and face-centered cubic structures (using PEO–polystyrene) [80] with pore diameter up to 31 nm, were also obtained.

One of the essential features of the micelle templating of OMMs is the tunability of their pore size using micelle-swelling agents (expanders) that enlarge the micelles and consequently, the pore dimensions in the templated materials. This pore enlargement pathway was known from the inception of the micelle-templating field [14], and is very attractive, as it presents an opportunity to use a single surfactant to attain a wide range of pore sizes tunable continuously by the adjustment of the quantity of the swelling agent in the synthesis mixture. Numerous compounds (typically hydrophobic ones) were found suitable as swelling agents, which includes benzene and its alkyl-substituted derivatives [14,44,47,74,81–84] (with 1,3,5-trimethylbenzene, TMB, being the most commonly used one), linear hydrocarbons [85–88], and long-chain amines [89]. Micelle expanders indeed increase the pore diameter and pore volume of micelle-templated materials [44,90], but their action commonly results in the decrease or loss of structural ordering [90]. The change in the structure type, for instance from SBA-15 to a mesocellular foam [76] may also take place.

In 2005, combinations of Pluronic block copolymer surfactants with appropriate swelling agents at the initial synthesis temperature of around 15 °C, instead of typically used 25–40 °C, afforded ordered mesoporous silicas with uncommonly large pore diameters [81,86], which in some cases were controlled by adjusting the initial temperature in the subambient range. In particular, the decrease in temperature of the first step of the synthesis from 25–40 to 14–15 °C in the case of FDU-12 silica with face-centered cubic structure of spherical mesopores templated by Pluronic F127 swollen by TMB led the unit-cell parameter increase from ~30 to 44 nm, and about twofold pore diameter increase up to 27 nm [81]. The resulting pore size was exceptionally large as for a well-ordered mesoporous silica. The pore diameter of resulting large-pore FDU-12 (LP-FDU-12) was found to be dependent on the initial synthesis temperature and additionally tunable by adjusting the temperature of the postsynthesis treatment in acid solution [81]. It was also shown that the pore diameter of highly ordered SBA-15 silica templated by Pluronic P123 can be increased up to ~15 nm (with (100) interplanar spacing up to ~14 nm) as the alkyl chain length of the linear hydrocarbon swelling agent decreased [86,88]. It was emphasized [86] that the order in which the pore sizes were increased followed the increasing extent of solubilization of alkanes in micelles of Pluronics [91]. The identification of the relation between the extent of solubilization of compounds and their swelling performance in micelle-templated OMM synthesis was a major new development. Until then, the focus was on the control of the pore size through the adjustment of the amount of the swelling agent in the synthesis mixture, which often resulted in the loss of structural ordering even after a modest pore size increase.

Further expansion of the pore size range attainable for OMMs in swelling-agent-based syntheses was achieved through the combination of subambient conditions and judicious selection of swelling agents for a particular surfactant template [74]. The concept was to apply a swelling agent that would expand micelles appreciably, but not excessively, so that the pore diameter can be increased while maintaining high degree of structural ordering [74,82]. To implement this concept, series of swelling agent candidates were identified [74] (on the basis of the literature data [91]) that would exhibit a systematically increasing extent of solubilization in a particular family of surfactants. In particular, the extent of solubilization of alkyl-substituted benzenes in Pluronics was reported to decrease with the increase in the number or size of alkyl substituents [91].

The above considerations were used to identify xylene, ethylbenzene, and toluene as superior swelling agents in the synthesis of LP-FDU-12 with face-centered cubic structure of spherical mesopores templated by Pluronic F127 [71,82], affording unit-cell parameters up to 69 nm and pore diameters of up to 47 nm. The use of xylene and toluene in the synthesis of large-pore periodic mesoporous organosilicas (PMOs) led to major increases of their unit-cell parameters and pore sizes [92] when compared to the TMB-based synthesis [93]. On the other hand, 1,3,5-triisopropylbenzene (TIPB) was used to obtain ultra-large-pore SBA-15 (ULP-SBA-15) with (100) interplanar spacings and pore diameters up to 30 nm [74,94]. The pore diameter of ULP-SBA-15 was controlled primarily through the adjustment of the initial synthesis temperature.

The suitability of swelling agents for particular block copolymer surfactant used can be predicted. For instance, Pluronic P123 ($EO_{20}PO_{70}EO_{20}$) surfactant typically employed in the synthesis of SBA-15 and other materials with cylindrical mesopores is composed of ~70 wt.% of hydrophobic PPO blocks and consequently solubilizes more alkylbenzenes per unit mass than Pluronic F127 ($EO_{106}PO_{70}EO_{106}$), which has only 30 wt.% PPO [82]. As a result, Pluronic P123 needs to be combined with a micelle expander that solubilizes to a moderate extent in Pluronics (e.g., TIPB) [74], but Pluronic F127 or F108 (with 20 wt.% PPO and 80 wt.% PEO) is preferably combined with a micelle expander that has a strong tendency to solubilize in Pluronics (e.g., xylene, ethylbenzene, or toluene) [71,82,95]. If this guideline is not followed, the swelling action may be excessive (for Pluronic P123 and similar surfactants), leading to large pore materials without any appreciable ordering, or not strong enough (for Pluronic F127 and similar surfactants), affording no appreciable pore size enlargement. While the above swelling-agent-based strategy proved to be successful at subambient temperatures, the use of suitable surfactants allows one to implement it also around room temperature [95]. Finally, it needs to be noted that in many successful swelling-agent-based synthesis procedures, the surfactant micelles appear to take up only a fraction (often small fraction) of the swelling agent that is present in the synthesis mixture [74]. Apparently, the swollen micelles are in equilibrium with the excess swelling agent, and perhaps this state is responsible for their uniform size, and consequently, for the well-defined structures of the templated materials.

So far, the pore size control based on the micelle size adjustment was discussed. However, it is important to keep in mind that the size of mesopores is not only governed by the size of the micelles (or their hydrophobic cores), but also by the extent of shrinkage that the framework undergoes when the surfactant is removed [96,97]. While a solvent extraction or low-temperature calcination (~300 °C) [98] often results in negligibly small shrinkage for silicas, the unit cell parameter decrease may be significant for common calcination conditions (being up to ~20% for some silica powders calcined at 550 °C) [96]. High calcination temperatures (up to ~1000 °C for silicas) result in even more significant shrinkage [96,99]. In the case of thin films or monoliths, the shrinkage may be high, and may be anisotropic (for films on a support) even if the calcination temperature is moderate [85,100]. Consequently, the increase in the calcination temperature can be used to systematically decrease the unit-cell size and pore diameter [97]. However, different synthesis procedures render materials with different propensity to shrink [71], and moreover, this strategy may result in low pore volumes and may render closed-pore materials [96,101]. The shrinkage can be further facilitated, if the calcination involves burning out of framework or pendent organic groups. In particular, the alkylammonium-templated synthesis of silicas with pendent vinyl groups followed by calcination rendered silicas with pore diameter in the upper part of the micropore range (1.3–2 nm) controlled by increasing the content of vinyl groups in the precursor [35]. Pores of that size are difficult to otherwise achieve through the surfactant templating. In the above case, the formation of silicas with narrow pores and small unit-cell sizes was additionally promoted by the decrease in these two structural features as the loading of vinyl groups in the precursor decreased.

6.7
Pore Connectivity

In the preceding discussion, it was implied that the mesopores of surfactant-templated silicas are accessible, and that the surfactant template can be effectively removed from the mesopores. While this is often the way that the surfactant templating is understood, there are number of cases where the surfactant-templating strategy offers an opportunity to adjust [47,49,63,94,102] or even suppress [99,101] the pore connectivity, depending on the nature of the surfactant template used and on the nanoscale structure of the templated material.

The silicate-surfactant charge density matching mechanism [23] operational in the formation of silicas templated by cationic surfactants does not provide an inherent way for the pore connectivity adjustment, so it is not surprising that the resulting silicas (MCM-41, MCM-48) have well-accessible mesopores. However, constrictions in the pore system may still be present if surfactant-templated mesopores are spherical. Indeed, the molecular sieving was observed in such cases, suggesting that the size of the connections between the mesopores is in the micropore range (diameter <2 nm), even as small as ~0.4 in diameter, which is on the order of magnitude smaller than the pore diameter [103]. Still, the general approach to adjust the size of these connections did not seem to emerge and moreover, these silicas are likely to be prone to water vapor adsorption from the air, perhaps followed by rehydroxylation, so the dimensional stability of the micropore connections is uncertain.

In contrast to their ionic counterparts, nonionic surfactants with hydrophilic poly(ethylene oxide) moieties create an inherent connectivity adjustment pathway in the templated silicas due to the occlusion of the poly(ethylene oxide) chains in the silica wall and the resulting microporosity in the wall [102]. If the silica frameworks feature spherical mesopores and the synthesis involves low hydrothermal treatment temperatures, the mesopores may be connected by narrow micropores (one or two orders of magnitude more narrow than the mesopore diameter) [102], which could possibly be eliminated by calcination at an appropriately high temperature [101]. The increase in the hydrothermal treatment temperature leads to a systematic enlargement of the connections, whose diameter may eventually exceed the micropore range (upper limit of 2 nm) and be well within the mesopore range (Scheme 6.3) [47,49]. If the hydrothermal treatment temperature is on the order of 140 °C, a very open pore structure may develop, in which pores are accessible through pathways of widths approaching the pore

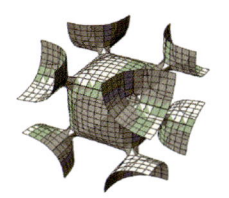

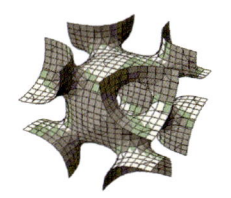

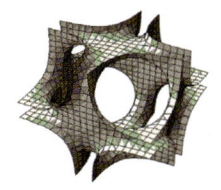

Hydrothermal treatment temperature or time

Scheme 6.3 The tailoring of pore connectivity in silicas with spherical mesopores templated by block-copolymer surfactants through the adjustment of the hydrothermal treatment temperature or time. The surface of the approximately spherical pores connected through openings of different size is shown. The actual openings are likely to have some degree of size and shape heterogeneity. Reprinted with permission from Ref. [63]. Copyright 2004 American Chemical Society.

diameter [47,49]. It should be noted that the connections may be appreciably irregular in shape and may have an appreciable size distribution, even if the sizes of the largest connection(s) per mesopore appear to be well defined [63].

The pore connectivity control is likely to go far beyond materials with spherical mesopores. While cylindrical mesopores templated by PEO-based surfactants often appear to be very open, they are accessible through narrow constrictions in some cases [104] or even they can be closed [71,99]. In the case of such surfactant templates, the silica wall primarily forms in the PEO corona of the micelles and around the PPO core, whose size and shape primarily determines the mesopore dimensions. This formation mechanism would imply that the cylindrical mesopores templated by cylindrical micelles are surrounded by the silica wall not only on the sides but at the ends as well. At present, intact rounded caps at the mesopore ends were observed in a few cases of silicas and organosilicas with very large mesopores and particularly thick walls [99,105]. Also, a limited size of entrances to cylindrical mesopores in some preparations of SBA-15, which are referred to as plugged hexagonal templated silicas (PHTSs), has been recognized [104]. However, most SBA-15 silicas have well accessible mesopores, which suggests that if the caps at pore ends exist at some stage of the synthesis, they do not persist till the final step(s) of the synthesis. Possibly, the silica framework at the surfactant micelle ends is not reinforced by the framework coming from another silica/micelle building block and thus is thinner that at the side of the pore and thus more prone to restructuring or even dissolution [99]. Other factors may be involved, especially as silicas with gyroidal pore arrangements templated by PEO-based surfactants do not appear to have any constrictions in the pore system [98].

Other surfactant-templated materials compositions may share some of the above pore entrance adjustment opportunities. However, there may be significant inherent limitations. To understand them, one needs to consider that the opportunities in the adjustment of the connectivity in the silica framework appear to stem from its noncrystallinity on atomic scale and reversibility of condensation with the formation of covalent siloxane (Si—O—Si) linkages. Moreover, the degree of framework condensation, that is the average number of siloxane linkages per silicon atom, systematically changes with synthesis conditions (including temperature, which tends to increase it). Therefore, the framework may adapt to different shapes, may uniformly shrink in three dimensions, and may seal micropores or potential defects. Even bridged organosilicas, which differ from silicas only in having a fraction of Si—O—Si linkages replaced by Si—R—Si linkages, where R is an organic groups, pose major problems in the pore connectivity adjustment, because usually their frameworks do not restructure much during hydrothermal treatment [106]. The latter behavior may be due to a large number of cross-links (up to six for typical framework precursors) per each bis(silylated) organic unit. The attempt to induced shrinkage thermally may lead to the decomposition of the organic groups in the framework, which may either lead to a higher shrinkage [92] or to the development of voids in the framework. To make the situation even more complex, many nonsilica frameworks can crystallize [31], in which case the wall may become discontinuous between the atomic-scale ordered domains, and the wall integrity cannot be ascertained.

References

1. Sing, K.S.W., Everett, D.H., Haul, R.A.W., Moscou, L., Pierotti, R.A., Rouquerol, J., and Siemieniewska, T. (1985) Reporting physisorption data for gas/solid systems with special reference to the determination of surface area and porosity (recommendations 1984). *Pure Appl. Chem.*, **57**, 603–619.
2. Kruk, M. and Jaroniec, M. (2001) Gas adsorption characterization of ordered organic-inorganic nanocomposite materials. *Chem. Mater.*, **13**, 3169–3183.
3. Gregg, S.J. and Sing, K.S.W. (1982) *Adsorption, Surface Area and Porosity*, Academic Press, London.
4. Unger, K.K. (1979) *Porous Silica*, Elsevier.
5. Trueba, M. and Trasatti, S.P. (2005) γ-Alumina as a support for catalysts: a review of fundamental aspects. *Eur. J. Inorg. Chem.*, **2005**, 3393–3403.
6. Ryoo, R., Joo, S.H., Kruk, M., and Jaroniec, M. (2001) Ordered mesoporous carbons. *Adv. Mater.*, **13**, 677–681.
7. Wan, Y. and Zhao, D. (2007) On the controllable soft-templating approach to mesoporous silicates. *Chem. Rev.*, **107**, 2821–2860.
8. Cote, A.P., Benin, A.I., Ockwig, N.W., O'Keeffe, M., Matzger, A.J., and Yaghi, O.M. (2005) Porous, crystalline, covalent organic frameworks. *Science*, **310**, 1166–1170.
9. Deng, H., Grunder, S., Cordova, K.E., Valente, C., Furukawa, H., Hmadeh, M., Gándara, F., Whalley, A.C., Liu, Z., Asahina, S. et al. (2012) Large-pore apertures in a series of metal-organic frameworks. *Science*, **336**, 1018–1023.
10. Li, A.P., Muller, F., Birner, A., Nielsch, K., and Gosele, U. (1998) Hexagonal pore arrays with a 50–420 nm interpore distance formed by self-organization in anodic alumina. *J. Appl. Phys.*, **84**, 6023–6026.
11. Yokoi, T., Sakamoto, Y., Terasaki, O., Kubota, Y., Okubo, T., and Tatsumi, T. (2006) Periodic arrangement of silica nanospheres assisted by amino acids. *J. Am. Chem. Soc.*, **128**, 13664–13665.
12. Zalusky, A.S., Olayo-Valles, R., Wolf, J.H., and Hillmyer, M.A. (2002) Ordered nanoporous polymers from polystyrene-polylactide block copolymers. *J. Am. Chem. Soc.*, **124**, 12761–12773.
13. Hartmann, M. (2005) Ordered mesoporous materials for bioadsorption and biocatalysis. *Chem. Mater.*, **17**, 4577–4593.
14. Beck, J.S., Vartuli, J.C., Roth, W.J., Leonowicz, M.E., Kresge, C.T., Schmitt, K.D., Chu, C.T.W., Olson, D.H., Sheppard, E.W., McCullen, S.B. et al. (1992) A new family of mesoporous molecular sieves prepared with liquid crystal templates. *J. Am. Chem. Soc.*, **114**, 10834–10843.
15. Corma, A. (1997) From microporous to mesoporous molecular sieve materials and their use in catalysis. *Chem. Rev.*, **97**, 2373–2419.
16. Davis, M.E. (2002) Ordered porous materials for emerging applications. *Nature*, **417**, 813–821.
17. Alexandridis, P. and Hatton, A.T. (1995) Poly(ethylene oxide)-poly(propylene oxide)-poly(ethylene oxide) block copolymer surfactants in aqueous solutions and at interfaces: thermodynamics, structure, dynamics, and modeling. *Colloids Surf. A Physicochem. Eng. Asp.*, **96**, 1–46.
18. Monnier, A., Schuth, F., Huo, Q., Kumar, D., Margolese, D., Maxwell, R.S., Stucky, G.D., Krishnamurty, M., Petroff, P., Firouzi, A. et al. (1993) Cooperative formation of inorganic-organic interfaces in the synthesis of silicate mesostructures. *Science*, **261**, 1299–1303.

19 Alexandridis, P., Olsson, U., and Lindman, B. (1998) A record nine different phases (four cubic, two hexagonal, and one lamellar lyotropic liquid crystalline and two micellar solutions) in a ternary isothermal system of an amphiphilic block copolymer and selective solvents (water and oil). *Langmuir*, **14**, 2627–2638.

20 Yanagisawa, T., Shimizu, T., Kuroda, K., and Kato, C. (1990) The preparation of alkyltrimethylammonium-kanemite complexes and their conversion to microporous materials. *Bull. Chem. Soc. Jpn.*, **63**, 988–992.

21 Inagaki, S., Fukushima, Y., and Kuroda, K. (1993) Synthesis of highly ordered mesoporous materials from a layered polysilicate. *J. Chem. Soc., Chem. Commun.*, 680–682.

22 Hitz, S. and Prins, R. (1997) Influence of template extraction on structure, activity, and stability of MCM-41 catalysts. *J. Catal.*, **168**, 194–206.

23 Huo, Q., Margolese, D.I., and Stucky, G.D. (1996) Surfactant control of phases in the synthesis of mesoporous silica-based materials. *Chem. Mater.*, **8**, 1147–1160.

24 Sayari, A., Liu, P., Kruk, M., and Jaroniec, M. (1997) Characterization of large-pore MCM-41 molecular sieves obtained via hydrothermal restructuring. *Chem. Mater.*, **9**, 2499–2506.

25 Kruk, M., Jaroniec, M., Yang, Y., and Sayari, A. (2000b) Determination of the lamellar phase content in MCM-41 using X-ray diffraction, nitrogen adsorption, and thermogravimetry. *J. Phys. Chem. B.*, **104**, 1581–1589.

26 Jaroniec, M., Kruk, M., Shin, H.J., Ryoo, R., Sakamoto, Y., and Terasaki, O. (2001) Comprehensive characterization of highly ordered MCM-41 silicas using nitrogen adsorption, thermogravimetry, X-ray diffraction and transmission electron microscopy. *Microporous Mesoporous Mater.*, **48**, 127–134.

27 Attard, G.S., Glyde, J.C., and Goltner, C.G. (1995) Liquid-crystalline phases as templates for the synthesis of mesoporous silica. *Nature*, **378**, 366–368.

28 Huo, Q., Margolese, D.I., Ciesla, U., Feng, P., Gier, T.E., Sieger, P., Leon, R., Petroff, P.M., Schueth, F., and Stucky, G.D. (1994) Generalized synthesis of periodic surfactant/inorganic composite materials. *Nature*, **368**, 317–321.

29 Huo, Q., Leon, R., Petroff, P.M., and Stucky, G.D. (1995) Mesostructure design with gemini surfactants: supercage formation in a three-dimensional hexagonal array. *Science*, **268**, 1324–1327.

30 Lu, Y., Ganguli, R., Drewien, C.A., Anderson, M.T., Brinker, C.J., Gong, W., Guo, Y., Soyez, H., Dunn, B., Huang, M.H. *et al.* (1997) Continuous formation of supported cubic and hexagonal mesoporous films by sol-gel dip-coating. *Nature*, **389**, 364–368.

31 de Soler-Illia, G.J., Sanchez, C., Lebeau, B., and Patarin, J. (2002) Chemical strategies to design textured materials: from microporous and mesoporous oxides to nanonetworks and hierarchical structures. *Chem. Rev.*, **102**, 4093–4138.

32 Schueth, F. (2001) Non-siliceous mesostructured and mesoporous materials. *Chem. Mater.*, **13**, 3184–3195.

33 Burkett, S.L., Sims, S.D., and Mann, S. (1996) Synthesis of hybrid inorganic-organic mesoporous silica by co-condensation of siloxane and organosiloxane precursors. *Chem. Commun.*, 1367–1368.

34 Macquarrie, D.J. (1996) Direct preparation of organically modified MCM-type materials. Preparation and characterization of aminopropyl-MCM and 2-cyanoethyl-MCM. *Chem. Commun.*, 1961–1962.

35 Kruk, M., Asefa, T., Jaroniec, M., and Ozin, G.A. (2002b) Metamorphosis of ordered mesopores to micropores: periodic silica with unprecedented loading of pendant reactive organic groups transforms to periodic microporous silica with tailorable pore size. *J. Am. Chem. Soc.*, **124**, 6383–6392.

36 Yu, K., Wu, X., Brinker, C.J., and Ripmeester, J. (2003) Mesostructured MTES-derived silica thin film with spherical voids investigated by TEM: 1.

Mesostructure Determination. *Langmuir*, **19**, 7282–7288.

37 Asefa, T., MacLachlan, M.J., Coombs, N., and Ozin, G.A. (1999) Periodic mesoporous organosilicas with organic groups inside the channel walls. *Nature*, **402**, 867–871.

38 Inagaki, S., Guan, S., Fukushima, Y., Ohsuna, T., and Terasaki, O. (1999) Novel mesoporous materials with a uniform distribution of organic groups and inorganic oxide in their frameworks. *J. Am. Chem. Soc.*, **121**, 9611–9614.

39 Melde, B.J., Holland, B.T., Blanford, C.F., and Stein, A. (1999) Mesoporous sieves with unified hybrid inorganic/organic frameworks. *Chem. Mater.*, **11**, 3302–3308.

40 Inagaki, S., Guan, S., Ohsuna, T., and Terasaki, O. (2002) An ordered mesoporous organosilica hybrid material with a crystal-like wall structure. *Nature*, **416**, 304–307.

41 Tanev, P.T. and Pinnavaia, T.J. (1995) A neutral templating route to mesoporous molecular sieves. *Science*, **267**, 865–867.

42 Bagshaw, S.A., Prouzet, E., and Pinnavaia, T.J. (1995) Templating of mesoporous molecular-sieves by nonionic polyethylene oxide surfactants. *Science*, **269**, 1242–1244.

43 Templin, M., Franck, A., Du Chesne, A., Leist, H., Zhang, Y., Ulrich, R., Schadler, V., and Wiesner, U. (1997) Organically modified aluminosilicate mesostructures from block copolymer phases. *Science*, **278**, 1795–1798.

44 Zhao, D., Huo, Q., Feng, J., Chmelka, B.F., and Stucky, G.D. (1998) Nonionic triblock and star diblock copolymer and oligomeric surfactant syntheses of highly ordered, hydrothermally stable, mesoporous silica structures. *J. Am. Chem. Soc.*, **120**, 6024–6036.

45 Melosh, N.A., Lipic, P., Bates, F.S., Wudl, F., Stucky, G.D., Fredrickson, G.H., and Chmelka, B.F. (1999) Molecular and mesoscopic structures of transparent block copolymer-silica monoliths. *Macromolecules*, **32**, 4332–4342.

46 Ryoo, R., Ko, C.H., Kruk, M., Antochshuk, V., and Jaroniec, M. (2000) Block-copolymer-templated ordered mesoporous silica: array of uniform mesopores or mesopore-micropore network? *J. Phys. Chem. B*, **104**, 11465–11471.

47 Fan, J., Yu, C., Gao, F., Lei, J., Tian, B., Wang, L., Luo, Q., Tu, B., Zhou, W., and Zhao, D. (2003) Cubic mesoporous silica with large controllable entrance sizes and advanced adsorption properties. *Angew. Chem., Int. Ed.*, **42**, 3146–3150.

48 Kleitz, F., Liu, D., Anilkumar, G.M., Park, I.-S., Solovyov, L.A., Shmakov, A.N., and Ryoo, R. (2003b) Large cage face-centered-cubic Fm3m mesoporous silica: synthesis and structure. *J. Phys. Chem. B*, **107**, 14296–14300.

49 Matos, J.R., Kruk, M., Mercuri, L.P., Jaroniec, M., Zhao, L., Kamiyama, T., Terasaki, O., Pinnavaia, T.J., and Liu, Y. (2003) Ordered mesoporous silica with large cage-like pores: structural identification and pore connectivity design by controlling the synthesis temperature and time. *J. Am. Chem. Soc.*, **125**, 821–829.

50 Yu, C., Yu, Y., and Zhao, D. (2000) Highly ordered large caged cubic mesoporous silica structures templated by triblock PEO-PBO-PEO copolymer. *Chem. Commun.*, 575–576.

51 Liu, X., Tian, B., Yu, C., Gao, F., Xie, S., Tu, B., Che, R., Peng, L.-M., and Zhao, D. (2002) Room-temperature synthesis in acidic media of large-pore three-dimensional bicontinuous mesoporous silica with Ia3d symmetry. *Angew. Chem., Int. Ed.*, **41**, 3876–3878.

52 Kleitz, F., Choi, S.H., and Ryoo, R. (2003) Cubic Ia3d large mesoporous silica: synthesis and replication to platinum nanowires, carbon nanorods and carbon nanotubes. *Chem. Commun.*, 2136–2137.

53 Chen, D., Li, Z., Yu, C., Shi, Y., Zhang, Z., Tu, B., and Zhao, D. (2005) Nonionic block copolymer and anionic mixed surfactants directed synthesis of highly ordered mesoporous silica with bicontinuous cubic structure. *Chem. Mater.*, **17**, 3228–3234.

54 Cho, E.-B., Kim, D., Gorka, J., and Jaroniec, M. (2009) Periodic mesoporous benzene- and thiophene-silicas prepared using aluminum chloride as an acid

catalyst: effect of aluminum salt/organosilane ratio and stirring time. *J. Phys. Chem. C*, **113**, 5111–5119.

55 Goto, Y. and Inagaki, S. (2002) Synthesis of large-pore phenylene-bridged mesoporous organosilica using triblock copolymer surfactant. *Chem. Commun.*, 2410–2411.

56 Guo, W., Kim, I., and Ha, C.-S. (2003) Highly ordered three-dimensional large-pore periodic mesoporous organosilica with Im3m symmetry. *Chem. Commun.*, 2692–2693.

57 Matos, J.R., Kruk, M., Mercuri, L.P., Jaroniec, M., Asefa, T., Coombs, N., Ozin, G.A., Kamiyama, T., and Terasaki, O. (2002) Periodic mesoporous organosilica with large cagelike pores. *Chem. Mater.*, **14**, 1903–1905.

58 Zhu, H., Jones, D.J., Zajac, J., Roziere, J., and Dutartre, R. (2001) Periodic large mesoporous organosilicas from lyotropic liquid crystal polymer templates. *Chem. Commun.*, 2568–2569.

59 Tanaka, S., Nishiyama, N., Egashira, Y., and Ueyama, K. (2005) Synthesis of ordered mesoporous carbons with channel structure from an organic-organic nanocomposite. *Chem. Commun.*, 2125–2127.

60 Wan, Y., Shi, Y., and Zhao, D. (2007) Designed synthesis of mesoporous solids via nonionic-surfactant-templating approach. *Chem. Commun.*, 897–926.

61 Yang, P., Zhao, D., Margolese, D.I., Chmelka, B.F., and Stucky, G.D. (1998) Generalized syntheses of large-pore mesoporous metal oxides with semicrystalline frameworks. *Nature*, **396**, 152–155.

62 Meng, Y., Gu, D., Zhang, F., Shi, Y., Yang, H., Li, Z., Yu, C., Tu, B., and Zhao, D. (2005) Ordered mesoporous polymers and homologous carbon frameworks: amphiphilic surfactant templating and direct transformation. *Angew. Chem., Int. Ed.*, **44**, 7053–7059.

63 Kim, T.W., Ryoo, R., Kruk, M., Gierszal, K.P., Jaroniec, M., Kamiya, S., and Terasaki, O. (2004) Tailoring the pore structure of SBA-16 silica molecular sieve through the use of copolymer blends and control of synthesis temperature and time. *J. Phys. Chem. B*, **108**, 11480–11489.

64 Jun, S., Joo, S.H., Ryoo, R., Kruk, M., Jaroniec, M., Liu, Z., Ohsuna, T., and Terasaki, O. (2000) Synthesis of new, nanoporous carbon with hexagonally ordered mesostructure. *J. Am. Chem. Soc.*, **122**, 10712–10713.

65 Tian, B., Liu, X., Yang, H., Xie, S., Yu, C., Tu, B., and Zhao, D. (2003) General synthesis of ordered crystallized metal oxide nanoarrays replicated by microwave-digested mesoporous silica. *Adv. Mater.*, **15**, 1370–1374.

66 Joo, S.H., Choi, S.J., Oh, I., Kwak, J., Liu, Z., Terasaki, O., and Ryoo, R. (2001) Ordered nanoporous arrays of carbon supporting high dispersions of platinum nanoparticles. *Nature*, **412**, 169–172.

67 Flodstrom, K., Wennerstrom, H., and Alfredsson, V. (2004a) Mechanism of mesoporous silica formation. A time-resolved NMR and TEM study of silica–block copolymer aggregation. *Langmuir*, **20**, 680–688.

68 Flodstrom, K., Wennerstrom, H., Teixeira, C.V., Amenitsch, H., Linden, M., and Alfredsson, V. (2004b) Time-resolved *in situ* studies of the formation of cubic mesoporous silica formed with triblock copolymers. *Langmuir*, **20**, 10311–10316.

69 Romero, A.A., Alba, M.D., Zhou, W., and Klinowski, J. (1997) Synthesis and characterization of the mesoporous silicate molecular sieve MCM-48. *J. Phys. Chem. B*, **101**, 5294–5300.

70 Kim, J.M., Sakamoto, Y., Hwang, Y.K., Kwon, Y.-U., Terasaki, O., Park, S.-E., and Stucky, G.D. (2002) Structural design of mesoporous silica by micelle-packing control using blends of amphiphilic block copolymers. *J. Phys. Chem. B*, **106**, 2552–2558.

71 Huang, L. and Kruk, M. (2015) Versatile surfactant/swelling-agent template for synthesis of large-pore ordered mesoporous silicas and related hollow nanoparticles. *Chem. Mater.*, **27**, 679–689.

72 Kruk, M., Jaroniec, M., Sakamoto, Y., Terasaki, O., Ryoo, R., and Ko, C.H. (2000) Determination of pore size and

pore wall structure of MCM-41 by nitrogen adsorption, transmission electron microscopy, and X-ray diffraction. *J. Phys. Chem. B*, **104**, 292–301.

73 Kruk, M. (2012) Access to ultralarge-pore ordered mesoporous materials through selection of surfactant/swelling-agent micellar templates. *Acc. Chem. Res.*, **45**, 1678–1687.

74 Cao, L., Man, T., and Kruk, M. (2009) Synthesis of ultra-large-pore SBA-15 silica with two-dimensional hexagonal structure using triisopropylbenzene as micelle expander. *Chem. Mater.*, **21**, 1144–1153.

75 Zhao, D., Yang, P., Chmelka, B.F., and Stucky, G.D. (1999) Multiphase assembly of mesoporous-macroporous membranes. *Chem. Mater.*, **11**, 1174–1178.

76 Lettow, J.S., Han, Y.J., Schmidt-Winkel, P., Yang, P., Zhao, D., Stucky, G.D., and Ying, J.Y. (2000) Hexagonal to mesocellular foam phase transition in polymer-templated mesoporous silicas. *Langmuir*, **16**, 8291–8295.

77 Vinu, A., Murugesan, V., Bohlmann, W., and Hartmann, M. (2004) An optimized procedure for the synthesis of AlSBA-15 with large pore diameter and high aluminum content. *J. Phys. Chem. B.*, **108**, 11496–11505.

78 Lin, C.-F., Lin, H.-P., Mou, C.-Y., and Liu, S.-T. (2006) Periodic mesoporous silicas via templating of new triblock amphiphilic copolymers. *Microporous Mesoporous Mater.*, **91**, 151–155.

79 Tattershall, C.E., Aslam, S.J., and Budd, P.M. (2002) Dimethylamino- and trimethylammonium-tipped oxyethylene-oxybutylene diblock copolymers and their use as structure-directing agents in the preparation of mesoporous silica. *J. Mater. Chem.*, **12**, 2286–2291.

80 Deng, Y., Yu, T., Wan, Y., Shi, S., Meng, Y., Gu, D., Zhang, L., Huang, Y., Liu, C., Wu, X. et al. (2007) Ordered mesoporous silicas and carbons with large accessible pores templated from amphiphilic diblock copolymer poly(ethylene oxide)-b-polystyrene. *J. Am. Chem. Soc.*, **129**, 1690–1697.

81 Fan, J., Yu, C., Lei, J., Zhang, Q., Li, T., Tu, B., Zhou, W., and Zhao, D. (2005) Low-temperature strategy to synthesize highly ordered mesoporous silicas with very large pores. *J. Am. Chem. Soc.*, **127**, 10794–10795.

82 Huang, L., Yan, X., and Kruk, M. (2010) Synthesis of ultra-large-pore FDU-12 silica with face-centered cubic structure. *Langmuir*, **26**, 14871–14878.

83 Jana, S.K., Nishida, R., Shindo, K., Kugita, T., and Namba, S. (2004) Pore size control of mesoporous molecular sieves using different organic auxiliary chemicals. *Microporous Mesoporous Mater.*, **68**, 133–142.

84 Namba, S. and Mochizuki, A. (1998) Effect of auxiliary chemicals on preparation of silica MCM-41. *Res. Chem. Intermediates*, **24**, 561–570.

85 Feng, P., Bu, X., Stucky, G.D., and Pine, D.J. (2000) Monolithic mesoporous silica templated by microemulsion liquid crystals. *J. Am. Chem. Soc.*, **122**, 994–995.

86 Sun, J., Zhang, H., Ma, D., Chen, Y., Bao, X., Klein-Hoffmann, A., Pfaender, N., and Su, D.S. (2005) Alkanes-assisted low temperature formation of highly ordered SBA-15 with large cylindrical mesopores. *Chem. Commun.*, 5343–5345.

87 Ulagappan, N. and Rao, C.N.R. (1996) Evidence of supramolecular organization of alkane and surfactant molecules in the process of forming mesoporous silica. *Chem. Commun.*, 2759–2760.

88 Zhang, H., Sun, J., Ma, D., Weinberg, G., Su, D.S., and Bao, X. (2006) Engineered complex emulsion system: toward modulating the pore length and morphological architecture of mesoporous silicas. *J. Phys. Chem. B*, **110**, 25908–25915.

89 Sayari, A., Kruk, M., Jaroniec, M., and Moudrakovski, I.L. (1998) New approaches to pore size engineering of mesoporous silicates. *Adv. Mater.*, **10**, 1376–1379.

90 Sayari, A., Yang, Y., Kruk, M., and Jaroniec, M. (1999) Expanding the pore size of MCM-41 silicas: use of amines as expanders in direct synthesis and postsynthesis procedures. *J. Phys. Chem. B*, **103**, 3651–3658.

91 Nagarajan, R. (1999) Solubilization of hydrocarbons and resulting aggregate shape transitions in aqueous solutions of pluronic (PEO-PPO-PEO) block copolymers. *Colloids Surf. B Biointerfaces*, **16**, 55–72.

92 Mandal, M. and Kruk, M. (2010a) Large-pore ethylene-bridged periodic mesoporous organosilicas with face-centered cubic structure. *J. Phys. Chem. C*, **114**, 20091–20099.

93 Zhou, X., Qiao, S., Hao, N., Wang, X., Yu, C., Wang, L., Zhao, D., and Lu, G.Q. (2007) Synthesis of ordered cubic periodic mesoporous organosilicas with ultra-large pores. *Chem. Mater.*, **19**, 1870–1876.

94 Cao, L. and Kruk, M. (2014) Short synthesis of ordered silicas with very large mesopores. *RSC Adv.*, **4**, 331–339.

95 Li, Y., Yi, J., and Kruk, M. (2015) Tuning of temperature window for unit-cell and pore size enlargement in face-centered-cubic large-mesopore silicas templated by swollen block copolymer micelles. *Chem. Eur. J.*, **21**, 12747–12754.

96 Kruk, M. and Hui, C.M. (2008) Synthesis and characterization of large-pore FDU-12 silica. *Microporous Mesoporous Mater.*, **114**, 64–73.

97 Naono, H., Hakuman, M., and Shiono, T. (1997) Analysis of nitrogen adsorption isotherms for a series of porous silicas with uniform and cylindrical pores: a new method of calculating pore size distribution of pore radius 1–2 nm. *J. Colloid. Interface Sci.*, **186**, 360–368.

98 Yi, J. and Kruk, M. (2015) Pluronic-P123-templated synthesis of silica with cubic Ia3d structure in the presence of micelle swelling agent. *Langmuir*, **31**, 7623–7632.

99 Mandal, M. and Kruk, M. (2012) Surfactant-templated synthesis of ordered silicas with closed cylindrical mesopores. *Chem. Mater.*, **24**, 149–154.

100 Wu, C.-W., Yamauchi, Y., Ohsuna, T., and Kuroda, K. (2006) Structural study of highly ordered mesoporous silica thin films and replicated Pt nanowires by high-resolution scanning electron microscopy (HRSEM). *J. Mater. Chem.*, **16**, 3091–3098.

101 Kruk, M. and Hui, C.M. (2008) Thermally induced transition between open and closed spherical pores in ordered mesoporous silicas. *J. Am. Chem. Soc.*, **130**, 1528–1529.

102 Kruk, M., Antochshuk, V., Matos, J.R., Mercuri, L.P., and Jaroniec, M. (2002) Determination and tailoring the pore entrance size in ordered silicas with cage-like mesoporous structures. *J. Am. Chem. Soc.*, **124**, 768–769.

103 Garcia-Bennett, A.E., Williamson, S., Wright, P.A., and Shannon, I.J. (2002) Control of structure, pore size and morphology of three-dimensionally ordered mesoporous silicas prepared using the dicationic surfactant $[CH_3(CH_2)_{15}N(CH_3)_2(CH_2)_3N(CH_3)_3]Br_2$. *J. Mater. Chem.*, **12**, 3533–3540.

104 Van Der Voort, P., Ravikovitch, P.I., De Jong, K.P., Benjelloun, M., Van Bavel, E., Janssen, A.H., Neimark, A.V., Weckhuysen, B.M., and Vansant, E.F. (2002) A new templated ordered structure with combined micro- and mesopores and internal silica nanocapsules. *J. Phys. Chem. B*, **106**, 5873–5877.

105 Mandal, M. and Kruk, M. (2010) Versatile approach to synthesis of 2-D hexagonal ultra-large-pore periodic mesoporous organosilicas. *J. Mater. Chem.*, **20**, 7506–7516.

106 Manchanda, A.S. and Kruk, M. (2016) Synthesis of large-pore face-centered-cubic periodic mesoporous organosilicas with unsaturated bridging groups. *Microporous Mesoporous Mater.*, **222**, 153–159.

7
Porous Coordination Polymers/Metal–Organic Frameworks

Ohtani Ryo[1] and Kitagawa Susumu[2]

[1]*Kumamoto University, Department of Chemistry, Graduate School of Science and Technology, 2-39-1 Kurokami, Chuo-ku, 860-8555, Kumamoto, Japan*
[2]*Kyoto University, Institute for Integrated Cell-Material Sciences (WPI-iCeMS), Yoshida, Sakyo-ku, 606-8501, Kyoto, Japan*

7.1
Introduction and Fundamentals of PCPs/MOFs

Porous frameworks that consist of metal-complex components and adsorbed gas molecules were first observed in 1997, when porous coordination polymer/metal–organic framework (PCP/MOF) chemistry began [1]. Since then, PCP/MOF compounds have been investigated as a new generation of porous materials besides conventional zeolites, activated carbons, and mesoporous silica, and the functional properties and characteristics of their cavities and frameworks have been reported [2–4]. PCPs/MOFs have three important features that are lacking in conventional porous materials. The first is "designability." Porous frameworks of PCPs/MOFs are constructed via self-assembly of metal ions, which act as nodes, and organic ligands, which act as spacers. Therefore, one can select the combination of metal ions and organic ligands to construct frameworks exhibiting desirable properties such as pore size, shape, surface functionality, and metal-complex properties (Figure 7.1a). The second is the "regularity of pore." PCPs/MOFs are crystalline materials with long-range (infinite) structural ordering in which micropores are aligned regularly. In such pores (or channels), accommodated guest molecules are also arranged regularly, and their transportation is associated with a characteristic mobility that is governed by a confinement effect (Section 7.5). Moreover, the infinite structures of PCPs/MOFs have high cooperativity, which is essential for the development of stimuli-responsive materials exhibiting high sensitivity (Section 7.4). The third is a "flexible framework," which arises from the moderate bond strength of coordination bonds. The structures of crystalline PCPs/MOFs are sometimes transformed in response to external physical and chemical stimuli, such as thermal treatment, light irradiation, and guest adsorption/desorption [5–8]. In particular, the guest-

Handbook of Solid State Chemistry, First Edition. Edited by Richard Dronskowski, Shinichi Kikkawa, and Andreas Stein.
© 2017 Wiley-VCH Verlag GmbH & Co. KGaA. Published 2017 by Wiley-VCH Verlag GmbH & Co. KGaA.

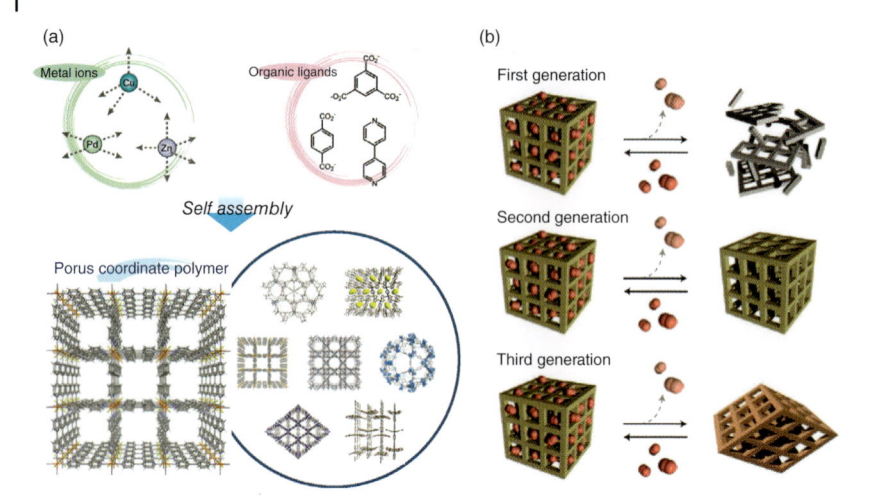

Figure 7.1 (a) Scheme for the self-assembly of metal ions and organic ligands producing porous frameworks of various structures. (b) The three categories of PCPs/MOFs, which differ in their porosity and structural flexibility. (Reprinted with permission from Ref. [9]. Copyright 2009, Macmillan Publishers Ltd: (Nature Chemistry).)

responsive structural transformation of flexible and dynamic frameworks gives rise to "molecular recognition," where substrates are selectively captured, as also occurs in enzymes. There have been many reports of selective adsorption, separation, and catalytic abilities resulting from such flexibility (Section 7.3).

In terms of flexibility, PCPs/MOFs are classified into three categories (Figure 7.1b) [10]. First-generation materials have frameworks without permanent porosity, that is, their porosity collapses irreversibly after the removal of guest molecules. Second-generation materials have robust frameworks with porosity, which is maintained before and after guest adsorption. Third-generation materials possess both highly ordered porous network and structural flexibility, and are referred to as "soft porous crystals" [9].

PCP/MOF materials serve two important functions: a porous function and a framework function (Figure 7.2). Porosity is available for guest adsorption, for carrying and releasing guests, as a reaction field for molecular conversion, and so on (Section 7.3). The framework provides physical properties that arise from its metal ion and organic ligand constituents, for example, magnetism, luminescence, conductivity, and dielectric property (Section 7.4). Synergistic and cooperative functions in host–guest systems have also attracted much attention.

Since the origin of PCP/MOF chemistry, this area of investigation has continuously expanded. Here, fundamental synthetic procedures for PCP/MOF compounds are described (Section 7.2), and subsequently, hot topics relating to both "porous" and "framework" functions, including important subjects such as "crystal engineering" (Section 7.6) and "physical chemistry" (Section 7.7) for PCPs/MOFs, are introduced with reference to representative works.

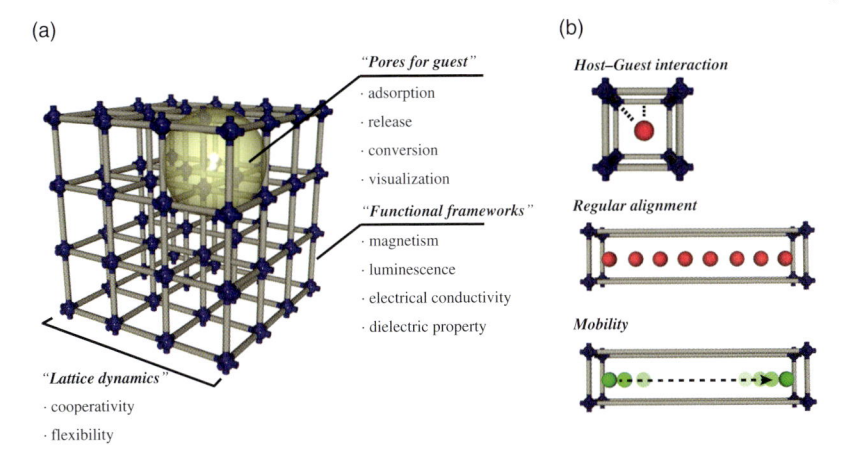

Figure 7.2 (a) Functional characteristics of PCP/MOF materials involving "lattice dynamics," "pores for guest," and "functional frameworks." (b) Activities of guests confined in the pores of PCPs/MOFs.

7.2 Synthetic Procedures

Since PCPs/MOFs have infinite structures and are not soluble, these compounds cannot be recrystalized to obtain their single crystals, unlike conventional molecular-metal complexes. Therefore, one-pot synthesis methods based on self-assembly, which is a diffusion method, solvothermal methods, sonochemical methods, microwave methods, and so on, have been applied for preparing single crystals of PCPs/MOFs. In the self-assembly process, crystal structures or morphologies are sometimes influenced by reaction conditions such as temperature, time, solvent, pH, and concentration. Therefore, trial and error is essential for acquiring the desired compounds for further research. In order to perform such syntheses rapidly and effectively, high-throughput systems have been utilized [11–13]. In these one-pot synthesis methods, insoluble organic ligands with complex structures and large molecular weights cannot be used. Sometimes, a mechanochemical method without solvents is utilized to overcome such limitations, thus increasing the variety of applicable organic ligands [14]. Generally, first-row transition metal ions such as Ni(II), Cu(II), and Zn(II) are utilized as metal sources for the construction of PCPs/MOFs. Recently, Zr(IV) [15–17] and La(III) [18,19] ions have been selected because they form chemically robust and thermally stable frameworks that are suitable for industrial applications. Moreover, inexpensive and low-toxicity compounds have been obtained using Group II Mg(II) and Ca(II) ions [20].

As a sequential technique, a "post-synthesis" method has been demonstrated for further functionalization of a parent framework [21,22]. This method is a valuable alternative for the synthesis of decorated frameworks by the chemical modification of organic ligands in PCP/MOF crystals, through subsequent chemical reactions. For the use of this "post-synthesis" method, reactive and accessible functional groups in the ligands are required, such as amino and carbonyl groups. Recently, "click chemistry" has been applied as a "post-synthesis" method that is effective for azide-decorated frameworks [23]. Moreover, in order to enhance the framework performance, not only chemical modifications (as mentioned above) but also crystal engineering approaches and crystal particle structuring are used (Section 7.6).

7.3
How to use "Pores" in PCPs/MOFs

The pores and channels that exist in PCPs/MOFs have four functions: "adsorption," "release," "conversion," and "visualization" of substrates. In each case, pore size, surface environment, and responsivity, which are determined by the flexibility of the framework, affect these functional characteristics strongly.

7.3.1
Adsorption

Typical functions of PCPs/MOFs are storage and separation of a target gas. Second-generation compounds incorporating high porosity and specific surface area are suitable for storage. In order to synthesize such compounds, in principle, the use of long pillar ligands is the simplest idea. However, this approach has issues because the extended long ligands prefer to produce interpenetrating dense networks with small porosity or fragile frameworks. To overcome these issues, Zaworotko and coworkers investigated the self-assembly conditions leading to a twofold interpenetrated dense structure or a noninterpenetrated framework of [Cd(bdc)(bpy)] (bpy = 4,4'-bipyridine, bdc = 1,4-benzenedicarboxylic acid), and reported that high-temperature, high-concentration conditions favored the dense form and low-temperature, low-concentration conditions favored a noninterpenetrated framework possessing high free volume (Figure 7.3a) [24]. Wöll and coworkers reported a methodology to obtain a noninterpenetrated framework of MOF-508 [Zn_2(bdc)$_2$(bpy)] [25] by using layer-by-layer growth on the substrate (a surface-mounted metal–organic framework, SURMOF) to prevent the interpenetration of its frameworks (Figure 7.3b) [26].

Yaghi and coworkers established a means of mixing two different linkers to prevent the formation of interpenetrating networks [27]. They synthesized a MOF-210 using 4,4',4''-[benzene-1,3,5-triyl-tris(ethyne-2,1-diyl)]tribenzoate (BTE) and biphenyl-4,4'-dicarboxylate (BPDC), which exhibits exceptional porosity and high Brunauer–Emmett–Teller (BET) surface area (6240 m^2/g) (Figure 7.4a). This value is near the ultimate adsorption limit for solid materials. This compound has high storage capacity for N_2, CO_2, H_2, and CH_4. Moreover, they constructed noninterpenetrating frameworks in an isoreticular (having the same topology) series of MOF-74 [28] using Mg ions and extended pillar ligands

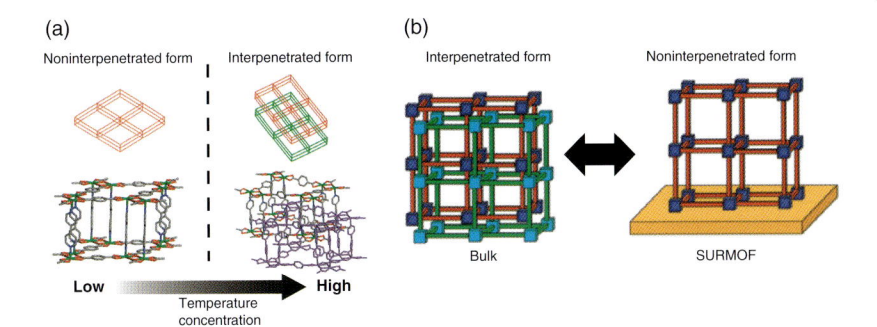

Figure 7.3 Synthetic approaches for controlling the formation of interpenetrated and noninterpenetrated frameworks. (a) Synthetic conditions of temperature and reagent concentration determine the resulting structures. (Adapted with permission from Ref. [24]. Copyright 2009, American Chemical Society.) (b) The SURMOF technique prevents interpenetration and a noninterpenetrated structure is obtained. (Adapted with permission from Ref. [26]. Copyright 2009, Macmillan Publishers Ltd: (Nature Materials).)

based on 2,5-dioxidoterephthalate, which incorporate 2–11 benzene rings (Figure 7.4b) [29]. IRMOF-74-XI, with a pore aperture of 98 Å, has the lowest density (0.195 g/cm^3). These extremely large pores were visualized by low-voltage high-resolution scanning electron microscopy (LV-HRSEM) and high-resolution transmission electron microscopy (HRTEM). Through the large pore apertures, not only gas molecules but also large molecules such as vitamin B12, metal–organic polyhedron-18, myoglobin, and green fluorescent protein were passed and accommodated. In frameworks with such large pores, supercritical CO_2 is often used to remove guest molecules without decomposing the framework, thus affording stable porous structures that have various applications.

In order to develop gas separation abilities in porous materials, selective and targeted adsorption behavior or different substrate-dependent adsorption speeds are required. Up to now, two approaches for improving the gas separation

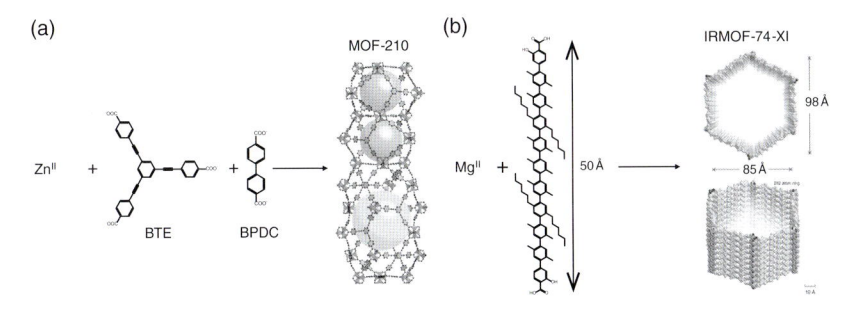

Figure 7.4 Representative compounds exhibiting ultrahigh porosity. (a) Structure of MOF-210 incorporates $Zn_4O(CO_2)_6$ clusters, BTE, and BPDC ligands [27]. (b) Structure of IRMOF-74-XI with hexagonal channels [29].

abilities of PCPs/MOFs have been studied: one is a means of host (pore walls)–guest interaction, such as van der Waals interaction and charge transfer, and the other is a means of structural transformation (dynamics) of frameworks in response to guest accommodation. In the former approach, open metal sites on pore walls or redox active ligands have been utilized as active sites to impart selectivity. Long and coworkers reported hydrocarbon separation of olefin/paraffin mixtures using $Fe_2(dobdc)$ (dobdc = 2,5-dioxido-1,4-benzenedicarboxylate) (Figure 7.5a) [30]. Open metal sites consisting of Fe^{II} ions interacted more strongly with unsaturated hydrocarbon olefins, through their π-electron clouds, than with saturated hydrocarbon. These different strengths of host–guest interaction contribute to the permeation selectivity of $Fe_2(dobdc)$, leading to its excellent performance in hydrocarbon separation. Matsuda and coworkers synthesized an electronically dynamic framework [Zn(TCNQ-TCNQ)bpy] incorporating tetracyanoquinodimethane (TCNQ), and its selective adsorption of O_2 and NO was reported [31]. This unique adsorption behavior arises from the charge-transfer interaction between TCNQ and these gas molecules, as corroborated by *in situ* Raman spectroscopy. A class of compounds whose gas separation abilities arise from gas-responsive structural transformations is coordination polymers with an interdigitated structure (CIDs) [7]. Their frameworks are composed of two-dimensional grid motifs, stacked through van der Waal interactions, and show open-gate phenomena in gas adsorption isotherms for specific gas species (Figure 7.5b). For example, CID-3 [Zn(ndc)(bpy)] (ndc = 2,7-naphthalene dicarboxylate) adsorbed CO_2 at 195 K with a hysteresis

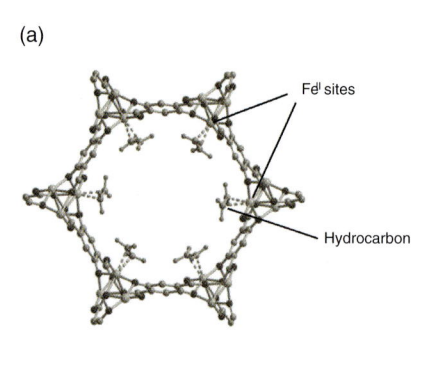

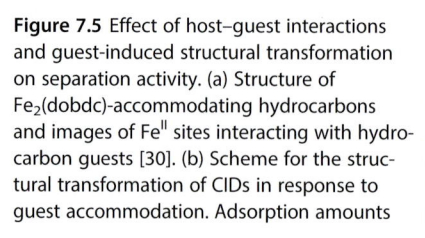

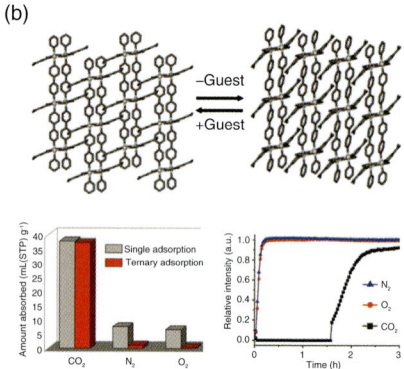

Figure 7.5 Effect of host–guest interactions and guest-induced structural transformation on separation activity. (a) Structure of $Fe_2(dobdc)$-accommodating hydrocarbons and images of Fe^{II} sites interacting with hydrocarbon guests [30]. (b) Scheme for the structural transformation of CIDs in response to guest accommodation. Adsorption amounts estimated from ternary adsorption experiments with gas mixtures in the ratio CO_2: O_2: N_2 = 1: 1: 1 at a total pressure of 1 MPa (red bar) and those from single-component adsorption isotherms (gray bar). Breakthrough curve for CO_2: O_2: N_2 = 1: 21: 78 at a total pressure of 0.5 MPa. (Adapted from Ref. [32] with permission from Royal Society of Chemistry.)

loop, but did not adsorb O_2 and N_2, resulting in highly selective adsorption of CO_2 from an O_2 and N_2 mixture [32].

More recently, separation of CO from mixtures with N_2 was achieved with a Cu^{II} framework, [Cu(aip)] (aip = 5-azidoisophthalate), which has combined characteristics of guest recognition and reversible structural transformation in response to guest accommodation [33]. Such cooperative behavior is defined as "self-accelerating adsorption" and gives rise to highly effective trapping of targeted gas molecules in a mixture.

7.3.2
Release

The releasing of guest molecules from pores in PCP/MOF containers has important biochemical and medical applications [34], such as drug delivery systems. Guest release has been demonstrated by two different mechanisms: slow release through equilibration and active release using external stimuli such as light irradiation.

Horcajada *et al.* reported the application of Materials of Institute Lavoisier (MIL) series compounds as nontoxic nanocarriers for ibuprofen [35,36], antitumoral, and retroviral drugs, which are used to combat cancer and AIDS (Figure 7.6a) [37]. The drug's loading capacity and release speed were tuned by changing the pore environments, shapes, and sizes. The potential of the iron compound MIL-88A as a contrast agent for MRI has also been investigated [37].

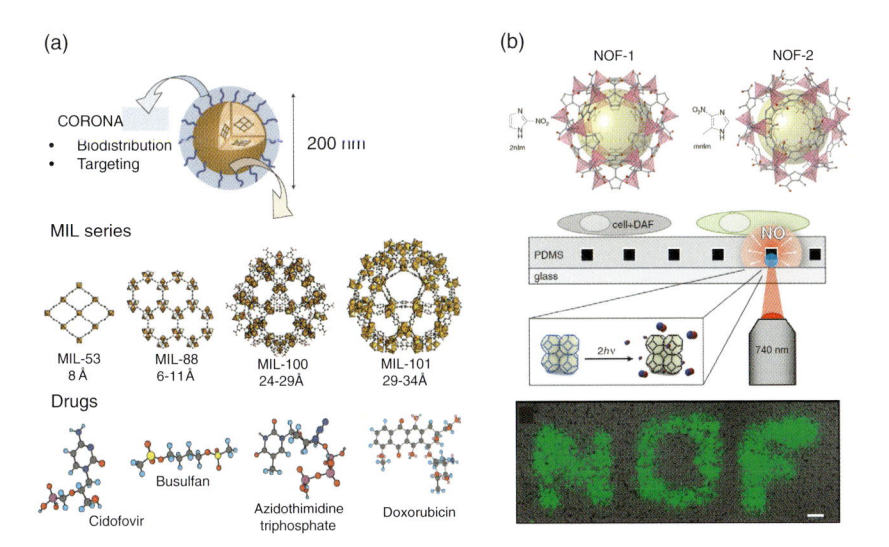

Figure 7.6 (a) Drug delivery systems using MIL series compounds and four challenging drugs. (Adapted with permission from Ref. [37]. Copyright 2009, Macmillan Publishers Ltd: (Nature Materials).) (b) Crystal structures of NOF-1 and NOF-2. Light irradiation-driven NO-releasing systems using NOFs, and demonstration of spatiotemporal control of cell activation by releasing NO molecules. (Reprinted with permission from Ref. [38]. Copyright 2013, Macmillan Publishers Ltd: (Nature Communications).

NO plays a significant role in numerous signaling events such as proliferation and vasodilatation. Therefore, controllable NO-releasing scaffolds are required to investigate signaling paths and to develop healing approaches. In order to obtain NO-adsorbing/releasing materials, frameworks incorporating NO-active open metal sites or imidazolate ligands have been synthesized [39–41]. Morris and coworkers synthesized a copper sulfoisophthalate MOF, Cu-SIP-3·3 H_2O, that showed a structural transformation upon dehydration [42]. This framework has copper sites to which water molecules coordinate, and it transforms to a nonporous structure incorporating active open metal sites upon dehydration. This dehydrated nonporous compound did not adsorb conventional gases up to at least 10 bar. On the other hand, NO gas induced gate-opening adsorption behavior arising from the interaction between NO and the copper sites at room temperature. Although the structure was not determined, NO is expected to coordinate to the copper sites. Subsequent exposure of the material to water vapor after NO coordination gave rise to a slow release of NO gas through guest exchange, and the hydrated structure was reformed. Furukawa and coworkers reported NO frameworks (NOFs) exhibiting light-induced spatiotemporal NO-releasing abilities (Figure 7.6b) [38]. They constructed two photoactive NOFs incorporating the imidazole-based ligands 2-nitroimidazole (2nIm) and 5-methyl-4-nitroimidazole (mnIm), and achieved a precisely localized stimulation of cells with NO released by near-infrared two-photon laser activation (chosen to reduce the risk of damaging the cells with harmful ultraviolet irradiation). In these systems, the porous framework contributes to the rapid and accurate release of NO molecules through its channels.

7.3.3
Conversion

Pores (channels) in PCPs/MOFs act as a reaction field for molecular conversion [43,44]. The pore surface environment, which is tuned by changing the composition of the framework, adds selectivity for targeted substrates to chemical reactions and controls final products. For example, homochiral frameworks, in which the chirality of organic ligands induces structural chirality, have enabled enantioselective catalytic reactions [45].

In the frameworks, metal sites or organic ligands on pore surfaces act as catalytic sites for chemical reactions. An outstanding framework that acts as a Lewis acid catalyst is HKUST-1 [$Cu_3(TMA)_2(H_2O)_3$] (TMA = benzene-1,3,5-tricarboxylate), which incorporates binuclear Cu_2 paddlewheel secondary building units [46]. The Cu Lewis acid sites of these units can directly access a substrate in the pores, and various catalytic reactions have been demonstrated [47]. Chemical conversion of gas substrates is also an important application of porous frameworks. For example, methane activation was demonstrated using a [VO(dmbdc)] (MOF-48, dmbdc: 2,5-dimethylbenzenedicarboxylate) MOF-incorporating vanadium(IV) ions, in which methane molecules were converted to acetic acid [48].

Base catalysis of PCPs/MOFs is still rare because the nitrogen lone pairs on the ligands, which are catalytic sites, usually coordinate to metal ions and are consequently unavailable for substrate activation. [Cd(4-btapa)$_2$(NO$_3$)$_2$] (4-btapa: 1,3,5-benzene tricarboxylic acid tris[N-(4-pyridyl)amide]) promoted the Knoevenagel condensation reaction, which is a well-known base-catalyzed model reaction [49]. In this framework, active amide groups are aligned uniformly on the channel surfaces. The yields of final products were influenced by the size of the substrate, demonstrating that selectivity depends on the relationship between the size of the substrate and the aperture size of the pores.

Composite catalysts consisting of PCPs/MOFs and different supports such as metal particles or organometallic complexes have been developed. They could possibly reduce the number of steps required in multistep reactions, resulting in an epoch-making one-pot reaction [50]. Recently, structural defects in crystals were also demonstrated to be catalytic sites [51]. Development of heterogeneous catalysts based on PCPs/MOFs is accelerating worldwide.

Polymerization of monomers confined in PCP/MOF channels has been demonstrated. Primary and secondary polymer structures, molecular weight, and functional characteristics were controlled by utilizing designable nanochannels, which may be regarded as nanosized flasks with structural regularity, controllable size, shape, surface environment, and flexibility (Figure 7.7) [52]. Uemura and coworkers demonstrated radical polymerization of vinyl monomers such as styrene and methyl methacrylate in various channels of [M$_2$(L)$_2$(dabco)] (M; Cu, Zn, L; biphenyl-4,4′-dicarboxylate, terephthalate, 1,4-ndc, etc.), and reported on the impact of channel size on the polydispersities (Mw/Mn) of the resulting polymers [53]. The molecular weight distributions of polystyrene and poly(methyl

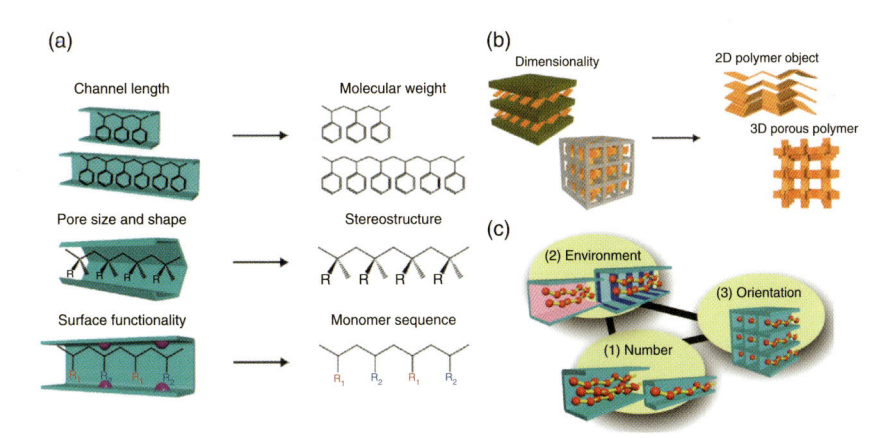

Figure 7.7 Polymer chemistry in PCP/MOF channels. (a) Polymer molecular weight and primary structure, such as stereostructure and monomer sequence, are controlled by the channel environment. (b) Polymer secondary structure is controlled using PCP/MOF templates. (c) The properties of polymers confined in PCP/MOF channels are controlled by changing the channel size and surface environment. (Adapted from Ref. [52] with permission from Royal Society of Chemistry.)

methacrylate) were narrower than those of the corresponding bulk-synthesized polymers, and became narrower as the channel size decreased, resulting in a Mw/Mn value of 1.5 for polystyrene synthesized in [Zn_2(1,4-ndc)$_2$(dabco)]. The stereoregularity (tacticity) was also controlled in these polymerization systems. Polymerization in restricted channels gave rise to a relative increase in the number of isotactic units with less steric bulk, compared to polymers obtained by conventional bulk synthesis [53].

The properties of polymers confined in the nanochannels of PCPs/MOFs are different to those of the corresponding bulk-synthesized polymers because they have different chain number, orientation, and conformation compared with polymers in the bulk state; thus, host–guest interactions should appear. For example, thermal transitions [54], dielectric properties [55], and conductivities [56] of confined polymers have been investigated. Moreover, Uemura and coworkers reported unique host–guest copolymerizations [57] as well as mixing of immiscible polymers [58] utilizing PCPs/MOFs as templates.

7.3.4
Visualization of Guest Species

Ordered guest molecules in PCP/MOF channels have been characterized (visualized) by X-ray structural analyses. Kitagawa and coworkers used *in situ* X-ray powder diffraction to demonstrate regular alignments of O_2 and C_2H_2 gas molecules in PCP/MOF channels [59,60] (details in Section 7.5). Recently, Fujita and coworkers reported an improved method for structural analysis of organic molecules, which they named "crystalline sponge method" (Figure 7.8) [61]. The

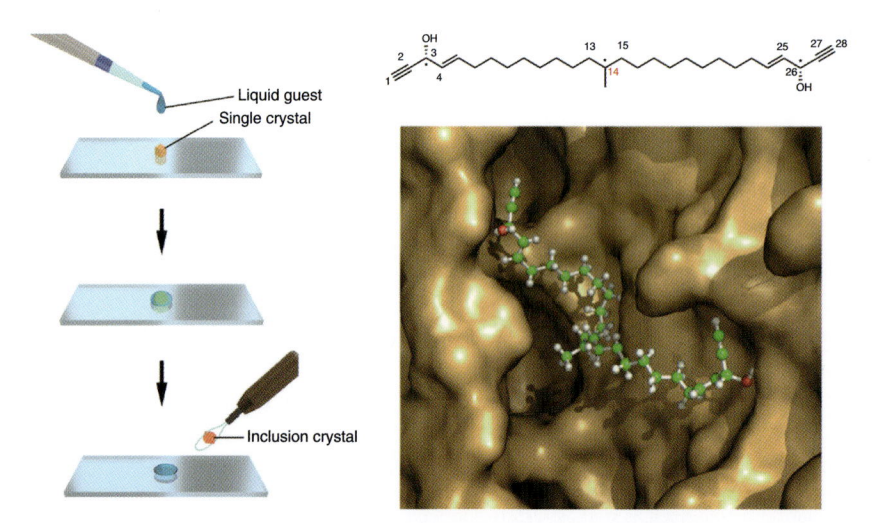

Figure 7.8 Scheme for the crystalline sponge method and a miyakosyne A guest visualized in the pores of a PCP/MOF. (Reprinted with permission from Ref. [61]). Copyright 2013, Macmillan Publishers Ltd: (Nature).)

crystalline sponge adsorbs target sample molecules from their solution, resulting in regularly ordered molecules in a crystal, and their structures are solved using X-ray analysis. Using this method, we can obtain molecular structures without crystallization of the targeted samples themselves. Therefore, even noncrystalline compounds and nanogram or microgram amounts of sample are suitable for the method. They succeeded in determining the absolute configuration of a chiral center in a scarce marine natural product, miyakosyne A, for the first time. This method will enable chemists to obtain an easy and precise characterization of many natural and synthetic compounds. Moreover, intermediates in chemical reactions can also be visualized by the same method [62].

7.4
Functional Framework of PCPs/MOFs

Framework functions based on metal complexes in PCPs/MOFs have attracted much attention for the development of sensing devices, in which porous properties are combined with metal-complex properties such as magnetism, luminescence, conductivity, and dielectric properties. In such devices, these properties provide the outputs, while guest molecules work as chemical stimuli. The infinite structures of PCPs/MOFs cause high stimuli sensitivity, reflecting their high cooperativity. In order to obtain functions that are suitable for devices, appropriate metal ions and linker ligands should be selected.

7.4.1
Magnetism

Generally, magnetic properties in frameworks include both magnetic ordering and the spin crossover (SCO) phenomenon (or spin transition (ST)). In terms of magnetic ordering, porous magnets, or magnet sponges have been synthesized, and switching of their magnetic interactions by guests has been investigated [63–67]. Principally, such changes in magnetic properties arise from structural transformations induced by adsorption/desorption of solvents and gas molecules. Long and coworkers reported magnetic exchange couplings between paramagnetic O_2 guest molecules and Prussian blue-type microporous magnetic frameworks of $CsNi[Cr(CN)_6]$ and $Cr_3[Cr(CN)_6]_2$ [68]. In the case of $CsNi[Cr(CN)_6]$, ferromagnetic interactions between O_2 and the framework was observed, whereas an antiferromagnetic interaction occurred in $Cr_3[Cr(CN)_6]_2$ upon O_2 absorption. In many cases, the distances between metal ions in PCPs/MOFs incorporating pillar ligands are long, leading to their weak magnetic interaction. It seems, therefore, that magnetic ordering and porous properties conflict with each other. From this viewpoint, in the design and synthesis of difunctional materials with direct coupling between magnetic and porous properties, SCO (or ST) has been focused on because its transition temperature is near room temperature, which is much higher than that of magnetic ordering. SCO is well

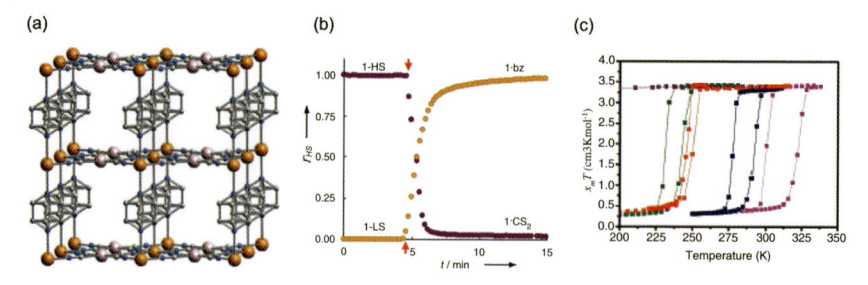

Figure 7.9 (a) Crystal structure of Hofmann-type ST framework [Fe(pz)M(CN)$_4$] (M = Pt, Pd, Ni). (b) Guest-induced spin state switching of [Fe(pz)Pt(CN)$_4$]. The HS state is switched to the LS state by introducing CS$_2$ at 298 K, whereas bz induces switching from the LS state to the HS state. (Reprinted with permission from Ref. [71]. Copyright 2009, Wiley-VCH Verlag GmbH & Co. KGaA, Weinheim.) (c) Control of the ST behavior of [Fe(pz)Ni(CN)$_4$] by various solvent molecules. (Reprinted with permission from Ref. [74]. Copyright 2009, American Chemical Society.)

known in Fe^{II} complexes, whose spin states are switched between high-spin (HS) and low-spin (LS) states by external stimuli such as temperature, pressure, and light irradiation, producing magnetic, optical, and structural changes. So far, several SCO-PCPs/MOFs have been synthesized and their guest-modulated magnetic behavior has been reported [69]. Hofmann-type SCO-PCPs/MOFs, [Fe(pz)M(CN)$_4$] (M = Pt, Pd, Ni), are the most outstanding and useful compounds because they are stable and show abrupt ST with a hysteresis of about 20 K around room temperature ($T_c^{\downarrow} = 285$ K, $T_c^{\uparrow} = 309$ K) (Figure 7.9a) [70]. They provide dynamic frameworks in which a 30% change in pore size occurs in response to the spin state switching, with large pores in the HS state and small pores in the LS state. Ohba et al. reported a bidirectional chemoswitching of the spin states of [Fe(pz)Pt(CN)$_4$], with coupling of guest adsorption and memory effects at room temperature using benzene and CS$_2$. The LS state is switched to the HS state in response to benzene adsorption, whereas the LS state is stabilized in a CS$_2$-adsorbed form at room temperature (Figure 7.9b) [71]. They also reported precise control of ST temperature by framework modifications through oxidative addition of halogen guest molecules to Pt metal sites [72,73]. Kepert and coworkers reported that various solvents modulated the ST behavior in [Fe(pz)Ni(CN)$_4$] (Figure 7.9c) [74].

7.4.2
Luminescence

Luminescent frameworks are constructed by incorporating a chromophore consisting of luminescent lanthanide (Ln) ions (such as Tb^{III} and Eu^{III}), organic ligands (such as anthracene and pyrene derivatives), and guest molecules (such as distyryl-benzene; DSB) to form a PCP/MOF. Energy transfer between the host and guest and tuning of the luminescent behavior via guest adsorption/desorption have been investigated [75]. [Zn$_2$(adc)$_2$(dabco)] (adc = 9,10-anthracenedicarboxylate, dabco = 1,4-diazabicyclo[2.2.2]octane) gave blue emission at

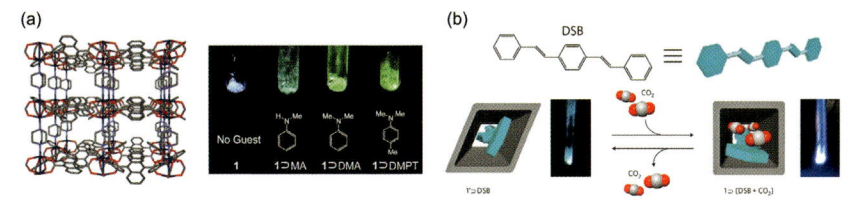

Figure 7.10 (a) Crystal structure of [Zn$_2$(adc)$_2$(dabco)] and changes in the color of fluorescence induced by host–guest charge transfer. (Reprinted from Ref. [76] with permission from Royal Society of Chemistry.) (b) CO$_2$ detection by monitoring changes in the fluorescence intensity of a host–guest composite of [Zn$_2$(terephthalate)$_2$(dabco)] and DSB. The conformation of DSB changes in response to the structural transformation of the host induced by CO$_2$ adsorption. (Reprinted with permission from Ref. [77]. Copyright 2011, Macmillan Publishers Ltd: (Nature Materials).)

415 nm, which was assigned to an emission from anthracene monomers. Its fluorescence behavior was altered by incorporating N-methylaniline (MA), N,N-dimethylaniline (DMA), and N,N-dimethyl-*p*-toluidine (DMPT), which induced host–guest charge transfer. In their host–guest systems, the fluorescence wavelengths were related to the ionization potentials of the incorporated guest species (Figure 7.10a) [76].

A host–guest composite of [Zn$_2$(terephthalate)$_2$(dabco)] and fluorescent DSB enabled the detection of gas molecules by changes in luminescence (Figure 7.10b) [77]. The composite had a rhombic net structure in which DSB had a distorted (twisted) conformation with weak emission. Upon CO$_2$ adsorption, the distorted DSB adopted a planar conformation and exhibited stronger emission, combined with structural transformation of the host from the rhombic net structure to a square grid structure. This was the first example of fluorescence detection of specific gases without any chemical interaction or energy transfer.

7.4.3
Electrical Conductivity

PCP/MOF compounds are typically poor electrical conductors because of the poor overlap between the orbitals of metal ions and ligands in their porous frameworks. Electrically conducting PCPs/MOFs could directly lead to electronic devices and reconfigurable electronics; however, at present there are few examples. Long and coworkers reported the first example of a conducting framework incorporating permanent porosity, Cu[Ni(pdt)$_2$] (pdt = pyrazine-2,3-dithiolate) (Figure 7.11a) [78]. This compound showed a type I N$_2$ adsorption isotherm and a BET surface area of 385 m^2/g. The conductivity was 1×10^{-4} S/cm after an I$_2$ doping treatment, implying that the framework was a p-type semiconductor. Allendorf and coworkers reported a host–guest composite of [Cu$_3$(MTA)$_2$] (HKUST-1) and TCNQ that exhibited a tunable air-stable electrical conductivity of 7 S/m [79]. TCNQ bound to the open metal sites in the framework and bridged copper dimer units, producing a new charge-transfer band and enabling

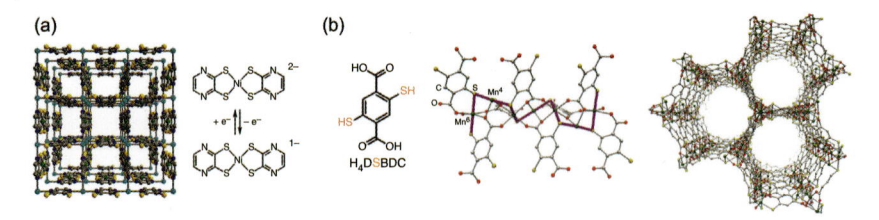

Figure 7.11 (a) Crystal structure of Cu[Ni(pdt)$_2$] incorporating the redox active ligand pdt. (Reprinted with permission from Ref. [78]. Copyright 2010, American Chemical Society.) (b) Crystal structure of Mn$_2$(DSBDC) and the conduction path of infinite metal–sulfur chains. (Reprinted with permission from Ref. [80]. Copyright 2013, American Chemical Society.)

electronic coupling between the framework and TCNQ. Recently, Dinca and coworkers synthesized MOF-74 analogs incorporating 2,5-disulfhydrylbenzene-1,4-dicarboxylic acid (H$_4$DSBDC), the thiolated analog of H$_4$DOBDC, and Mn (or Fe) ions. These frameworks consist of infinite metal–sulfur chains with M-thiophenoxide linkages. Mn$_2$(DSBDC) exhibited a type I N$_2$ adsorption isotherm, a BET surface area of 978 m^2/g, and a high charge mobility of 0.01 cm^2/(V s), as measured by flash-photolysis time-resolved microwave conductivity (FP-TRMC) and direct-current time-of-flight (TOF) methods (Figure 7.11b) [80,81].

7.4.4
Dielectric Properties

Porous frameworks sometimes induce dynamic motions such as rotation of pillar ligands, as confirmed by solid-state NMR and neutron spectroscopies [82,83]. Up to now, two or four site flipping motions of pillar ligands have been reported, whose activation energies are impacted by the substituents and pore size (Figure 7.12). Such ligand motions are considered to be involved in the uptake and

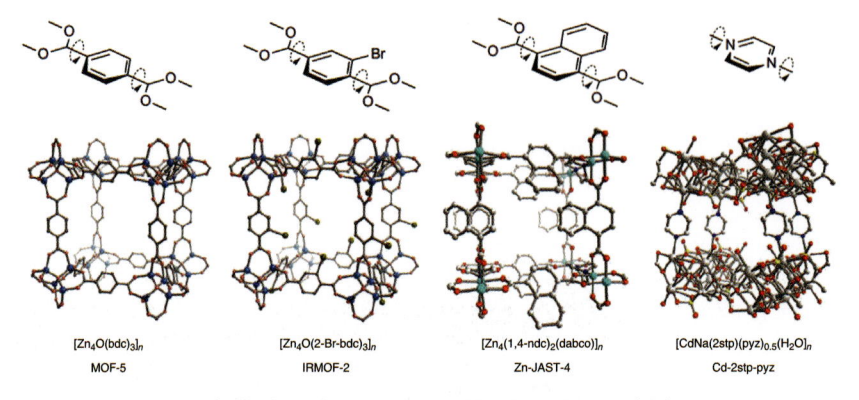

Figure 7.12 Rotation of pillar ligands in MOF-5, IRMOF-2, Zn-JAST-4, and Cd-2stp-pyz. Activation energies were 11.3, 7.3, 12.7, and 1.8 kcal/mol, respectively. (Reprinted with permission from Ref. [9]. Copyright 2009, Macmillan Publishers Ltd: (Nature Chemistry).

diffusion of guests; moreover, such dynamic systems showing order–disorder phase transitions are associated with the dielectric properties of the crystals [84,85]. Recently, Miyasaka and coworkers reported dielectric responses associated with structural transformations that accompany gas adsorption [86].

7.5
Guests in the Pores of PCPs/MOFs

Confinement effects arising in restricted pores or channels cause guests to acquire characteristic alignments and mobilities that are different to those in their bulk states. The confinement effects can be tuned by changing the pore sizes and surface environments.

Kitagawa and coworkers, for the first time, reported a pseudohigh-pressure effect on O_2 molecules accommodated in channels of CPL-1 [$Cu_2(pzdc)_2(pyz)$] (pzdc = 2,3-pyrazinedicarboxylate, pyz = pyrazine) [59]. O_2 molecules were aligned in one-dimensional chains in which intermolecular antiferromagnetic interactions occurred. The vibrational mode of the O_2 corresponded to the mode of O_2 in the alpha phase, which usually occurs under 2 GPa pressure, as evidenced by Raman spectroscopy. Moreover, they found a high-density phase of adsorbed C_2H_2 in CPL-1 at room temperature – for which the density was estimated to be 0.434 g/cm^3, which is 200 times larger than the compression limit for the safe use of C_2H_2 (Figure 7.13a and b) [60]. Although reports on the pseudo-high-pressure effect are still rare, such behavior is very interesting from a fundamental scientific viewpoint, and will also contribute to safe (stable), large-volume storage applications.

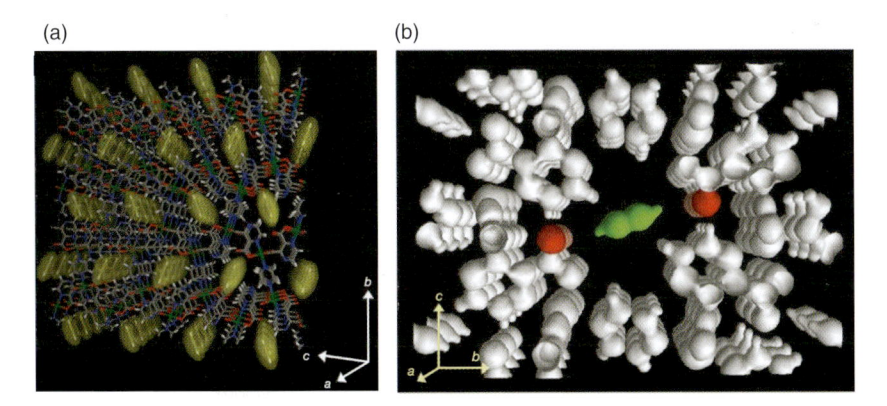

Figure 7.13 (a) Crystal structure of CPL-1 and regular alignments of C_2H_2 molecules confined in the channels of CPL-1. (b) Electron densities of CPL-1 adsorbing C_2H_2. The C_2H_2 molecule and oxygen atoms are represented by green and red, respectively. (Reprinted with permission from Ref. [60]. Copyright 2005, Macmillan Publishers Ltd: (Nature).)

The guests in the nanospace of PCPs/MOFs sometimes show unusual mobility compared with that in the bulk phase because their assembly depends on the nature of the channels such as size, shape, and chemical environment. In terms of guest migration in the channels, protons have received much attention as a carrier species because proton conductors are important for fuel cell technology. In order to develop proton conductors based on PCPs/MOFs, a key concept is to construct proton-conducting pathways in the channels [87]. For this purpose, water and heterocyclic organic molecules such as imidazole have been used as guest proton carriers. Kitagawa and coworkers have synthesized proton conductors in which conducting paths were formed via water molecules. For example, they demonstrated that the acidity of the functional groups —COOH, —OH, and —NH$_2$ on the pore surfaces of [Al(OH)(bdc)] (MIL-53) impacted proton conductivity and activation energy, and reported that a framework modified with a carboxy group with the lowest pK_a showed the highest proton conductivity and the lowest activation energy (Figure 7.14) [88].

For the case of water–guest systems, the water molecules play an important role in the construction of conducting paths; therefore, working temperatures are limited to below 373 K because of the low stability of the water in the channels. The conductivity is also sensitive to humidity. In order to obtain applicable proton conductors over 373 K, which are anhydrous proton conductors, amphoteric heterocycles were used as proton sources confined in PCP/MOF channels. Kitagawa and coworkers constructed a host–guest proton conductor, [Al(OH)(1,4-ndc)] (1,4-ndc = 1,4-naphthalenedicarboxylate), which incorporates imidazole molecules, exhibiting a high proton conductivity of 2.2×10^{-5} S/cm at 393 K (Figure 7.15) [89]. By means of solid-state NMR spectroscopy, they revealed that high-frequency imidazole rotations, which appear to be similar to

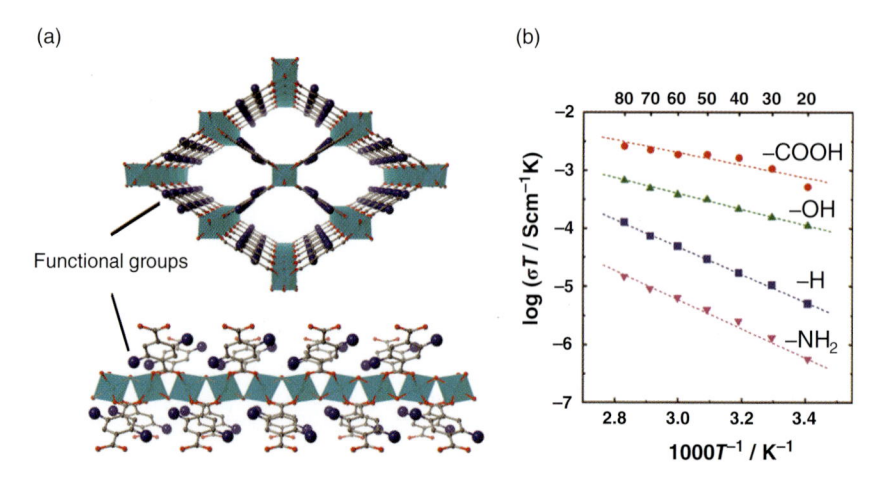

Figure 7.14 (a) Crystal structure of functionalized MIL-53. (b) Proton conductivities of MIL-53 with its framework modified by the functional group —COOH, —OH, or —NH$_2$. (Reprinted with permission from Ref. [88]. Copyright 2011, American Chemical Society.)

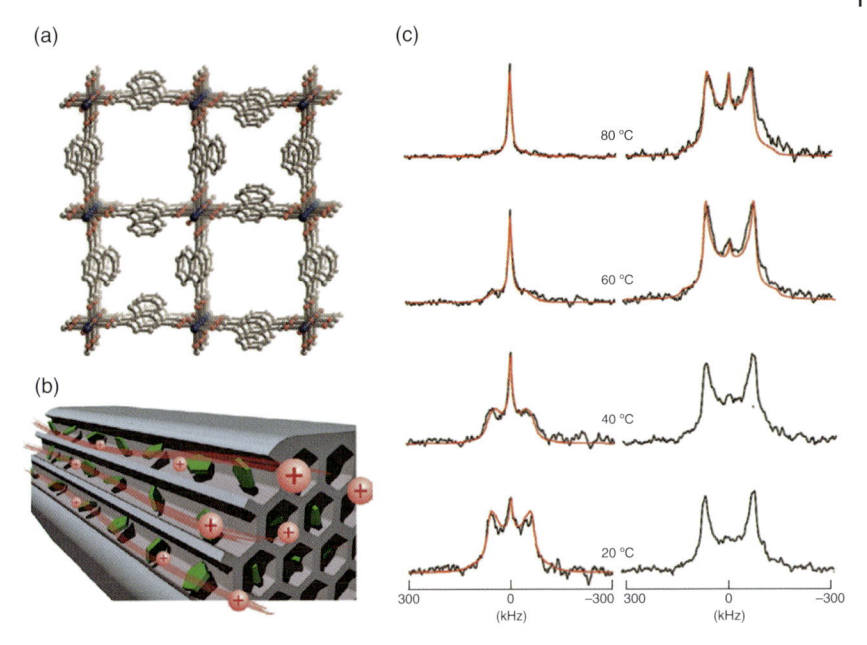

Figure 7.15 (a) Crystal structure of [Al(OH)(1,4-ndc)]. (b) Scheme for proton conduction via imidazole molecules in one-dimensional channels. (Reprinted with permission from Ref. [87]. Copyright 2013, American Chemical Society.) (c) Variable-temperature solid-state NMR spectra for deuterated imidazole molecules confined in the channels of [Al(OH)(1,4-ndc)]. (Reprinted with permission from Ref. [89]. Copyright 2009, Macmillan Publishers Ltd: (Nature Materials).)

those occurring in the liquid state, are associated with a Grotthus mechanism for proton conductivity in the channels (Figure 7.15c). At the same time, Shimizu and coworkers demonstrated that Na_3(2,4,6-trihydroxy-1,3,5-benzenetrisulfonate)-incorporating 1H-1,2,4-triazole guests exhibited a proton conductivity of 5×10^{-4} S/cm at 423 K [90].

7.6
Crystal Engineering of PCPs/MOFs

The performance of PCPs/MOFs is improved not only by changing their composition but also by constructing macroscopic/mesoscopic superstructures through crystal engineering techniques (Figure 7.16) [91,92]. Crystal engineering of PCPs/MOFs includes the preparation of hybrid crystals such as core–shell-type and solid–solution-type hybrid crystals, control of crystal sizes or crystal morphologies, crystal growth on substrates (to produce the so-called SURMOFs), and control of crystal assembly (aggregate) structures.

Hybrid PCP/MOF crystals such as core–shell or solid–solution-type crystals have been obtained through one-step and multi-step procedures using different

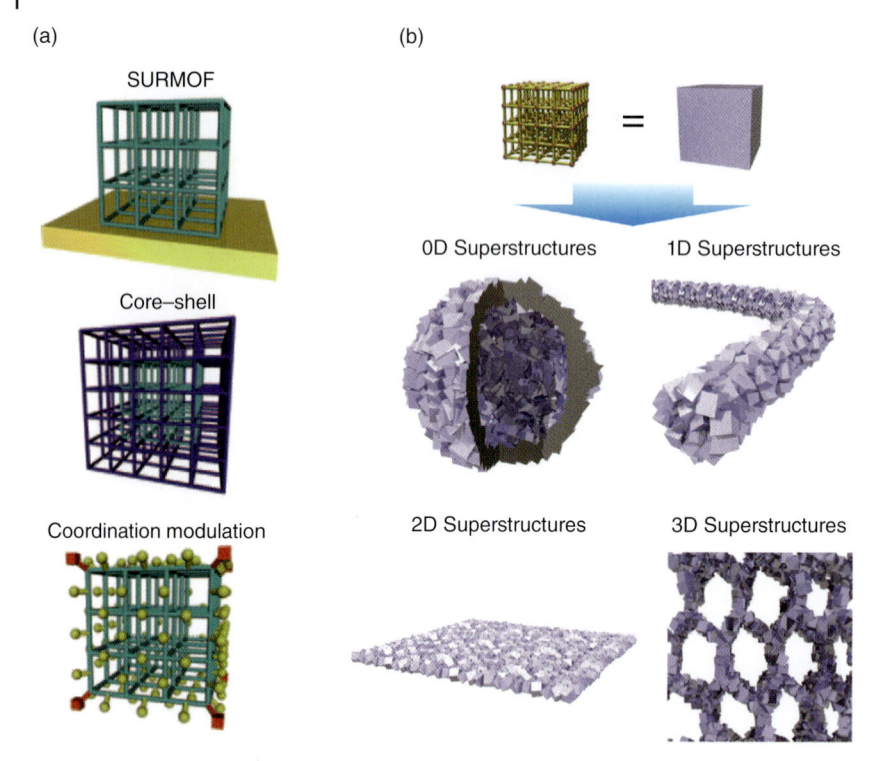

Figure 7.16 (a) Surface engineering of PCP/MOF crystals involving layer-by-layer growth of thin films (SURMOF), core–shell-type heterostructures, and the coordination modulation method for controlling the crystal growth. (Adapted with permission from Ref. [92]. Copyright 2010, Wiley-VCH Verlag GmbH & Co. KGaA, Weinheim.) (b) Structuring of PCP/MOF crystals leads to zero-dimensional to three-dimensional superstructures. (Adapted from Ref. [91] with permission from Royal Society of Chemistry.)

frameworks with similar lattice parameters [93]. In core–shell crystals, epitaxial growth of a framework on another framework occurs, resulting in the formation of unique crystal interfaces. These interface designs are useful to enhance the selective adsorption of substrates. Approaches to solid–solution-type crystals that tune cooperative and overall dynamics in the crystals, which is achieved by incorporating different metal ions or ligands in arbitrary mixing ratios, result in the control of gate pressures in adsorption isotherms [94,95].

Crystal sizes and morphologies can be controlled by changing the reaction conditions for crystal growth (the coordination modulation method). In particular, preparation of PCP/MOF nanoparticles is important for many applications and industrial processes. For example, catalytic activity was enhanced due to the increase in surface area, and the speed of guest migrations in the channels was increased. More recently, the impact of crystal size on crystal structure and cooperative dynamics was reported [96–98]. By downsizing the crystals of an interpenetrated framework of $[Cu_2(bdc)_2(bpy)]$, characteristic dynamic behavior, such as the "shape memory effect," was discovered in the nanoparticles. Moreover, morphology control of crystals was demonstrated; using similar approaches for controlling crystal growth, the desired crystal faces were displayed [99].

Structuring of PCPs/MOFs has been demonstrated using particles as building units to construct superstructures. Various aggregate structures such as zero-dimensional spheres, one-dimensional chains (tubes), two-dimensional sheets, and three-dimensional architectures have been constructed using appropriate templates [100] or colloidal approaches [101,102]. Research into such structuring also produced new methodologies for tuning the framework functional characteristics, and revealed the importance of the crystal morphology of PCPs/MOFs in their various applications.

7.7
Physical Chemistry in PCPs/MOFs

The strength and stability of coordination networks are related to the durability and mechanical properties of PCPs/MOFs. In order to probe these properties in PCPs/MOFs, structural transformation, thermal expansion, crystal–amorphous transition, and melting by heat and pressure have been investigated (Figure 7.17) [103]. For example, Cheetham and coworkers reported that network glasses of three-dimensional zeolitic imidazolate frameworks (ZIFs) transformed on heating under pressure [104–106]. Horike and coworkers demonstrated that proton-conductive PCPs/MOFs that consist of zinc ions, phosphates, and azoles undergo a reversible transformation from solid to liquid (melting) [107]. These works afforded intrinsic information about PCPs/MOFs with respect to their lattice vibrations and thermodynamic diagrams. Very recently, an insulator–semiconductor conversion associated with a transformation between crystalline and amorphous phases was demonstrated in $CuCl(ttcH_3)$ ($ttcH_3$: trithiocyanuric acid) [108].

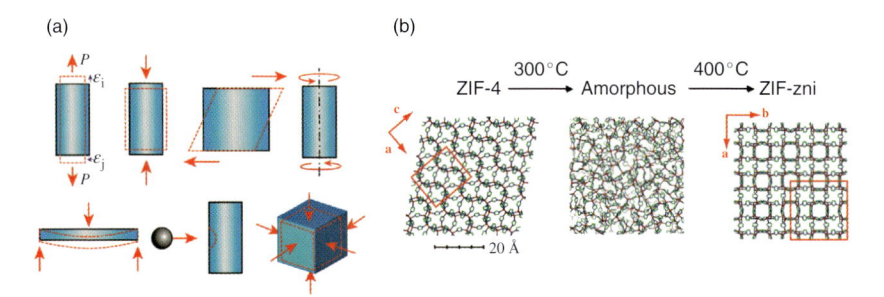

Figure 7.17 (a) Images of the common modes of mechanical loading such as tension, compression, shear, torsion, bending, impact, and hydrostatic compression. (b) Crystalline–amorphous–crystalline transition of ZIF-4 on heating. (Adapted from Ref. [103] with permission from Royal Society of Chemistry.)

7.8
Outlook

An "age of gas" is dawning in the twenty-first century because depletion of petroleum has become a critical issue and gases such as natural gas, biogas, and air will play important roles [109]. PCP/MOF chemistry, which deals with these gases using next-generation porous materials, is now expanding from being a purely scientific field of inquiry to being a field with real industrial applications. Here, the "porous" and "framework" functions of PCPs/MOFs are described. The porous function of incorporating guests (gases), the framework functions involving coordination chemistry, and their combinations are all fascinating subjects. One of the ultimate goals of research in this area is to develop a process for the selective adsorption of N_2 and H_2, and their subsequent conversion to NH_3 with high efficiency under ambient conditions, resulting in a PCP/MOF-based technology that supersedes the conventional Haber–Bosch process. Moreover, a reductive conversion of CO_2 to methanol is significant. In order to approach such goals, many scientists are investigating the nature of both pores and frameworks, and are creating a new area of science associated with PCP/MOF chemistry.

References

1 Kondo, M., Yoshitomi, T., Seki, K., Matsuzaka, H., and Kitagawa, S. (1997) *Angew. Chem., Int. Ed.*, **36**, 1725.
2 Yaghi, O.M., O'Keeffe, M., Ockwig, N.W., Chae, H.K., Eddaoudi, M., and Kim, J. (2003) *Nature*, **423**, 705.
3 Kitagawa, S., Kitaura, R., and Noro, S. (2004) *Angew. Chem., Int. Ed.*, **43**, 2334.
4 Ferey, G. (2008) *Chem. Soc. Rev.*, **37**, 191.
5 Henke, S., Schneemann, A., and Fischer, R.A. (2013) *Adv. Funct. Mater.*, **23**, 5990.
6 Liu, Y., Her, J.-H., Dailly, A., Ramirez-Cuesta, A.J., Neumann, D.A., and Brown, C.M. (2008) *J. Am. Chem. Soc.*, **130**, 11813.
7 Kitaura, R., Seki, K., Akiyama, G., and Kitagawa, S. (2003) *Angew. Chem., Int. Ed.*, **42**, 428.
8 Yanai, N., Uemura, T., Inoue, M., Matsuda, R., Fukushima, T., Tsujimoto, M., Isoda, S., and Kitagawa, S. (2012) *J. Am. Chem. Soc.*, **134**, 4501.
9 Horike, S., Shimomura, S., and Kitagawa, S. (2009) *Nat. Chem.*, **1**, 695.
10 Kitagawa, S. and Kondo, M. (1998) *Bull. Chem. Soc. Jpn.*, **71**, 1739.
11 Stock, N. (2010) *Microporous Mesoporous Mater.*, **129**, 287.
12 Banerjee, R., Phan, A., Wang, B., Knobler, C., Furukawa, H., O'Keeffe, M., and Yaghi, O.M. (2008) *Science*, **319**, 939.
13 Sumida, K., Horike, S., Kaye, S.S., Herm, Z.R., Queen, W.L., Brown, C.M., Grandjean, F., Long, G.J., Dailly, A., and Long, J.R. (2010) *Chem. Sci.*, **1**, 184.
14 Sakamoto, H., Matsuda, R., and Kitagawa, S. (2012) *Dalton Trans.*, **41**, 3956.
15 Cavka, J.H., Jakobsen, S., Olsbye, U., Guillou, N., Lamberti, C., Bordiga, S., and Lillerud, K.P. (2008) *J. Am. Chem. Soc.*, **130**, 13850.
16 Kalidindi, S.B., Nayak, S., Briggs, M.E., Jansat, S., Katsoulidis, A.P., Miller, G.J., Warren, J.E., Antypov, D., Cora, F., Slater, B., Prestly, M.R., Marti-Gastaldo, C., and Rosseinsky, M.J. (2015) *Angew. Chem., Int. Ed.*, **54**, 221.
17 Liu, T.-F., Feng, D., Chen, Y.-P., Zou, L., Bosch, M., Yuan, S., Wei, Z., Fordham, S., Wang, K., and Zhou, H.-C. (2015) *J. Am. Chem. Soc.*, **137**, 413.

18 Duan, J., Higuchi, M., Horike, S., Foo, M.L., Rao, K.P., Inubushi, Y., Fukushima, T., and Kitagawa, S. (2013) *Adv. Funct. Mater.*, **23**, 3525.

19 Duan, J., Higuchi, M., Krishna, R., Kiyonaga, T., Tsutsumi, Y., Sato, Y., Kubota, Y., Takata, M., and Kitagawa, S. (2014) *Chem. Sci.*, **5**, 660.

20 Noro, S.-i., Mizutani, J., Hijikata, Y., Matsuda, R., Sato, H., Kitagawa, S., Sugimoto, K., Inubushi, Y., Kubo, K., and Nakamura, T. (2015) *Nat. Commun.*, **6**, 5851.

21 Tanabe, K.K., Wang, Z., and Cohen, S.M. (2008) *J. Am. Chem. Soc.*, **130**, 8508.

22 Sato, H., Matsuda, R., Sugimoto, K., Takata, M., and Kitagawa, S. (2010) *Nat. Mater.*, **9**, 661.

23 Goto, Y., Sato, H., Shinkai, S., and Sada, K. (2008) *J. Am. Chem. Soc.*, **130**, 14354.

24 Zhang, J., Wojtas, L., Larsen, R.W., Eddaoudi, M., and Zaworotko, M.J. (2009) *J. Am. Chem. Soc.*, **131**, 17040.

25 Chen, B.L., Liang, C.D., Yang, J., Contreras, D.S., Clancy, Y.L., Lobkovsky, E.B., Yaghi, O.M., and Dai, S. (2006) *Angew. Chem., Int. Ed.*, **45**, 1390.

26 Shekhah, O., Wang, H., Paradinas, M., Ocal, C., Schuepbach, B., Terfort, A., Zacher, D., Fischer, R.A., and Wöll, C. (2009) *Nat. Mater.*, **8**, 481.

27 Furukawa, H., Ko, N., Go, Y.B., Aratani, N., Choi, S.B., Choi, E., Yazaydin, A.O., Snurr, R.Q., O'Keeffe, M., Kim, J., and Yaghi, O.M. (2010) *Science*, **329**, 424.

28 Rosi, N.L., Kim, J., Eddaoudi, M., Chen, B.L., O'Keeffe, M., and Yaghi, O.M. (2005) *J. Am. Chem. Soc.*, **127**, 1504.

29 Deng, H., Grunder, S., Cordova, K.E., Valente, C., Furukawa, H., Hmadeh, M., Gandara, F., Whalley, A.C., Liu, Z., Asahina, S., Kazumori, H., O'Keeffe, M., Terasaki, O., Stoddart, J.F., and Yaghi, O.M. (2012) *Science*, **336**, 1018.

30 Bloch, E.D., Queen, W.L., Krishna, R., Zadrozny, J.M., Brown, C.M., and Long, J.R. (2012) *Science*, **335**, 1606.

31 Shimomura, S., Higuchi, M., Matsuda, R., Yoneda, K., Hijikata, Y., Kubota, Y., Mita, Y., Kim, J., Takata, M., and Kitagawa, S. (2010) *Nat. Chem.*, **2**, 633.

32 Nakagawa, K., Tanaka, D., Horike, S., Shimomura, S., Higuchi, M., and Kitagawa, S. (2010) *Chem. Commun.*, **46**, 4258.

33 Sato, H., Kosaka, W., Matsuda, R., Hori, A., Hijikata, Y., Belosludov, R.V., Sakaki, S., Takata, M., and Kitagawa, S. (2014) *Science*, **343**, 167.

34 Huxford, R.C., Della Rocca, J., and Lin, W. (2010) *Curr. Opin. Chem. Biol.*, **14**, 262.

35 Horcajada, P., Serre, C., Vallet-Regi, M., Sebban, M., Taulelle, F., and Ferey, G. (2006) *Angew. Chem., Int. Ed.*, **45**, 5974.

36 Horcajada, P., Serre, C., Maurin, G., Ramsahye, N.A., Balas, F., Vallet-Regi, M., Sebban, M., Taulelle, F., and Ferey, G. (2008) *J. Am. Chem. Soc.*, **130**, 6774.

37 Horcajada, P., Chalati, T., Serre, C., Gillet, B., Sebrie, C., Baati, T., Eubank, J.F., Heurtaux, D., Clayette, P., Kreuz, C., Chang, J.-S., Hwang, Y.K., Marsaud, V., Bories, P.-N., Cynober, L., Gil, S., Ferey, G., Couvreur, P., and Gref, R. (2010) *Nat. Mater.*, **9**, 172.

38 Diring, S., Wang, D.O., Kim, C., Kondo, M., Chen, Y., Kitagawa, S., Kamei, K.-i., and Furukawa, S. (2013) *Nat. Commun.*, **4**, 2684.

39 Xiao, B., Wheatley, P.S., Zhao, X., Fletcher, A.J., Fox, S., Rossi, A.G., Megson, I.L., Bordiga, S., Regli, L., Thomas, K.M., and Morris, R.E. (2007) *J. Am. Chem. Soc.*, **129**, 1203.

40 Morris, R.E. and Wheatley, P.S. (2008) *Angew. Chem., Int. Ed.*, **47**, 4966.

41 Hinks, N.J., McKinlay, A.C., Xiao, B., Wheatley, P.S., and Morris, R.E. (2010) *Microporous Mesoporous Mater.*, **129**, 330.

42 Xiao, B., Byrne, P.J., Wheatley, P.S., Wragg, D.S., Zhao, X., Fletcher, A.J., Thomas, K.M., Peters, L., Evans, J.S.O., Warren, J.E., Zhou, W., and Morris, R.E. (2009) *Nat. Chem.*, **1**, 289.

43 Farrusseng, D., Aguado, S., and Pinel, C. (2009) *Angew. Chem., Int. Ed.*, **48**, 7502.

44 Lee, J., Farha, O.K., Roberts, J., Scheidt, K.A., Nguyen, S.T., and Hupp, J.T. (2009) *Chem. Soc. Rev.*, **38**, 1450.

45 Seo, J.S., Whang, D., Lee, H., Jun, S.I., Oh, J., Jeon, Y.J., and Kim, K. (2000) *Nature*, **404**, 982.

46 Chui, S.S.Y., Lo, S.M.F., Charmant, J.P.H., Orpen, A.G., and Williams, I.D. (1999) *Science*, **283**, 1148.

47 Alaerts, L., Seguin, E., Poelman, H., Thibault-Starzyk, F., Jacobs, P.A., and De Vos, D.E. (2006) *Chem. Eur. J.*, **12**, 7353.

48 Phan, A., Czaja, A.U., Gandara, F., Knobler, C.B., and Yaghi, O.M. (2011) *Inorg. Chem.*, **50**, 7388.

49 Hasegawa, S., Horike, S., Matsuda, R., Furukawa, S., Mochizuki, K., Kinoshita, Y., and Kitagawa, S. (2007) *J. Am. Chem. Soc.*, **129**, 2607.

50 Pan, Y., Yuan, B., Li, Y., and He, D. (2010) *Chem. Commun.*, **46**, 2280.

51 Fang, Z., Bueken, B., De Vos, D.E., and Fischer, R.A. (2015) *Angew. Chem., Int. Ed.*, **54**, 7234.

52 Uemura, T., Yanai, N., and Kitagawa, S. (2009) *Chem. Soc. Rev.*, **38**, 1228.

53 Uemura, T., Ono, Y., Kitagawa, K., and Kitagawa, S. (2008) *Macromolecules*, **41**, 87.

54 Uemura, T., Yanai, N., Watanabe, S., Tanaka, H., Numaguchi, R., Miyahara, M.T., Ohta, Y., Nagaoka, M., and Kitagawa, S. (2010) *Nat. Commun.*, **1**, 83.

55 Yanai, N., Uemura, T., Kosaka, W., Matsuda, R., Kodani, T., Koh, M., Kanemura, T., and Kitagawa, S. (2012) *Dalton Trans.*, **41**, 4195.

56 Uemura, T., Uchida, N., Asano, A., Saeki, A., Seki, S., Tsujimoto, M., Isoda, S., and Kitagawa, S. (2012) *J. Am. Chem. Soc.*, **134**, 8360.

57 Distefano, G., Suzuki, H., Tsujimoto, M., Isoda, S., Bracco, S., Comotti, A., Sozzani, P., Uemura, T., and Kitagawa, S. (2013) *Nat. Chem.*, **5**, 335.

58 Uemura, T., Kaseda, T., Sasaki, Y., Inukai, M., Toriyama, T., Takahara, A., Jinnai, H., and Kitagawa, S. (2015) *Nat. Commun.*, **6**, 7473.

59 Kitaura, R., Kitagawa, S., Kubota, Y., Kobayashi, T.C., Kindo, K., Mita, Y., Matsuo, A., Kobayashi, M., Chang, H.C., Ozawa, T.C., Suzuki, M., Sakata, M., and Takata, M. (2002) *Science*, **298**, 2358.

60 Matsuda, R., Kitaura, R., Kitagawa, S., Kubota, Y., Belosludov, R.V., Kobayashi, T.C., Sakamoto, H., Chiba, T., Takata, M., Kawazoe, Y., and Mita, Y. (2005) *Nature*, **436**, 238.

61 Inokuma, Y., Yoshioka, S., Ariyoshi, J., Arai, T., Hitora, Y., Takada, K., Matsunaga, S., Rissanen, K., and Fujita, M. (2013) *Nature*, **495**, 461.

62 Inokuma, Y., Kawano, M., and Fujita, M. (2011) *Nat. Chem.*, **3**, 349.

63 Navarro, J.A.R., Barea, E., Rodriguez-Dieguez, A., Salas, J.M., Ania, C.O., Parra, J.B., Masciocchi, N., Galli, S., and Sironi, A. (2008) *J. Am. Chem. Soc.*, **130**, 3978.

64 Milon, J., Daniel, M.-C., Kaiba, A., Guionneau, P., Brandes, S., and Sutter, J.-P. (2007) *J. Am. Chem. Soc.*, **129**, 13872.

65 Yanai, N., Kaneko, W., Yoneda, K., Ohba, M., and Kitagawa, S. (2007) *J. Am. Chem. Soc.*, **129**, 3496.

66 Wang, Z.M., Zhang, B., Fujiwara, H., Kobayashi, H., and Kurmoo, M. (2004) *Chem. Commun.*, 416.

67 Maspoch, D., Ruiz-Molina, D., Wurst, K., Domingo, N., Cavallini, M., Biscarini, F., Tejada, J., Rovira, C., and Veciana, J. (2003) *Nat. Mater.*, **2**, 190.

68 Kaye, S.S., Choi, H.J., and Long, J.R. (2008) *J. Am. Chem. Soc.*, **130**, 16921.

69 Halder, G.J., Kepert, C.J., Moubaraki, B., Murray, K.S., and Cashion, J.D. (2002) *Science*, **298**, 1762.

70 Niel, V., Martinez-Agudo, J.M., Munoz, M.C., Gaspar, A.B., and Real, J.A. (2001) *Inorg. Chem.*, **40**, 3838.

71 Ohba, M., Yoneda, K., Agusti, G., Munoz, M. Carmen, Gaspar, A.B., Real, J.A., Yamasaki, M., Ando, H., Nakao, Y., Sakaki, S., and Kitagawa, S. (2009) *Angew. Chem., Int. Ed.*, **48**, 4767.

72 Agusti, G., Ohtani, R., Yoneda, K., Gaspar, A.B., Ohba, M., Sanchez-Royo, J.F., Munoz, M.C., Kitagawa, S., and Real, J.A. (2009) *Angew. Chem., Int. Ed.*, **48**, 8944.

73 Ohtani, R., Yoneda, K., Furukawa, S., Horike, N., Kitagawa, S., Gaspar, A.B., Munoz, M. Carmen., Real, J.A., and Ohba, M. (2011) *J. Am. Chem. Soc.*, **133**, 8600.

74 Southon, P.D., Liu, L., Fellows, E.A., Price, D.J., Halder, G.J., Chapman, K.W., Moubaraki, B., Murray, K.S., Letard, J.F., and Kepert, C.J. (2009) *J. Am. Chem. Soc.*, **131**, 10998.

75 Allendorf, M.D., Bauer, C.A., Bhakta, R.K., and Houk, R.J.T. (2009) *Chem. Soc. Rev.*, **38**, 1330.

76 Tanaka, D., Horike, S., Kitagawa, S., Ohba, M., Hasegawa, M., Ozawa, Y., and Toriumi, K. (2007) *Chem. Commun.*, 3142.

77 Yanai, N., Kitayama, K., Hijikata, Y., Sato, H., Matsuda, R., Kubota, Y., Takata, M., Mizuno, M., Uemura, T., and Kitagawa, S. (2011) *Nat. Mater.*, **10**, 787.

78 Kobayashi, Y., Jacobs, B., Allendorf, M.D., and Long, J.R. (2010) *Chem. Mater.*, **22**, 4120.

79 Talin, A.A., Centrone, A., Ford, A.C., Foster, M.E., Stavila, V., Haney, P., Kinney, R.A., Szalai, V., El Gabaly, F., Yoon, H.P., Leonard, F., and Allendorf, M.D. (2014) *Science*, **343**, 66.

80 Sun, L., Miyakai, T., Seki, S., and Dinca, M. (2013) *J. Am. Chem. Soc.*, **135**, 8185.

81 Sun, L., Hendon, C.H., Minier, M.A., Walsh, A., and Dinca, M. (2015) *J. Am. Chem. Soc.*, **137**, 6164.

82 Horike, S., Matsuda, R., Tanaka, D., Matsubara, S., Mizuno, M., Endo, K., and Kitagawa, S. (2006) *Angew. Chem., Int. Ed.*, **45**, 7226.

83 Rodriguez-Velamazan, J. Alberto, Gonzalez, M.A., Real, J.A., Castro, M., Carmen Munoz, M., Gaspar, A.B., Ohtani, R., Ohba, M., Yoneda, K., Hijikata, Y., Yanai, N., Mizuno, M., Ando, H., and Kitagawa, S. (2012) *J. Am. Chem. Soc.*, **134**, 5083.

84 Jain, P., Ramachandran, V., Clark, R.J., Zhou, H.D., Toby, B.H., Dalal, N.S., Kroto, H.W., and Cheetham, A.K. (2009) *J. Am. Chem. Soc.*, **131**, 13625.

85 Jain, P., Dalal, N.S., Toby, B.H., Kroto, H.W., and Cheetham, A.K. (2008) *J. Am. Chem. Soc.*, **130**, 10450.

86 Kosaka, W., Yamagishi, K., Zhang, J., and Miyasaka, H. (2014) *J. Am. Chem. Soc.*, **136**, 12304.

87 Horike, S., Umeyama, D., and Kitagawa, S. (2013) *Acc. Chem. Res.*, **46**, 2376.

88 Shigematsu, A., Yamada, T., and Kitagawa, H. (2011) *J. Am. Chem. Soc.*, **133**, 2034.

89 Bureekaew, S., Horike, S., Higuchi, M., Mizuno, M., Kawamura, T., Tanaka, D., Yanai, N., and Kitagawa, S. (2009) *Nat. Mater.*, **8**, 831.

90 Hurd, J.A., Vaidhyanathan, R., Thangadurai, V., Ratcliffe, C.I., Moudrakovski, I.L., and Shimizu, G.K.H. (2009) *Nat. Chem.*, **1**, 705.

91 Furukawa, S., Reboul, J., Diring, S., Sumida, K., and Kitagawa, S. (2014) *Chem. Soc. Rev.*, **43**, 5700.

92 Zacher, D., Schmid, R., Wöll, C., and Fischer, R.A. (2011) *Angew. Chem., Int. Ed.*, **50**, 176.

93 Furukawa, S., Hirai, K., Nakagawa, K., Takashima, Y., Matsuda, R., Tsuruoka, T., Kondo, M., Haruki, R., Tanaka, D., Sakamoto, H., Shimomura, S., Sakata, O., and Kitagawa, S. (2009) *Angew. Chem., Int. Ed.*, **48**, 1766.

94 Fukushima, T., Horike, S., Inubushi, Y., Nakagawa, K., Kubota, Y., Takata, M., and Kitagawa, S. (2010) *Angew. Chem. Int. Ed.*, **49**, 4820.

95 Horike, S., Inubushi, Y., Hori, T., Fukushima, T., and Kitagawa, S. (2012) *Chem. Sci.*, **3**, 116.

96 Tanaka, D., Henke, A., Albrecht, K., Moeller, M., Nakagawa, K., Kitagawa, S., and Groll, J. (2010) *Nat. Chem.*, **2**, 410.

97 Hijikata, Y., Horike, S., Tanaka, D., Groll, J., Mizuno, M., Kim, J., Takata, M., and Kitagawa, S. (2011) *Chem. Commun.*, **47**, 7632.

98 Sakata, Y., Furukawa, S., Kondo, M., Hirai, K., Horike, N., Takashima, Y., Uehara, H., Louvain, N., Meilikhov, M., Tsuruoka, T., Isoda, S., Kosaka, W., Sakata, O., and Kitagawa, S. (2013) *Science*, **339**, 193.

99 Umemura, A., Diring, S., Furukawa, S., Uehara, H., Tsuruoka, T., and Kitagawa, S. (2011) *J. Am. Chem. Soc.*, **133**, 15506.

100 Reboul, J., Furukawa, S., Horike, N., Tsotsalas, M., Hirai, K., Uehara, H., Kondo, M., Louvain, N., Sakata, O., and Kitagawa, S. (2012) *Nat. Mater.*, **11**, 717.

101 Yanai, N. and Granick, S. (2012) *Angew. Chem., Int. Ed.*, **51**, 5638.

102 Yanai, N., Sindoro, M., Yan, J., and Granick, S. (2013) *J. Am. Chem. Soc.*, **135**, 34.

103 Tan, J.C. and Cheetham, A.K. (2011) *Chem. Soc. Rev.*, **40**, 1059.

104 Cao, S., Bennett, T.D., Keen, D.A., Goodwin, A.L., and Cheetham, A.K. (2012) *Chem. Commun.*, **48**, 7805.

105 Bennett, T.D., Goodwin, A.L., Dove, M.T., Keen, D.A., Tucker, M.G., Barney, E.R., Soper, A.K., Bithell, E.G., Tan, J.-C., and Cheetham, A.K. (2010) *Phys. Rev. Lett.*, **104**, 115503.

106 Du, Y., Wooler, B., Nines, M., Kortunov, P., Paur, C.S., Zengel, J., Weston, S.C., and Ravikovitch, P.I. (2015) *J. Am. Chem. Soc.*, **137**, 13603.

107 Umeyama, D., Horike, S., Inukai, M., Itakura, T., and Kitagawa, S. (2015) *J. Am. Chem. Soc.*, **137**, 864.

108 Tominaka, S., Hamoudi, H., Suga, T., Bennett, T.D., Cairns, A.B., and Cheetham, A.K. (2015) *Chem. Sci.*, **6**, 1465.

109 Kitagawa, S. (2015) *Angew. Chem., Int. Ed.*, **54**, 10686.

8
Metal–Organic Frameworks: An Emerging Class of Solid-State Materials

Joseph E. Mondloch,[1] Rachel C. Klet,[2] Ashlee J. Howarth,[2] Joseph T. Hupp,[2] and Omar K. Farha[2,3]

[1]University of Wisconsin–Stevens Point, Department of Chemistry, 2001 Fourth Avenue, Stevens Point, WI 54482, USA
[2]Northwestern University, Department of Chemistry, 2145 Sheridan Road Evanston, IL 60208, USA
[3]Department of Chemistry, Faculty of Science, King Abdulaziz University, Jeddah, Saudi Arabia

8.1
Introduction

Over the past two decades metal–organic frameworks (MOFs) have emerged as an intriguing class of solid-state materials. This has been fueled by the synthesis of thousands of discrete examples of these materials through the judicious choice of their components, inorganic nodes (metal ions or polymetallic complexes) and organic linkers [1]. MOFs are held together by coordinate covalent bonds, which is in stark contrast to many other solid-state materials such as zeolites, metal oxides, and activated carbons that are held together by covalent or ionic bonds. In the most favorable instances, this bestows MOFs with a highly attractive set of characteristics, including: (i) crystallinity, leading to precise structural determination; (ii) synthetic tunability – pore size, shape, and functionality are all readily accessible by modifying both the inorganic nodes and organic linkers; (iii) porosity, MOFs possess some of the highest known surface areas and pore volumes reported to date, and can be ultramicroporous, microporous, and/or mesoporous; and (iv) flexibility, MOFs are dynamic materials which can allow for adsorbate restructuring as well as gating effects [2].

These features give rise to new and unique physical and chemical properties that are driving researchers to investigate the suitability of MOFs in a handful of practical applications. Examples include gas storage [3], chemical separations [4], catalysis [5], sensing [6], conductivity [7], solar fuels conversion [8], light harvesting and energy conversion [9], and magnetic materials among others [10]. Given that MOFs are a rather young class of materials, their true potential in

Handbook of Solid State Chemistry, First Edition. Edited by Richard Dronskowski, Shinichi Kikkawa, and Andreas Stein.
© 2017 Wiley-VCH Verlag GmbH & Co. KGaA. Published 2017 by Wiley-VCH Verlag GmbH & Co. KGaA.

application driven chemistry is only starting to blossom. A recent feature article has highlighted the most promising industrial developments [11]. Several of these discoveries center around the need to reduce the pressure required to store gases in tanks on either the large scale, such as natural gas storage for use as alternative fuel supplies [12], or smaller specialized markets [13]. In other examples, MOFs are being put to the test against industrial sorbents and solid-state materials for gas separation [14] and chemical catalysis [15].

Herein we give a brief overview of MOF chemistry. We focus on examples that highlight important concepts in the field of MOF chemistry, starting with the toolbox that MOF chemists use for the synthesis of these materials. Examples include the *de novo* synthesis of MOFs, post-synthesis modification (PSM) of the inorganic nodes and organic linkers, and building-block replacements strategies such as transmetalation and solvent-assisted linker exchange (SALE). Many of the potential applications for MOFs, such as gas storage and separation, as well as gas-phase catalysis involve the removal of guest molecules from their pores (i.e., activation). Accordingly, we briefly discuss the activation of MOFs given its importance throughout much of the field. Finally, we discuss some emerging areas of application driven MOF chemistry. Our choice of coverage here is largely driven by our expertise in the field and by examples that we find most interesting.

8.2
Synthesis of MOFs

8.2.1
De Novo Synthesis

In a typical procedure (i.e., a *de novo* synthesis), MOFs are synthesized under thermal conditions in a one-pot reaction with a soluble metal precursor or salt and a multitopic organic linker in a high boiling point solvent (Figure 8.1) [16]. The metal source, reaction temperature and time, solvent, reagent concentration, heat source, and pH can all strongly influence the phase purity, crystal size, and morphology of the product [16a,17]. In some cases, a monotopic modulator molecule is employed which often has the same functional group as the linker. Modulators can help control crystal size and shape, or may terminate a structure leading to more reproducible syntheses [18]; modulators have been particularly critical for the development of Zr_6-based MOFs (Section 8.4.4) [19].

Two additional and important aspects of MOF synthesis include the preparation of phase-pure MOFs and the control of catenation (*vide infra*). Phase-pure MOFs are most often obtained via synthetic optimization, a process that can be accelerated via high-throughput screening [20]. When product mixtures are obtained, density separation using combinations of organic solvents can be utilized to purify mixtures of MOFs [21]. Catenation of MOFs, in which two or more frameworks self-assemble within each other, is a well-known phenomenon in MOF synthesis [22]. While control of catenation can be challenging, synthesis

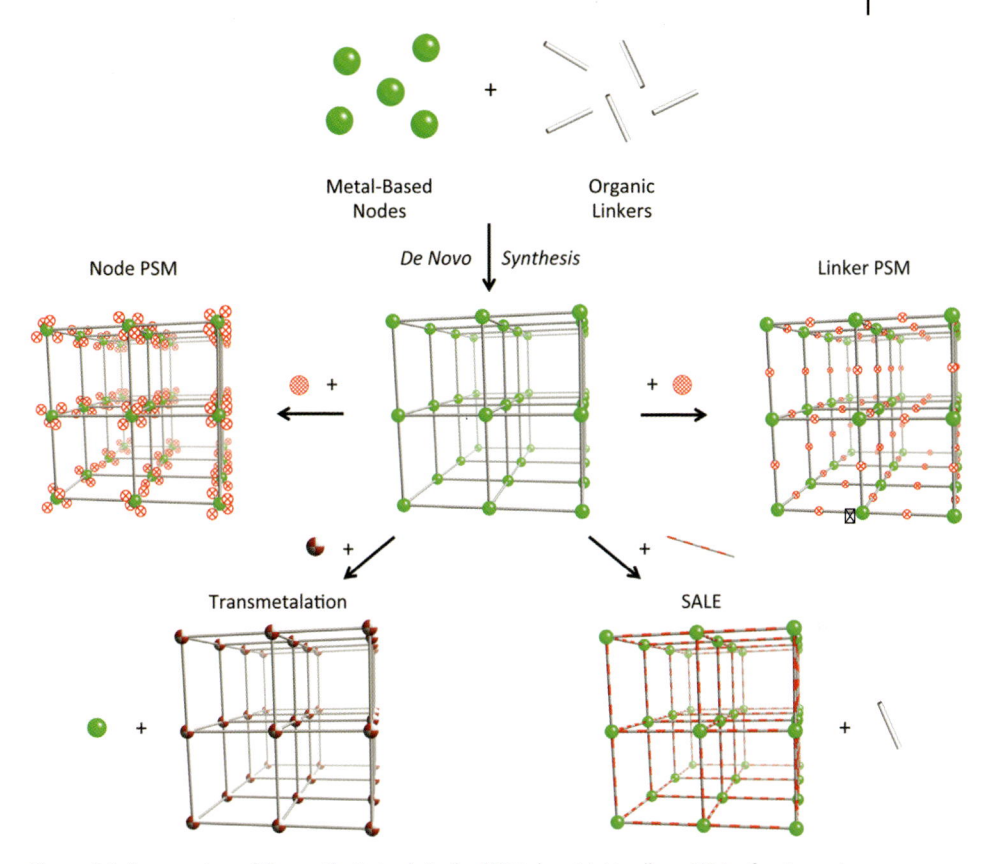

Figure 8.1 An overview of the synthetic tools in the MOF chemists' toolbox. PSM refers to post-synthesis modificiation and SALE refers to solvent-assisted linker exchange.

parameters such as temperature or concentration can strongly impact catenation [16a]. Careful design of MOF building blocks has been shown to be effective to prevent catenation [23].

Interest in the commercialization of MOFs has driven the development of alternative synthetic methods [24]. BASF first reported the electrochemical synthesis of Cu- and Zn-based MOFs in 2005. Electrochemical synthesis uses bulk metal plates as the anode in an electrochemical cell with the linker dissolved in a suitable solvent. This route avoids the use of metal salts as starting materials, which may include anions such as nitrate, perchlorate, or chloride, which could be challenging for commercial processes [25]. Microwave-assisted syntheses of MOFs have also been reported and often lead to the production of smaller MOF particles in shorter reaction times compared to conventional thermal routes [26]. Mechanochemical routes to MOFs are known and have the advantage of typically operating at room temperature under solvent-free or low-solvent conditions [27], and include ball milling [28], manual grinding with a pestle and mortar [29], and extrusion [30]. The addition of small amounts of solvent, known as liquid-assisted grinding (LAG), often greatly improves mechanochemical syntheses [28c,31]. MOF synthesis by sonochemistry has also been reported [26a]. Notably, several well-known MOFs, including HKUST-1, ZIF-8, and MIL-53(Al), have successfully been scaled up and are commercially available from BASF.

In addition to bulk powders, thin films of MOFs have been synthesized by various strategies, including direct growth using conventional thermal methods or microwave-assisted heating; layer-by-layer or liquid phase epitaxial growth, in which the substrate is immersed in reactant solutions in a sequential stepwise fashion; and seeded growth, in which a substrate is first coated with a seed layer, then immersed in a MOF reaction solution [32].

8.2.2
Post-Synthesis Modification

While many MOFs have been prepared via *de novo* synthetic strategies, in certain instances it can be difficult to access a designed MOF for a specific application through *de novo* routes. For example, incorporation of organic linkers with desirable functional groups, or installation of a second metal ion in a MOF framework can be difficult to achieve through a one-pot syntheses. Post-synthesis modification (PSM) allows for access to desired materials through modification of the nodes and/or linkers of a preconstructed MOF (Figure 8.1). PSM can overcome problems associated with linker solubility, undesirable metal-ion coordination, formation of undesirable topologies, and thermal stability among others to achieve otherwise inaccessable framework materials. Below we highlight several important examples of PSM.

8.2.2.1 Metal-Based Node Modification

Larabi and Quadrelli have supported a $Au(PMe_3)$ species on the Zr-based MOF UiO-67 ($Zr_6(\mu_3-O)_4(\mu_3-OH)_4(bpdc)_6$, bpdc = biphenyldicarboxylate) using an ethereal solution of the organometallic precursor $AuMe(PMe_3)$. The Au precursor reacts with hydroxy groups on the MOF node to release methane, which was quantified by GC analysis. Based on elemental analysis, approximately one Au (PMe_3) species was grafted per $Zr_6(OH)_4O_4$ node yielding $Zr_6(OH)_3O_4(bpdc)_6(OAuPMe_3)$. Notably, the authors also attempted reaction of UiO-67 with the organometallic precursors trineopentyl(neopentylidene)tantalum^V ($Ta(=CH^tBu)(CH_2^tBu)_3$), cyclooctadiene(cyclooctatriene)ruthenium(0) ($Ru(c-C_8H_{10})(c-C_8H_8)$), and cyclooctadiene(naphthalene)ruthenium(0) ($Ru(c-C_8H_{10})(C_{10}H_8)$), but did not observe any reactivity most likely due to the large dimensions of these complexes compared to the relatively small pore apertures of UiO-67 [33]. Nguyen and coworkers have subsequently demonstrated that the node of a very similar MOF UiO-66, which contains 1,4-benzene dicarboxylate (bdc) linkers in place of bpdc linkers, could be modified with vanadyl acetylacetonate to give V-UiO-66 [34]. Yang *et al.* have supported $Ir(C_2H_4)_2$ and/or $Ir(CO)_2$ species on UiO-66 and NU-1000 ($Zr_6(\mu_3-O)_4(\mu_3-OH)_4(H_2O)_4(OH)_4(TBAPy)_2$) where TBAPy =

1,3,6,8-tetrakis(*p*-benzoate)pyrene) by reacting pentane solutions of Ir(CO)$_2$ (*acac*) (*acac* = acetylacetonate) or Ir(C$_2$H$_4$)$_2$(*acac*) with each MOF [35]. The nodes of the Hf congener of NU-1000 (Hf$_6$(μ$_3$-O)$_4$(μ$_3$-OH)$_4$(H$_2$O)$_4$(OH)$_4$ (TBAPy)$_2$) have also been metallated with the organometallic precursor ZrBn$_4$(Bn = benzyl) to yield a supported monobenzylzirconium species Hf-NU-1000-ZrBn. This species was shown to be an active single-component catalyst (i.e., not requiring a cocatalyst/activator) for both ethylene and stereoregular 1-hexene polymerization; however, formation of a thermodynamically favored Zr-oxo species, resulting from a TBAPy linker carboxylate group shifting from the Hf$_6$ node to the embedded Zr ion, led to inconsistent catalytic activity [36].

Post-synthesis metalation from the vapor-phase is also emerging as an attractive synthetic strategy [37]. In comparison to solution-based techniques, the use of vapor-phase precursors circumvents the need for washing and purification steps while eliminating the use of solvents that can block sites of coordinative unsaturation. One particularly attractive strategy is to utilize MOFs with spatially isolated and chemically reactive functional groups (e.g., —OH containing groups). This serves as a platform for generating well-defined site isolated metal complexes within the framework. Because the chemistry is self-limiting – that is, chemical reactions can only occur at the reactive functional groups – the process is akin to atomic layer deposition (ALD) which has traditionally been utilized to deposit thin films (Figure 8.2a) [38]. Given the similarities to traditional ALD, Mondloch *et al.* coined this synthetic technique atomic layer deposition in a metal–organic framework (AIM), Figure 8.2b [37].

A key aspect to realizing AIM was to simultaneously develop a MOF containing large mesopores (facilitating diffusion of vapor-phase precursors throughout the framework) and chemical functionality (e.g., —OH containing groups), all

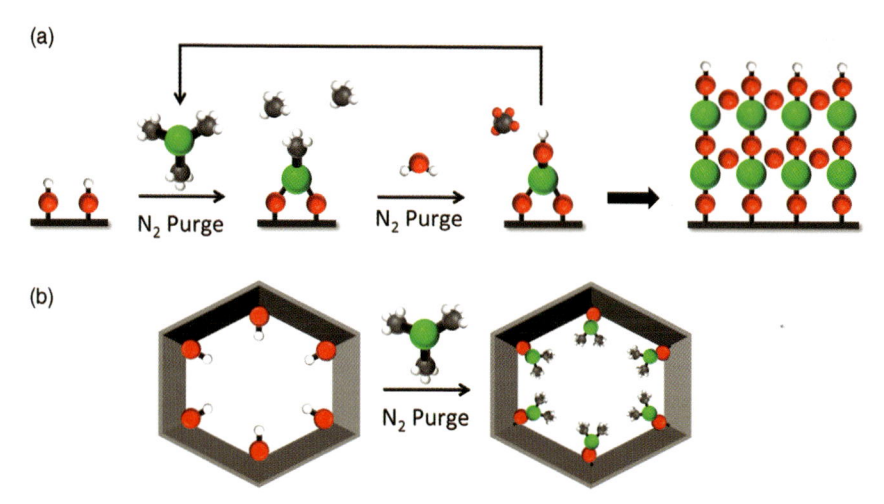

Figure 8.2 An illustration of (a) ALD on a flat substrate and (b) AIM within a MOF scaffold using the well defined precursor Al(Me)$_3$. Al: Green; O: Red; C: Black; H: White.

while being thermally and hydrolytically robust. To meet the aforementioned criteria, Mondloch and Bury et al. synthesized the MOF NU-1000 (Figure 8.3a) from tetracarboxylate linkers (TBAPy) and $Zr_6(\mu_3\text{-O})_4(\mu_3\text{-OH})_4(H_2O)_4(OH)_4$ nodes [37]. By making use of AIM, trimethylaluminum ($AlMe_3$) and diethyl zinc ($ZnEt_2$) infiltrated the entire MOF. Inductively coupled plasma-optical emission spectroscopy (ICP-OES) demonstrated that eight Al atoms and three Zn atoms were deposited per Zr_6 node, while the structural integrity of the MOF was confirmed via powder X-ray diffraction (PXRD) and N_2 adsorption analysis. Incorporation of Al and Zn ions throughout the MOF crystals was demonstrated by scanning electron microscopy-energy dispersive X-ray spectroscopy (SEM-EDX) analysis. Finally, diffuse reflectance infrared Fourier transform spectroscopy (DRIFTS) demonstrated that the $AlMe_3$ and $ZnEt_2$ metal complexes were indeed deposited on the chemically reactive $Zr_6(\mu_3\text{-O})_4(\mu_3\text{-OH})_4(H_2O)_4(OH)_4$ nodes.

In a follow-up study Kim et al. demonstrated that trimethyl indium ($InMe_3$) could also be deposited within NU-1000 [39]. Experimental X-ray pair distribution function (PDF) analysis coupled with density functional theory (DFT) modeling led to precise structural analysis for both the $AlMe_3$ and $InMe_3$ modified versions of NU-1000. These analyses indicated that up to eight tetrahedrally coordinated $M(Me)(OR)_3$ (M = Al or In) species are bound to the Zr_6-based nodes (Figure 8.3b). In addition, Kim et al. demonstrated that modulating the reaction temperature and dehydrating the Zr_6-based node could allow for control over metal loading. Peters et al. demonstrated that the precursor bis(N,N'-di-i-propylacetamidinato)cobalt(II) ($Co(amd)_2$) could be deposited within NU-1000 and subsequently reacted with dihydrogen sulfide (H_2S) [40]. The resultant material, a CoS-based species within the pores of NU-1000, is an active and selective hydrogenation catalyst for the conversion of nitroaromatic compounds into amines. Kung et al. deposited the same $Co(amd)_2$ precursor into thin films of NU-1000 on conductive glass [41]. The Co modified nodes of

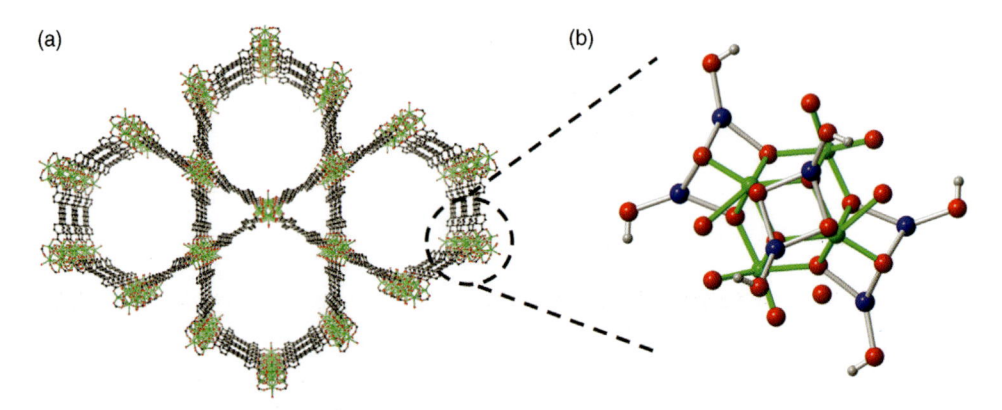

Figure 8.3 (a) the MOF NU-1000 and (b) and its InOMe metallated analog. Zr: Green; O: red; C: black; H: gray; In: blue.

NU-1000 were electrochemically addressable and able to catalytically oxidize water. Single-site Ni^{II} modified versions of the NU-1000 platform were also found to be efficient gas-phase hydrogenation catalysts [42]. Finally, it is notable that microporous MOFs can also be functionalized with metal complexes via vapor-phase infiltration [43], however much longer infiltration times (days to weeks in comparison to less than an hour) are needed and the precise spatial distribution of the metal complexes within the MOF has not been demonstrated.

8.2.2.2 Organic-Based Node Modification

Metal node functionalization with organic moieties has been demonstrated using both ionic and neutral ligands. This functionalization method is of particular interest for modulating the adsorption capacity of MOFs (e.g., for CO_2 capture [3a,44]) and as a method to install homogenous catalysts [45]. Deria et al. have incorporated fluoroalkyl carboxylic acids into the MOF NU-1000 through ionic bonding of the carboxylate groups to the Zr_6 node via a method called solvent-assisted ligand incorporation (SALI). The carboxylate groups of the fluoroalkyl carboxylic acids displace the $-OH/-OH_2$ groups that constitute the node of NU-1000 ($Zr_6(\mu_3\text{-}O)_4(\mu_3\text{-}OH)_4(H_2O)_4(OH)_4$) (Figure 8.4). The fluoroalkyl carboxylate chain lengths range from one (CF_3CO_2H) to nine ($CF_3(CF_2)_8CO_2H$) and approximately 3.4–4 perfluoroalkyl carboxylates were incorporated per Zr_6 node [44]. In follow-up studies, Deria and coworkers have demonstrated that SALI is a versatile synthetic technique compatible with a wide range of carboxylate-based alkyl and aromatic moieties with secondary functional groups [46], as well as phosphonate-based moieties [47]. This work has been expanded on by Madrahimov et al. who used SALI to incorporate a phosphonate-modified 2,2′-bipyridine (bpy) ligand into NU-1000 that could then be metallated with $NiCl_2$ to yield NU-1000-bpy-$NiCl_2$. Upon activation with Et_2AlCl, NU-1000-bpy-$NiCl_2$ was found to be an active catalyst for ethylene dimerization that could be reused three times (competitive formation of

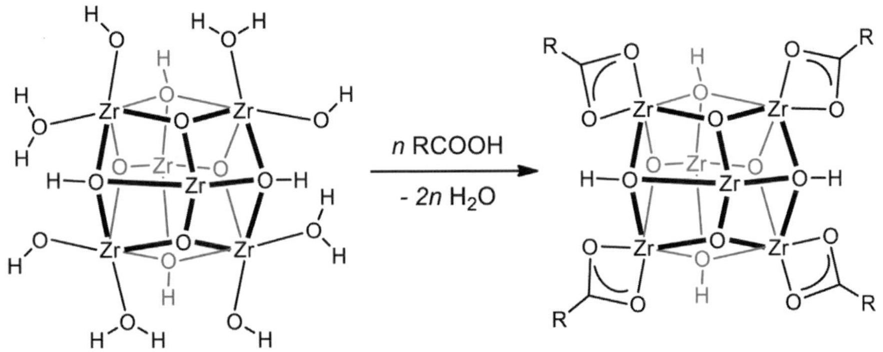

Figure 8.4 Representative functionalization of the Zr_6-node of NU-1000 with carboxylic acid moieties via SALI (MOF carboxylate linkers have been omitted for clarity).

polyethylene eventually blocks access to active sites and decreases dimerization activity) [45].

Unsaturated metal sites in MOF nodes can also be functionalized with organic moieties through coordination (dative bonding) of various functional groups (typically nitrogen-based). Hwang and coworkers demonstrated grafting of alkylamines (ethylenediamine, diethylenetriamine, and 3-aminopropyltrialkoxysilane) to the open metal sites in the Cr-based MOF MIL-101 $Cr_3(OH)(H_2O)_2O$ $(bdc)_3$ [48]. In related work, Farha et al. generated unsaturated Zn^{II} sites in the MOF $Zn_2(L)$ (L = 4,4′,4″,4‴-benzene-1,2,4,5-tetrayltetrabenzoate) by removal of coordinated solvent molecules, and then elaborated the open metal sites with a series of pyridine derivatives [49]. Long and co-workers also utilized this methodology as a post-synthesis modification strategy for enhancing CO_2 binding in MOFs. Specifically, they have investigated incorporation of the alkylamines into the MOFs CuBTTri ($H_3[(Cu_4Cl)_3(BTTri)_8]$ H_3BTTri = 1,3,5-tri(1H-1,2,3-triazol-4-yl)benzene), which has unsaturated Cu^{II} sites [50], and $Mg_2(dobpdc)$ (dobpdc = 4,4′-dioxido-3,3′-biphenyldicarboxylate), with coordinatively unsaturated Mg^{II} sites [51]. In general, grafting alkylamines into MOF pores was found to enhance CO_2 uptake through chemisorption at the amine functionalities [52].

8.2.2.3 Linker Modification

Post-synthesis modification of MOF organic linkers has been extensively reviewed [53]. Here we highlight several main themes in this widely studied area. Lin et al. designed a pyridyl-BINOL (BINOL = 1,1′-bi-naphthol) ligand that was used to synthesize a new Cd^{II} MOF. The pyridyl groups of the ligand coordinate to Cd^{II} ions to form the MOF backbone, while the hydroxy groups remain uncoordinated. After obtaining the crystalline MOF material, the authors treated the MOF with $Ti(O^iPr)_4$, which presumably coordinates to the MOF through the available hydroxyl functional groups. This material was then shown to catalyze asymmetric diethylzinc addition to aromatic aldehydes [54]. This strategy has since been expanded to include a wide variety of functional groups that have all been successfully metallated post-synthesis, including salen [55], bpy [56], porphyrin [57], phosphine [58], aminopyridinato [59], and thiol [60].

An alternative strategy to install functional groups is to chemically modify the crystalline MOF material after its synthesis. For example, Cohen et al. demonstrated covalent modification of the linkers in IRMOF-3 ($Zn_4O(bdc-NH_2)_3$) via reaction of linker amino groups with anhydrides [61]. In related work, Rosseinsky and coworkers derivatized the linker amino group in IRMOF-3 to a salicylidene through a condensation reaction with salicylaldehyde. The salicylidene could then be metalated with $O=V(acac)_2$ to afford a supported vanadyl complex (Figure 8.5) [62]. Other linker modification reactions include: click chemistry on linkers with pendant azide [63] or alkyne [64] groups; oxidation of sulfur-containing linker functional groups [65]; and bromination of alkenyl- [66] or alkynyl-linkers [67]. Recently, post-synthesis polymerization of MOF crystallites has been demonstrated through reaction of appropriately functionalized

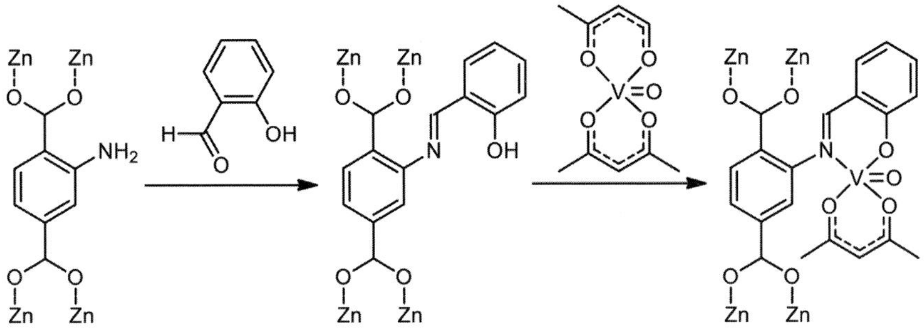

Figure 8.5 Functionalization of the bdc-NH$_2$ linker in IRMOF-3 with salicylaldehyde and subsequent metallation with vanadyl *acac*.

linkers [68]; in one example, the hybrid polymer-MOF material was then utilized as an effective membrane for separating Cr^{VI} ions from water [68b].

8.2.3
Building Block Replacement

Building block replacement is a fundamentally different approach to the synthesis of MOFs that has recently emerged [69]. Here key structural components of the MOF (i.e., the nodes and the linkers) are simply exchanged for new and desired components in a solution (Figure 8.1). Exchange of the metal-based nodes has been termed transmetalation or cation exchange [70], while exchange of the organic linkers has been coined solvent-assisted linker exchange (SALE) or post-synthetic exchange [71]. This building block replacement phenomenon is thought to occur via single-crystal-to-single-crystal transformations and is attractive given that isostructural MOFs that are not accessible via traditional *de novo* synthetic strategies can be readily generated from a parent MOF.

Transmetalation occurs between a parent MOF and a solution containing a secondary metal salt. The resultant MOF can contain partially or completely transmetalated metal nodes. One of the most well studied systems is the transmetalation of Zn^{II} ions in the Zn_4O nodes of MOF-5 ($Zn_4O(bdc)_3$) (Figure 8.6). In an early study, Brozek et al. demonstrated partial transmetalation by substituting one of four tetrahedral Zn^{II} ions with Ni^{II} [72]. Subsequent studies demonstrated that this partial substitution is governed by a dissociative solvent-mediated pathway [73]. While the Ni-modified version of MOF-5 can be prepared under *de novo* synthetic conditions, several other mixed-metal variants of MOF-5, including the Ti^{III}, V^{II}, V^{III}, Cr^{II}, Cr^{III}, Mn^{II}, Fe^{II}, and Co^{II} transmetalated versions, could not [74]. This example highlights one of the most attractive aspects of transmetalation and post-synthesis modification in general – MOFs which are not accessible via *de novo* synthetic strategies can indeed be accessed via post-synthesis modification strategies. It is notable that complete substitution via transmetalation has also been experimentally demonstrated in several MOFs [69,70b]. Together partial and complete transmetalation have led to

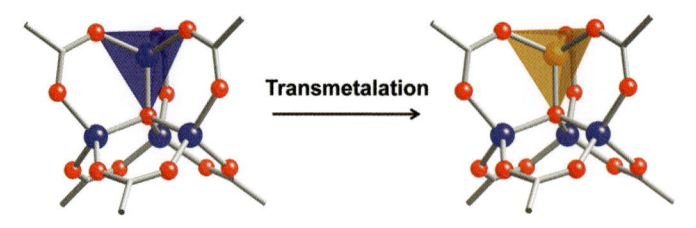

Figure 8.6 Transmetalation at the Zn$_4$O node of MOF-5. Zn: blue; O: red; transmetalated metal ion: orange.

MOFs with improved gas adsorption properties [75], increased stability [76], and new coordination environments that are giving rise to new and unique electron transfer [74] and catalytic chemistry [15].

Solvent-assisted linker exchange (SALE) occurs between a parent MOF and a solution of protonated organic linker. Burnett et al. first demonstrated the feasibility of SALE in the pillared-paddlewheel MOF PPF-20 by exchanging the 15.4 Å pillars for smaller 7.0 Å pillars [77]. Subsequently, Karagiaridi et al. demonstrated that larger linkers (up to 17 Å) could be substituted for shorter linkers (9 Å) within a pillared-paddlewheel MOF [78], a finding confirmed by Li et al. [79]. Surprisingly, even some of the most robust known MOFs, such as ZIF-8 (Zn(mim)$_2$, where mim is 2-methylimidazolate) and UiO-66, are amenable to SALE. Karagiaridi et al. demonstrated that the methylimidazole linker of ZIF-8 could be exchanged for imidazole, effectively opening up a secondary aperture within the structure of ZIF-8 [80]. Kim et al. demonstrated that benzene dicarboxylate (bdc) could be exchanged for a series of functionalized bdc ligands including Br-bdc, NH$_2$-bdc, N$_3$-bdc, and HO-bdc [81]. SALE has also successfully been utilized to synthesize new MOF-based catalysts [80,82], proton conductive MOFs [83], mixed-linker MOFs, defect-controlled MOFs [84], and thin films [85] as well as membranes [86]. Clearly, SALE is an attractive strategy for controlling pore dimensions, aperture size, and pore functionality. In almost all of the examples reported to date, the MOFs accessed via SALE are not accessible under *de novo* synthetic conditions.

8.3
Activation of MOFs

The extraordinary surface areas, pore volumes, and permanent porosity exhibited by MOFs can only be accessed by gas molecules after the solvent or other guest molecules (present during the synthesis of these materials) are removed from their pores. This process is often termed "activation" [87], and can be particularly cumbersome for MOFs. Unlike many other solid-state materials, such as metal oxides, zeolites, and activated carbons which often utilize a thermal annealing step during their syntheses, thermal annealing of MOFs can lead to pore collapse and loss of permanent porosity in all but the most stable MOFs. This loss of permanent porosity can be a result of capillary-force-driven channel collapse especially when high boiling point solvents are present within the pores of the MOF [88].

Several strategies for minimizing MOF-solvent intermolecular forces have been developed to mitigate capillary-force-driven channel collapse. The classic approach is solvent exchange; a high boiling-point solvent (such as dimethylformamide or water) is exchanged for a lower boiling-point solvent such as chloroform or acetone [89]. The lower boiling-point solvent is then removed and collapse is avoided. Supercritical carbon dioxide ($scCO_2$) has more recently evolved as a general strategy for activating MOFs [90]. Here, solvents such as ethanol and other solvents that are miscible with liquid CO_2 are exchanged at high pressure (i.e., >73 atm). The sample is then brought above the supercritical temperature (i.e., 31 °C) and slowly vented; avoiding the liquid-to-gas phase transition and going directly to the supercritical phase circumvents capillary forces. This strategy has been effective in activating some of the most porous MOFs to date [87]. A similar, but less utilized strategy is benzene freeze drying (lyophilization) [91]. Here the sample is frozen in benzene at 0 °C at a temperature and pressure below the triple point. The sample is then brought to room temperature under reduced pressure to avoid the liquid-to-gas phase transition and its associated capillary forces. There are also strategies being developed to rapidly assess the activation conditions and porosity of MOFs simultaneously. For example, McDonald *et al.* demonstrated that thermogravimetric analysis (TGA) coupled to isothermal propane adsorption could be utilized to screen newly synthesized MOFs for porosity [92].

8.4
The Quest for Increasingly "Stable" MOFs

When considering the potential applications of MOFs, particularly in a large-scale industrial or commercial setting, the stability of these materials is of utmost importance. There are many different ways to define "stability" as it relates to MOFs. The most straightforward definition is *the resistance of a structure to degradation when placed under mechanical, thermal, chemical, and/or hydrothermal stress* [93]. More specific definitions relating to water stability have been proposed by Walton and coworkers that take into account the differences between thermodynamic and kinetic stability of MOFs. For example, Walton and coworkers describe MOF thermodynamic stability as *the inertness of the metal cluster* that is largely attributed to metal-ligand bond strength and/or the lability of the metal cluster toward water [94]. On the other hand, MOF kinetic stability is attributed to the *hydrophobicity* of the framework and/or *steric factors* that block water from coordinating to the metal cluster and/or replacing labile linkers within the framework. It is important to note that depending on the application envisioned, different types of stability may be more desirable than others and in some instances, predictable instability may actually be required. For example, some potential industrial applications of MOFs include gas separation and storage [95], water desalination [96], ion exchange [97] and catalysis [5a] in which varying degrees of thermal, chemical, hydrothermal, and hydrolytic stability are required. Mechanical stability is also important given that many of these applications require MOF powders to be compacted into pellets before loading into fuel tanks or ion exchange columns [98]. MOF instability, or more importantly predictable or controlled instability, can also be desirable when considering applications in drug storage, delivery, and release [99].

There are, unfortunately, no standardized methods currently being used to measure MOF stability. PXRD experiments of MOF samples before and after exposure to stress gives information about the bulk crystallinity of a sample, but unfortunately the pattern could look the same before and after applied stress even if the MOF has partially degraded or lost some or all porosity and functionality. Instead, MOFs should undergo multiple stability tests relevant to the candidate application. While PXRD patterns are helpful for screening (the retention of crystallinity may be all that is required of certain applications), retention of porosity should also be confirmed by testing inert gas adsorption before and after applied stress. For some applications, for example, catalytic transformations, retention of chemical functionality should also be tested by determining catalytic efficiency before and after the applied stress. Methods of applying stress to MOFs such as temperature, pressure, humidity, and pH are also yet to be standardized, in part due to the extreme variation in conditions that are required for many potential applications. It has, however, been recognized that some type of standardization or labeling system should be implemented to describe the many variations of MOF stability [100]. In this section, different families of MOFs will be discussed and important features related to each family that help to impart framework stability will be described.

8.4.1
MOFs Containing M–N Bonds

Zeolitic imidazolate frameworks (ZIFs) and metal-azolate frameworks (MAFs) contain divalent metals connected by N-donor linkers and can be particularly stable. As the name suggests, ZIF topologies resemble those of aluminosilicate zeolites owing in part to a match between the M—N—M bond angle in ZIFs and the Si—O—Si or the Al—O—Al bond angles in zeolites (Figure 8.7a). ZIF-8 (Figure 8.7b), comprised of Zn^{II} nodes and 2-methylimidazolate linkers (mim), is a well-studied example that is stable (retains crystallinity and porosity) in refluxing water, alkaline solutions and various organic solvents [101]. ZIF-8 is also stable up to 300 °C in air or up to 500 °C under inert conditions [102]. The stability of ZIFs arises, in part, from the strong bonds formed between the divalent metal ions and nitrogen-donor imidazolate linkers [101]. Zn^{II} imidazolate bonds are among the strongest of all transition metal–nitrogen-donor ligand bonds.

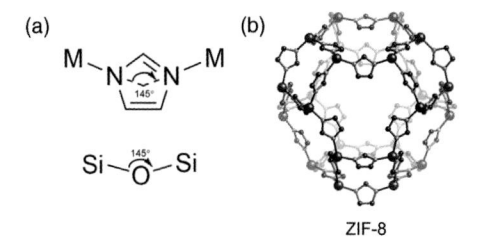

Figure 8.7 (a) The M–N–M angle in ZIFs compared to the Si—O—Si angle in zeolites. (b) Structure of ZIF-8. Zn: purple; N: light blue; C: black.

MAFs contain pyrazolate, triazolate, or tetrazolate-based linkers [103]. Similar to ZIFs, the stability of MAFs arises primarily from the strength of the metal–nitrogen (M—N) bond, which in general correlates with the basicity of the azole linker [103]. The higher the basicity of the azole linker, the stronger the M—N bond and the more robust the framework. Frameworks comprised of pyrazole ($pK_a = 19.8$) linkers have thus been shown to be more stable than those comprised of 1,2,3-triazole ($pK_a = 13.9$) linkers, which in turn are more stable than analogues comprised of tetrazole ($pK_a = 4.9$) linkers [104]. Taking advantage of this concept, Long and coworkers reported a very robust Ni-azolate framework using 1,3,5-tris(1H-pyrazol-4-yl)benzene (H_3BTP). $Ni_3(BTP)_2$ was shown to be stable upon heating in air up to 430 °C and also toward exposure to boiling solutions ranging from pH 2–14 for 2 weeks [104].

8.4.2
MOFs Containing Metal–Carboxylate Bonds

MILs (Materials Institute Lavoisier), rare earth MOFs and Zr_6-based MOFs are all linked together via metal–carboxylate bonds and are exceptionally stable. The stability of these families of MOFs can be traced to the strength of metal-carboxylate bonds which can be correlated to the charge density of the metal ion involved, meaning that smaller higher valent metals tend to form very strong bonds with carboxylate-terminated linkers (i.e., hard acids form strong bonds with hard bases). MILs are primarily comprised of trivalent metal ions such as Cr^{III}, Fe^{III}, Al^{III}, V^{III}, Ga^{III}, and In^{III} (Figure 8.8, some examples) [105]. One example from the MIL-family is MIL-125-NH_2, which is comprised of tetravalent Ti^{IV} metal nodes and 2-amino-terephthalate-based linkers. MIL-125-NH_2 is stable to 300 °C in air, 100 °C in the presence of water vapor, and is stable toward the corrosive gas, H_2S, at 30 °C [106]. A number of rare earth MOFs, comprised of trivalent metal ions such as Eu^{III}, Tb^{III}, and Y^{III} and carboxylic acid linkers, have also been found to be very stable [107]. A Eu-based MOF containing 1,4-naphthalenedicarboxylate linkers is stable in boiling water for 24 h and in aqueous solutions from pH 3.5–10 at room temperature. The stability of this MOF is not only attributed to the strong Eu^{III}—O bonds in the framework

MIL-101 (Cr) **MIL-53(Al)** **MIL-125(Ti)**

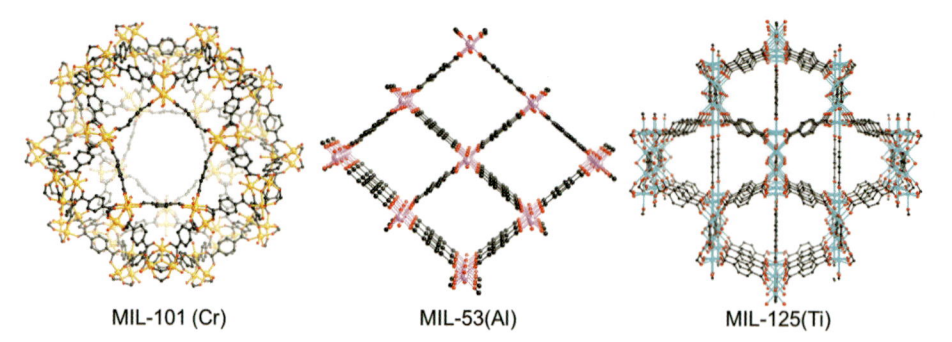

Figure 8.8 Structure of MIL-101(Cr), MIL-53(Al), and MIL-125(Ti). O: red; C: black; Cr: yellow; Al: light pink; Ti: teal.

but also to the use of short, hydrophobic linkers which physically shield the Eu^{III}-cluster from water [107a].

Zr_6-based MOFs are those comprised of Zr_6 metal nodes with the general core formula $Zr_6(\mu_3\text{-}O)_x(\mu_3\text{-}OH)_y$ and carboxylate-based linkers [108]. Examples include MOFs from the UiO-series [109] (University of Oslo), PCN-series [108] (porous coordination network), DUT-series [110] (Dresden University of Technology), NU-series [37] (Northwestern University), and MOF-series [1] (Figure 8.9, examples of each). Similar to the MOFs discussed above, the stability of Zr_6-MOFs arises primarily from the strong metal-ligand bonds (Zr^{IV}-O) in these frameworks. The first Zr_6-based MOF, UiO-66, was reported in 2008 [109]

NU-1000 **UiO-66**

MOF-841

PCN-225 **DUT-67**

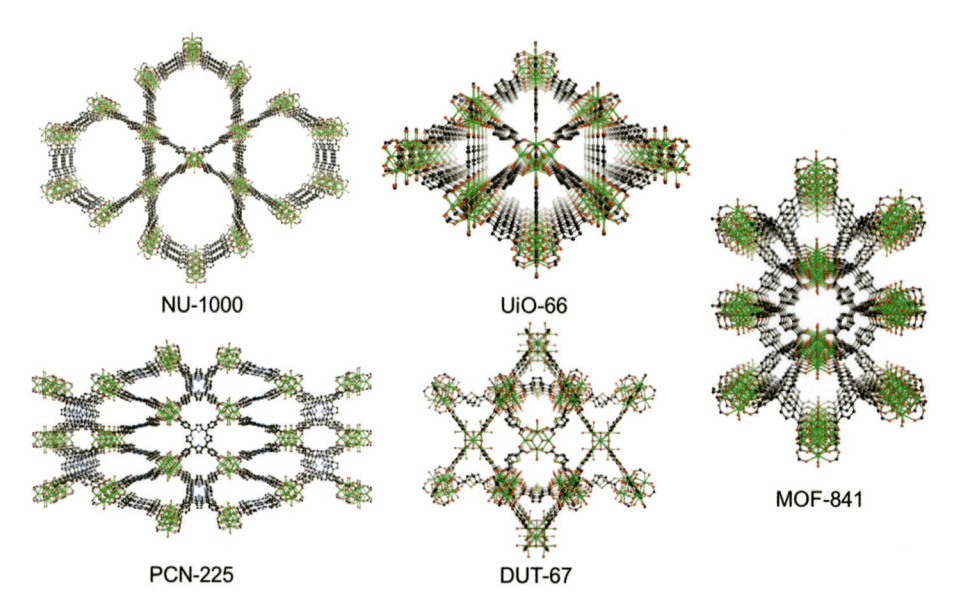

Figure 8.9 Structure of Zr-based MOFs NU-1000, UiO-66, PCN-225, DUT-67, and MOF-841. Zr: green; O: red; C: black; N: light blue; S: dark yellow.

and it has since been shown to be stable in strong acid (HCl), as well as heating in air up to 375 °C, in boiling water [111] and toward 10 tons/cm^2 of external pressure [98].

Although strong bonding is desirable for obtaining robust MOFs, it can also make the synthesis of MOFs rather challenging. This is because metals and linkers that form increasingly strong bonds also tend to form amorphous materials that rapidly precipitate. Significant effort has been put forth into precisely controlling the synthetic conditions (e.g., with the use of competing modulators that presumably slow the nucleation and growth of these materials) of MOFs with strong bonds. Moving forward in the quest for increasingly stable MOFs, the synthesis of frameworks with even stronger metal-ligand bonds will continue to be a challenge. The selection of strong metal-ligand bonds with adequate lability for structure propagation and/or the judicious choice of modulators will continue to be of importance in this regard.

8.5
Select Potential Applications of MOFs

8.5.1
Nerve Agent Degradation

A recent area of active research has been the hydrolysis of organophosphrous compounds (phosphate and phosphonate esters) found in chemical warfare agents (CWAs) as well as pesticides (often used as simulants to CWAs in academic laboratories due to safety concerns) [112]. These compounds are among the most toxic chemicals known to mankind and strategies are needed for bulk chemical destruction, filtration from airborne agents, and protection of clothing and military vehicles among others [113]. MOFs may play a unique role in addressing some or all of these aspects of chemical threat protection [114]. In an early demonstration, Wang *et al.* utilized dialkylaminopyridine-modified derivatives of the Cr-based MOF MIL-101 to hydrolyze the nerve agent simulant diethyl 4-nitrophenylphosphate, albeit with rather long half-lives (5 h). Katz *et al.* subsequently demonstrated that the Zr_6-based MOF UiO-66 could selectively hydrolyze dimethyl 4-nitrophenylphosphate (DMNP) with moderate half-lives (45 min) [115]. The Zr_6-based nodes of UiO-66 have a Zr—OH—Zr moiety that is similar to the active site in the enzyme phosphotriesterase that hydrolyzes phosphonate ester bonds in nature. Unfortunately, the small apertures in UiO-66 prohibit diffusion of the bulky organophosphorous substrates to the Zr_6-based nodes throughout the framework.

To promote diffusion of the DMNP simulant throughout the entire framework, Mondloch *et al.* turned to the mesoporous MOF NU-1000 (30 Å channels). Half-lives of 15 min (hydrated node) were observed suggesting that access to the entire framework enhances hydrolysis [116]. Computational efforts demonstrated that hydrolysis of DMNP was thermodynamically downhill by

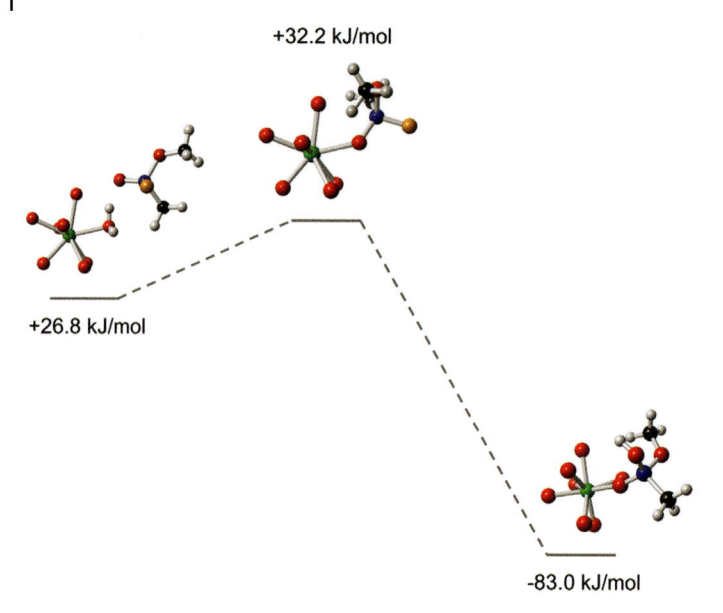

Figure 8.10 The thermodynamics of GD simulant hydrolysis in NU-1000. Zr: green; O: red; P: blue; C: black; and H: gray.

48.1 kJ/mol. The reaction likely proceeds through a ligand exchange reaction at Zr^{IV} centers (on the node) followed by Lewis acid activation of the substrate. To support the proposed mechanism, NU-1000 was thermally treated to remove water and hydroxide ligands from the node of NU-1000 (NU-1000-dehyd). Removal of the water and hydroxide ligands from the node promoted hydrolysis, dropping the half-life of DMNP to 1.5 min. Computational screening also suggested that hydrolysis of CWAs such as O-pinacolyl methylphosphonofluoridate (GD, also known as soman) and O-ethyl S-[2-(diisopropylamino)ethyl] methylphosphonothioate (VX) were also thermodynamically favorable, Figure 8.10. Indeed, NU-1000 selectively hydrolyzes GD with a half-life of 3 min, 80-fold faster than the Cu-based MOF HKUST-1 [117]. Subsequently, it has been found that the node (and therefore framework) topology [118], linker [119], and MOF crystallite size [120] can all have significant effects on the hydrolysis rates of DMNP. In addition, these Zr_6-based catalysts have been effective for destroying VX [121] and creating textile-framework composites that exhibit self-detoxifying properties [122].

8.5.2
Gas-Phase Catalysis with Alkenes

Catalysis with MOFs has been explored and reviewed extensively with particular interest paid to the potential of MOFs as uniform periodic supports for single-site catalysis in condensed phase media [36,45,54,58,62,5a]. Gas-phase catalytic reactions with heterogeneous MOF catalysts, however, is a nascent area within

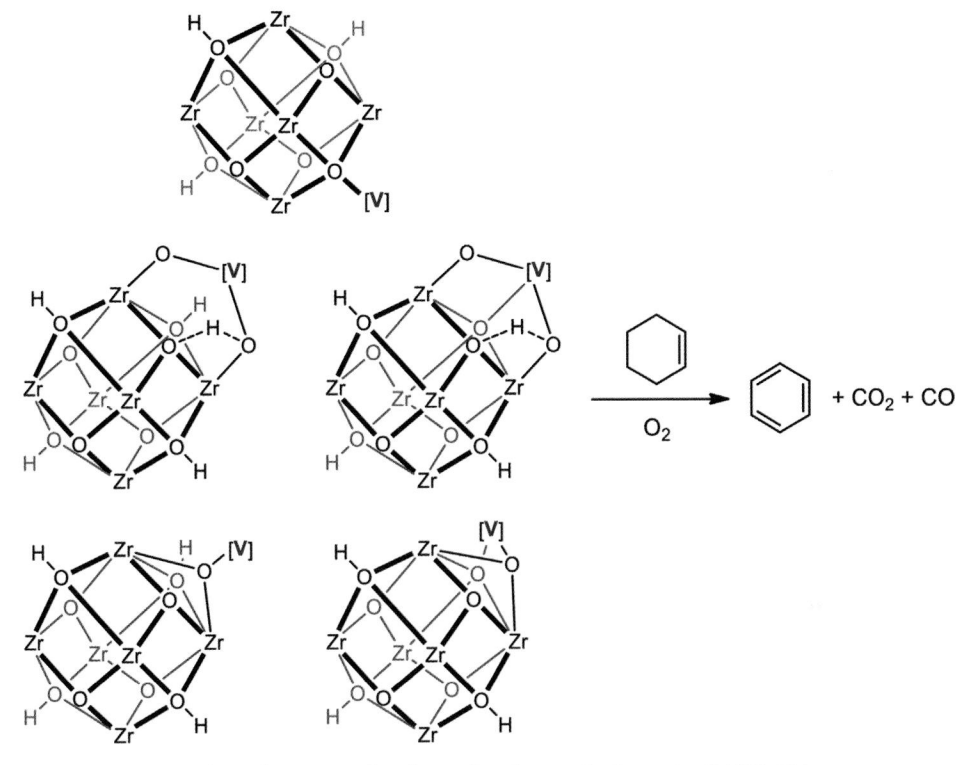

Figure 8.11 Proposed coordination modes of vanadium ions to the Zr_6-node of V-UiO-66 (carboxylate linkers have been omitted for clarity) and oxidative dehydrogenation of cyclohexene.

this field. Nguyen *et al.* have demonstrated metallation of the Zr_6-based MOF UiO-66 ($Zr_6(\mu_3\text{-}O)_4(\mu_3\text{-}OH)_4(bdc)_6$) using vanadyl acetylacetonate ($O=V(acac)_2$) in methanol to yield V-UiO-66. V-UiO-66 contains isolated V^V species immobilized in the MOF through defect node —OH sites (Figure 8.11). The authors employed the new material V-UiO-66 as a precatalyst for the gas phase oxidative dehydrogenation of cyclohexene using O_2 as the oxidant. Under low-conversion conditions (<2%), the MOF catalyst showed excellent selectivity (100 mol%) for the formation of benzene [34].

Yang *et al.* have supported $Ir(C_2H_4)_2$ and/or $Ir(CO)_2$ species on UiO-66 and NU-1000 ($Zr_6(\mu_3\text{-}O)_4(\mu_3\text{-}OH)_4(H_2O)_4(OH)_4(TBAPy)_2$ where TBAPy = 1,3,6,8-tetrakis(*p*-benzoate)pyrene) by reacting pentane solutions of $Ir(CO)_2(acac)$ (acac = acetylacetonate) or $Ir(C_2H_4)_2(acac)$ with each MOF. On the basis of IR spectroscopy, EXAFS, and DFT calculations, the authors propose that Ir coordinates in two ways to defect sites in UiO-66. Based on the dimensions of the Ir precursor compared to the pore apertures, the Ir species are likely on the exterior of the UiO-66 MOF particles. In contrast, coordination of Ir complexes to NU-1000 leads to a single type of chemisorbed species throughout the entire MOF. Upon activation with H_2, the supported $Ir(C_2H_4)_2$ species were determined to be catalysts for ethylene hydrogenation (with minor ethylene dimerization activity also observed) [35b]. In a follow-up study, Yang et al. employed acetic acid or HCl as modulators in the synthesis of UiO-66 or UiO-67 to control the number of defect sites resulting from missing linkers and thereby influence the Ir binding sites [35a]. Gas-phase ethylene hydrogenation [42] and dimerization/oligomerization [45] with metal ions/complexes installed in NU-1000 have also been described.

Kaskel and coworkers have reported incipient wetness impregnation of MIL-101 using Pd($acac$)$_2$. The impregnated material was heated under flowing hydrogen to form dispersed Pd nanoparticles in the MOF, Pd/MIL-101, which was then tested for catalytic hydrogenation activity with a mixture of acetylene, ethylene, and hydrogen at elevated temperatures. Pd/MIL-101 showed superior initial catalytic activity compared to Pd/ZnO and Pd/Al$_2$O$_3$, which the authors attribute to the highly dispersed nature of Pd nanoparticles in Pd/MIL-101 [123].

Long and coworkers have demonstrated oligomerization of propene with Ni$_2$(dobdc) (dobdc = 2,5-dioxodo-1,4-benzenedicarboxylate) and Ni$_2$(dobpdc) drawing an analogy to nickel-exchanged aluminosilicate (zeolite) materials. Ni$_2$(dobdc) and Ni$_2$(dobpdc) were investigated for gas-phase propene oligomerization activity in a fixed bed reactor at 453 K and 5 bar propene. Both materials were found to be >99% selective for oligomerization, with dimers composing >95% of the product mixture. For comparison, Mg$_2$(dobdc) was also tested, but showed no oligomerization activity. Based on this observation, the authors conclude that the coordinatively unsaturated Ni^{2+} sites in Ni$_2$(dobdc) and Ni$_2$(dobpdc) give rise to their catalytic activity [124].

8.5.3
Environmental Pollution Remediation

Environmental pollution is one of the most problematic issues facing our planet today. The efficient capture, removal, and/or degradation of pollutants from our air and water is therefore extremely important for the continued existence of the Earth's diverse flora and fauna [125]. MOFs have been studied for the capture of gases such as NH$_3$, H$_2$S, CO, NO$_x$, and SO$_x$ [112,126] as well as for the removal of water pollutants such as organic solvents, dyes, insecticides/herbicides/pesticides [127], and various metal and metalloid-based species [128]. In a number of studies on the capture of gaseous or aqueous analytes in MOFs, it has been shown that open metal sites (metals that are coordinatively unsaturated or have labile ligands available for substitution) are an important design feature for efficient capture. Open metal sites, which are Lewis acidic in M-MOF-74/CPO-27 (where M = Cu^{II} or Zn^{II}) and HKUST-1, for example, have been shown to coordinate the Lewis base NH$_3$ [129]. The capture of SeO$_4^{2-}$ and SeO$_3^{2-}$ from aqueous solutions using NU-1000 also occurs through interactions between the metal node and analyte, where labile, terminal –OH/H$_2$O ligands on the Zr$_6$-node are displaced by the selenium oxyanions [130]. Analyte interactions with MOF linkers can also be important, such as hydrogen bonding or electrostatic interactions. MIL-125(Ti)-NH$_2$, for example, is shown to capture H$_2$S through hydrogen bonding interactions between the amino group on the linker and H_2S [106]. MOFs can also be used as an ion exchange medium which has been demonstrated by the uptake of $Cr_2O_7^{2-}$ in ZJU-101(Zr) [131], as well as the capture of anionic organic dyes in P-MOF(In) [97]. While for some applications in analyte capture, a "one-time-use" MOF-based material is adequate, in other cases the stability of the candidate MOF toward the analyte of interest, as well as the conditions under which the analyte must be captured, is important to consider. For this reason, the continued design and synthesis of MOFs that are able to withstand harsh conditions (Section 8.4) will continue to drive this promising potential application of MOFs forward in the future.

8.6 Conclusions

In this short overview of MOF chemistry, we have highlighted some of the most important concepts in this rapidly expanding field of solid-state chemistry. The MOF community has devoted significant effort to the synthesis, purification, and activation of these materials. This in turn has given rise to an abundant number of solid-state materials exhibiting a wide range of characteristics that can be readily manipulated including stability, flexibility, porosity, and chemical reactivity. Given these unique characteristics it is our opinion that MOFs will inevitably be harnessed for a handful of practical applications.

References

1 Furukawa, H., Cordova, K.E., O'Keeffe, M., and Yaghi, O.M. (2013) The chemistry and applications of metal-organic frameworks. *Science*, **341** (6149), 1230444.

2 (a) Yaghi, O.M., O'Keefee, M., Ockwig, N.W., Chae, H.K., Eddaoudi, M., and Kim, J. (2003) Reticular synthesis and the design of new materials. *Nature*, **423**, 705–714. (b) Férey, G. (2008) Hybrid porous solids: past, present, future. *Chem. Soc. Rev.*, **37**, 191–214. (c) Horike, S., Shimomura, S. and Kitagawa, S. (2009) Soft porous crystals. *Nat. Chem.*, **1**, 695–704.

3 (a) Sumida, K., Rogow, D.L., Mason, J.A., McDonald, T.M., Bloch, E.D., Herm, Z.R., Bae, T.-H., and Long, J.R. (2012) Carbon dioxide capture in metal-organic frameworks. *Chem. Rev.*, **112** (2), 724–781. (b) Suh, M.P., Park, H.J., Prasad, T.K., and Lim, D.-W. (2012) Hydrogen storage in metal-organic frameworks. *Chem. Rev.*, **112** (2), 782–835.

4 Li, J.-R., Sculley, J., and Zhou, H.-C. (2012) Metal-organic frameworks for separations. *Chem. Rev.*, **112** (2), 869–932.

5 (a) Lee, J., Farha, O.K., Roberts, J., Scheidt, K.A., Nguyen, S.T., and Hupp, J.T. (2009) Metal-organic framework materials as catalysts. *Chem. Soc. Rev.*, **38**, 1450–1459. (b) Corma, A., García, H. and Llabrés i Xamena, F.X. (2010) Engineering metal organic frameworks for heterogeneous catalysis. *Chem. Rev.*, **110** (8), 4606–4655.

6 Kreno, L.E., Leong, K., Farha, O.K., Allendorf, M., Van Duyne, R.P., and Hupp, J.T. (2012) Metal-organic

framework materials as chemical sensors. *Chem. Rev.*, **112** (2), 1105–1125.
7 Sun, L., Campbell, M.G., and Dincă, M. (2016) Electrically conductive porous metal–organic frameworks. *Angew. Chem., Int. Ed.*, **55**, 3566–3579.
8 Hod, I., Deria, P., Bury, W., Mondloch, J.E., Kung, C.-W., So, M., Sampson, M.D., Peters, A.W., Kubiak, C.P., Farha, O.K., and Hupp, J.T. (2015) A porous proton-relaying metal-organic framework material that accelerates electrochemical hydrogen evolution. *Nat. Commun.*, **6**, 8304.
9 So, M.C., Wiederrecht, G.P., Mondloch, J.E., Hupp, J.T., and Farha, O.K. (2015) Metal-organic framework materials for light-harvesting and energy transfer. *Chem. Commun.*, **51** (17), 3501–3510.
10 Darago, L.E., Aubrey, M.L., Yu, C.J., Gonzalez, M.I., and Long, J.R. (2015) Electronic conductivity, ferrimagnetic ordering, and reductive insertion mediated by organic mixed-valence in a ferric semiquinoid metal–organic framework. *J. Am. Chem. Soc.*, **137** (50), 15703–15711.
11 Peplow, M. (2015) Materials science: the hole story. *Nature*, **520**, 148–150.
12 Mason, J.A., Oktawiec, J., Taylor, M.K., Hudson, M.R., Rodriguez, J., Bachman, J.E., Gonzalez, M.I., Cervellino, A., Guagliardi, A., Brown, C.M., Llewellyn, P.L., Masciocchi, N., and Long, J.R. (2015) Methane storage in flexible metal–organic frameworks with intrinsic thermal management. *Nature*, **527** (7578), 357–361.
13 Weston, M.H., Morris, W., Siu, P.W., Hoover, W.J., Cho, D., Richardson, R.K., and Farha, O.K. (2015) Phosphine gas adsorption in a series of metal–organic frameworks. *Inorg. Chem.*, **54** (17), 8162–8164.
14 McDonald, T.M., Mason, J.A., Kong, X., Bloch, E.D., Gygi, D., Dani, A., Crocellà, V., Giordanino, F., Odoh, S.O., Drisdell, W.S., Vlaisavljevich, B., Dzubak, A.L., Poloni, R., Schnell, S.K., Planas, N., Lee, K., Pascal, T., Wan, L.F., Prendergast, D., Neaton, J.B., Smit, B., Kortright, J.B., Gagliardi, L., Bordiga, S., Reimer, J.A., and Long, J.R. (2015) Cooperative insertion of CO_2 in diamine-appended metal–organic frameworks. *Nature*, **519** (7543), 303–308.
15 Metzger, E.D., Brozek, C.K., Comito, R.J., and Dincă, M. (2016) Selective dimerization of ethylene to 1-butene with a porous catalyst. *ACS Cent. Sci*, **2** (3), 148–153.
16 (a) Stock, N. and Biswas, S. (2012) Synthesis of metal–organic frameworks (MOFs): routes to various MOF topologies, morphologies, and composites. *Chem. Rev.*, **112** (2), 933–969. (b) Lee, Y.-R., Kim, J. and Ahn, W.-S. (2013) Synthesis of metal–organic frameworks: a mini review. *Korean J. Chem. Eng.*, **30** (9), 1667–1680. (c) Meek, S.T., Greathouse, J.A. and Allendorf, M.D. (2011) Metal–organic frameworks: a rapidly growing class of versatile nanoporous materials. *Adv. Mater.*, **23** (2), 249–267.
17 Millange, F., El Osta, R., Medina, M.E., and Walton, R.I. (2011) A time-resolved diffraction study of a window of stability in the synthesis of a copper carboxylate metal–organic framework. *CrystEngComm*, **13** (1), 103–108.
18 (a) Hermes, S., Witte, T., Hikov, T., Zacher, D., Bahnmüller, S., Langstein, G., Huber, K., and Fischer, R.A. (2007) Trapping metal–organic framework nanocrystals: an *in situ* time-resolved light scattering study on the crystal growth of MOF-5 in solution. *J. Am. Chem. Soc.*, **129** (17), 5324–5325. (b) Tsuruoka, T., Furukawa, S., Takashima, Y., Yoshida, K., Isoda, S., and Kitagawa, S. (2009) Nanoporous nanorods fabricated by coordination modulation and oriented attachment growth. *Angew. Chem., Int. Ed.*, **48** (26), 4739–4743. (c) Diring, S., Furukawa, S., Takashima, Y., Tsuruoka, T., and Kitagawa, S. (2010) Controlled multiscale synthesis of porous coordination polymer in nano/micro regimes. *Chem. Mater.*, **22** (16), 4531–4538. (d) Cravillon, J., Nayuk, R., Springer, S., Feldhoff, A., Huber, K., and Wiebcke, M. (2011) Controlling zeolitic imidazolate framework nano- and microcrystal formation: insight into crystal growth by time-resolved *in situ*

static light scattering. *Chem. Mater.*, **23** (8), 2130–2141.

19 (a) Schaate, A., Roy, P., Godt, A., Lippke, J., Waltz, F., Wiebcke, M., and Behrens, P. (2011) Modulated synthesis of zr-based metal–organic frameworks: from nano to single crystals. *Chem. Eur. J.*, **17** (24), 6643–6651. (b) Vermoortele, F., Bueken, B., Le Bars, G., Van de Voorde, B., Vandichel, M., Houthoofd, K., Vimont, A., Daturi, M., Waroquier, M., Van Speybroeck, V., Kirschhock, C., and De Vos, D.E. (2013) Synthesis modulation as a tool to increase the catalytic activity of metal–organic frameworks: the unique case of UiO-66 (Zr). *J. Am. Chem. Soc.*, **135** (31), 11465–11468. (c) Wu, H., Chua, Y.S., Krungleviciute, V., Tyagi, M., Chen, P., Yildirim, T., and Zhou, W. (2013) Unusual and highly tunable missing-linker defects in zirconium metal–organic framework UiO-66 and their important effects on gas adsorption. *J. Am. Chem. Soc.*, **135** (28), 10525–10532.

20 (a) Banerjee, R., Phan, A., Wang, B., Knobler, C., Furukawa, H., O'Keeffe, M., and Yaghi, O.M. (2008) High-throughput synthesis of zeolitic imidazolate frameworks and application to CO_2 capture. *Science*, **319** (5865), 939–943. (b) Bauer, S., Serre, C., Devic, T., Horcajada, P., Marrot, J., Férey, G., and Stock, N. (2008) High-throughput assisted rationalization of the formation of metal organic frameworks in the iron (III) aminoterephthalate solvothermal system. *Inorg. Chem.*, **47** (17), 7568–7576. (c) Volkringer, C., Loiseau, T., Guillou, N., Férey, G., Haouas, M., Taulelle, F., Elkaim, E., and Stock, N. (2010) High-throughput aided synthesis of the porous metal–organic framework-type aluminum pyromellitate, MIL-121, with extra carboxylic acid functionalization. *Inorg. Chem.*, **49** (21), 9852–9862.

21 Farha, O.K., Mulfort, K.L., Thorsness, A.M., and Hupp, J.T. (2008) Separating solids: purification of metal-organic framework materials. *J. Am. Chem. Soc.*, **130** (27), 8598–8599.

22 (a) Chen, B., Eddaoudi, M., Hyde, S.T., O'Keeffe, M., and Yaghi, O.M. (2001) Interwoven metal-organic framework on a periodic minimal surface with extra-large pores. *Science*, **291** (5506), 1021–1023. (b) Maji, T.K., Matsuda, R. and Kitagawa, S. (2007) A flexible interpenetrating coordination framework with a bimodal porous functionality. *Nat. Mater.*, **6** (2), 142–148.

23 Farha, O.K., Malliakas, C.D., Kanatzidis, M.G., and Hupp, J.T. (2010) Control over catenation in metal–organic frameworks via rational design of the organic building block. *J. Am. Chem. Soc.*, **132** (3), 950–952.

24 Czaja, A.U., Trukhan, N., and Müller, U. (2009) Industrial applications of metal–organic frameworks. *Chem. Soc. Rev.*, **38** (5), 1284–1293.

25 Al-Kutubi, H., Gascon, J., Sudhölter, E.J.R., and Rassaei, L. (2015) Electrosynthesis of metal–organic frameworks: challenges and opportunities. *ChemElectroChem*, **2** (4), 462–474.

26 (a) Khan, N.A. and Jhung, S.H. (2015) Synthesis of metal–organic frameworks (MOFs) with microwave or ultrasound: rapid reaction, phase-selectivity, and size reduction. *Coord. Chem. Rev.*, **285**, 11–23. (b) Klinowski, J., Almeida Paz, F.A., Silva, P., and Rocha, J. (2011) Microwave-assisted synthesis of metal–organic frameworks. *Dalton Trans.*, **40** (2), 321–330.

27 James, S.L., Adams, C.J., Bolm, C., Braga, D., Collier, P., Friščic, T., Grepioni, F., Harris, K.D.M., Hyett, G., Jones, W., Krebs, A., Mack, J., Maini, L., Orpen, A.G., Parkin, I.P., Shearouse, W.C., Steed, J.W., and Waddell, D.C. (2012) Mechanochemistry: opportunities for new and cleaner synthesis. *Chem. Soc. Rev.*, **41** (1), 413–447.

28 (a) Klimakow, M., Klobes, P., Thünemann, A.F., Rademann, K., and Emmerling, F. (2010) Mechanochemical synthesis of metal–organic frameworks: a fast and facile approach toward quantitative yields and high specific surface areas. *Chem. Mater.*, **22** (18), 5216–5221. (b) Beldon, P.J., Fábián, L.,

Stein, R.S., Thirumurugan, A., Cheetham, A.K., and Friščić, T. (2010) Rapid room-temperature synthesis of zeolitic imidazolate frameworks by using mechanochemistry. *Angew. Chem., Int. Ed.*, **49** (50), 9640–9643. (c) Friščić, T., Reid, D.G., Halasz, I., Stein, R.S., Dinnebier, R.E., and Duer, M.J. (2010) Ion- and liquid-assisted grinding: improved mechanochemical synthesis of metal–organic frameworks reveals salt inclusion and anion templating. *Angew. Chem., Int. Ed.*, **49** (4), 712–715. (d) Uzarevic, K., Wang, T.C., Moon, S.-Y., Fidelli, A.M., Hupp, J.T., Farha, O.K., and Friščic, T. (2016) Mechanochemical and solvent-free assembly of zirconium-based metal–organic frameworks. *Chem. Commun.*, **52** (10), 2133–2136.

29 Willans, C.E., French, S., Anderson, K.M., Barbour, L.J., Gertenbach, J.-A., Lloyd, G.O., Dyer, R.J., Junk, P.C., and Steed, J.W. (2011) Tripodal imidazole frameworks: reversible vapour sorption both with and without significant structural changes. *Dalton Trans.*, **40** (3), 573–582.

30 Crawford, D., Casaban, J., Haydon, R., Giri, N., McNally, T., and James, S.L. (2015) Synthesis by extrusion: continuous, large-scale preparation of MOFs using little or no solvent. *Chem. Sci.*, **6** (3), 1645–1649.

31 Friščić, T. (2012) Supramolecular concepts and new techniques in mechanochemistry: cocrystals, cages, rotaxanes, open metal–organic frameworks. *Chem. Soc. Rev.*, **41** (9), 3493–3510.

32 (a) Zacher, D., Shekhah, O., Wöll, C., and Fischer, R.A. (2009) Thin films of metal–organic frameworks. *Chem. Soc. Rev.*, **38** (5), 1418–1429. (b) Shekhah, O., Liu, J., Fischer, R.A., and Wöll, C. (2011) MOF thin films: existing and future applications. *Chem. Soc. Rev.*, **40** (2), 1081–1106.

33 Larabi, C. and Quadrelli, E.A. (2012) Titration of $Zr_3(\mu\text{-OH})$ hydroxy groups at the cornerstones of bulk MOF UiO-67, $[Zr_6O_4(OH)_4(\text{biphenyldicarboxylate})_6]$, and their reaction with $[\text{AuMe}(PMe_3)]$. *Eur. J. Inorg. Chem.*, **2012** (18), 3014–3022.

34 Nguyen, H.G.T., Schweitzer, N.M., Chang, C.-Y., Drake, T.L., So, M.C., Stair, P.C., Farha, O.K., Hupp, J.T., and Nguyen, S.T. (2014) Vanadium-node-functionalized UiO-66: a thermally stable MOF-supported catalyst for the gas-phase oxidative dehydrogenation of cyclohexene. *ACS Catal.*, **4** (8), 2496–2500.

35 (a) Yang, D., Odoh, S.O., Borycz, J., Wang, T.C., Farha, O.K., Hupp, J.T., Cramer, C.J., Gagliardi, L., and Gates, B.C. (2016) Tuning Zr_6 metal–organic framework (MOF) nodes as catalyst supports: site densities and electron-donor properties influence molecular iridium complexes as ethylene conversion catalysts. *ACS Catal.*, **6** (1), 235–247. (b) Yang, D., Odoh, S.O., Wang, T.C., Farha, O.K., Hupp, J.T., Cramer, C.J., Gagliardi, L., and Gates, B.C. (2015) Metal–organic framework nodes as nearly ideal supports for molecular catalysts: NU-1000- and UiO-66-supported iridium complexes. *J. Am. Chem. Soc.*, **137** (23), 7391–7396.

36 Klet, R.C., Tussupbayev, S., Borycz, J., Gallagher, J.R., Stalzer, M.M., Miller, J.T., Gagliardi, L., Hupp, J.T., Marks, T.J., Cramer, C.J., Delferro, M., and Farha, O.K. (2015) Single-site organozirconium catalyst embedded in a metal–organic framework. *J. Am. Chem. Soc.*, **137** (50), 15680–15683.

37 Mondloch, J.E., Bury, W., Fairen-Jimenez, D., Kwon, S., DeMarco, E.J., Weston, M.H., Sarjeant, A.A., Nguyen, S.T., Stair, P.C., Snurr, R.Q., Farha, O.K., and Hupp, J.T. (2013) Vapor-phase metalation by atomic layer deposition in a metal–organic framework. *J. Am. Chem. Soc.*, **135** (28), 10294–10297.

38 George, S.M. (2010) Atomic layer deposition: an overview. *Chem. Rev.*, **110** (1), 111–131.

39 Kim, I.S., Borycz, J., Platero-Prats, A.E., Tussupbayev, S., Wang, T.C., Farha, O.K., Hupp, J.T., Gagliardi, L., Chapman, K.W., Cramer, C.J., and Martinson, A.B.F. (2015) Targeted single-site MOF node modification: trivalent metal loading via atomic layer deposition. *Chem. Mater.*, **27** (13), 4772–4778.

40 Peters, A.W., Li, Z., Farha, O.K., and Hupp, J.T. (2015) Atomically precise growth of catalytically active cobalt sulfide on flat surfaces and within a metal–organic framework via atomic layer deposition. *ACS Nano*, **9** (8), 8484–8490.

41 Kung, C.-W., Mondloch, J.E., Wang, T.C., Bury, W., Hoffeditz, W., Klahr, B.M., Klet, R.C., Pellin, M.J., Farha, O.K., and Hupp, J.T. (2015) Metal–organic framework thin films as platforms for atomic layer deposition of cobalt ions to enable electrocatalytic water oxidation. *ACS Appl. Mater. Interfaces*, **7** (51), 28223–28230.

42 Li, Z., Schweitzer, N.M., League, A.B., Bernales, V., Peters, A.W., Getsoian, A., Wang, T.C., Miller, J.T., Vjunov, A., Fulton, J.L., Lercher, J.A., Cramer, C.J., Gagliardi, L., Hupp, J.T., and Farha, O.K. (2016) Sintering-resistant single-site nickel catalyst supported by metal–organic framework. *J. Am. Chem. Soc.*, **138** (6), 1977–1982.

43 Meilikhov, M., Yusenko, K., Esken, D., Turner, S., Van Tendeloo, G., and Fischer, R.A. (2010) Metals@MOFs – loading MOFs with metal nanoparticles for hybrid functions. *Eur. J. Inorg. Chem.*, **2010** (24), 3701–3714.

44 Deria, P., Mondloch, J.E., Tylianakis, E., Ghosh, P., Bury, W., Snurr, R.Q., Hupp, J.T., and Farha, O.K. (2013) Perfluoroalkane functionalization of NU-1000 via solvent-assisted ligand incorporation: synthesis and CO_2 adsorption studies. *J. Am. Chem. Soc.*, **135** (45), 16801–16804.

45 Madrahimov, S.T., Gallagher, J.R., Zhang, G., Meinhart, Z., Garibay, S.J., Delferro, M., Miller, J.T., Farha, O.K., Hupp, J.T., and Nguyen, S.T. (2015) Gas-phase dimerization of ethylene under mild conditions catalyzed by MOF materials containing (bpy)Ni^{II} complexes. *ACS Catal.*, **5** (11), 6713–6718.

46 Deria, P., Bury, W., Hupp, J.T., and Farha, O.K. (2014) Versatile functionalization of the NU-1000 platform by solvent-assisted ligand incorporation. *Chem. Commun.*, **50** (16), 1965–1968.

47 Deria, P., Bury, W., Hod, I., Kung, C.-W., Karagiaridi, O., Hupp, J.T., and Farha, O.K. (2015) MOF Functionalization via solvent-assisted ligand incorporation: phosphonates vs carboxylates. *Inorg. Chem.*, **54** (5), 2185–2192.

48 Hwang, Y.K., Hong, D.-Y., Chang, J.-S., Jhung, S.H., Seo, Y.-K., Kim, J., Vimont, A., Daturi, M., Serre, C., and Férey, G. (2008) Amine grafting on coordinatively unsaturated metal centers of MOFs: consequences for catalysis and metal encapsulation. *Angew. Chem., Int. Ed.*, **47** (22), 4144–4148.

49 Farha, O.K., Mulfort, K.L., and Hupp, J.T. (2008) An example of node-based postassembly elaboration of a hydrogen-sorbing, metal–organic framework material. *Inorg. Chem.*, **47** (22), 10223–10225.

50 (a) Demessence, A., D'Alessandro, D.M., Foo, M.L., and Long, J.R. (2009) Strong CO_2 binding in a water-stable, triazolate-bridged metal–organic framework functionalized with ethylenediamine. *J. Am. Chem. Soc.*, **131** (25), 8784–8786. (b) McDonald, T.M., D'Alessandro, D.M., Krishna, R., and Long, J.R. (2011) Enhanced carbon dioxide capture upon incorporation of N,N′-dimethylethylenediamine in the metal-organic framework CuBTTri. *Chem. Sci.*, **2** (10), 2022–2028.

51 McDonald, T.M., Lee, W.R., Mason, J.A., Wiers, B.M., Hong, C.S., and Long, J.R. (2012) Capture of carbon dioxide from air and flue gas in the alkylamine-appended metal–organic framework mmen-Mg$_2$(dobpdc). *J. Am. Chem. Soc.*, **134** (16), 7056–7065.

52 Planas, N., Dzubak, A.L., Poloni, R., Lin, L.-C., McManus, A., McDonald, T.M., Neaton, J.B., Long, J.R., Smit, B., and Gagliardi, L. (2013) The mechanism of carbon dioxide adsorption in an alkylamine-functionalized metal–organic framework. *J. Am. Chem. Soc.*, **135** (20), 7402–7405.

53 (a) Wang, Z. and Cohen, S.M. (2009) Postsynthetic modification of metal–organic frameworks. *Chem. Soc. Rev.*, **38** (5), 1315–1329. (b) Tanabe, K.K. and Cohen, S.M. (2011) Postsynthetic modification of metal–organic frameworks-a progress report. *Chem. Soc.*

Rev., **40** (2), 498–519. (c) Cohen, S.M. (2012) Postsynthetic methods for the functionalization of metal–organic frameworks. *Chem. Rev.*, **112** (2), 970–1000. (d) Evans, J.D., Sumby, C.J. and Doonan, C.J. (2014) Post-synthetic metalation of metal-organic frameworks. *Chem. Soc. Rev.*, **43** (16), 5933–5951.

54 Wu, C.-D., Hu, A., Zhang, L., and Lin, W. (2005) A homochiral porous metal–organic framework for highly enantioselective heterogeneous asymmetric catalysis. *J. Am. Chem. Soc.*, **127** (25), 8940–8941.

55 Shultz, A.M., Sarjeant, A.A., Farha, O.K., Hupp, J.T., and Nguyen, S.T. (2011) Post-synthesis modification of a metal–organic framework to form metallosalen-containing MOF materials. *J. Am. Chem. Soc.*, **133** (34), 13252–13255.

56 (a) Bloch, E.D., Britt, D., Lee, C., Doonan, C.J., Uribe-Romo, F.J., Furukawa, H., Long, J.R., and Yaghi, O.M. (2010) Metal insertion in a microporous metal–organic framework lined with 2,2′-bipyridine. *J. Am. Chem. Soc.*, **132** (41), 14382–14384. (b) Jacobs, T., Clowes, R., Cooper, A.I., and Hardie, M.J. (2012) A chiral, self-catenating and porous metal–organic framework and its post-synthetic metal uptake. *Angew. Chem., Int. Ed.*, **51** (21), 5192–5195. (c) Manna, K., Zhang, T. and Lin, W. (2014) Postsynthetic metalation of bipyridyl-containing metal–organic frameworks for highly efficient catalytic organic transformations. *J. Am. Chem. Soc.*, **136** (18), 6566–6569. (d) Fei, H. and Cohen, S.M. (2014) A robust, catalytic metal–organic framework with open 2,2[prime or minute]-bipyridine sites. *Chem. Commun.*, **50** (37), 4810–4812. (e) Toyao, T., Miyahara, K., Fujiwaki, M., Kim, T.-H., Dohshi, S., Horiuchi, Y., and Matsuoka, M. (2015) Immobilization of Cu complex into Zr-based MOF with bipyridine units for heterogeneous selective oxidation. *J. Phys. Chem. C*, **119** (15), 8131–8137. (f) Gonzalez, M.I., Bloch, E.D., Mason, J.A., Teat, S.J., and Long, J.R. (2015) Single-crystal-to-single-crystal metalation of a metal–organic framework: a route toward structurally well-defined catalysts. *Inorg. Chem.*, **54** (6), 2995–3005.

57 (a) Morris, W., Volosskiy, B., Demir, S., Gándara, F., McGrier, P.L., Furukawa, H., Cascio, D., Stoddart, J.F., and Yaghi, O.M. (2012) Synthesis, structure, and metalation of two new highly porous zirconium metal–organic frameworks. *Inorg. Chem.*, **51** (12), 6443–6445. (b) Kung, C.-W., Chang, T.-H., Chou, L.-Y., Hupp, J.T., Farha, O.K., and Ho, K.-C. (2015) Post metalation of solvothermally grown electroactive porphyrin metal–organic framework thin films. *Chem. Commun.*, **51** (12), 2414–2417.

58 Falkowski, J.M., Sawano, T., Zhang, T., Tsun, G., Chen, Y., Lockard, J.V., and Lin, W. (2014) Privileged phosphine-based metal–organic frameworks for broad-scope asymmetric catalysis. *J. Am. Chem. Soc.*, **136** (14), 5213–5216.

59 Fang, Q.-R., Yuan, D.-Q., Sculley, J., Li, J.-R., Han, Z.-B., and Zhou, H.-C. (2010) Functional mesoporous metal–organic frameworks for the capture of heavy metal ions and size-selective catalysis. *Inorg. Chem.*, **49** (24), 11637–11642.

60 Yee, K.-K., Reimer, N., Liu, J., Cheng, S.-Y., Yiu, S.-M., Weber, J., Stock, N., and Xu, Z. (2013) Effective mercury sorption by thiol-laced metal–organic frameworks: in strong acid and the vapor phase. *J. Am. Chem. Soc.*, **135** (21), 7795–7798.

61 (a) Wang, Z. and Cohen, S.M. (2007) Postsynthetic covalent modification of a neutral metal–organic framework. *J. Am. Chem. Soc.*, **129** (41), 12368–12369. (b) Wang, Z. and Cohen, S.M. (2008) Tandem modification of metal–organic frameworks by a postsynthetic approach. *Angew. Chem., Int. Ed.*, **47** (25), 4699–4702.

62 Ingleson, M.J., Perez Barrio, J., Guilbaud, J.-B., Khimyak, Y.Z., and Rosseinsky, M.J. (2008) Framework functionalisation triggers metal complex binding. *Chem. Commun.*, (23), 2680–2682.

63 (a) Goto, Y., Sato, H., Shinkai, S., and Sada, K. (2008) "Clickable" metal–organic framework. *J. Am. Chem. Soc.*, **130** (44), 14354–14355. (b) Jiang, H.-L., Feng, D., Liu, T.-F., Li, J.-R., and Zhou, H.-C. (2012) Pore surface

engineering with controlled loadings of functional groups via click chemistry in highly stable metal–organic frameworks. *J. Am. Chem. Soc.*, **134** (36), 14690–14693.

64 (a) Gadzikwa, T., Farha, O.K., Malliakas, C.D., Kanatzidis, M.G., Hupp, J.T., and Nguyen, S.T. (2009) Selective bifunctional modification of a non-catenated metal–organic framework material via "Click" chemistry. *J. Am. Chem. Soc.*, **131** (38), 13613–13615. (b) Li, B., Gui, B., Hu, G., Yuan, D., and Wang, C. (2015) Postsynthetic modification of an alkyne-tagged zirconium metal–organic framework via a "Click" reaction. *Inorg. Chem.*, **54** (11), 5139–5141.

65 (a) Burrows, A.D., Frost, C.G., Mahon, M.F., and Richardson, C. (2009) Sulfur-tagged metal-organic frameworks and their post-synthetic oxidation. *Chem. Commun.* (28), 4218–4220. (b) Phang, W.J., Jo, H., Lee, W.R., Song, J.H., Yoo, K., Kim, B., and Hong, C.S. (2015) Superprotonic conductivity of a UiO-66 framework functionalized with sulfonic acid groups by facile postsynthetic oxidation. *Angew. Chem., Int. Ed.*, **54** (17), 5142–5146.

66 Jones, S.C. and Bauer, C.A. (2009) Diastereoselective heterogeneous bromination of stilbene in a porous metal–organic framework. *J. Am. Chem. Soc.*, **131** (35), 12516–12517.

67 Marshall, R.J., Griffin, S.L., Wilson, C., and Forgan, R.S. (2015) Single-crystal to single-crystal mechanical contraction of metal–organic frameworks through stereoselective postsynthetic bromination. *J. Am. Chem. Soc.*, **137** (30), 9527–9530.

68 (a) Zhang, Z., Nguyen, H.T.H., Miller, S.A., and Cohen, S.M. (2015) polyMOFs: a class of interconvertible polymer-metal-organic-framework hybrid materials. *Angew. Chem., Int. Ed.*, **54** (21), 6152–6157. (b) Zhang, Y., Feng, X., Li, H., Chen, Y., Zhao, J., Wang, S., Wang, L., and Wang, B. (2015) Photoinduced postsynthetic polymerization of a metal–organic framework toward a flexible stand-alone membrane. *Angew. Chem., Int. Ed.*, **54** (14), 4333–4337.

69 Deria, P., Mondloch, J.E., Karagiaridi, O., Bury, W., Hupp, J.T., and Farha, O.K. (2014) Beyond post-synthesis modification: evolution of metal–organic frameworks via building block replacement. *Chem. Soc. Rev.*, **43** (16), 5896–5912.

70 (a) Lalonde, M., Bury, W., Karagiaridi, O., Brown, Z., Hupp, J.T., and Farha, O.K. (2013) Transmetalation: routes to metal exchange within metal–organic frameworks. *J. Mater. Chem. A*, **1** (18), 5453–5468. (b) Brozek, C.K. and Dinca, M. (2014) Cation exchange at the secondary building units of metal-organic frameworks. *Chem. Soc. Rev.*, **43** (16), 5456–5467.

71 (a) Karagiaridi, O., Bury, W., Mondloch, J.E., Hupp, J.T., and Farha, O.K. (2014) Solvent-assisted linker exchange: an alternative to the *de novo* synthesis of unattainable metal-organic frameworks. *Angew. Chem., Int. Ed.*, **53**, 4530–4540. (b) Cohen, S.M. (2012) Postsynthetic methods for the functionalization of metal–organic frameworks. *Chem. Rev.*, **112** (2), 970–1000.

72 Brozek, C.K. and Dincă, M. (2012) Lattice-imposed geometry in metal-organic frameworks: lacunary Zn4O clusters in MOF-5 serve as tripodal chelating ligands for Ni^{2+}. *Chem. Sci.*, **3** (6), 2110–2113.

73 Brozek, C.K., Michaelis, V.K., Ong, T.-C., Bellarosa, L., López, N., Griffin, R.G., and Dincă, M. (2015) Dynamic DMF binding in MOF-5 enables the formation of metastable cobalt-substituted MOF-5 analogues. *ACS Cent. Sci.*, **1** (5), 252–260.

74 Brozek, C.K. and Dincă, M. (2013) Ti^{3+}-, $V^{2+/3+}$-, $Cr^{2+/3+}$-, Mn^{2+}-, and Fe^{2+}-substituted MOF-5 and redox reactivity in Cr- and Fe-MOF-5. *J. Am. Chem. Soc.*, **135** (34), 12886–12891.

75 Dincă, M. and Long, J.R. (2007) High-enthalpy hydrogen adsorption in cation-exchanged variants of the microporous metal–organic framework $Mn_3[(Mn_4Cl)_3(BTT)_8(CH_3OH)_{10}]_2$. *J. Am. Chem. Soc.*, **129** (36), 11172–11176.

76 Karagiaridi, O., Bury, W., Fairen-Jimenez, D., Wilmer, C.E., Sarjeant, A.A., Hupp,

J.T., and Farha, O.K. (2014) Enhanced gas sorption properties and unique behavior toward liquid water in a pillared-paddlewheel metal–organic framework transmetalated with Ni(II). *Inorg. Chem.*, **53** (19), 10432–10436.

77 Burnett, B.J., Barron, P.M., Hu, C., and Choe, W. (2011) Stepwise synthesis of metal–organic frameworks: replacement of structural organic linkers. *J. Am. Chem. Soc.*, **133** (26), 9984–9987.

78 Karagiaridi, O., Bury, W., Tylianakis, E., Sarjeant, A.A., Hupp, J.T., and Farha, O.K. (2013) Opening metal–organic frameworks vol. 2: inserting longer pillars into pillared-paddlewheel structures through solvent-assisted linker exchange. *Chem. Mater.*, **25** (17), 3499–3503.

79 Li, T., Kozlowski, M.T., Doud, E.A., Blakely, M.N., and Rosi, N.L. (2013) Stepwise ligand exchange for the preparation of a family of mesoporous MOFs. *J. Am. Chem. Soc.*, **135** (32), 11688–11691.

80 Karagiaridi, O., Lalonde, M.B., Bury, W., Sarjeant, A.A., Farha, O.K., and Hupp, J.T. (2012) Opening ZIF-8: a catalytically active zeolitic imidazolate framework of sodalite topology with unsubstituted linkers. *J. Am. Chem. Soc.*, **134** (45), 18790–18796.

81 Kim, M., Cahill, J.F., Su, Y., Prather, K.A., and Cohen, S.M. (2012) Postsynthetic ligand exchange as a route to functionalization of 'inert' metal-organic frameworks. *Chem. Sci.*, **3** (1), 126–130.

82 Pullen, S., Fei, H., Orthaber, A., Cohen, S.M., and Ott, S. (2013) Enhanced photochemical hydrogen production by a molecular diiron catalyst incorporated into a metal–organic framework. *J. Am. Chem. Soc.*, **135** (45), 16997–17003.

83 Kim, S., Dawson, K.W., Gelfand, B.S., Taylor, J.M., and Shimizu, G.K.H. (2013) Enhancing proton conduction in a metal–organic framework by isomorphous ligand replacement. *J. Am. Chem. Soc.*, **135** (3), 963–966.

84 Karagiaridi, O., Vermeulen, N.A., Klet, R.C., Wang, T.C., Moghadam, P.Z., Al-Juaid, S.S., Stoddart, J.F., Hupp, J.T., and Farha, O.K. (2015) Functionalized defects through solvent-assisted linker exchange: synthesis, characterization, and partial postsynthesis elaboration of a metal–organic framework containing free carboxylic acid moieties. *Inorg. Chem.*, **54** (4), 1785–1790.

85 So, M.C., Beyzavi, M.H., Sawhney, R., Shekhah, O., Eddaoudi, M., Al-Juaid, S.S., Hupp, J.T., and Farha, O.K. (2015) Post-assembly transformations of porphyrin-containing metal–organic framework (MOF) films fabricated via automated layer-by-layer coordination. *Chem. Commun.*, **51** (1), 85–88.

86 Denny, M.S. Jr. and Cohen, S.M. (2015) In situ modification of metal–organic frameworks in mixed-matrix membranes. *Angew. Chem., Int. Ed.*, **54**, 9029–9032.

87 Mondloch, J.E., Karagiaridi, O., Farha, O.K., and Hupp, J.T. (2013) Activation of metal–organic framework materials. *CrystEngComm*, **15** (45), 9258–9264.

88 Mondloch, J.E., Katz, M.J., Planas, N., Semrouni, D., Gagliardi, L., Hupp, J.T., and Farha, O.K. (2014) Are Zr_6-based MOFs water stable? Linker hydrolysis vs. capillary-force-driven channel collapse. *Chem. Commun.*, **50** (64), 8944–8946.

89 Li, H., Eddaoudi, M., O'Keeffe, M., and Yaghi, O.M. (1999) Design and synthesis of an exceptionally stable and highly porous metal-organic framework. *Nature*, **402** (6), 276–279.

90 Nelson, A.P., Farha, O.K., Mulfort, K.L., and Hupp, J.T. (2009) Supercritical processing as a route to high internal surface areas and permanent microporosity in metal–organic framework materials. *J. Am. Chem. Soc.*, **131** (2), 458–460.

91 Ma, L., Jin, A., Xie, Z., and Lin, W. (2009) Freeze drying significantly increases permanent porosity and hydrogen uptake in 4,4-connected metal–organic frameworks. *Angew. Chem., Int. Ed.*, **48** (52), 9905–9908.

92 McDonald, T.M., Bloch, E.D., and Long, J.R. (2015) Rapidly assessing the activation conditions and porosity of metal-organic frameworks using thermogravimetric analysis. *Chem. Commun.*, **51** (24), 4985–4988.

93 Howarth, A.J., Liu, Y., Li, P., Li, Z., Wang, T.C., Hupp, J.T., and Farha, O.K. (2016)

Chemical, thermal and mechanical stabilities of metal–organic frameworks. *Nat. Rev. Mater.*, **1**, 15018.

94 Burtch, N.C., Jasuja, H., and Walton, K.S. (2014) Water stability and adsorption in metal–organic frameworks. *Chem. Rev.*, **114** (20), 10575–10612.

95 Li, J.-R., Kuppler, R.J., and Zhou, H.-C. (2009) Selective gas adsorption and separation in metal–organic frameworks. *Chem. Soc. Rev.*, **38** (5), 1477–1504.

96 Liu, X., Demir, N.K., Wu, Z., and Li, K. (2015) Highly water-stable zirconium metal–organic framework UiO-66 membranes supported on alumina hollow fibers for desalination. *J. Am. Chem. Soc.*, **137** (22), 6999–7002.

97 Zhao, X., Bu, X., Wu, T., Zheng, S.-T., Wang, L., and Feng, P. (2013) Selective anion exchange with nanogated isoreticular positive metal-organic frameworks. *Nat. Commun.*, **4**, 2344.

98 Wu, H., Yildirim, T., and Zhou, W. (2013) Exceptional mechanical stability of highly porous zirconium metal–organic framework UiO-66 and its important implications. *J. Phys. Chem. Lett.*, **4** (6), 925–930.

99 Horcajada, P., Gref, R., Baati, T., Allan, P.K., Maurin, G., Couvreur, P., Férey, G., Morris, R.E., and Serre, C. (2012) Metal–organic frameworks in biomedicine. *Chem. Rev.*, **112** (2), 1232–1268.

100 Gelfand, B.S. and Shimizu, G.K.H. (2016) Parameterizing and grading hydrolytic stability in metal-organic frameworks. *Dalton Trans*, **45** (9), 3668–3678.

101 Park, K.S., Ni, Z., Côté, A.P., Choi, J.Y., Huang, R., Uribe-Romo, F.J., Chae, H.K., O'Keeffe, M., and Yaghi, O.M. (2006) Exceptional chemical and thermal stability of zeolitic imidazolate frameworks. *Proc. Natl. Acad. Sci. USA*, **103** (27), 10186–10191.

102 Yin, H., Kim, H., Choi, J., and Yip, A.C.K. (2015) Thermal stability of ZIF-8 under oxidative and inert environments: a practical perspective on using ZIF-8 as a catalyst support. *Chem. Eng. J.*, **278**, 293–300.

103 Zhang, J.-P., Zhang, Y.-B., Lin, J.-B., and Chen, X.-M. (2012) Metal azolate frameworks: from crystal engineering to functional materials. *Chem. Rev.*, **112** (2), 1001–1033.

104 Colombo, V., Galli, S., Choi, H.J., Han, G.D., Maspero, A., Palmisano, G., Masciocchi, N., and Long, J.R. (2011) High thermal and chemical stability in pyrazolate-bridged metal-organic frameworks with exposed metal sites. *Chem. Sci.*, **2** (7), 1311–1319.

105 (a) Férey, G., Mellot-Draznieks, C., Serre, C., Millange, F., Dutour, J., Surblé, S., and Margiolaki, I. (2005) A chromium terephthalate-based solid with unusually large pore volumes and surface area. *Science*, **309** (5743), 2040–2042. (b) Loiseau, T., Lecroq, L., Volkringer, C., Marrot, J., Férey, G., Haouas, M., Taulelle, F., Bourrelly, S., Llewellyn, P.L., and Latroche, M. (2006) MIL-96, a porous aluminum trimesate 3D structure constructed from a hexagonal network of 18-membered rings and μ_3-Oxo-centered trinuclear units. *J. Am. Chem. Soc.*, **128** (31), 10223–10230. (c) Surblé, S., Millange, F., Serre, C., Düren, T., Latroche, M., Bourrelly, S., Llewellyn, P.L., and Férey, G. (2006) Synthesis of MIL-102, a chromium carboxylate metal–organic framework, with gas sorption analysis. *J. Am. Chem. Soc.*, **128** (46), 14889–14896.

106 Vaesen, S., Guillerm, V., Yang, Q., Wiersum, A.D., Marszalek, B., Gil, B., Vimont, A., Daturi, M., Devic, T., Llewellyn, P.L., Serre, C., Maurin, G., and De Weireld, G. (2013) A robust amino-functionalized titanium(IV) based MOF for improved separation of acid gases. *Chem. Commun.*, **49** (86), 10082–10084.

107 (a) Xue, D.X., Belmabkhout, Y., Shekhah, O., Jiang, H., Adil, K., Cairns, A.J., and Eddaoudi, M. (2015) Tunable rare earth fcu-MOF platform: access to adsorption kinetics driven gas/vapor separations via pore size contraction. *J. Am. Chem. Soc.*, **137** (15), 5034–5040. (b) Alezi, D., Peedikakkal, A.M., Weselinski, L.J., Guillerm, V., Belmabkhout, Y., Cairns, A.J., Chen, Z., Wojtas, L., and Eddaoudi, M. (2015) Quest for highly connected metal–organic framework platforms: rare-earth polynuclear clusters versatility

meets net topology needs. *J. Am. Chem. Soc.*, **137** (16), 5421–5430.

108 Bai, Y., Dou, Y., Xie, L.-H., Rutledge, W., Li, J.-R., and Zhou, H.-C. (2016) Zr-based metal-organic frameworks: design, synthesis, structure, and applications. *Chem. Soc. Rev.*, **45**, 2327–2367.

109 Cavka, J.H., Jakobsen, S., Olsbye, U., Guillou, N., Lamberti, C., Bordiga, S., and Lillerud, K.P. (2008) A new zirconium inorganic building brick forming metal organic frameworks with exceptional stability. *J. Am. Chem. Soc.*, **130** (42), 13850–13851.

110 Bon, V., Senkovska, I., Weiss, M.S., and Kaskel, S. (2013) Tailoring of network dimensionality and porosity adjustment in Zr- and Hf-based MOFs. *CrystEngComm*, **15** (45), 9572–9577.

111 Valenzano, L., Civalleri, B., Chavan, S., Bordiga, S., Nilsen, M.H., Jakobsen, S., Lillerud, K.P., and Lamberti, C. (2011) Disclosing the complex structure of UiO-66 metal organic framework: a synergic combination of experiment and theory. *Chem. Mater.*, **23** (7), 1700–1718.

112 DeCoste, J.B. and Peterson, G.W. (2014) Metal–organic frameworks for air purification of toxic chemicals. *Chem. Rev.*, **114** (11), 5695–5727.

113 Rosseinsky, M.J., Smith, M.W., and Timperley, C.M. (2015) Metal–organic frameworks: breaking bad chemicals down. *Nat. Mater.*, **14** (5), 469–470.

114 Wang, S., Bromberg, L., Schreuder-Gibson, H., and Hatton, T.A. (2013) Organophophorous ester degradation by chromium(III) terephthalate metal–organic framework (MIL-101) chelated to N,N-dimethylaminopyridine and related aminopyridines. *ACS Appl. Mater. Interfaces*, **5** (4), 1269–1278.

115 Katz, M.J., Mondloch, J.E., Totten, R.K., Park, J.K., Nguyen, S.T., Farha, O.K., and Hupp, J.T. (2013) Simple and compelling biomimetic metal–organic framework catalyst for the degradation of nerve agent simulants. *Angew. Chem., Int. Ed.*, **126** (2), 507–511.

116 Mondloch, J.E., Katz, M.J., Isley, W.C. III, Ghosh, P., Liao, P., Bury, W., Wagner, G.W., Hall, M.G., DeCoste, J.B., Peterson, G.W., Snurr, R.Q., Cramer, C.J., Hupp, J.T., and Farha, O.K. (2015) Destruction of chemical warfare agents using metal–organic frameworks. *Nat. Mater.*, **14** (5), 512–516.

117 Peterson, G.W. and Wagner, G.W. (2013) Detoxification of chemical warfare agents by CuBTC. *J. Porous Mater.*, **21** (2), 121–126.

118 Moon, S.-Y., Liu, Y., Hupp, J.T., and Farha, O.K. (2015) Instantaneous hydrolysis of nerve-agent simulants with a six-connected zirconium-based metal-organic framework. *Angew. Chem., Int. Ed.*, **54**, 6795–6799.

119 Katz, M.J., Moon, S.-Y., Mondloch, J.E., Beyzavi, M.H., Stephenson, C.J., Hupp, J.T., and Farha, O.K. (2015) Exploiting parameter space in MOFs: a 20-fold enhancement of phosphate-ester hydrolysis with UiO-66-NH$_2$. *Chem. Sci.*, **6** (4), 2286–2291.

120 Li, P., Klet, R.C., Moon, S.-Y., Wang, T.C., Deria, P., Peters, A.W., Klahr, B.M., Park, H.-J., Al-Juaid, S.S., Hupp, J.T., and Farha, O.K. (2015) Synthesis of nanocrystals of Zr-based metal–organic frameworks with csq-net: significant enhancement in the degradation of a nerve agent simulant. *Chem. Commun.*, **51** (54), 10925–10928.

121 Moon, S.-Y., Wagner, G.W., Mondloch, J.E., Peterson, G.W., DeCoste, J.B., Hupp, J.T., and Farha, O.K. (2015) Effective, facile, and selective hydrolysis of the chemical warfare agent VX using Zr$_6$-based metal–organic frameworks. *Inorg. Chem.*, **54** (22), 10829–10833.

122 López-Maya, E., Montoro, C., Rodriguez-Albelo, L.M., Aznar Cervantes, S.D., Luis, Lozaono-Peréz, A.A. Ceníc, J., Barea, E., and Navarro, J.A.R. (2015) Textile/metal-organic framework composites as self-detoxifying filters for chemical-warfare agents. *Angew. Chem., Int. Ed.*, **54** (23), 6790–6794.

123 Henschel, A., Gedrich, K., Kraehnert, R., and Kaskel, S. (2008) Catalytic properties of MIL-101. *Chem. Commun.*, (35), 4192–4194.

124 Mlinar, A.N., Keitz, B.K., Gygi, D., Bloch, E.D., Long, J.R., and Bell, A.T. (2014) Selective propene oligomerization with

nickel(II)-based metal–organic frameworks. *ACS Catal.*, **4** (3), 717–721.

125 Middleton, N. (2013) *The Global Casino: An Introduction to Environmental Issues*, 5th edn, Routledge, New York, NY.

126 Barea, E., Montoro, C., and Navarro, J.A.R. (2014) Toxic gas removal – metal-organic frameworks for the capture and degradation of toxic gases and vapours. *Chem. Soc. Rev.*, **43** (16), 5419–5430.

127 Khan, N.A., Hasan, Z., and Jhung, S.H. (2013) Adsorptive removal of hazardous materials using metal-organic frameworks (MOFs): a review. *J. Hazard. Mater.*, **244–245**, 444–456.

128 Howarth, A.J., Liu, Y., Hupp, J.T., and Farha, O.K. (2015) Metal-organic frameworks for applications in remediation of oxyanion/cation-contaminated water. *CrystEngComm*, **17** (38), 7245–7253.

129 (a) Britt, D., Tranchemontagne, D. and Yaghi, O.M. (2008) Metal-organic frameworks with high capacity and selectivity for harmful gases. *Proc. Natl. Acad. Sci. USA*, **105** (33), 11623–11627. (b) Katz, M.J., Howarth, A.J., Moghadam, P.Z., DeCoste, J.B., Snurr, R.Q., Hupp, J.T., and Farha, O.K. (2016) High volumetric uptake of ammonia using Cu-MOF-74/Cu-CPO-27. *Dalton Trans.*, **45** (10), 4150–4153.

130 Howarth, A.J., Katz, M.J., Wang, T.C., Platero-Prats, A.E., Chapman, K.W., Hupp, J.T., and Farha, O.K. (2015) High efficiency adsorption and removal of selenate and selenite from water using metal–organic frameworks. *J. Am. Chem. Soc.*, **137** (23), 7488–7494.

131 Zhang, Q., Yu, J., Cai, J., Zhang, L., Cui, Y., Yang, Y., Chen, B., and Qian, G. (2015) A porous Zr-cluster-based cationic metal-organic framework for highly efficient $Cr_2O_7^{2-}$ removal from water. *Chem. Commun.*, **51** (79), 14732–14734.

9
Sol–Gel Processing of Porous Materials

Kazuki Nakanishi,[1] Kazuyoshi Kanamori,[1] Yasuaki Tokudome,[2] George Hasegawa,[3] and Yang Zhu[1]

[1]Kyoto University, Department of Chemistry, Graduate School of Science, Kitashirakawa, Sakyo-ku, Kyoto 606-8502, Japan
[2]Osaka Prefecture University, Department of Materials Science, Graduate School of Engineering, Sakai, Osaka 599-8531, Japan
[3]Osaka University, The Institute of Scientific and Industrial Research, Mihogaoka, Ibaraki, Osaka 567-0047, Japan

9.1 Introduction

In general sol–gel processing, porous gels obtained after removing the solvents have long been regarded as intermediates to fully sintered pore-free glasses or ceramics. Exactly the same was true also for Vycor® as well as other silica gels based on colloidal processes [1], and all of them were intended just to make the production process of high-melting-temperature materials easier. In most of these materials, the size of controlled continuous pores was smaller than submicrometers, which was desirable for sintering/densification but too small for efficient transport of liquids through the specimen. In 1991, the first paper was published that described the structure control method of macroporous silica through a sol–gel process accompanied by phase separation [2]. The incorporation of a water-soluble polymer into an alkoxysilane-based sol–gel process made it possible to fabricate pure silica gels having well-defined interconnected macropores in the size range of micrometers. This method has been gradually extended, using various kinds of water-soluble polymers, surfactants, or other additives, to siloxane-based organic–inorganic hybrids and metal oxides such as titania, zirconia, and alumina [3]. The starting materials include metal alkoxides, polyoxometallates, water glass, colloidal dispersions, and metal salts. Even fully organic networks such as cross-linked polystyrenes and polyacrylates could be fabricated into well-defined macroporous monoliths based on the same principle as above.

Since most chemically cross-linked rigid (tri- or tetra-functional) networks possess their inherent porosity, the assemblage of them into macroporous framework typically in micrometer range naturally results in the formation of hierarchically porous materials. Depending on the nature of the network, additional reorganization of the inherent micro- or mesoporosity into a better controlled structure is possible. In this chapter, methods to prepare macroporous frameworks using the polymerization-induced phase separation paralleled with the sol–gel transition are described. In combination, several practical ways of tailoring mesopore structure, frequently independent of the preformed macroporous frameworks, are explained. Based on the designed pore structure, several applications of such hierarchically porous monoliths to separation sciences are briefly introduced.

9.2
Background and Concepts

9.2.1
Polymerization-Induced Phase Separation in Oxide Sol–Gels

Starting from systems containing metal alkoxides and appropriate additives, the polymerization-induced phase separation, especially the spinodal decomposition, has been extensively utilized to generate well-defined heterogeneous structures [3]. The variation of precursors and polymerization scheme has been extended to include metal salts and organic monomers. Typical examples of applicable precursors are classified and summarized in Table 9.1. Further details of respective compositions will be described in the following sections. In most cases, the hydrolysis is conducted under acidic conditions where relatively narrow distribution of growing oligomers can be obtained [4].

The chemical compatibility in a system containing at least one kind of polymeric species can be estimated by the thermodynamic treatment known as the Flory–Huggins formulation [37]. The Gibbs free energy change of mixing for binary system can be expressed as

$$\Delta G = -T\Delta S + \Delta H = RT\left\{\left(\frac{\Phi_1}{P_1}\right)\ln\Phi_1 + \left(\frac{\Phi_2}{P_2}\right)\ln\Phi_2 + \chi_{12}\Phi_1\Phi_2\right\}.$$

(9.1)

Here, Φ_i and P_i ($i = 1, 2$) denote the volume fraction and the degree of polymerization of each component, and χ_{12} is the interaction parameter. The former two terms in the bracket express the entropic contribution, and the last term the enthalpic contribution. Since the decrease in absolute values of the negative entropic terms destabilizes the system, it is evident that an increase in the degree of polymerization of either component makes the mixture less compatible. When the sign of ΔG turns from negative to positive, a driving force of phase separation arises. In other words, an initially single-phase solution containing a

Table 9.1 Precursors and their variations of sol–gel systems accompanied by phase separation.

General classification	Precursor type/applicable element	Possible variations	Possible combinations	References
Tetrafunctional alkoxide	Si, Ti, Zr, Al alkoxide	$Si(OMe)_4$, $Si(OEt)_4$, $Ti(Oi\text{-}Pr)_4$, $Ti(On\text{-}Pr)_4$, $Zr(Oi\text{-}Pr)_4$, $Al(Osec\text{-}Bu)_3$	Si–Ti, Si–Zr, Si–Al, Si–colloidal particles	[5–7]
Tri- or difunctional alkoxide	Alkyltrialkoxysilane	$RSi(OR')_3$, $R_2Si(OR')_2$	R = methyl, ethyl, vinyl, allyl, and others; R' = methyl, ethyl	[8,9]
Bridged alkoxide	Bis(trialkoxysilyl) alkane	$(R'O)_3\text{-}Si\text{-}R\text{-}Si\text{-}(OR')_3$	R = methylene, ethylene, propylene, diethylbenzene, phenylene, biphenylene and others; R' = methyl, ethyl	[10–13]
Colloidal dispersion	Silica, titania	Acid or base stabilized		[14]
Polysilicate solution	Water glass (alkaline silicate)	—		[15,16]
Metal salt	Aluminum chloride/nitrate	Vanadium(V), chromium(III), manganese(II), iron(III), cobalt(II), nickel(II), copper(II), zinc(II)	Additions of yttrium for YAG, magnesium for spinel; doping rare earth ions for luminescent properties	[17–29]
Metal salt	Calcium nitrate/phosphoric acid	Zirconium, titanium ,oxysulfate)	Additions of second metals to form NASICON phase	[30–33]
Tetrafunctional organic monomer	Divinylbenzene/acrylamide/1,3-glycerol dimethacrylate			[17,34–36]

polymerizable component becomes less stable with the progress of polymerization reaction, finally resulting in the separation into different phases. Exactly the same occurs if the positive enthalpic contribution increases as the polymerization proceeds. A polycondensation reaction that consumes polar parts of molecules, for example, that between silanol groups mediated in a polar solvent, is a possible case for the substantial change in the enthalpic term during the polymerization.

Reviewing Eq. (9.1), the decrease either in T or ΔS results in the increase of ΔG, that is, the system becomes destabilized against homogeneous mixing. The decrease in T corresponds to ordinary cooling, while that in ΔS to the polymerization, which decreases the degree of freedom among the polymerizing components. The equation implies that decreases in T and ΔS equally contribute to the phase separation of a mixture. The decreases in T and ΔS are respectively termed "physical cooling" and "chemical cooling" (Figure 9.1). Irrespective of the mode of "cooling," once the phase separation is induced, the process of domain formation follows an identical path described in the next section. In the rest of this chapter, the phase separation induced by polymerization is extensively described in relation to the principle of macropore control of monolithic gels. An important difference between physical and chemical cooling is that the former is usually reversible and can be easily controlled artificially, but the latter is often irreversible and only the rate of cooling (polymerization) can be adjusted by the experimental parameters.

Sol–gel systems depicted here undergo a phase separation to generate micrometer-range heterogeneity composed of "gel-phase" and "fluid-phase." After the solidification (gelation) of the whole system, the fluid-phase can be removed relatively easily to leave vacant spaces in the length scale of micrometers

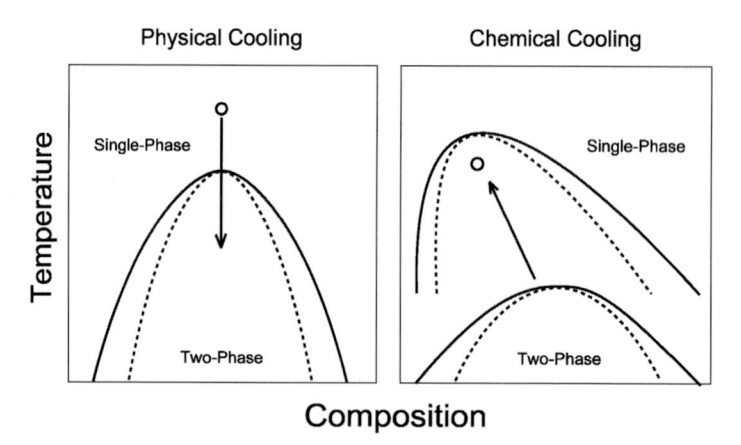

Figure 9.1 Comparison between physical and chemical coolings of the systems with miscibility windows. In the chemical cooling, the composition and temperature of initially homogeneous mixture becomes included in the miscibility gap with the progress of polymerization reaction.

(macropores). In many cases of thermally induced phase separation in metallic alloys, polymer blends, and multicomponent glasses, the kinetics of phase separation can be externally controlled through temperature. One can quench the shape and size of the developing phase domains simply by cooling the system down. On the other hand, the structure formation process has more or less spontaneous nature in the chemical sol–gel systems. The onsets of both phase separation and sol–gel transition are governed by the kinetics of essentially irreversible chemical bond formation. With a predetermined composition, the homogeneously dissolved starting constituents are just left to react at a constant temperature under closed conditions (to avoid evaporation of volatile components). As shown in Table 9.1, it is noteworthy that quite a few gel-forming systems exhibit common features of concurrent phase separation and sol–gel transition, irrespective of the origins of their gel-forming reactions.

9.2.2
Structure Formation in Parallel with Sol–Gel Transition

When the phase separation is induced in the unstable region of a phase diagram (a temperature–composition region within the spinodal curve), a specific process called spinodal decomposition occurs. With comparable volume fractions of conjugate phase domains without crystallographic or mechanical anisotropy, a sponge-like structure called a cocontinuous structure forms (Figure 9.2, top). The cocontinuous structure is characterized by mutually continuous conjugate domains and hyperbolic interfaces.

Development of cocontinuous structure

Self-similar coarsening 1

Self-similar coarsening 2

Fragmentation of domains

Spheroidization and sedimentation

Figure 9.2 Time evolution of spinodally decomposed isotropic phase domains driven by surface energy. After the domains grow only in characteristic size while maintaining the connectivity (self-similar growth), fragmentation and spheroidization follow to minimize the interfaces with energetically unfavorable curvatures.

The final morphology of the spinodally decomposed phase domains is strongly governed by the dynamics driven by the interfacial energy [38]. As shown in the figure, the well-defined cocontinuous structure of the spinodal decomposition is a transient one, which coarsens self-similarly for a limited duration of time and then breaks up into fragments. In order to reduce the total interfacial energy, the system reorganizes the domain structure toward that with less interfacial area and less local interfacial energy. Within the regime of self-similar coarsening (Figure 9.2, bottom), the geometrical features of the developing domains remain unchanged except the characteristic size. Then it is followed by the fragmentation of either of the continuous domains, which results in the dispersion of one phase within the other continuous phase.

9.2.3
Macropore Control

The sol–gel transition is a dynamical freezing process by cross-linking reactions. If any transient (dynamic) heterogeneity is present in a gelling solution, it will be arrested in a gel network if the timescale of a sol–gel transition is short enough to take "snap-shot" of the transient heterogeneity. The "frozen" structure depends, therefore, on the onset of phase separation relative to the "freezing" point by sol–gel transition. The earlier the phase separation is initiated relative to the sol–gel transition, the coarser the resultant structure becomes, and vice versa. For example, a higher reaction temperature normally increases the mutual solubility of the constituents and hence suppresses the phase separation tendency, and in parallel it accelerates the hydrolysis/polycondensation reactions. As a result, the onset of phase separation is retarded and the solution is solidified earlier by the sol–gel transition. Due to these duplicate effects, gels with drastically finer phase-separated domains are obtained at higher temperatures. With an appropriate choice of the reaction parameters such as starting composition and temperature, the pore size (domain size) and pore volume of the gels can be designed in a broad range [5].

9.2.4
Mesopore Control

The sol–gel process accompanied by phase separation provides monolithic gels having phase-separated bicontinuous micrometer-domains, one already solidified as a wet gel and the other still remaining as a fluid. Most of the gel-forming components described in the rest of this chapter are rigid inorganic or organic–inorganic hybrid networks. The continuous gel framework contains inherent vacant spaces on a length scale of nanometers that are filled with solvents in the wet state. On evaporation drying, however, most of these spaces are collapsed by capillary forces. Drying shrinkage is explained by the yield of pore walls against the tensile force exerted by the menisci (curved gas–liquid interface) formed on the individual pore walls. For the purpose of preserving porosity in the micro- to mesopore regimes, enlargement of the inherent pore size should be performed by additional treatments that reorganize the gel network without breaking the existing macroporous framework. Micropores in most amorphous gels in the wet state can be converted into larger mesopores by aging in an appropriate solvent.

In the case of pure silica, weakly basic aqueous solutions in the temperature range up to 100 °C can reorganize micropores into mesopores larger than 20 nm in diameter. The mechanism of pore coarsening by aging under basic conditions is explained by a classical Ostwald ripening theory based on the difference in solubility of the solid (hydrated silica in this case) as a function of the surface roughness. That is, the dissolution is most enhanced on the sharp points with the smallest positive curvature, whereas the reprecipitation is most pronounced at the cavities with the smallest negative curvature. As a result, with an elapse of aging, finer roughness is removed and the whole surface is reorganized into that with only coarser points and cavities. If this process occurs in the three-dimensional network of silica gels, smaller pores are eliminated and the whole pore system is reorganized into one with larger pores.

In the case of less water-soluble solids such as titania, zirconia, and alkyl-modified silsesquioxanes, aging under severe conditions such as hydrothermal conditions is required to tailor the mesopore structure. In these systems, the classical explanation based on an appreciable solubility of solid into water seems not adequate. Additional mechanisms including cooperative reorganization of partially cross-linked metalloxane network under a strong hydrothermal condition should be considered. In any case, the rate of pore coarsening is accelerated by increasing the temperature. Details of aging in specific compositions are described in the following sections.

9.3
Silica

9.3.1
Typical Synthesis Conditions

Pure silica formulations have been exploited to give every possible morphology, material shape, and doped compositions. In the presence of a limited amount of water, using especially tetramethoxysilane (TMOS) as a precursor, the phase separation is induced just by adding a polar solvent, as reported earlier by Kaji et al. [39]. Tetraethoxysilane (TEOS) can also be used as a major precursor often combined with a variety of phase separation inducers. Recent reports by Kajihara et al. show the method of preparing macroporous monolithic silica from TEOS by the simple two-step reactions without additives [40–42]. In the presence of a higher molar ratio of water to silicon, the phase separation is necessarily induced by polymeric or amphiphilic additives. Water glass (alkaline silicate solution), reported by Takahashi et al., as well as colloidal dispersions of silica can be the precursors with lower costs [15,16]. Polymers or surfactants having no specific

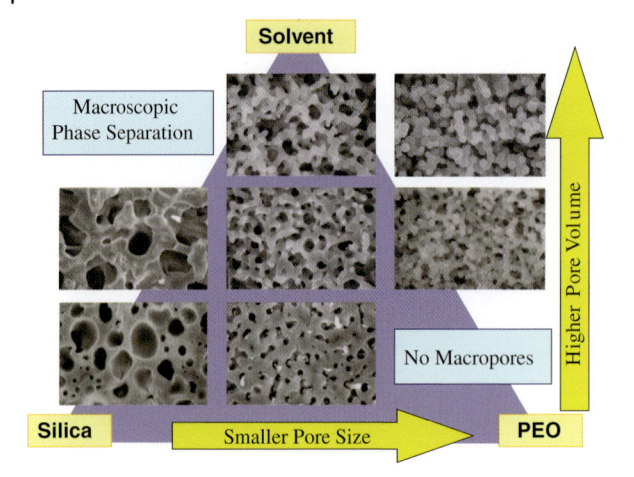

Figure 9.3 Relation between starting composition and resultant gel morphology in TMOS-PEO-solvent pseudo-ternary system. Pore size is controlled by PEO/Si ratio and pore volume by the fraction of solvent.

attractive interaction with silanol surfaces, for example, poly(acrylic acid) or anionic polymers and surfactants, tend to be distributed to the fluid phase, so that the amount of additives directly relates to the volume fraction of macropores. On the other hand, due to the strong hydrogen bonds between silanols and polyoxyethylenes, additives having —CH_2—CH_2—O— repeating units, poly(ethylene oxide) (PEO), and Pluronic or Brij family surfactants are always distributed to the gel-phase, while the solvent mixture becomes a majority in the fluid phase. Similarly, cationic surfactants are preferentially distributed to the gel phase under acidic conditions. In these cases, the volume fraction of the fluid phase can be controlled mostly by the amount of solvent, while the domain size is determined by the additive concentration that dominantly governs the phase separation tendency. This implies that one can independently design the volume and size of macropores by the concentrations of solvent and additive, respectively (Figure 9.3). For gels prepared in a macroscopic mold followed by the evaporation drying and heat treatment at 600 °C, the typical porosity (vol/vol) covers 40–80% with the median pore size ranging from 0.1 to 50 μm [5]. In addition, the strong attractive interaction between silica and additive molecules makes it possible to template the mesoscale structures of the structural unit of the gels by surfactants.

9.3.2
Additional Mesopore Formation by Aging

Since the interconnected macropores enhance the material transport within the bulk gel sample, the exchange of pore liquid with an external solvent can be performed much faster than the case with gels having only meso- to micropores. Conventional methods of tailoring mesopore structure by aging wet silica gels under basic and/or hydrothermal conditions can be suitably applied to the monolithic macroporous silica gels without essentially disturbing the preformed macroporous structure. Experimentally, the as-gelled wet monolithic specimen is immersed in an excess amount of an external solvent such as aqueous ammonia solution. Alternatively, one can add urea in the starting composition of the gel preparation, and subsequently heat the wet gel in a closed vessel to generate aqueous ammonia *in situ*. The preferential dissolution of gel network sites with small positive curvature and subsequent reprecipitation onto those with small negative curvature results in the reorganization of smaller pores into larger pores (the so-called Ostwald ripening mechanism). In the case of pure silica, NMR and SAXS (small-angle X-ray scattering) measurements proved that the chemical reorganization of an initially microporous network into that with sharply distributed mesopores takes place on the timescale of a few hours [43].

9.3.3
Hierarchically Porous Monoliths

The above-mentioned mesopore formation processes take place within the preformed micrometer-sized gel skeletons, so that the size of mesopores can be controlled independent of the macropore size unless the local dissolution of the gel skeletons causes significant deformation of the whole macroporous framework during the solvent exchange. For easier and quicker solvent removal without breaking the monolithic materials shape, mesopores larger than 10 nm in diameter are favored. After evaporation–drying of the wet gels at ambient or elevated temperatures up to 80 °C, the monolithic gel pieces are heat-treated typically in the temperature range between 600 and 800 °C to strengthen the network and yet to preserve the mesopores and appreciable surface area. An example of pore size distribution curve of the finally obtained hierarchically porous monolith is shown in Figure 9.4, where sharply distributed macropores

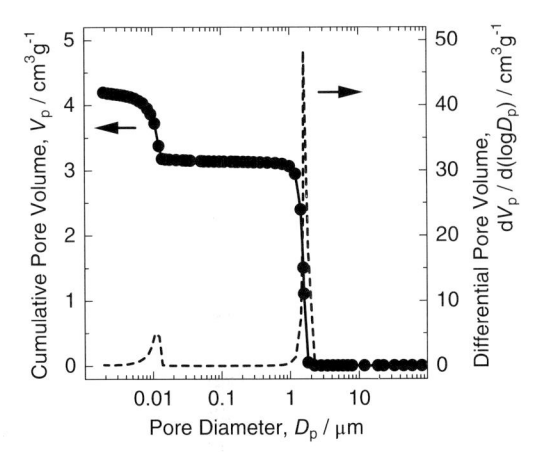

Figure 9.4 Cumulative and differential pore size distribution of hierarchically porous silica monolith designed for HPLC column.

around 1.5 μm and mesopores around 10 nm are clearly evidenced. While the enlargement of mesopores can be carried out up to 100 nm in diameter, the approximate range of macropores homogeneously formed throughout the specimen is between 100 nm and 50 μm in diameter. In addition, the maximum size of columnar monoliths industrially manufactured has been 25 mm in diameter and 200 mm in length. Since the specimens in smaller dimensions are generally easier to be manufactured, the above size range well covers the dimensions of most practical liquid chromatography columns.

9.3.4
Supramolecular Templating of Mesopores

Supramolecular templating is an attractive alternative to the post-gelation aging process to obtain mesopores with a higher degree of order in pore size, shape, and spatial arrangement. It has been found that several kinds of surfactants can be used to induce the phase separation concurrently with the sol–gel transition [44–46]. With an appropriate choice of surfactants suitable also to the supramolecular templating of mesopores, materials have been prepared with crystal-like long-range ordered mesopores homogeneously embedded in the micrometer-sized well-defined gel skeletons.

The key to combine phase separation/gelation and supramolecular templating/precipitation is that both processes include a kind of polymerization-induced phase separation. It has been established that cooperative assembly between surfactant micelles and oligomeric oxides enhances the ordered arrangement of the micelles. Highly ordered mesostructures are organized by such cooperative assembly mechanism in generally amorphous oxide networks. Due to relatively strong attractive interactions between micelles and oxides, submicrometer- to micrometer-sized particles are precipitated out of the solution in dilute systems under closed conditions.

Starting from a composition favorable for the formation of cocontinuous macroporous structure containing a triblock copolymer Pluronic P123 (EO_{20}-PO_{70}-EO_{20}, EO: ethylene oxide, PO: propylene oxide), an additive, 1,3,5-trimethylbenzene (TMB), known to preferentially distributed to the hydrophobic cores of micelles, was introduced to enhance long-range ordering of mesophases in TMOS-derived system [47,48]. Alternatively, a relatively large amount of water together with high concentration of P123 can be used. In heat-treated gels, the long-range ordering of cylindrical pores in 2D hexagonal symmetry has been confirmed by XRD measurements that indicate sharp multiple peaks comparable to single-crystal-like SBA-15. Furthermore, the real-space observations performed by SEM and TEM (Figure 9.5) evidenced the long-range ordering of the cylindrical pores. It is noteworthy that the shape of gel skeletons is affected by the anisotropy of mesopores contained in the skeletons; that is, those with cylindrical mesopores exhibit fibrous features.

With an appropriate post-gelation aging that reorganizes the micropore structure within the frameworks comprising 2D hexagonal mesopores, large

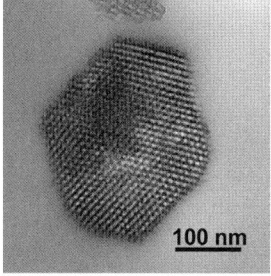

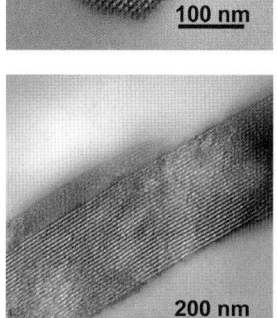

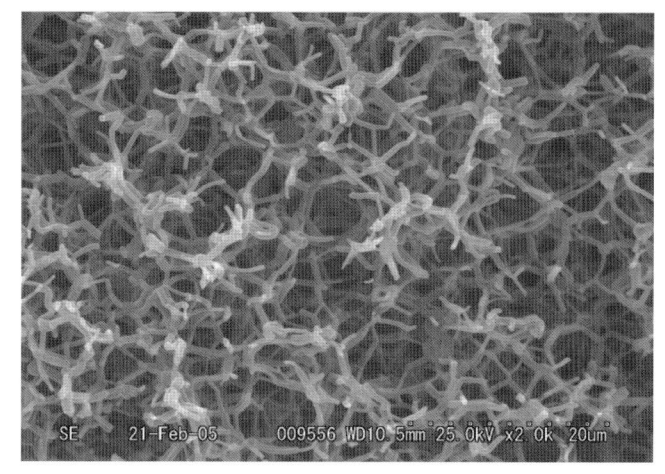

Figure 9.5 TEM (right, upper: cross section, lower: longitudinal cross section) and SEM (left) photographs of high-porosity macroporous silica with thin frameworks embedded with fully templated 2D hexagonal cylindrical pores that are running parallel to the length of the frameworks.

monolithic pieces of hierarchically porous silica can be fabricated. The porosity offered by macropores reaches 90% by volume, whereas about 50% of the silica framework remains porous mainly with sharply defined templated mesopores.

9.3.5
Applications

With high-purity silica composition, the most prominent application of hierarchically porous monolith is a novel type of separation media for liquid chromatography [49]. A single piece of silica gel in columnar shape is embedded with continuous macropores ($\sim$1 µm) and fully accessible mesopores ($\sim$10 nm) that is readily used as a chromatographic column with its sidewall sealed with pressure-tight clad. The novel type of column is now called "monolithic" columns in contrast with the conventional "particle-packed" columns. The monolithic columns have higher macropore volume than particle-packed columns, which increases the permeability of the columns by a factor of at least 2–3. In addition, monolithic columns are characterized by thinner silica gel skeletons compared with the diameters of silica gel particles packed in conventional columns, which suppresses the decreased efficiency at higher mobile phase velocity. Combined with these features, monolithic columns are suitably used in the following two extreme cases: (i) High-throughput analysis with moderate plate numbers (efficiency) and very short analysis time [50,51]. (ii) Ultrahigh-performance analysis with a column generating plate number of 1 million even with extended analysis time (a few to few tens of hours). Since no other practical liquid chromatography columns can exceed the plate number of the long monolithic capillary column, its potential will further be explored in bioanalysis where the separation of thousands of compounds is required in connection with mass spectrometric detections [52,53].

9.4
Silsesquioxane and Other Silicone-Like Systems: Hybrid Aerogels and Low-Density Materials

9.4.1
Network Formation and Pore Control in Methylsilsesquioxane (MSQ) Systems

Silsesquioxanes ($RSiO_{1.5}$) derived from trialkoxysilanes ($RSi(OR')_3$) through sol–gel process offer attractive features especially in surface characters and mechanical properties [54,55]. Apart from particles and films, however, the monolithic form of organosilsesquioxane networks from organotrialkoxysilanes with a hydrophobic substituent group such as methyl has been limited because of the difficulties in monolithic gel formation in typical aqueous sol–gel systems [55]. In acid-catalyzed conditions with low concentration of organotrialkoxysilane, in particular, the formation of cyclic species (two examples of polyhedral oligomeric silsesquioxanes (POSS) are exhibited in Figure 9.6a) rather than cross-linked networks (Figure 9.6b) would become dominant to result in a stable sol or separation of oligomeric oils/polymeric resins [56,57]. In addition, since those organosilsesquioxane networks are hydrophobic and provide lower concentration of remaining silanol group, there is high tendency to form hydrophobic precipitates in an uncontrolled manner [55]. Since the random networks, not cyclic species, are the basis of monolithic gel formation, it is important to promote the random cross-links through adequate sol–gel design strategies. In our study, careful control over fundamental sol–gel parameters (pH, solvent, additives, etc.) has been proved to be effective in obtaining monolithic porous silsesquioxane materials with the length scale of the porous structures from several tens of nanometers (typical aerogels) to micrometers (macroporous gels) by enhancing the network formation and controlling the phase separation tendency [54].

In acidic one-step systems, where both hydrolysis and polycondensation of the precursor are conducted under acidic conditions, monolithic macroporous materials are obtained only when [water]/[alkoxysilanes] ratio (r) is low (such as $r \sim 2.0$) [58]. In the case where methyltrimethoxysilane (MTMS) and vinyltrimethoxysilane (VTMS) are used as the precursor and formamide (FA) as the solvent, macroporous monolithic gels with cocontinuous structure are obtained as a result of sol–gel transition accompanied by phase separation (Figure 9.7) [59,60]. Enthalpy-driven spinodal decomposition takes place in the

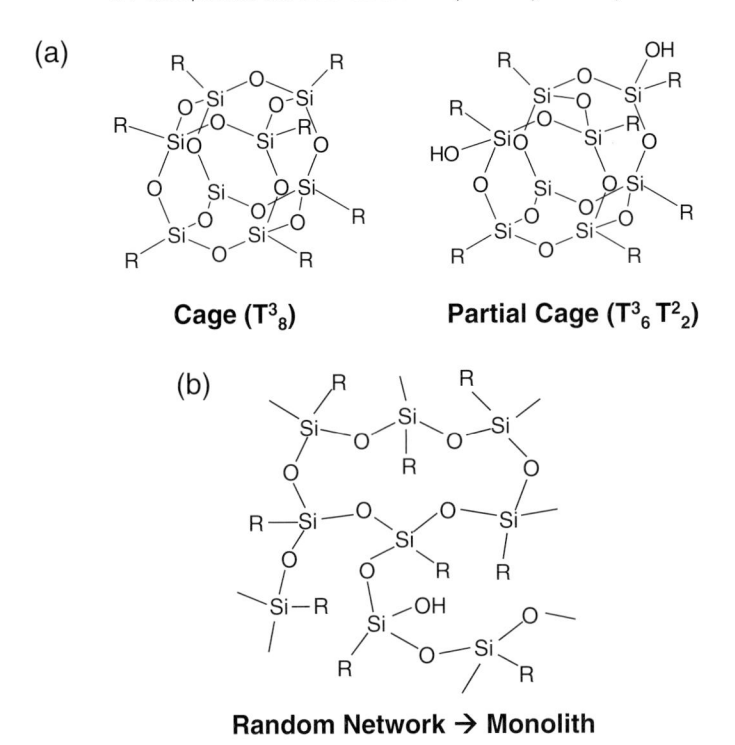

9.4 Silsesquioxane and Other Silicone-Like Systems: Hybrid Aerogels and Low-Density Materials | 207

Figure 9.6 (a) Silsesquioxanes with closed-ring structure; two examples of polyhedral oligomeric silsesquioxanes (POSS). (b) Random network structure crucial for the formation of monoliths.

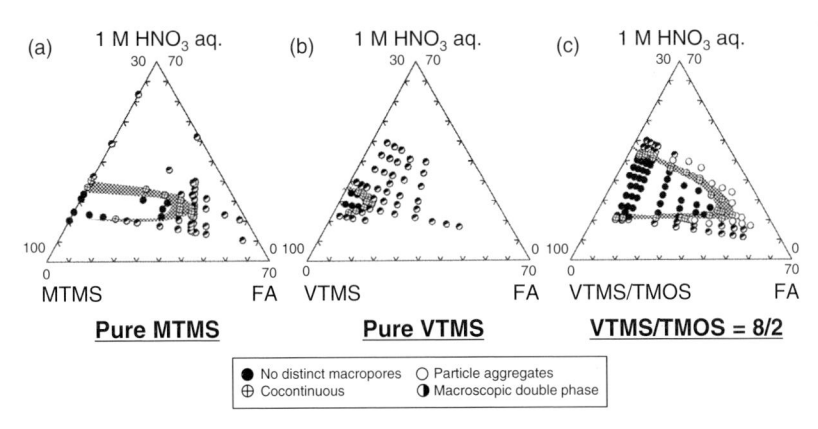

Figure 9.7 Relationship between starting composition and resultant microstructure in (a) MTMS, (b) VTMS, and (c) VTMS + TMOS systems represented in mass%. An acid one-step process has been employed to prepare macroporous monoliths.

course of gelation in water-based solvents, due to the chemical incompatibility between hydrophobic silsesquioxane networks and polar solvents. Methylsilsesquioxane gels with well-defined macropores form in a wider range of starting composition compared to vinylsilsesquioxane, because the hydrophobicity by the methyl groups is lower and the controllability of phase separation is higher. The lower steric effect also contributes to the enhanced network formation in MSQ system. Cocondensation with tetramethoxysilane (TMOS) extends the compositional region where well-defined cocontinuous structure is obtained as the average polarity of the network becomes higher, which allows higher controllability of the phase separation process.

The MSQ monoliths thus obtained have been applied as monolithic capillary columns for high-performance liquid chromatography (HPLC) [60]. Moderate retention of polar and nonpolar compounds to the MSQ surface enables efficient separations both in the normal- and reversed-phase modes in a single column [61] because of the simultaneous presence of hydrophobic methyl groups and hydrophilic silanol groups. The highest theoretical plate number reaches $N = 100\,000\text{ m}^{-1}$ in the normal-phase mode, although the retention volume is not sufficient because the macropore skeletons have only micropores as evidenced from the type-I isotherm of nitrogen adsorption, and the micropores are hardly accessible for eluent molecules.

The introduction of mesoporosity is crucial for various applications to increase the contact between guest molecules and the surface. Mesoporosity, in addition to the macroporosity, has been for the first time imparted to MSQ by employing a different synthetic strategy in sol–gel processes [62]. A combination of acid-catalyzed hydrolysis and base-catalyzed polycondensation enhances the random network formation, which enables wider control over pore size and porosity when combined with an appropriate surfactant for better control of phase separation. Figure 9.8 demonstrates the control of phase separation using a nonionic surfactant Pluronic F127, poly(ethylene oxide)-block-poly(propylene oxide)-

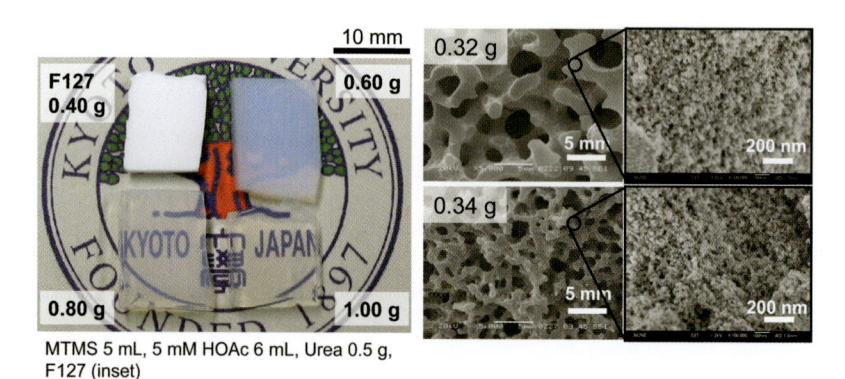

MTMS 5 mL, 5 mM HOAc 6 mL, Urea 0.5 g, F127 (inset)

Figure 9.8 Appearance of MSQ monolithic gels prepared with different amounts of Pluronic F127 via acid–base two-step sol–gel process, and SEM images of the typical cocontinuous macroporous structure. Mesopores are also found in the macropore skeletons.

block-poly(ethylene oxide). The pH swing for the second polycondensation step is conducted by hydrolysis of urea at >60 °C. The appearance of the MSQ gel changes from opaque to transparent with increasing amount of F127. With appropriate amounts of F127, a well-defined cocontinuous macroporous structure is obtained and the macropore skeletons are found to contain mesopores. These mesopores with the size range of 10–20 nm are formed as a result of aggregation of colloidal MSQ condensates in a higher amount of solvent (water) under basic conditions. These MSQ monoliths have also been confirmed to be applicable to HPLC separation media in the normal phase mode [63]. Similar structural controls using different series of precursors for silsesquioxane, those with multiple alkoxysilane units bridged by hydrocarbon chains, can also be performed to result in monolithic gels with well-defined macropores [10,48].

9.4.2
Methylsilsesquioxane Aerogels and Xerogels

As mentioned in the previous section, the acid–base two-step process is found to be suitable for wider controls over pore size and porosity in MSQ system. Here an approach to MSQ aerogels and aerogel-like xerogels is demonstrated.

Aerogels [64,65] are porous materials distinguished by exceptionally high porosity (>90%). Typical silica aerogels are, in particular, differentiated by visible light transparency with well-defined mesopores with a few tens of nanometers, in which gas molecules cannot effectively transfer kinetic momentum, resulting in the severely restricted contribution to total thermal conductivity of the material. Silica aerogels are thus known as the most insulating solid material and expected as the transparent thermal superinsulator, although a fatal problem in the mechanical friability and the need of a supercritical drying process remain.

We have for the first time developed transparent aerogels with MSQ composition by employing the acid–base two-step process under the presence of surfactants (Figure 9.9) [8,66,67]. The one-pot acid–base two-step process using urea as the base generator, as described above, contributes to the homogeneous formation of three-dimensional random networks over cyclic species. In addition, appropriate cationic surfactant (such as n-hexadecyltrimethylammonium bromide/chloride, CTAB/CTAC) [68] and nonionic triblock copolymer (such as Pluronic, poly(ethylene oxide)-*block*-poly(propylene oxide)-*block*-poly(ethylene oxide)) [69] are effective to render the hydrophobic MSQ condensates more hydrophilic in the aqueous sol–gel system. Phase separation of the MSQ condensates/networks is thus effectively suppressed in aqueous media. The obtained transparent wet gels can be transformed into aerogels by supercritical drying. It has been found that these aerogels are mechanically strong and flexible against compression (Figure 9.10), and MSQ xerogel monoliths with aerogel-like properties can be prepared by ambient pressure drying, because the drying gels show reversible shrinkage–re-expansion (the so-called spring-back) behavior during drying. Large-area MSQ xerogel plates can be successfully obtained by optimizing the starting composition and drying process (Figure 9.11). These MSQ

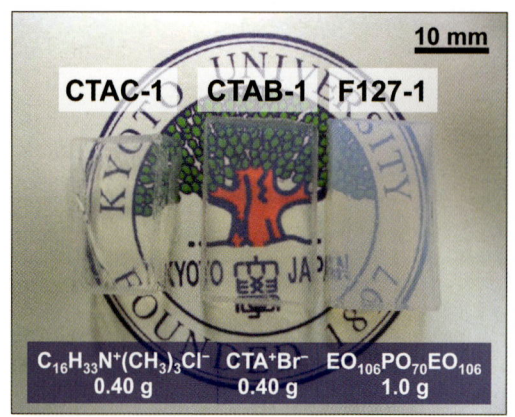

Figure 9.9 Appearance of MSQ aerogels prepared with different surfactant via acid–base two-step sol–gel process. Pore size, porosity, and light transmittance are confirmed to be close to standard silica aerogels.

xerogels are confirmed to show thermal conductivity comparable (~0.015 W/(m K) at an ambient condition) with a typical silica aerogel [70,71]. This material would be the breakthrough for practical application to thermal superinsulators.

9.4.3
Marshmallow-Like Gels

Co-condensation of MTMS with diorganodialkoxysilane such as dimethyldimethoxysilane (DMDMS) yields bendable, opaque, low-density gels, to which we coined a name of "marshmallow-like gels" (Figure 9.12). Starting from a solution similar to that for making transparent MSQ aerogels/xerogels as described above, an introduction of DMDMS with a fixed total amount of alkoxysilanes results in the coarsened pore size toward micrometer range, since

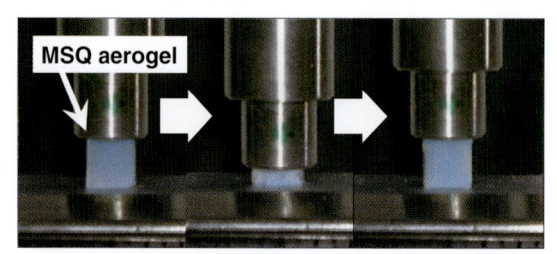

Figure 9.10 Spring-back behavior observed during uniaxial compression–decompression test on an MSQ aerogel. The specimen does not break up to 80% compression, and recovers original size after unloading. This characteristic compression flexibility enables ambient pressure drying toward aerogel-like xerogels.

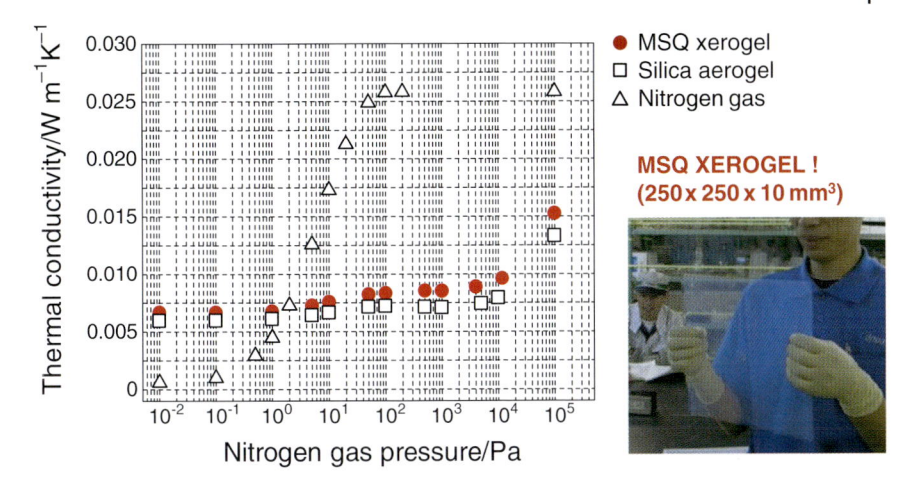

9.4 Silsesquioxane and Other Silicone-Like Systems: Hybrid Aerogels and Low-Density Materials | 211

Figure 9.11 Dependence of thermal conductivity on nitrogen gas pressure (measurement ambient), and a photo of a large-area MSQ xerogel tile. Thermal conductivity shows the comparable behavior with a standard silica aerogel, promising an applicability to transparent thermal superinsulators.

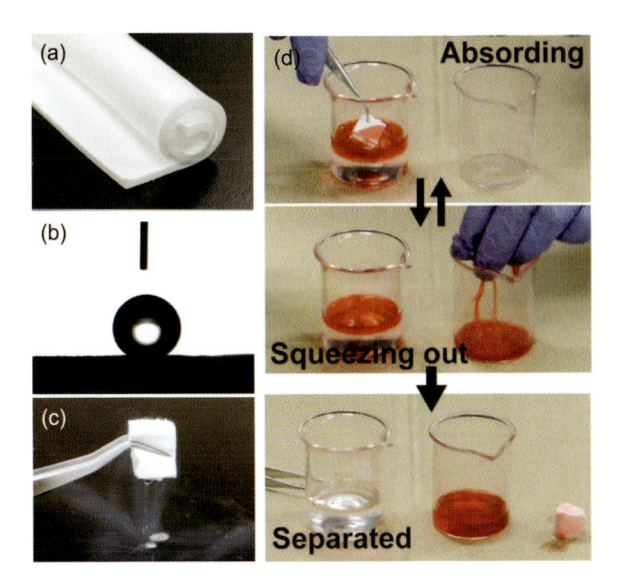

Figure 9.12 A marshmallow-like gel prepared from MTMS and DMDMS with (a) bending flexibility and (b) superhydrophobicity. (c) The marshmallow-like gel shows flexibility even at liquid nitrogen temperature, demonstrating squeezing out liquid nitrogen. (d) Separating n-hexane (colored with Oil-Red O) from an n-hexane–water two-phase system by hand.

hydrophobicity of the polysiloxane network increases with increasing DMDMS ratio and phase separation tendency increases [72].

The marshmallow-like gels show bending flexibility (Figure 9.12a) and superhydrophobicity with water contact angle >150° on the surface of the monolith (Figure 9.12b), due to the combined effect of hydrophobic nature of the polydimethylsiloxane (PDMS)-like surface and physical roughness on the surface by the macropores [9]. A piece of marshmallow-like gel has been used for demonstrating oil–water separation in a two-phase layered system consisting of oil (such as n-hexane, colored by Oil-Red O) and water (Figure 9.12d). The marshmallow-like gel absorbs only oil and the oil can be squeezed out in a different place by mechanical compression, thus leading to efficient separation. It is worth noting that this behavior can only be seen in materials with high hydrophobicity (a similar material such as kitchen sponge can absorb and squeeze out liquid, but cannot selectively absorb a liquid out of others). Another unique property includes maintained flexibility even at liquid nitrogen temperature (77 K) (Figure 9.12c). This is presumably due to the low cross-linking density of the polysiloxane network with high mobility, the thin macropore skeletons, and low density. Some variations of the marshmallow-like gels have been demonstrated by using alkoxysilanes having vinyl, thiol, phenyl, and fluorohydrocarbon substituent groups.

A more advanced surface design can be performed when alkoxysilane precursors with reactive groups are employed to prepare the marshmallow-like gel [73]. Figure 9.13a exhibits an example of cocondensation between VTMS and vinylmethyldimethoxysilane (VMDMS), followed by the thiol-ene reaction on the pore surface. A perfluoroalkylthiol compound is employed to the surface modification in this specific case in order to add oleophobicity while keeping the flexibility of the marshmallow-like gel. While the gel without surface modification (MG1) shows superhydrophobicity and absorbs oil (1,3,5-trimethylbenzene, colored by Oil-Red O), the modified MG2 does not absorb the oil as well as water, showing superoleophobicity (Figure 9.13b). The hydro- and oleophobicity are demonstrated on MG2 by repelling liquids with a wide range of polarity from water to hydrocarbon (Figure 9.13c). These examples of flexible material design would be beneficial for developing multifunctional soft porous materials.

9.5
Titania and Zirconia

9.5.1
Choice of Starting Compounds

Alkoxides of titanium and zirconium generally exhibit much higher reactivity toward hydrolysis/polycondensation than those of silicon, mainly due to the difference in partial charge of the metals in their respective oxygen-coordinated environments [14]. Slower gelation is essentially required to obtain monolithic gels even without special pore structures. Chelating agents such as

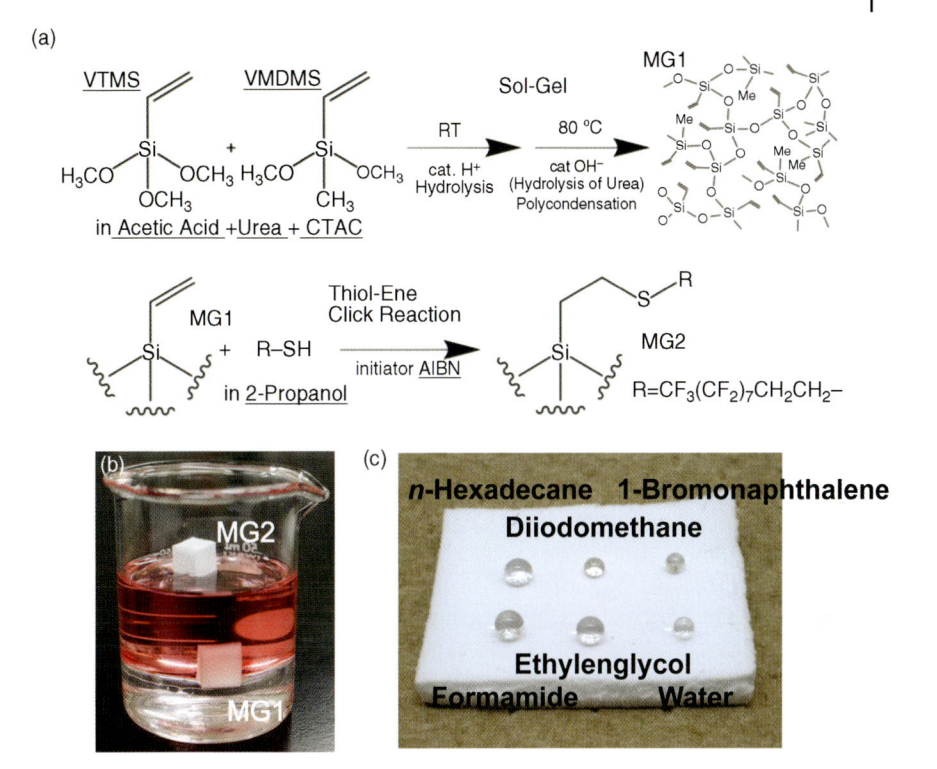

Figure 9.13 (a) Preparation of vinyl-modified marshmallow-like gels (MG1) from VTMS and VMDMS, followed by surface modification by the thiol-ene reaction on the surface to add perfluoroalkyl groups (MG2). (b) MG1 absorbing 1,3,5-trimethylbenzene (colored with Oil-Red O) and floating on the water layer. MG2 does not absorb both liquids. (c) MG2 repelling liquids with a wide range of polarity from water to hydrocarbon.

ethylacetoacetates can effectively block some of the available coordination sites of titanium or zirconium, and thus slow down the overall kinetics of hydrolysis/polycondensation. Dilution with a large amount of parent alcohol also works to make homogeneous gels, frequently followed by a significant syneresis due to the continuing cross-linking after gelation. In the case of preparing macroporous monolithic gels by arresting the transient micrometer-scale structure of the spinodal decomposition, the sol–gel transition should take place rapidly enough to prevent undesired coarsening of the phase-separating solutions. With this regard, options to slow down the gelation kinetics described above do not always give satisfactory results. When started especially from i-propoxides, well-defined macroporous titania and zirconia had been very hard to be synthesized in monolithic form. Another option of preparing titania and zirconia gels with moderate gelation kinetics is to destabilize the colloidal dispersion by pH swing. In the case of titania, acid-stabilized high-purity colloidal dispersions with a nominal particle size below 10 nm are commercially available, which are mixed with urea or acid amides as base generator with heating to obtain homogeneous gels. One advantage of using colloidal dispersions is that since the primary particles contain fewer hydroxyl-terminated parts than those obtained from alkoxides, the overall shrinkage on drying can be kept smaller.

9.5.2
Control over Reactivity

Konishi *et al.* first succeeded in preparing titania monoliths with well-defined macropores from colloidal dispersions [14]. It was one of the evidences that polymerization-induced phase separation can take place even in the case that the polymerizing units are not truly molecules or molecular-scale aggregates. Even though the nominal size of the titania particles was nearly an order larger than those of oligomers, the final morphology well resembled those obtained in many other systems described above. The compressive mechanical strength of the final product, however, was very low, reflecting that the particulate structural units physically aggregate on point-contact basis. The same group reported that the use of titanium *n*-propoxide instead of *i*-propoxide under high concentration of HCl enabled one to control the hydrolysis/polycondensation kinetics in an experimentally feasible timescale, where gelation occurs in a few tens of minutes to a few hours [6]. The reaction can be well controlled with a relatively limited amount of water against complete hydrolysis, so that titania oligomers rich in unreacted alkoxy groups separates from a polar solvent mixture containing acid amides. An appropriate amount of PEO can also be incorporated for an adjustment of phase separation dynamics to obtain a better defined macroporous structure. Macroporous titania gels thus obtained are composed of a microcrystalline anatase phase even in the wet-gel stage. By controlling the grain growth of the anatase crystallites during the aging and heat treatment, the interstices of the crystallites form sharply distributed mesopores in the size range of 5–10 nm (Figure 9.14). Monolithic pure zirconia with controlled macropores can also be prepared. Zirconia tends to form gels with amorphous structural

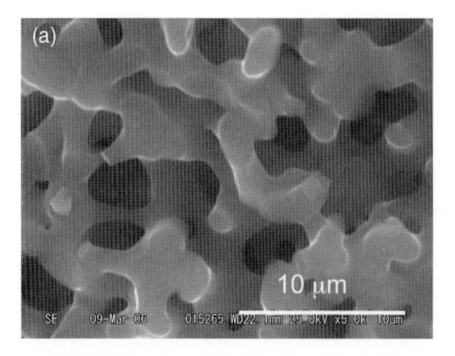

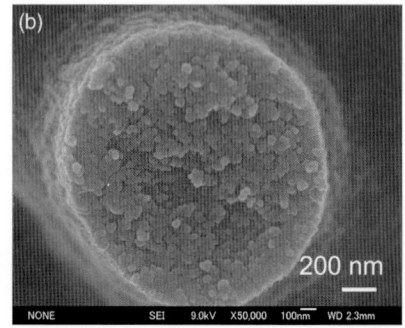

Figure 9.14 Fractured surfaces of monolithic titania columns with well-defined macropores (a) and mesopores (b) embedded in the continuous frameworks.

units and transforms via a monoclinic into a tetragonal phase as the heat-treatment temperature increases [7].

Hasegawa et al. recently reported that inclusion of various inorganic salts into the titania sol–gel system started from chelated titanium n-propoxide enables the control over an extended range of gelation time [74,75]. In terms of avoiding contaminations, the use of ammonium nitrate is a convenient choice. Since the chelating agents are incorporated in the gelled network, their gradual decomposition by successive solvent exchange steps using alcohol/water solutions with increasing water concentration makes it possible to finally obtain crack-free macroporous monoliths.

9.5.3
Applications

Titania is known to specifically adsorb phosphorus-containing compounds and is used as packing materials for commercial HPLC columns. As an extension of pure silica monolithic HPLC columns, separations of phosphorus-containing compounds have first been reported with monolithic columns with their inner surface coated by titania layers [76]. Although the potential of titania surface for separating a mixture of adenosine, AMP, ADP, and ATP was successfully demonstrated, the efficiency was not as high as that of existing particulate counterparts. Recently, further improvements of efficiency have been reported by Konishi et al. [77] and Hasegawa et al. [78] using titanium n-propoxide with concentrated HCl and chelating agent with inorganic salts, respectively. In both methods, mesopores are tailored under strong hydrothermal conditions typically up to 200 °C. Subsequent heat treatment between 400 and 600 °C gives mesopore surfaces composed of microcrystalline anatase with appreciable surface area. In addition to their use as separation media, the hierarchically porous titania monoliths are also useful as a catalyst support for Ni [79].

Making use of the high refractive index of titania (rutile) combined with well-controlled macroporous framework in the wavelength scale of visible light, the confinement of light can be realized. The degree of light confinement was evaluated by measuring coherent back scattering using an Ar^+ laser ($\lambda = 488$ nm) as a light source [80,81]. A scattering medium with a short-enough transport mean free path approaching the wavelength of the light can efficiently confine the light of corresponding wavelength. Compared with a similar macroporous matrix made of silica, the scattering from a macroporous titania monolith with rutile skeletons exhibited a shorter transport mean free path of the incident light. Owing to the precisely controllable uniform macropore structures, various interesting photonic properties are reported using titania and other oxide systems, including silica, a silica-alumina matrix doped with organic dyes, and rare earth ions [81–85].

9.6
Epoxide-Mediated System: (Oxy)hydroxides from Metal Salts

Hierarchically porous silica monoliths are typically synthesized from metal alkoxides at a relatively high silica concentration (10–20 wt%), allowing the formation of monolithic gels that possess well-defined porous architectures. The reactivity of alkoxide precursors significantly affects the feasibility to prepare uniform/crack-free monoliths, and moreover to control meso/macrostructures. The successful formation of hierarchically porous silica primarily takes advantage of slow hydrolysis and polycondensation kinetics of silicon alkoxides. A highly reactive alkoxide with a large partial positive charge on its central atom is basically not involved in the aforementioned alkoxide-based technique except the case of using chelating agents and strong acids as a retardant of the reactions. A more difficult challenge is to achieve a mixed metal oxide with a porous structure starting from a mixture of heterogeneous alkoxides. The difference of reactivities between heterogeneous alkoxides prevents the reactions from taking place homogeneously in these multicomponent systems. For instance, in titania–silica alkoxide systems, $Ti(OR)_4$ catalyzes the condensation of $Si(OR)_{4-x}(OH)_x$, promoting homocondensation of the silicate species rather than uniform incorporation of Ti at the molecular level through heterocondensation [4]. Developing an alternative reaction that is free from metal alkoxides is the key to the versatile synthesis of porous materials based on the phase separation route. This section is mainly dedicated to introduce (i) the synthesis of hierarchically porous monoliths from metal salts and acids instead of alkoxides, and (ii) the applicability of the procedure to the various multicomponent systems, especially mixed metal oxides.

9.6.1
Gelation from Metal Salts and Acids

A pioneering work by Gash et al. reported the synthesis of monolithic aerogels from metal salts and epoxides (Scheme 9.1) [86]. Using stable metal salts such as metal chlorides and nitrates as metal sources, homogeneous gelation occurs by the addition of epoxides, such as propylene oxide and trimethylene oxide. Since the epoxide-mediated reaction can take place in aqueous media, metal salts that form stable metal aqua complexes can be used as metal sources instead of using metal alkoxides. Products obtained by the reaction are generally metal hydroxide or oxyhydroxides, and importantly possess a monolithic form due to the high supersaturation achieved in the course of the reaction. Epoxides act as a alkalization agent through the protonation of the epoxide oxygen and the subsequent ring opening of epoxide by the nucleophilic attack of conjugate bases, such as H_2O or Cl^-. The successive reactions increase the pH of the reaction solution. The uniform increase of solution pH taking place in minute scale allows the homogeneous hydrolysis and polycondensation reactions to produce a monolithic gel. Figure 9.15a shows pH evolutions when propylene oxide is reacted

9.6 Epoxide-Mediated System: (Oxy)hydroxides from Metal Salts

$$[M(H_2O)_m]^{x+} + \triangle\!\!\!\!|^O \rightleftharpoons \triangle\!\!\!\!|^{O^+\!H} + [M(OH)(H_2O)_{m-1}]^{(x-1)+} \quad (1)$$

$$\triangle\!\!\!\!|^{O^+\!H} + A^- \xrightarrow[(A^-:Cl^-,\,NO_3^-,...)]{\text{ring opening}} \begin{array}{c} OH \\ \diagup\!\!\!\!\diagdown\!\!A \end{array} \left(+ \begin{array}{c} A \\ \diagup\!\!\!\!\diagdown\!\!OH \end{array} \right) \quad (\text{pH}\uparrow) \quad (2)$$

$$n[M(OH)(H_2O)_{m-1}]^{2+} \xrightarrow[\text{polymerization}]{\text{hydrolysis}} \text{metal hydroxides (or metal oxyhydroxides)} \quad (3)$$
(metal complex)

Scheme 18.1 Scheme showing the reaction between a metal salt and propylene oxide in aqueous media. Propylene oxide works as a alkalization agent; (1) propylene oxide is protonated: (2) the protonated propylene oxide causes ring-opening reaction by the nucleophilic attack of coexisting anionic species. pH increases as a result of the consumption of protons: (3) the increase of pH drives hydrolysis and polycondensation reactions of metal aqua complexes to form metal hydroxides (or metal oxyhydroxides).

with three different metal chloride aqueous solutions. A monolithic wet gel can be obtained typically in several to tens of minutes (Figure 9.15b) (no gelation in case of using alkali salts).

The epoxide-mediated sol–gel reaction has several advantages over the alkoxide-derived sol–gel reaction. First, the reaction can be applied to prepare various main and transition group metal oxides, starting from stable metal salts, in aqueous media, through green processes [34]. Second, heterogeneous metal cations are homogeneously and stably dissolved in a starting solution. Third, the reaction is readily combined with other sophisticated solution processes, such as

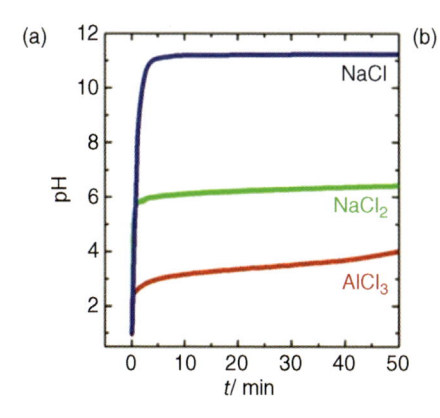

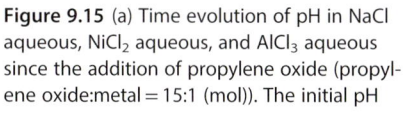

Figure 9.15 (a) Time evolution of pH in NaCl aqueous, NiCl$_2$ aqueous, and AlCl$_3$ aqueous since the addition of propylene oxide (propylene oxide:metal = 15:1 (mol)). The initial pH was set at 1 by adding HCl aqueous. (b) Appearance of wet aluminum hydroxide gel prepared by the epoxide-mediated route.

the inverse micelle method [35]. Indeed, the epoxide route can be successfully coupled with the technique of "phase-separation accompanied by sol–gel transition" toward hierarchically porous monoliths. Several highlighted porous materials in aluminate systems are closely reviewed in the following sections.

9.6.2
Hierarchically Porous Aluminum Hydroxide

Aluminum hydroxide is the first successful example of hierarchical porous materials via the epoxide-mediated reaction accompanied by phase separation [17]. Briefly, aluminum chloride hexahydrate ($AlCl_3 \cdot 6H_2O$) was used as an inorganic source, and a mixture of H_2O and ethanol (EtOH) as a solvent. Propylene oxide was used as an epoxide to increase the pH of the solution, and poly(ethylene oxide) (PEO) ($M_v = 1,000,000$) as a phase separation inducer.

The solution pH is <1 before PO addition and reaches 3 at the gelation in 10 min. The uniform increase of pH drives the hydrolysis and polycondensation of the metal aqua complexes to form the monolithic gel. The addition of PEO induces phase separation, although it does not impact the gelation time. Hierarchically porous alumina monoliths are synthesized from starting solutions containing 3–4 mol% of $AlCl_3 \cdot 6H_2O$, which is three times higher compared to the composition used by Gash and coworkers [36]. The metal salt solutions with a relatively high concentration enable one to achieve a high degree of supersaturation in the reaction solution, allowing one to form nano hydroxide particles (8.5 nm) [87] required for the structural control in nano- and macrosize scales. The phase separation in this system is driven by the entropy loss due to the polymerization of aluminate species, and a homogeneous starting mixture phase separates into alumina-rich solid phase and PEO-rich fluid phase [88].

Figure 9.16a–f shows macroporous structures of the dried aluminum hydroxides prepared with different amounts of PEO (W_{PEO}). Cocontinuous macropores form at $W_{PEO} > 0.05$ g and the size and the volume of macropore increase with W_{PEO}. Macropore size and volume respectively are 0.4 µm and 0.20 cm^3/g for $W_{PEO} = 0.06$ g, and 1.8 µm and 0.51 cm^3/g for $W_{PEO} = 0.10$ g. SAXS analysis revealed that aluminum hydroxides are composed of primary nanoparticles with a diameter of 8.5 as confirmed by TEM (Figure 9.16g). In addition to macropores, mesostructures are also observed in the skeletons of dried gels (Figure 9.16h and i) as interstices of the primary particles. Dried alumina gels possess mesopores with a median diameter of 2.6 nm and a BET surface area as high as 400 m^2/g. The mesopore characteristics highly depend on drying conditions because a large deformation occurs during ambient drying (Figure 9.17) [87].

The crystalline structure of the dried aluminum hydroxide is amorphous or pseudo-boehmite. The degree of crystallinity is generally affected by several factors, including concentration of metal salts, reaction pH, solvents, temperature, and the presence or absence of polymers in the starting solution. In the present reaction, the gelation takes place under acid conditions in a high ionic strength solution, and the crystallization triggered by PO is relatively rapid compared to

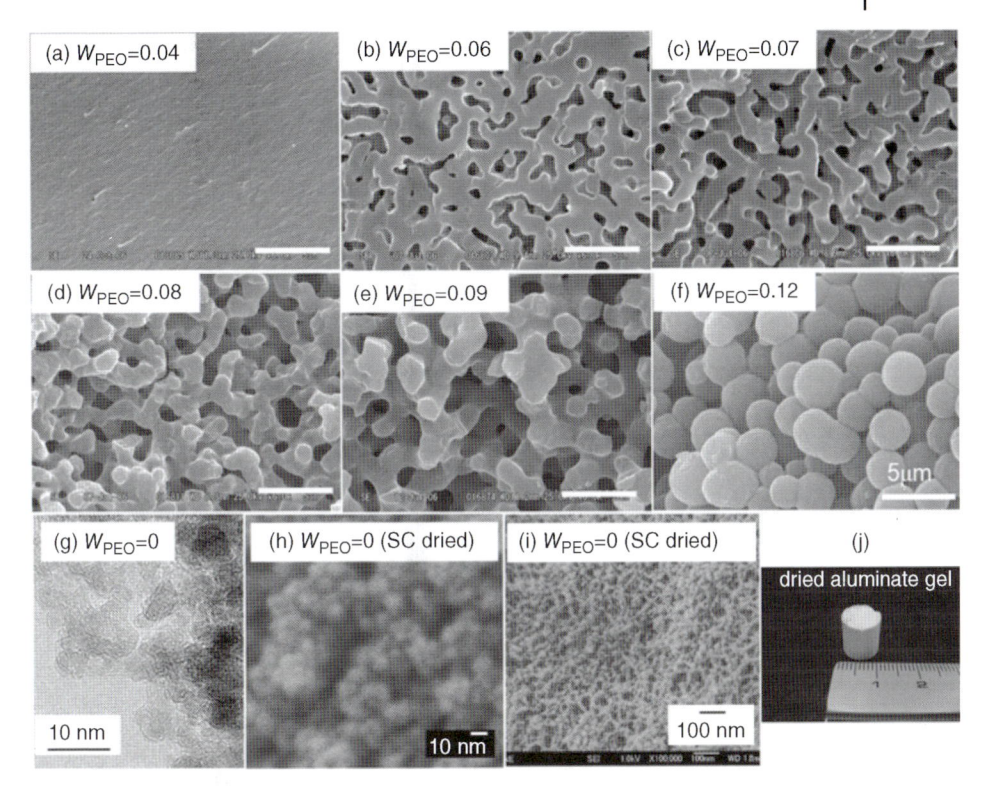

9.6 Epoxide-Mediated System: (Oxy)hydroxides from Metal Salts | 219

Figure 9.16 (a–f) SEM images of the hierarchically porous aluminum hydroxides prepared with varied PEO amounts (W_{PEO} (in part g)). (g) TEM image of the aluminum hydroxide ($W_{PEO} = 0$). (h and i) Magnified SEM images of aluminum hydroxides prepared with a supercritical drying process ($W_{PEO} = 0$). (j) Appearance of part (e).

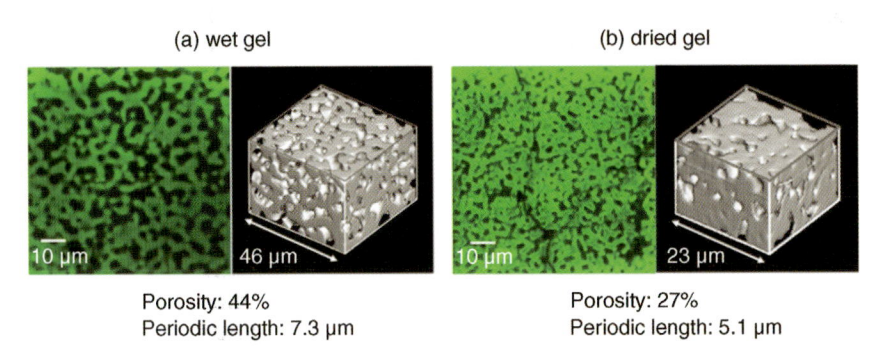

Figure 9.17 Two- and three-dimensional images of (a) wet gel and (b) dried gel observed by laser scanning confocal microscope (LSCM). A volume shrinkage of 66% takes place during drying at 40 °C.

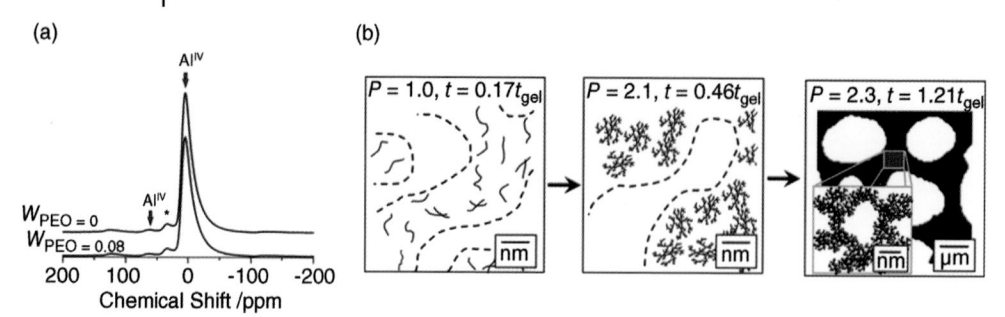

Figure 9.18 (a) ^{27}Al MAS NMR spectra for aluminum hydroxides prepared with $W_{PEO} = 0$ and 0.08. * typically observed for poor crystalline aluminum hydroxides that are usually assigned as Al^V. (b) Time evolution of macroporous structure. P: fractal dimension of oligomeric species; t: time since PO addition; t_{gel}: gelation time.

typical homogeneous crystallization with using urea. As a result, nanoparticulate aluminum hydroxide with a low crystallinity is formed, which is favorable to form a hierarchical structure (Figure 9.18). Phase transitions take place to form γ- and α-Al$_2$O$_3$ at 800 and 1100 °C, respectively. During the thermal process, hierarchical pores in respective size scales keep their sharp pore size distributions. The capability of preserving the porous structure even after the calcination at high temperatures allows one to prepare functional porous ceramics with a monolithic form. Fujita et al. reported the synthesis of porous ruby (Cr^{3+}-doped alumina) based on this scheme [18]. The synthesis of La^{3+}-doped alumina with a high surface area is also reported (BET surface area = 91 m^2/g after calcination at 1100 °C for 12 h) [19].

The present strategy toward hierarchically porous metal hydroxides has been reportedly extended to various systems such as iron hydroxide [20], nickel hydroxide [21], and copper hydroxide [22]. These monoliths are obtained as composites with polymers, and thereby carbon/oxide composites are also obtained after the calcination in an inert atmosphere.

9.6.3
Double Hydroxides, Mixed Metal Oxides, and Others

Metal salts are soluble and stable in an aqueous solvent, which allows homogeneous mixing of starting components. As a result, the present method based on the epoxide-mediated reaction allows the preparation of double hydroxide and mixed metal oxides. This capability toward mixed metal oxides was first demonstrated in Y-Al-O system, where Y^{3+} was incorporated into the above-mentioned aluminum hydroxide system. Y^{3+} ions are homogeneously distributed in the gel matrix and yttrium aluminum garnet (YAG) is obtained at as low as 800 °C [23]. The thus obtained macroporous YAG is applicable to a scattering media to achieve spectral hole burning [24]. Recently, mixed metal oxides with complex stoichiometries have also been reported; macroporous β-cordierite

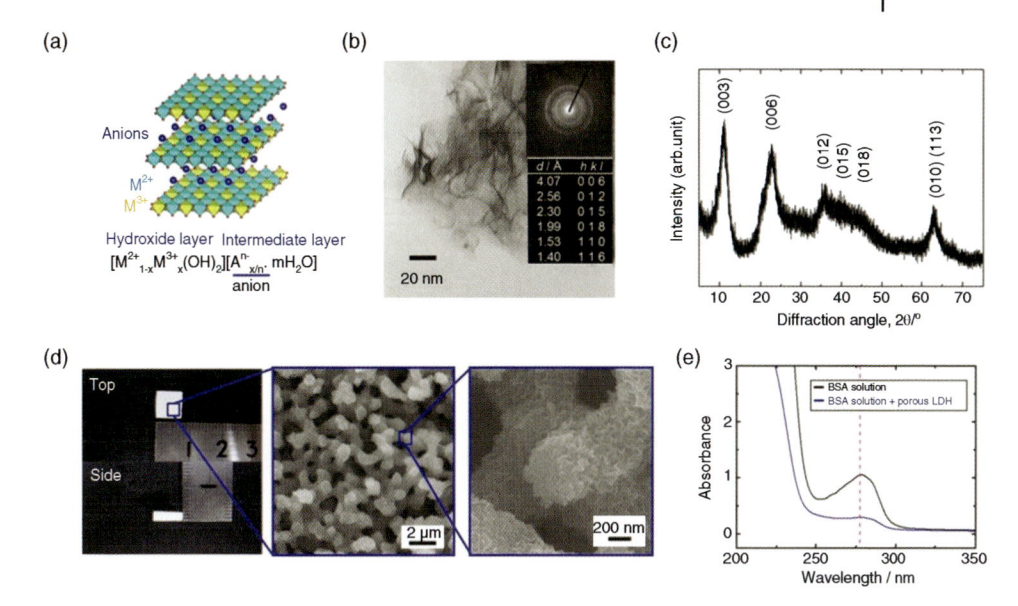

Figure 9.19 (a) Schematic illustration of crystalline structure of hydrotalcite-type LDH. (b) TEM, (c) XRD pattern, (d) photo and SEM images of hierarchically porous LDH. (e) UV-vis spectra of supernatants before and after bovine serum albumin (BSA) adsorption by hierarchically porous LDH. The absorption peak at 277.4 nm is derived from BSA.

($Mg_2Al_4Si_5O_{18}$) and mayenite ($Ca_{12}Al_{14}O_{32}Cl_2$) monoliths are obtained from the Mg-Al-Si ternary and Ca-Al binary systems, respectively [25,26]. Another important example that can be achieved by the present protocol is hierarchically porous layered double hydroxide (LDH) (Figure 9.19) [27]: $M^{2+}_{1-x}Al_x(OH)_2Cl_x$ (M^{2+}: Mg^{2+}, Mn^{2+}, Fe^{2+}, Co^{2+}, Ni^{2+}, Cu^{2+}, Zn^{2+}) with hierarchically porous structures [28]. The materials are obtained as a composite of LDH and aluminum hydroxide, both of which are known as biocompatible materials. For example, in the case of Mg^{2+} involved, the chemical composition is [$Mg_{0.66}Al_{0.33}(OH)_2Cl_{0.33}\cdot2.92H_2O$]·$2.0Al(OH)_3$. The tunable surface chemistry as well as hierarchically porous architectures allows applications of adsorption and separation with a high selectivity [28].

Posttreatments of these porous materials are promising to obtain another class of materials with hierarchical pores. The hierarchically porous aluminum hydroxide can be converted to MFI, ANA, PHI-zeolites [89] and metal organic frameworks (MOFs) [90] via pseudomorphological replication. The replication is simply performed by soaking the monolithic aluminum hydroxide (or alumina) in solutions of respective reactants. In the case of zeolites, hydrothermal treatments are applied to cause the reactions. This simple scheme is versatile and applied to the replication from $Cu(OH)_2$ to Cu_2(btc = 1,3,5-benzenetricarboxylate)$_2$(MeOH)$_2$ and Cu_2(btc)$_2$(bpy)$_2$ (bdc^{2-} = 1,4-benzenedicarboxylate, bpy = 4,4'-bipyridine) [91].

9.7
Metal Phosphate Systems: Layered Crystalline Phosphates and Related Compounds

One of the distinctive structural features of metal phosphates is the layered structure, a 2D network constructed by the connection between metal–oxygen polyhedra and phosphorous–oxygen tetrahedra. Similar to many other 2D materials (clays, etc.), such a crystal structure renders properties such as ion exchange and ion conductivity to the family of metal phosphate materials [30,92,93]. Great progress has been witnessed in the last three decades for the preparation and structurization of metal phosphates into different morphological forms from nanoparticles (0D) [94], nanorods (1D) [95], thin films (2D) [96], and porous materials (3D) [97,98] to composites [99]. Among all the methods so far developed, the sol–gel process has been proven as an efficient and versatile one for the introduction of pores into metal phosphate materials [100,101]. In this section, our recent progress in the preparation of hierarchically porous metal phosphate monoliths via sol–gel processes is introduced.

9.7.1
Calcium Phosphate

As perhaps the best-known biomineral, the morphological control during the synthesis of calcium phosphate materials is always of great importance. Starting from calcium chloride dihydrate ($CaCl_2 \cdot 2H_2O$) and phosphoric acid as both the calcium and phosphate sources, with a mixture of water and methanol as the solvent, poly(acrylic acid) (100 kDa) as the phase separation inducer, and propylene oxide as the pH adjuster, hierarchically porous dicalcium phosphate anhydrous ($CaHPO_4$) monolith is synthesized (Figure 9.20) [102]. The addition of poly(acrylic acid) efficiently suppress the excessive crystal growth, leading to a compositionally homogeneous gel. The median macropore size of the obtained calcium phosphate monolith is 0.5 μm and the mesopore distribution derived from interstices of primary particles is relatively broad. Calcination at 700 °C leads to the transformation of the crystal phase into hydroxyapatite and a small amount of β-tricalcium phosphate with the macroporous structure preserved. Both the crystals of dicalcium phosphate anhydrous and hydroxyapatite bear a

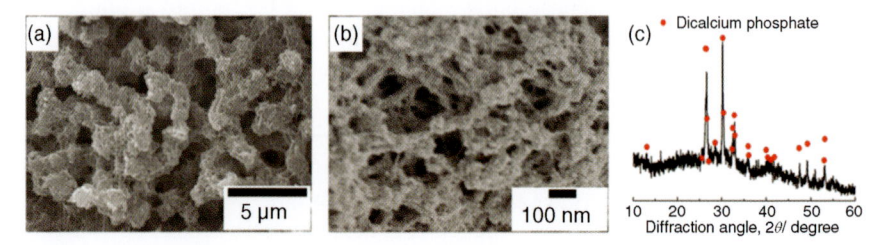

Figure 9.20 SEM image (a), FE-SEM image (b), and XRD pattern (c) of hierarchically porous calcium phosphate monolith.

9.7.2
Zirconium Phosphate and Its NaSICON-Type Derivatives

As a layered acidic compound, zirconium phosphate has found its applications in various fields such as adsorption [103], catalysis [104], and fuel cells [105]. Starting from $ZrOCl_2 \cdot 8H_2O$ and phosphoric acid as the metal and phosphate sources, with 1 M HCl aqueous and glycerol as the solvent and polyethylene oxide (35 kDa) and polyacrylamide (10 kDa) as the phase separation inducers, a hierarchically porous zirconium phosphate monolith is obtained (Figure 9.21a and b) [31]. Macropore size can be tuned by changing the amount of polymers from 0.5 to 5 μm. Instead of two-dimensional sheet-like morphology, the constituent zirconium phosphate nanoparticles are globular and are less than 10 nm in size, the interstitials of which lead to mesopores of 5 nm in average size. Compositionally, the obtained material is close to α-zirconium phosphate. However, a poorly crystalline structure with expanded interlayers is obtained due to the suppression of particle growth during the synthesis (Figure 9.21c). This material, because of its high surface area (600 m^2/g), hierarchical porosity, and layered crystal structure, is of particular interest for continuous flow adsorption and catalysis. A continuous flow setup is fabricated with zirconium phosphate

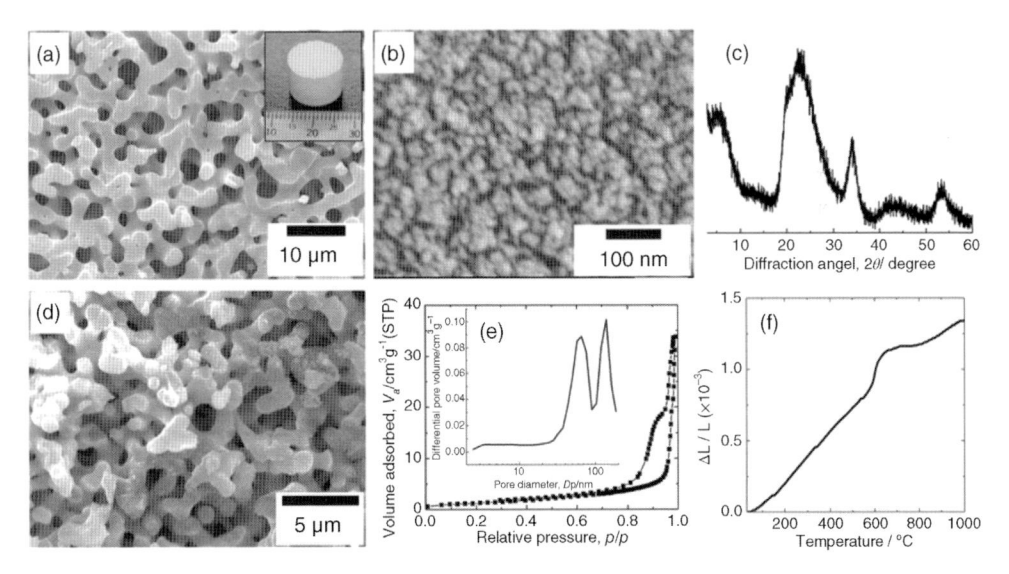

Figure 9.21 SEM image (a, inset is the appearance of the gel), FE-SEM image (b), and XRD pattern (c) of hierarchically porous zirconium phosphate monolith. SEM image (d), N$_2$ adsorption desorption isotherm (BJH pore size distribution obtained from the adsorption branch) (e), and temperature-dependent bulk thermal expansion curve (f) of hierarchically porous strontium zirconium phosphate monolith (after calcination at 1000 °C).

Table 9.2 Concentrations of metal ions in solutions before and after adsorption by hierarchically porous zirconium phosphate.

Metal ions	Ag^+	Cs^+	Sr^{2+}	Zn^{2+}	Cu^{2+}	Pb^{2+}	Cd^{2+}	Fe^{3+}
Before adsorption/ppm	1070	1280	1100	290	740	2200	1150	620
After adsorption	63 ppm	1.5 ppm	0.3 ppm	0.4 ppm	<1 ppb	<10 ppb	87 ppm	<1 ppb
Percentage of adsorption/%	94	99.9	>99.9	99.9	>99.9	>99.9	92	>99.9

Temperature: 25 °C, solution volume: 10 ml, flow rate: 0.2 ml/min.

monolith and subjected to the removal of different heavy metals in water. High efficiency for the removal of ions is confirmed for all the metals in Table 9.2. The concentrations of some metal ions (Cu^{2+}, Pb^{2+}, Fe^{3+}) in the solution after adsorption could not be detected even by the ICP-AES instrument (Varian 720-ES, Agilent Technology Inc.). Other applications of such hierarchically porous zirconium phosphate monoliths are as acidic catalysts for catalytic reactions, as catalyst supports for the immobilization of proteins or noble metal nanoparticles, and as proton conductor.

Introduction of other metal species into the crystal structure of zirconium phosphate leads to the formation of sodium superionic conductor (NaSICON)-type metal zirconium phosphates, a family of compounds well known for their ion conductivity and ultralow/negative thermal expansion property [106–108]. By adding the target metal salts in the starting composition for the synthesis of zirconium phosphate and following the same synthesis procedure, macroporous monolithic gels are achieved (Figure 9.21d) [32]. Instead of solvent exchange, direct drying is applied to avoid the leakage of metal salts. Glycerol as the cosolvent in the starting composition remains in the gel network due to its high boiling point and low vapor pressure, which coincidentally acts as the solvent for the stabilization of metal salts from precipitation. Therefore, the target metal species can be homogeneously distributed in the gel network after drying. Calcination at elevated temperatures (in most of the cases, 1000 °C) gives rise to the incorporation of metal ion into the crystallizing gel network, resulting in the formation of macroporous NaSICON-type zirconium phosphate monoliths. Successes have been achieved for systems of various metal species (alkaline metal, alkaline earth metal, transition metals, and rare earth metals) and some mixtures of them (both metals of same valence and metals of different valences). In the case of strontium zirconium phosphate, the presence of nanoscale pores even after the calcination at 1000 °C can act as the free space to buffer the anisotropic thermal expansion of each crystallite at elevated temperatures (Figure 9.21e). The coefficient of thermal expansion of our hierarchically porous strontium zirconium phosphate monolith is 1.4×10^{-6}/K (Figure 9.21f), lower than that of the reported dense bulky ceramics (2.2×10^{-6}/K) and falls right into the category of ultralow thermal expansion materials (coefficient of thermal expansion $\leq 2 \times 10^{-6}$/K) [109]. Other potential applications for such NaSICON-type metal zirconium phosphate porous monoliths are as solid-state electrolyte due to their high ion conductivity and for the solidification of radioactive nuclear waste due to their high structural and chemical stability.

9.7.3
Titanium Phosphate

Another important member of metal(IV) phosphates is titanium phosphate. Starting from $TiOSO_4 \cdot xH_2O$ as a cheaper titanium precursor and phosphoric acid as the phosphate source, with a mixture of H_2O, glycerol, and dimethyl sulfoxide as the solvent and polyethylene oxide (100 kDa) and polyvinylpyrrolidone (55 kDa) as the phase separation inducers, hierarchically porous titanium phosphate monolith is obtained (Figure 9.22a–c) [33]. Addition of dimethyl sulfoxide helps the suppression of particle growth. Therefore, similar to the case of zirconium phosphate, the skeleton of the monolith is constituted by globular amorphous nanoparticles (20–30 nm in size). However, due to the amorphous nature of crystal structure, the composition of the obtained monolith is somewhat complicated. Although no layered structure is observed in the dried titanium phosphate monolith, solvothermal treatment in ethylene glycol leads to the

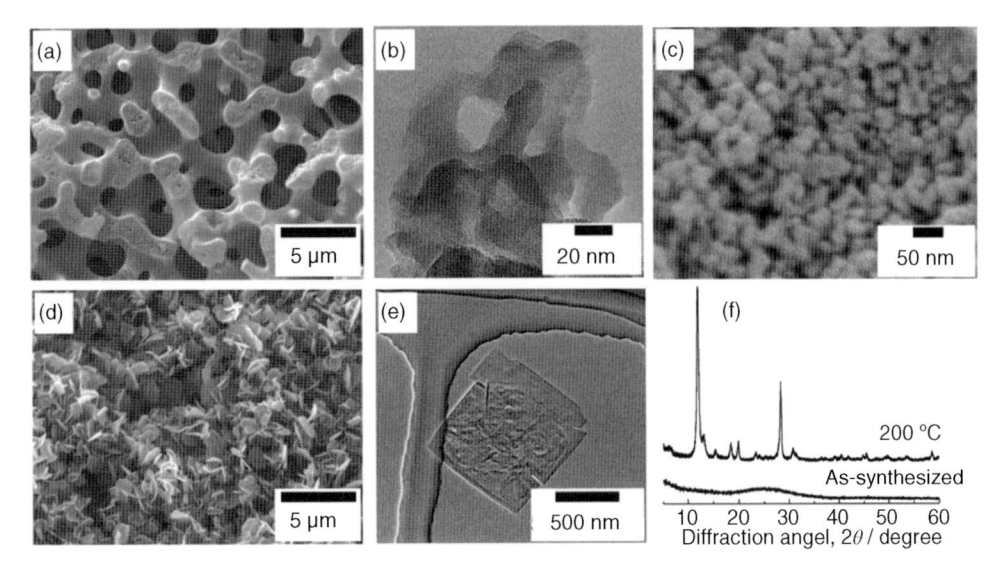

Figure 9.22 SEM image (a), TEM image (b), and FE-SEM image (c) of hierarchically porous titanium phosphate monolith. SEM image (d) and TEM image (e) of titanium phosphate monolith solvothermally treated in ethylene glycol at 200 °C for 24 h. XRD patterns (f) of both samples.

crystallization of the gel network and the subsequent transformation of the globular nanoparticles into thin nanosheets, resulting in a crystalline porous titanium phosphate monolith (Figure 9.22d–f). Temperature dependency of the crystal growth kinetics is observed, which affects the size of the nanosheets in the resultant monolith. The dried titanium phosphate monolith, due to its high surface area (320 m^2/g) and hierarchical porous structure with large mesopore size (21 nm) can readily be applied as adsorbent, catalyst/catalyst support and electrode materials for ion batteries. In the meantime, crystallization in ethylene glycol at high temperature is expected to lead to drastic changes in the properties from the starting titanium phosphate monolith and to bring enhancement for the applications such as ion adsorption.

9.8
Functionalization by Postreduction: Carbon, Reduced Oxides, Carbides, and Nitrides

9.8.1
Carbon

Porous carbon materials have been utilized since the dawn of human history. In addition to the traditional wood-derived carbons, the growing application fields currently require finely designed porous carbon materials with controlled pore properties over multiple length scales from subnanometer to micrometer order [110]. Hierarchically porous structures can improve various functionalities of carbon due to the high surface area arising from small pores (micropores) and the efficient mass transport stemming from large pores (meso- and macropores).

Fabrication of cross-linked polymer precursors with designed porous structures is a promising way to synthesize porous carbons in terms of feasibility and controllability. If a suitable polymer is employed, the tailored porous structure can be preserved through the carbonization process. The following shows a couple of examples of the porous carbon monoliths characterized by the interconnected macroporous structure, which are derived from the macroporous polymers obtained via the sol–gel reaction accompanied by spinodal decomposition.

The first example is the micro/meso/macroporous activated carbon monolith prepared from the poly(divinylbenzene) (PDVB) gel [111,112], as presented in Figure 9.23. The combination of the living radical polymerization of divinylbenzene and the phase separation method yields macroporous PDVB gels with well-defined mesopores. The carbonization of PDVB is, however, associated with large shrinkage, resulting in collapse of mesopores [113]. The sulfonation of PDVB effectively suppresses the shrinkage, which allows the PDVB-derived carbon monolith to retain meso- and macroporous structures [111,114]. The subsequent activation process by CO_2 provides the micro/meso/macroporous carbon

9.8 Functionalization by Postreduction: Carbon, Reduced Oxides, Carbides, and Nitrides | 227

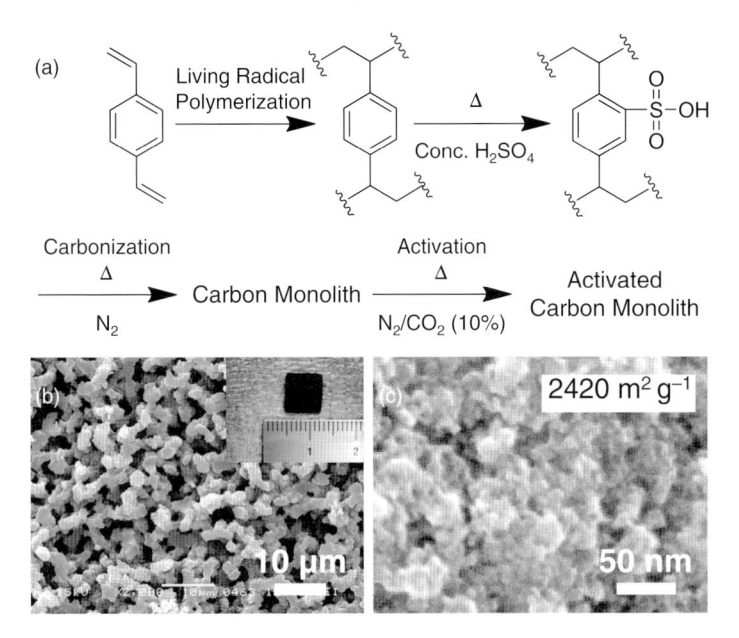

Figure 9.23 (a) Synthesis scheme showing the synthesis of activated carbon monoliths derived from porous PDVB monoliths. (b) Appearance (*inset*) and macroporous morphology of a representative activated carbon monolith. (c) Mesoporous morphology of the cross section of macropore skeleton in the activated carbon monolith.

monoliths with high specific surface area (~2400 m^2/g), which is a promising binder-free electrode for supercapacitors [112].

Resorcinol-formaldehyde (RF) gel is a good candidate for a carbon precursor due to relatively low shrinkage and high yield during carbonization [115]. Since the polymerization-induced phase separation of RF gels can be controlled in aqueous media simply by adjusting the polarity of solvent, the addition of metal salts into the sol provides macroporous RF gels complexed with the corresponding metal ions. The metal ions embedded in the polymer network act as a catalyst for graphitization at as low as 1000 °C [116]. Hence, the porous RF gels with metal ions end up with the graphitized carbon monoliths after the calcination at relatively low temperatures (<1500 °C) followed by washing with HCl aqueous solution [117], as depicted in Figure 9.24. The graphitized carbon monoliths show superior electrical conductivity of ~13 S/cm, which is one order of magnitude higher than that of the amorphous carbon monoliths with the similar pore properties.

Heteroatom doping into carbon matrix further broadens the horizons of carbon materials. The heteroatoms introduced in the polyaromatic frameworks and the heteroatom-containing surface functional groups can arouse or enhance a variety of capabilities of carbon materials, such as catalytic activities and electrochemical characteristics [118]. Figure 9.25 demonstrates an example of the

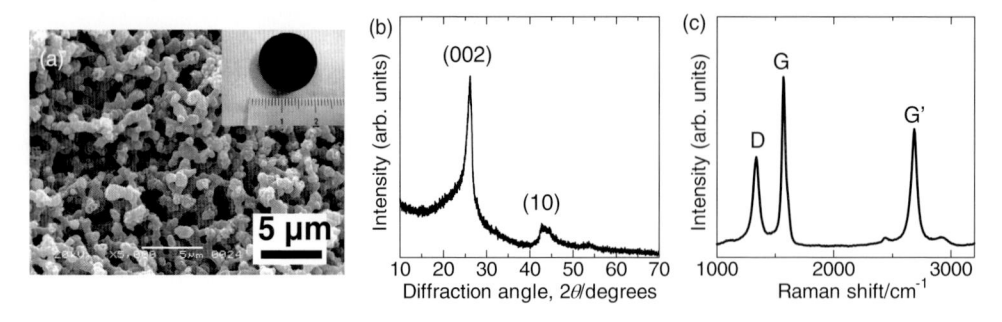

Figure 9.24 (a) Appearance (*inset*) and macroporous structure of a typical graphitized carbon monolith. (b) XRD pattern and (c) Raman spectrum of the graphitized carbon monolith.

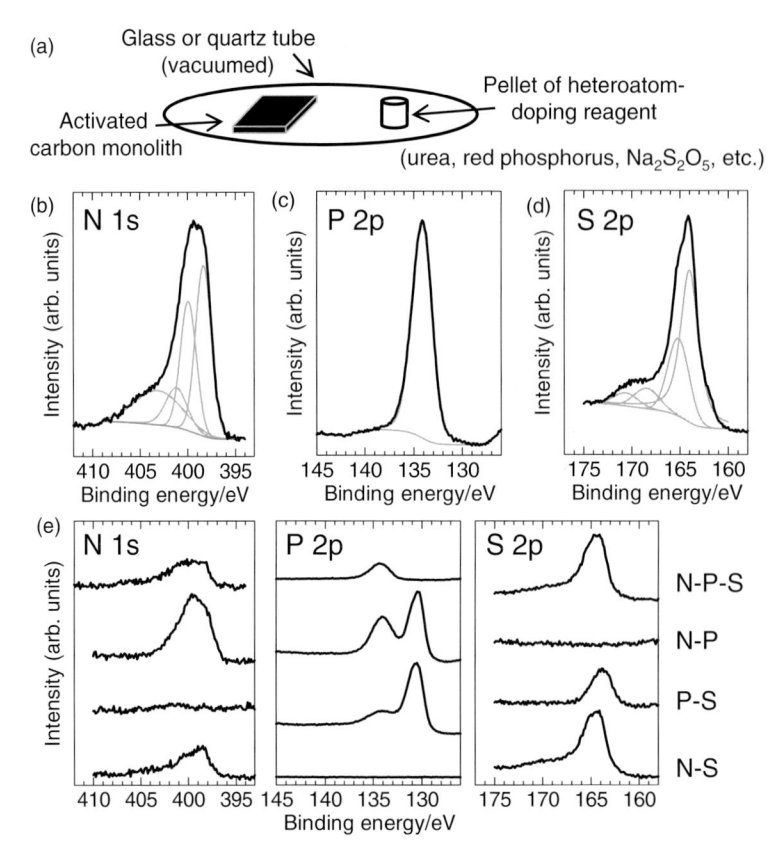

Figure 9.25 (a) Schematic illustration of the posttreatment of carbons for heteroatom doping. (b) N 1s XPS spectrum of the N-doped carbon treated with urea. (c) P 2p XPS spectrum of the P-doped carbon treated with red phosphorus. (d) S 2p XPS spectrum of the S-doped carbon treated with $Na_2S_2O_5$. (e) XPS spectra of the dual- (N–P, P–S, and N–S) and triple-doped carbons (N–P–S).

heteroatom doping of carbon by the facile and versatile posttreatment [119]. The heat treatment of carbon materials with a reagent, which is stable in an ambient atmosphere and evolves reactive gases on heating, in a vacuum-closed container allows the introduction of manifold heteroatom-containing functional groups into the carbon matrix. In addition, this procedure can also provide dual- and triple-heteroatom-doped carbons by the sequential doping reactions (Figure 9.25e). The pore properties of the precursor carbon materials are well preserved through the doping process, which indicates that the independent tuning of heteroatom doping and nanostructural design becomes possible in relation to various carbon materials.

9.8.2
Silicon Oxycarbide (SiO_xC_y) and Silicon Carbide (SiC)

Calcination of inorganic–organic hybrid networks under inert atmosphere gives rise to non-oxide ceramic materials, which is known as "preceramic polymer route" [120]. This methodology is to be incorporated with the structural design of precursors so as to control the porous morphology of various reduced ceramic materials. Herein, the preceramic polymer route to porous SiO_xC_y and SiC monoliths and the control of their pore properties are demonstrated.

The sol–gel reaction of arylene-bridged alkoxysilane accompanied by phase separation yields macroporous arylene-bridged poly(silsesquioxane) gels [11,12], as illustrated in Figure 9.26. The mesopore control by the posttreatment in a basic aqueous solution (Section 9.2.4) is also applicable to such poly(silsesquioxane)-based gels, leading to hierarchically porous inorganic–organic hybrid materials [13]. On calcining, the hybrid network changes to an amorphous matrix based on Si—O and Si—C bonds, that is, SiO_xC_y, together with carbonaceous species. Above 1200 ∼ 1300 °C, the carbothermal reduction of SiO_xC_y facilitates the crystallization into β-SiC (Figure 9.27a), resulting in SiC monoliths that retain the macroporous structure, as shown in Figure 9.27b. When the phenylene-bridged poly(silsesquioxane) is employed, the mesopore size of the SiC monoliths increases as the calcination temperature becomes higher because of the crystal growth of SiC (Figure 9.27c and d). On the other hand, starting from

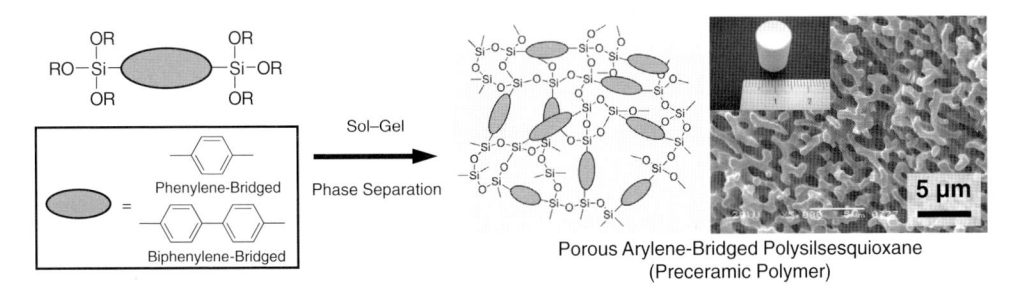

Figure 9.26 Synthesis pathway of the porous arylene-bridged polysilsesquioxane monoliths.

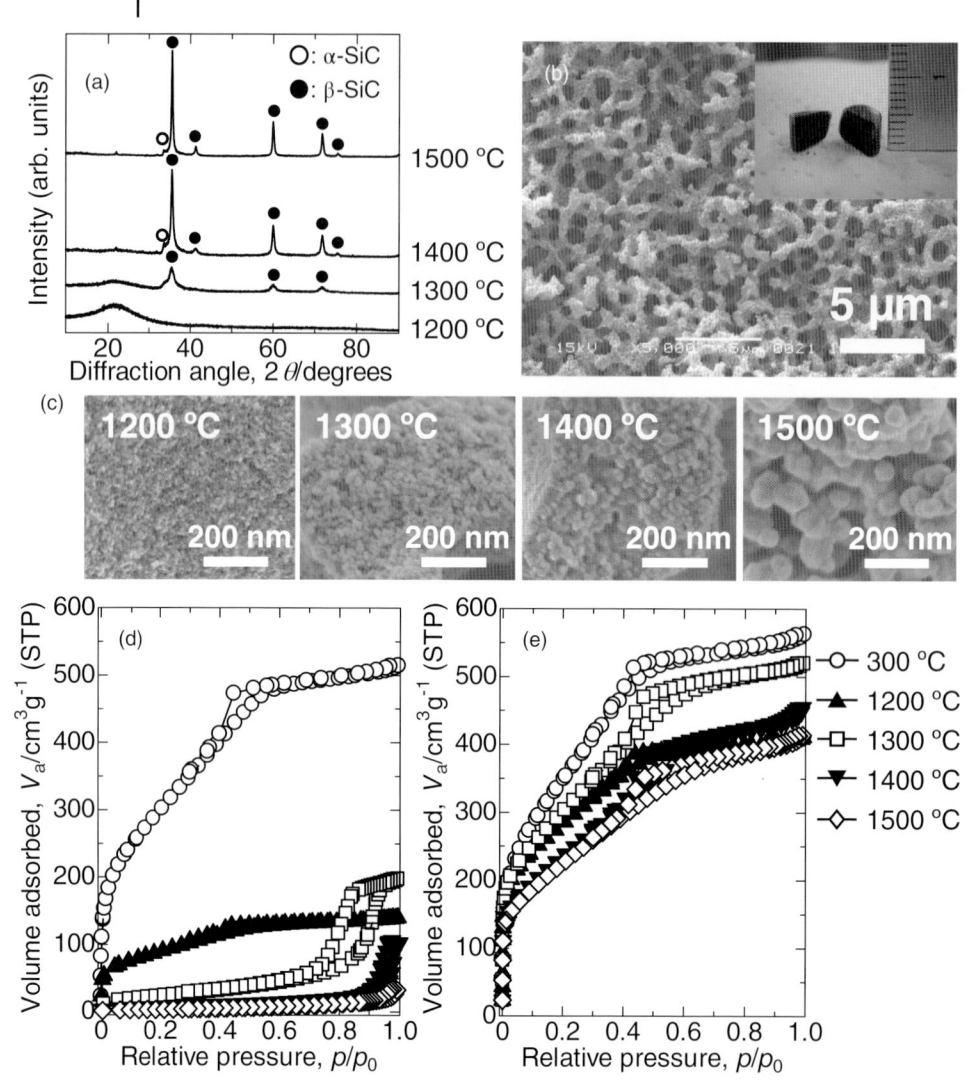

Figure 9.27 (a) XRD patterns of the porous SiO_xC_y and SiC monoliths derived from the phenylene-bridged polysilsesquioxanes calcined at different temperatures. (b) Appearance (inset) and macroporous structure of the SiC monolith calcined at 1500 °C. (c) Magnified images of the cross section of macropore skeleton in the SiO_xC_y and SiC monoliths calcined at different temperatures. (d and e) Comparison between the N_2 sorption isotherms of the SiO_xC_y and SiC monoliths derived from (d) phenylene-bridged and (e) biphenylene-bridged polysilsesquioxanes.

the biphenylene-bridged counterparts, the calcined sample contains a large quantity of carbon even after the calcination at 1500 °C, which allows maintaining the original mesoporous structure in the precursor, as indicated in Figure 9.27e.

9.8.3
Reduced Titanium Oxides (Ti_nO_{2n-1}) and Titanium Nitride (TiN)

Reduced titanium oxides are featured with a variety of compositions expressed as Ti_nO_{2n-1} ($n = 2-10$), and each of them exhibits unique magnetic and electric characteristics [121–123]. The control of crystal phase is therefore of great importance in addition to the architectural design. One approach is the reduction of TiO_2 materials with designed morphology. For instance, the macroporous reduced titanium oxide monoliths can be prepared via the reduction of the porous TiO_2 monoliths (Section 9.5) with Zr metal in a vacuum-closed container at 1050–1200 °C (Figure 9.28) [124]. The Zr metal acts as an oxygen scavenger according to the following reaction: $2n\ TiO_2 + Zr \rightarrow 2\ Ti_nO_{2n-1} + ZrO_2$. By adjusting the TiO_2/Zr ratio, a series of reduced titanium oxides can be selectively synthesized as a single phase, as shown in Figure 9.28b. Although the reduced titanium oxide monoliths with well-defined macroporous structure originated from the TiO_2 precursor show good electrical conductivity (Figure 9.28d), the high-temperature treatment destroys micro- and mesopores owing to the

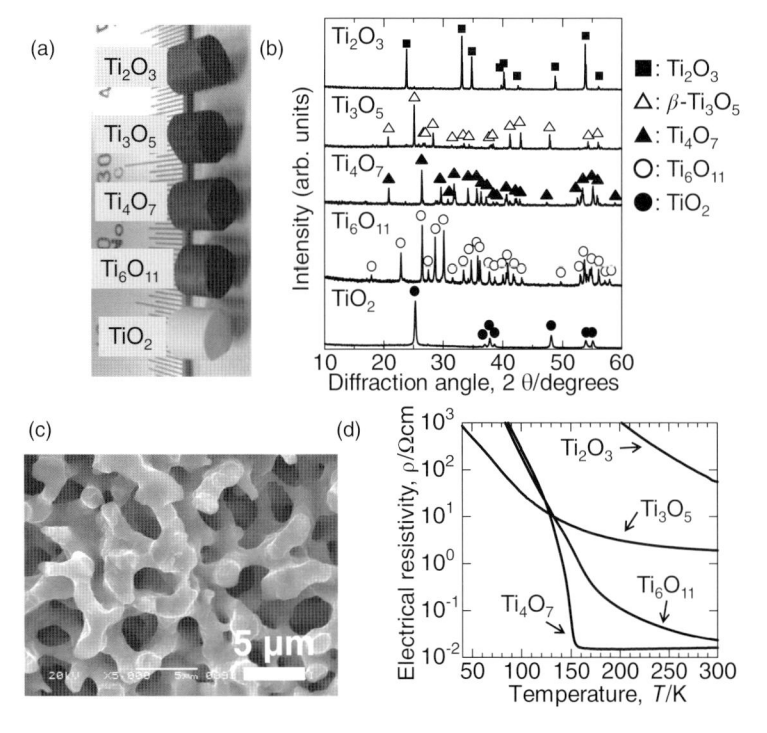

Figure 9.28 (a) Appearance and (b) XRD patterns of the porous monoliths based on a series of titanium oxides. (c) Macroporous structure of the Ti_4O_7 monolith. (d) Temperature dependence of the electrical resistivity for the porous monoliths based on a series of reduced titanium oxides.

sintering, resulting in small specific surface area. In the meantime, the reduction of porous TiO_2 with NH_3 gas at 1000 °C provides porous TiN monoliths with meso- and macropores, which show excellent electrical conductivity (about 320 S/cm at room temperature) along with the superconducting transition at $T_c \sim 5.0$ K [125].

Another strategy is the preceramic polymer route from inorganic–organic hybrid networks likewise the synthesis of SiC as described in Section 9.8.2. As the Ti—O bond is fairly strong, the titanium-based preceramic polymer needs to be created from the aspect of molecular design. When ethylenediamine is employed as a cross-linker, three-dimensionally cross-linked titanium-based networks are formed, leading to the preceramic polymer with macroporous morphology by applying the phase separation method (Figure 9.29a) [126,127]. The calcination of the preceramic polymer under inert atmosphere gives rise to the crystallization into Ti_4O_7, γ-Ti_3O_5, and Ti_2O_3, ending up in cubic TiO_xN_y, as shown in Figure 9.29b. It is noteworthy that even when the hybrid network initially crystallizes into TiO_2, the carbothermal reduction into the reduced phases takes place at relatively low temperatures ($\sim$800 °C). This is because the incorporation of N atoms distorts the O—Ti—O lattice, diminishing the lattice enthalpy [128]. In addition, the preceramic polymer route yields composites of

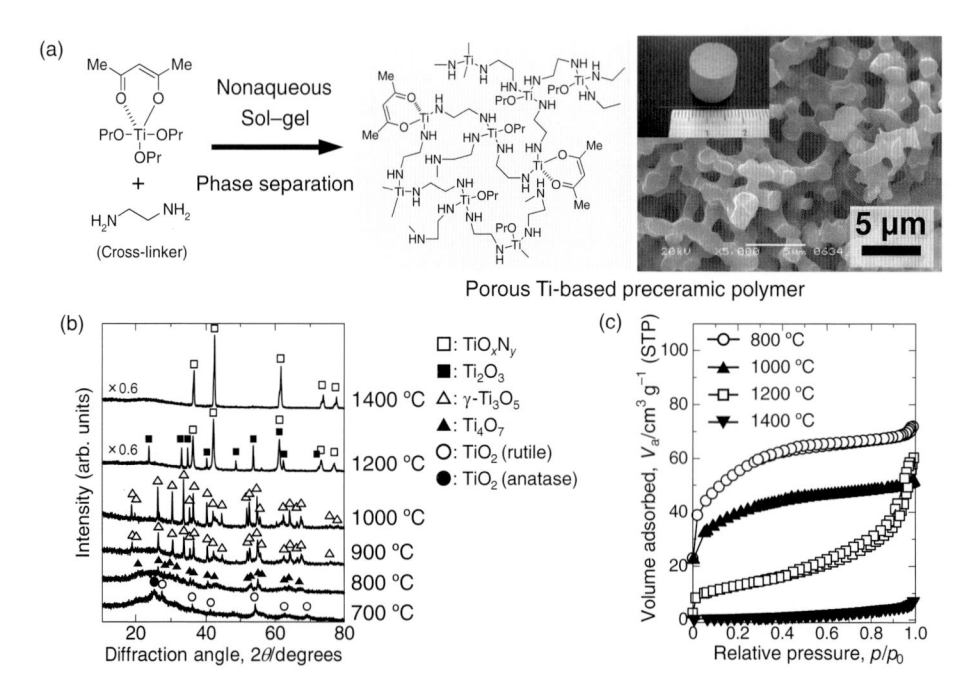

Figure 9.29 (a) Scheme of the nonaqueous sol–gel process to obtain the porous Ti-based preceramic polymer. (b) XRD patterns of the reduced titanium oxides prepared via the preceramic polymer route. (c) N_2 isotherms of the reduced titanium oxides calcined at different temperatures.

reduced titanium oxides and carbon, resulting in relatively high specific surface area (~200 m^2/g) with micro- and mesoporosities, as shown in Figure 9.29c.

9.8.4
Other Metal Carbides and Nitrides via "Urea Glass Route"

As already demonstrated, the preceramic polymer route is a fascinating methodology to obtain non-oxide ceramic materials. With respect to the versatility, however, the type of transition metal elements applicable to this technique is still limited because of the difficulty in controlling both gelation and phase separation. On the other hand, the epoxide-mediated sol–gel systems are introduced as a versatile means to fabricate the porous monoliths based on transition metals (Section 9.6). Meanwhile, it is reported that urea can be utilized as nitrogen and/or carbon sources for preparing various metal oxynitrides, nitrides, and carbides, which is known as the "urea glass route" [129,130]. The combination of these two strategies has the potential to be a versatile and facile method to obtain porous monolithic materials based on multiple non-oxide ceramics.

The meso- and macroporous chromium-based monoliths can be fabricated according to the aforementioned procedure. In this case, the precursor is a porous chromium (oxy)hydroxide/urea composite obtained simply by adding urea to the chromium-based sol followed by gelation and drying. On calcining the composite under inert atmosphere, the thermal decomposition of urea generates various reactive gases, such as NH_3 and HNCO [131], in the monolith, leading to the crystallization into CrN. The polymer added as a phase separation inducer changes into carbon, which supports the monolithic shape and macroporous morphology, resulting in the crack-free CrN monolith with hierarchical porosity [29]. The heat treatment above 900 °C encourages the crystal transition from CrN into Cr_3C_2 via the reaction with the surrounding carbon. This approach would be applicable not only to other metal nitrides and carbides but also to those with more complex composition with several metal elements.

9.9
Summary

The sol–gel processing of porous materials has been introduced. Especially, a focus was put on monolithic porous gels that can be prepared via sol–gel reaction accompanied by polymerization-induced phase separation. After continuous development over 20 years, the formation of well-defined macropores in sol–gel systems in a broad range of chemical compositions has been explored, and the versatility of the method was clearly demonstrated. Most metal oxide gels retain their inherent porosity within the gel framework comprising the continuously macroporous monoliths, so that additional aging treatments under basic and/or hydrothermal conditions worked efficiently to tailor the mesopores essentially independent of the preformed macroframeworks. Although this chapter focuses on demonstrating the formation of hierarchically porous structures, also included are some related topics such as low-density aerogel-like materials with unimodal well-defined porosity. The introduced applications were limited to those proven to work and/or commercialized in separation sciences and biopurification. Further details on basic concepts are best found in one of the reviews [5]. Regarding materials prepared from ionic precursors (metal salts) in aqueous media, mechanically stable aerogel-like transparent low-density hybrid materials, and those convertible to non-oxide ceramics and carbons, there remains a lot to be explored in the near future from both basic and application viewpoints.

Acknowledgments

Contributions of Prof. Koji Fujita and Dr. Shunsuke Murai, Kyoto University, are warmly acknowledged. All the research would have been impossible without full support of Emeritus Profs. Naohiro Soga, and Teiichi Hanada, Kyoto University. Heartfelt thanks also go to all the students, postdocs, and researchers from companies for their indispensable inputs throughout the research. Stable financial supports from Kakenhi, MEXT, Japan and JST-ALCA project, Japan are especially acknowledged.

References

1 Shoup, R.D. (1976) Controlled pore silica bodies gelled from silica–alkali silicate mixtures, in *Colloid and Interface Science*, vol. **3** (ed. M. Kerker), Academic Press, New York, pp. 63–69.

2 Nakanishi, K. and Soga, N. (1991) Phase separation in gelling silica–organic polymer solution: systems containing poly[sodium styrenesulfonate]. *J. Am. Ceram. Soc.*, **74**, 2518–2530.

3 Nakanishi, K. (2006) Sol–gel process of oxides accompanied by phase separation. *Bull. Chem. Soc. Jpn.*, **79**, 673–691.

4 Brinker, C.J. and Scherer, G.W. (1990) *Sol-Gel Science; The Physics and Chemistry of Sol-Gel Processing*, Academic Press, San Diego, CA.

5 Nakanishi, K. (1997) Pore structure control of silica gels based on phase separation. *J. Porous Mater.*, **4**, 67–112.

6 Konishi, J., Fujita, K., Nakanishi, K., and Hirao, K. (2006) Monolithic TiO_2 with controlled multiscale porosity via a template-free sol–gel process accompanied by phase separation. *Chem. Mater.*, **18**, 6069–6074.

7 Konishi, J., Fujita, K., Oiwa, S., Nakanishi, K., and Hirao, K. (2008) Crystalline ZrO_2 monoliths with well-defined macropores and mesostructured skeletons prepared by combining the alkoxy-derived sol–gel process accompanied by phase separation and the solvothermal process. *Chem. Mater.*, **20**, 2165–2173.

8 Kanamori, K., Aizawa, M., Nakanishi, K., and Hanada, T. (2007) New transparent methylsilsesquioxane aerogels and xerogels with improved mechanical properties. *Adv. Mater.*, **19**, 1589–1593.

9 Hayase, G., Kanamori, K., Fukuchi, M., Kaji, H., and Nakanishi, K. (2013) Facile synthesis of marshmallow-like macroporous gels usable under harsh conditions for the separation of oil and water. *Angew. Chem., Int. Ed.*, **52**, 1986–1989.

10 Kobayashi, Y., Amatani, T., Nakanishi, K., Hirao, K., and Kodaira, T. (2004) Spontaneous formation of hierarchical macro-mesoporous ethane-silica monolith. *Chem. Mater.*, **16**, 3652–5658.

11 Hasegawa, G., Kanamori, K., Nakanishi, K., and Hanada, T. (2009) Fabrication of macroporous silicon carbide ceramics by intramolecular carbothermal reduction of phenyl-bridged polysilsesquioxane. *J. Mater. Chem.*, **19**, 7716–7720.

12 Hasegawa, G., Kanamori, K., Nakanishi, K., and Hanada, T. (2010) A new route to monolithic macroporous SiC/C composites from biphenylene-bridged polysilsesquioxane gels. *Chem. Mater.*, **22**, 2541–2547.

13 Hasegawa, G., Kanamori, K., and Nakanishi, K. (2012) Pore properties of hierarchically porous carbon monoliths with high surface area obtained from bridged polysilsesquioxanes. *Microporous Mesoporous Mater.*, **155**, 265–273.

14 Konishi, J., Fujita, K., Nakanishi, K., and Hirao, K. (2006) Phase-separation-induced titania monoliths with well-defined macropores and mesostructured framework from colloid-derived sol–gel systems. *Chem. Mater.*, **18**, 864–866.

15 Yachi, A., Takahashi, R., Sato, S., Sodesawa, T., and Azuma, T. (2004) Silica with bimodal pores for solid catalysts prepared from water glass. *J. Sol-Gel Sci. Technol.*, **31**, 373–376.

16 Takahashi, R., Sato, S., Sodesawa, T., Oguma, K., Matsutani, K., and Mikami, N. (2005) Silica gel with continuous macropores prepared from water glass in the presence of poly(acrylic acid). *J. Non-Cryst. Solids*, **351**, 331–339.

17 Tokudome, Y., Fujita, K., Nakanishi, K., Miura, K., and Hirao, K. (2007) Synthesis of monolithic Al_2O_3 with well-defined macropores and mesostructured skeletons via the sol–gel process accompanied by phase separation. *Chem. Mater.*, **19**, 3393–3398.

18 Fujita, K., Tokudome, Y., Nakanishi, K., Miura, K., and Hirao, K. (2008) Cr^{3+}-doped macroporous Al_2O_3 monoliths prepared by the metal-salt-derived sol–gel method. *J. Non-Cryst. Solids*, **354**, 659–664.

19 Tokudome, Y., Nakanishi, K., and Hanada, T. (2009) Effect of La addition on thermal microstructural evolution of macroporous alumina monolith prepared from ionic precursors. *J. Ceram. Soc. Jpn.*, **117**, 351–355.

20 Kido, Y., Nakanishi, K., Miyasaka, A., and Kanamori, K. (2012) Synthesis of monolithic hierarchically porous iron-based xerogels from iron(III) salts via an epoxide-mediated sol–gel process. *Chem. Mater.*, **24**, 2071–2077.

21 Kido, Y., Nakanishi, K., Okumura, N., and Kanamori, K. (2013) Hierarchically porous nickel/carbon composite monoliths prepared by sol–gel method from an ionic precursor. *Microporous Mesoporous Mater.*, **176**, 64–70.

22 Fukumoto, S., Nakanishi, K., and Kanamori, K. (2015) Direct preparation and conversion of copper hydroxide-based monolithic xerogels with hierarchical pores. *New J. Chem.*, **39**, 6771–6777.

23 Tokudome, Y., Fujita, K., Nakanishi, K., Kanamori, K., Miura, K., Hirao, K., and Hanada, T. (2007) Sol–gel synthesis of macroporous YAG from ionic precursors via phase separation route. *J. Ceram. Soc. Jpn.*, **115**, 925–928.

24 Murai, S., Fujita, K., Iwata, K., and Tanaka, K. (2011) Scattering-based hole burning in $Y_3Al_5O_{12}$:Ce^{3+} monoliths with hierarchical porous structures prepared via the sol–gel route. *J. Phys. Chem. C*, **115**, 7676–7681.

25 Guo, X.Z., Nakanishi, K., Kanamori, K., Zhu, Y., and Yang, H. (2014) Preparation of macroporous cordierite monoliths via the sol–gel process accompanied by phase separation. *J. Eur. Ceram. Soc.*, **34**, 817–823.

26 Guo, X.Z., Cai, X.B., Song, J., Zhu, Y., Nakanishi, K., Kanamori, K., and Yang, H. (2014) Facile synthesis of monolithic mayenite with well-defined macropores via an epoxide-mediated sol–gel process accompanied by phase separation. *New J. Chem.*, **38**, 5832–5839.

27 Tokudome, Y., Tarutani, N., Nakanishi, K., and Takahashi, M. (2013) Layered double hydroxide (LDH)-based monolith with interconnected hierarchical

channels: enhanced sorption affinity for anionic species. *J. Mater. Chem. A*, **1**, 7702–7708.

28 Tarutani, N., Tokudome, Y., Fukui, M., Nakanishi, K., and Takahashi, M. (2015) Fabrication of hierarchically porous monolithic layered double hydroxide composites with tunable microcages for effective oxyanion adsorption. *RSC Adv.*, **5**, 57187–57192.

29 Kido, Y., Hasegawa, G., Kanamori, K., and Nakanishi, K. (2014) Porous chromium-based ceramic monoliths: oxides (Cr_2O_3), nitrides (CrN), and carbides (Cr_3C_2). *J. Mater. Chem. A*, **2**, 745–752.

30 Clearfield, A. (1988) Role of ion exchange in solid-state chemistry. *Chem. Rev.*, **88**, 125–148.

31 Zhu, Y., Shimizu, T., Kitashima, T., Morisato, K., Moitra, N., Brun, N., Kanamori, K., Takeda, K., Tafu, M., and Nakanishi, K. (2015) Synthesis of robust hierarchically porous zirconium phosphate monolith for efficient ion adsorption. *New J. Chem.*, **39**, 2444–2450.

32 Zhu, Y., Kanamori, K., Moitra, N., Kadono, K., Ohi, S., Shimobayashi, N., and Nakanishi, K. (2016) Metal zirconium phosphate macroporous monoliths: versatile synthesis, thermal expansion and mechanical properties. *Microporous Mesoporous Mater.*, **225**, 122–127.

33 Zhu, Y., Yoneda, K., Kanamori, K., Takeda, K., Kiyomura, T., Kurata, H., and Nakanishi, K. (2016) Hierarchically porous titanium phosphate monoliths and their crystallization behavior in ethylene glycol. *New J. Chem.* doi: 10.1039/C5NJ02820E

34 Gash, A.E., Tillotson, T.M., Satcher, J.H., Hrubesh, L.W., and Simpson, R.L. (2001) New sol–gel synthetic route to transition and main-group metal oxide aerogels using inorganic salt precursors. *J. Non-Cryst. Solids*, **285**, 22–28.

35 Woo, K., Lee, H.J., Ahn, J.P., and Park, Y.S. (2003) Sol–gel mediated synthesis of Fe_2O_3 nanorods. *Adv. Mater.*, **15**, 1761–1764.

36 Baumann, T.F., Gash, A.E., Chinn, S.C., Sawvel, A.M., Maxwell, R.S., and Satcher, J.H. (2005) Synthesis of high-surface-area alumina aerogels without the use of alkoxide precursors. *Chem. Mater.*, **17**, 395–401.

37 Flory, P.J. (1971) *Principles of Polymer Chemistry*, Cornell University Press, Ithaca, NY.

38 Hashimoto, T., Itakura, M., and Shimidzu, N. (1986) Late stage spinodal decomposition of a binary polymer mixture. II. Scaling analyses on $Q_m(\tau)$ and $I_m(\tau)$. *J. Chem. Phys.*, **85**, 6773–6786.

39 Kaji, H., Nakanishi, K., and Soga, N. (1993) Polymerization-induced phase separation in silica sol–gel systems containing formamide. *J. Sol-Gel Sci. Technol.*, **1**, 35–46.

40 Kuwatani, S., Maehana, R., Kajihara, K., and Kanamura, K. (2010) Amine-buffered phase separating tetraethoxysilane–water binary mixture: a simple precursor of sol–gel derived monolithic silica gels and glasses. *Chem. Lett.*, **39**, 712–713.

41 Kajihara, K., Kuwatani, S., Maehana, R., and Kanamura, K. (2009) Macroscopic phase separation in a tetraethoxysilane–water binary sol–gel system. *Bull. Chem. Soc. Jpn.*, **82**, 1470–1476.

42 Kajihara, K., Hirano, M., and Hosono, H. (2009) Sol–gel synthesis of monolithic silica gels and glasses from phase-separating tetraethoxysilane–water binary system. *Chem. Commun.*, 2580–2582.

43 Nakanishi, K., Takahashi, R., Nagakane, T., Kitayama, K., Koheiya, N., Shikata, H., and Soga, N. (2000) Formation of hierarchical pore structure in silica gel. *J. Sol-Gel Sci. Technol.*, **17**, 191–210.

44 Nakanishi, K., Nagakane, T., and Soga, N. (1998) Designing double pore structure in alkoxy-derived silica incorporated with nonionic surfactant. *J. Porous Mater.*, **5**, 103–110.

45 Sato, Y., Nakanishi, K., Hirao, K., Jinnai, H., Shibayama, M., Melnichenko, Y.B., and Wignall, G.D. (2001) Formation of ordered macropores and templated nanopores in silica sol–gel system incorporated with EO–PO–EO triblock copolymer. *Colloid Surf. A*, **187/188**, 117–122.

46 Nakanishi, K., Sato, Y., Ruyat, Y., and Hirao, K. (2003) Supramolecular

templating of mesopores in phase-separating silica sol–gels incorporated with cationic surfactant. *J. Sol-Gel Sci. Technol.*, **26**, 567–570.
47 Amatani, T., Nakanishi, K., and Hirao, K. (2005) Monolithic periodic mesoporous silica with well-defined macropores. *Chem. Mater.*, **17**, 2114–2119.
48 Nakanishi, K., Amatani, T., Yano, S., and Kodaira, T. (2008) Multiscale templating of siloxane gels via polymerization-induced phase separation. *Chem. Mater.*, **20**, 1108–1115.
49 Unger, K.K., Tanaka, N., and Machtejevas, E. (eds) (2011) *Monolithic Silicas in Separation Science: Concepts, Syntheses, Characterization, Modeling and Applications*, Wiley-VCH Verlag GmbH, Weinheim, Germany.
50 Ma, Y., Chassy, A.W., Miyazaki, S., Motokawa, M., Morisato, K., Uzu, H., Ohira, M., Furuno, M., Nakanishi, K., Minakuchi, H., Mriziq, K., Farkas, T., Fiehn, O., and Tanaka, N. (2015) Efficiency of short, small-diameter columns for reversed-phase liquid chromatography under practical operating conditions. *J. Chromatogr. A*, **1383**, 47–57.
51 Eghbali, H., Sandra, K., Detobel, F., Lynen, F., Nakanishi, K., Sandra, P., and Desmet, G. (2011) Performance evaluation of long monolithic silica capillary columns in gradient liquid chromatography using peptide mixtures. *J. Chromatogr. A*, **1218**, 3360–3366.
52 Iwasaki, M., Miwa, S., Ikegami, T., Tomita, M., Tanaka, N., and Ishihama, Y. (2010) One-dimensional capillary liquid chromatographic separation coupled with tandem mass spectrometry unveils the *Escherichia coli* proteome on a microarray scale. *Anal. Chem.*, **82**, 2616–2620.
53 Miyamoto, K., Hara, T., Kobayashi, H., Morisaka, H., Tokuda, D., Horie, K., Koduki, K., Makino, S., Núñez, O., Yang, C., Kawabe, T., Ikegami, T., Takubo, H., Ishihama, Y., and Tanaka, N. (2008) High-efficiency liquid chromatographic separation utilizing long monolithic silica capillary columns. *Anal. Chem.*, **80**, 8741–8750.

54 Kanamori, K. and Nakanishi, K. (2011) Controlled pore formation in organotrialkoxysilane-derived hybrids: from aerogels to hierarchically porous monoliths. *Chem. Soc. Rev.*, **40**, 754–770.
55 Loy, D.A., Baugher, B.M., Baugher, C.R., Schneider, D.A., and Rahimian, K. (2000) Substituent effects on the sol–gel chemistry of organotrialkoxysilanes. *Chem. Mater.*, **12**, 3624–3632.
56 Sugahara, Y., Okada, S., Sato, S., Kuroda, K., and Kato, C. (1994) ^{29}Si-NMR study of hydrolysis and initial polycondensation processes of organoalkoxysilanes. II. Methyltriethoxysilane. *J. Non-Cryst. Solids*, **167**, 21–28.
57 Dong, H., Lee, M., Thomas, R.D., Zhang, Z., Reidy, R.F., and Mueller, D.W. (2003) Methyltrimethoxysilane sol–gel polymerization in acidic ethanol solutions studied by ^{29}Si NMR spectroscopy. *J. Sol.-Gel. Sci. Technol.*, **28**, 5–14.
58 Nakanishi, K. and Kanamori, K. (2005) Organic–inorganic hybrid poly (silsesquioxane) monoliths with controlled macro- and mesopores. *J. Mater. Chem.*, **15**, 3776–3786.
59 Itagaki, A., Nakanishi, K., and Hirao, K. (2003) Phase separation in sol–gel system containing mixture of 3- and 4-functional alkoxysilanes. *J. Sol.-Gel. Sci. Technol.*, **26**, 153–156.
60 Kanamori, K., Yonezawa, H., Nakanishi, K., Hirao, K., and Jinnai, H. (2004) Structural formation of hybrid siloxane-based polymer monolith in confined spaces. *J. Sep. Sci.*, **27**, 874–886.
61 Kanamori, K., Nakanishi, K., and Hanada, T. (2006) Thick silica gel coatings on methylsilsesquioxane monoliths using anisotropic phase separation. *J. Sep. Sci.*, **29**, 2463–2470.
62 Kanamori, K., Kodera, Y., Hayase, G., Nakanishi, K., and Hanada, T. (2011) Transition from transparent aerogels to hierarchically porous monoliths in polymethylsilsesquioxane sol–gel system. *J. Colloid Interface Sci.*, **357**, 336–344.
63 Zhu, Y., Morimoto, Y., Shimizu, T., Morisato, K., Takeda, K., Kanamori, K., and Nakanishi, K. (2015) Synthesis of hierarchically porous

polymethylsilsesquioxane monoliths with controlled mesopores for HPLC separation. *J. Ceram. Soc. Jpn.*, **123**, 770–778.
64 Hüsing, N. and Schubert, U. (1998) Aerogels-airy materials: chemistry, structure, and properties. *Angew. Chem., Int. Ed.*, **37**, 22–45.
65 Pierre, A.C. and Pajonk, G.M. (2002) Chemistry of aerogels and their applications. *Chem. Rev.*, **102**, 4243–4265.
66 Kanamori, K., Aizawa, M., Nakanishi, K., and Hanada, T. (2008) Elastic organic–inorganic hybrid aerogels and xerogels. *J. Sol.-Gel. Sci. Technol.*, **48**, 172–181.
67 Kanamori, K., Nakanishi, K., and Hanada, T. (2009) Sol–gel synthesis, porous structure, and mechanical property of polymethylsilsesquioxane aerogels. *J. Ceram. Soc. Jpn.*, **117**, 1333–1338.
68 Hayase, G., Kanamori, K., and Nakanishi, K. (2012) Structure and properties of polymethylsilsesquioxane aerogels synthesized with surfactant *n*-hexadecyltrimethylammonium chloride. *Microporous Mesoporous Mater.*, **158**, 247–252.
69 Kurahashi, M., Kanamori, K., Takeda, K., Kaji, H., and Nakanishi, K. (2012) Role of block copolymer surfactant on the pore formation in methylsilsesquioxane aerogel systems. *RSC Adv.*, **2**, 7166–7173.
70 Hayase, G., Kugimiya, K., Ogawa, M., Kodera, Y., Kanamori, K., and Nakanishi, K. (2014) The thermal conductivity of polymethylsilsesquioxane aerogels and xerogels with varied pore sizes for practical application as thermal superinsulators. *J. Mater. Chem. A*, **2**, 6525–6531.
71 Kanamori, K. (2014) Monolithic silsesquioxane materials with well-defined pore structure. *J. Mater. Res.*, **29**, 2773–2786.
72 Hayase, G., Kanamori, K., and Nakanishi, K. (2011) New flexible aerogels and xerogels derived from methyltrimethoxysilane/dimethyldimethoxysilane co-precursors. *J. Mater. Chem.*, **21**, 17077–17079.
73 Hayase, G., Kanamori, K., Hasegawa, G., Maeno, A., Kaji, H., and Nakanishi, K. (2013) A superamphiphobic macroporous silicone monolith with marshmallow-like flexibility. *Angew. Chem., Int. Ed.*, **52**, 10788–10791.
74 Hasegawa, G., Kanamori, K., Nakanishi, K., and Hanada, T. (2010) Facile preparation of transparent monolithic titania gels utilizing a chelating ligand and mineral salts. *J. Sol.-Gel. Sci. Technol.*, **53**, 59–66.
75 Hasegawa, G., Kanamori, K., Nakanishi, K., and Hanada, T. (2010) Facile preparation of hierarchically porous TiO_2 monoliths. *J. Am. Ceram. Soc.*, **93**, 3110–3115.
76 Miyazaki, S., Miah, M.Y., Morisato, K., Shintani, Y., Kuroha, T., and Nakanishi, K. (2005) Titania-coated monolithic silica as separation medium for high performance liquid chromatography of phosphorus-containing compounds. *J. Sep. Sci.*, **28**, 39–44.
77 Konishi, J., Fujita, K., Nakanishi, K., Hirao, K., Morisato, K., Miyazaki, S., and Ohira, M. (2009) Sol–gel synthesis of macro-mesoporous titania monoliths and their applications to chromatographic separation media for organophosphate compounds. *J. Chromatogr. A*, **1216**, 7375–7378.
78 Hasegawa, G., Kanamori, K., Nakanishi, K., and Hanada, T. (2010) Facile preparation of hierarchically porous TiO_2 monoliths. *J. Am. Ceram. Soc.*, **93**, 3110–3115.
79 Numata, M., Takahashi, R., Yamada, I., Nakanishi, K., and Sato, S. (2010) Sol–gel preparation of Ni/TiO_2 catalysts with bimodal pore structures. *Appl. Catal. A*, **383**, 66–72.
80 Fujita, K., Konishi, J., Nakanishi, K., and Hirao, K. (2004) Strong light scattering in macroporous TiO_2 monoliths induced by phase separation. *Appl. Phys. Lett.*, **85**, 5595–5597.
81 Murai, S., Fujita, K., Konishi, J., Hirao, K., and Tanaka, K. (2010) Random lasing from localized modes in strongly scattering systems consisting of macroporous titania monoliths infiltrated with dye solution. *Appl. Phys. Lett*, **97**, 031118.

82 Fujita, K., Murai, S., Nakanishi, K., and Hirao, K. (2006) Formation of photonic structures in Sm^{2+}-doped aluminosilicate glasses through phase separation. *J. Non-Cryst. Solids*, **352**, 2496–2500.

83 Hirao, T., Fujita, K., Murai, S., Nakanishi, K., and Hirao, K. (2006) Fabrication of Sm^{2+}-doped macroporous aluminosilicate glasses with high alumina content. *J. Non-Cryst. Solids*, **352**, 2553–2557.

84 Murai, S., Fujita, K., Hirao, T., Nakanishi, K., and Hirao, K. (2007) Temperature-tunable scattering strength based on the phase transition of liquid crystal infiltrated in well-defined macroporous random media. *Opt. Mater.*, **29**, 949–954.

85 Meng, X., Fujita, K., Murai, S., Konishi, J., Mano, M., and Tanaka, K. (2010) Random lasing in ballistic and diffusive regimes for macroporous silica-based systems with tunable scattering strength. *Opt. Express*, **18**, 12153–12160.

86 Gash, A.E., Tillotson, T.M., Satcher, J.H., Poco, J.F., Hrubesh, L.W., and Simpson, R.L. (2001) Use of epoxides in the sol–gel synthesis of porous iron(III) oxide monoliths from Fe(III) salts. *Chem. Mater.*, **13**, 999–1007.

87 Tokudome, Y., Nakanishi, K., Kanamori, K., Fujita, K., Akamatsu, H., and Hanada, T. (2009) Structural characterization of hierarchically porous alumina aerogel and xerogel monoliths. *J. Colloid Interface Sci.*, **338**, 506–513.

88 Tokudome, Y., Nakanishi, K., Kanamori, K., and Hanada, T. (2010) *In situ* SAXS observation on metal-salt-derived alumina sol–gel system accompanied by phase separation. *J. Colloid Interface Sci.*, **352**, 303–308.

89 Tokudome, Y., Nakanishi, K., Kosaka, S., Kariya, A., Kaji, H., and Hanada, T. (2010) Synthesis of high-silica and low-silica zeolite monoliths with trimodal pores. *Microporous Mesoporous Mater.*, **132**, 538–542.

90 Reboul, J., Furukawa, S., Horike, N., Tsotsalas, M., Hirai, K., Uehara, H., Kondo, M., Louvain, N., Sakata, O., and Kitagawa, S. (2012) Mesoscopic architectures of porous coordination polymers fabricated by pseudomorphic replication. *Nat. Mater.*, **11**, 717–723.

91 Sumida, K., Moitra, N., Reboul, J., Fukumoto, S., Nakanishi, K., Kanamori, K., Furukawa, S., and Kitagawa, S. (2015) Mesoscopic superstructures of flexible porous coordination polymers synthesized via coordination replication. *Chem. Sci.*, **6**, 5938–5946.

92 Cao, G., Hong, H.G., and Mallouk, T.E. (1992) Layered metal phosphates and phosphonates: from crystals to monolayers. *Acc. Chem. Res.*, **25**, 420–427.

93 Colomban, P. (1992) *Proton Conductors: Solids, Membranes and Gels –Materials and Devices*, Cambridge University Press.

94 Díaz, A., Saxena, V., González, J., David, A., Casañas, B., Carpenter, C., Batteas, J.D., Colón, J.L., Clearfield, A., and Hussain, M.D. (2012) Zirconium phosphate nano-platelets: a novel platform for drug delivery in cancer therapy. *Chem. Commun.*, **48**, 1754–1756.

95 Zhang, F. and Wong, S.S. (2010) Ambient large-scale template-mediated synthesis of high-aspect ratio single-crystalline, chemically doped rare-earth phosphate nanowires for bioimaging. *ACS Nano*, **4**, 99–112.

96 Alberti, G., Casciola, M., Costantino, U., and Vivani, R. (1996) Layered and pillared metal(IV) phosphates and phosphonates. *Adv. Mater.*, **8**, 291–303.

97 Tian, B., Liu, X., Tu, B., Yu, C., Fan, J., Wang, L., Xie, S., Stucky, G.D., and Zhao, D. (2003) Self-adjusted synthesis of ordered stable mesoporous minerals by acid–base pairs. *Nat. Mater.*, **2**, 159–163.

98 Doherty, C.M., Caruso, R.A., Smarsly, B.M., Adelhelm, P., and Drummond, C.J. (2009) Hierarchically porous monolithic $LiFePO_4$/carbon composite electrode materials for high power lithium ion batteries. *Chem. Mater.*, **21**, 5300–5306.

99 Costamagna, P., Yang, C., Bocarsly, A.B., and Srinivasan, S. (2002) Nafion® 115/zirconium phosphate composite membranes for operation of PEMFCs above 100 °C. *Electrochim. Acta*, **47**, 1023–1033.

100 Ma, Y., Tong, W., Zhou, H., and Suib, S.L. (2000) A review of zeolite-like porous materials. *Microporous Mesoporous Mater.*, **37**, 243–252.

101 Lin, R. and Ding, Y. (2013) A review on the synthesis and applications of mesostructured transition metal phosphates. *Materials*, **6**, 217–243.

102 Tokudome, Y., Miyasaka, A., Nakanishi, K., and Hanada, T. (2011) Synthesis of hierarchical macro/mesoporous dicalcium phosphate monolith via epoxide-mediated sol–gel reaction from ionic precursors. *J. Sol-Gel Sci. Technol.*, **57**, 269–278.

103 Komarneni, S. and Roy, R. (1982) Use of γ-zirconium phosphate for Cs removal from radioactive waste. *Nature*, **299**, 707–708.

104 Ginestra, A.L., Patrono, P., Berardelli, M.L., Galli, P., Ferragina, C., and Massucci, M.A. (1987) Catalytic activity of zirconium phosphate and some derived phases in the dehydration of alcohols and isomerization of butenes. *J. Catal.*, **103**, 346–356.

105 Hogarth, W.H.J., da Costa, J.C.D., Drennan, J., and Lu, G.Q. (2005) Proton conductivity of mesoporous sol–gel zirconium phosphates for fuel cell applications. *J. Mater. Chem.*, **15**, 754–758.

106 Agrawal, D.K., Huang, C.Y., and McKinstry, H.A. (1991) NZP: a new family of low-thermal expansion materials. *Int. J. Thermophys.*, **12**, 697–710.

107 Hong, H.Y. and Goodenough, J.B. (1976) Crystal structures and crystal chemistry in the system $Na_{1+x}Zr_2Si_xP_{3-x}O_{12}$. *Mater. Res. Bull.*, **11**, 173–182.

108 Anantharamulu, N., Rao, K.K., Rambabu, G., Kumar, B.V., Radha, V., and Vithal, M. (2011) A wide-ranging review on Nasicon type materials. *J. Mater. Sci.*, **46**, 2821–2837.

109 Roy, R., Agrawal, D.K., and Mckinstry, H.A. (1989) Very low thermal expansion coefficient materials. *Annu. Rev. Mater. Sci.*, **19**, 59–81.

110 Stein, A., Wang, Z., and Fierke, M.A. (2009) Functionalization of porous carbon materials with designed pore architecture. *Adv. Mater.*, **21**, 265–293.

111 Hasegawa, G., Kanamori, K., Nakanishi, K., and Hanada, T. (2010) Fabrication of activated carbons with well-defined macropores derived from sulfonated poly (divinylbenzene) networks. *Carbon*, **48**, 1757–1766.

112 Hasegawa, G., Aoki, M., Kanamori, K., Nakanishi, K., Hanada, T., and Tadanaga, K. (2011) Monolithic electrode for electric double-layer capacitors based on macro/meso/microporous S-containing activated carbon with high surface area. *J. Mater. Chem.*, **21**, 2060–2063.

113 Hasegawa, J., Kanamori, K., Nakanishi, K., and Hanada, T. (2010) Macro- and microporous carbon monoliths with high surface areas pyrolyzed from poly (divinylbenzene) networks. *C. R. Chim.*, **13**, 207–211.

114 Neely, J.M. (1981) Characterization of polymer carbons derived from porous sulfonated polystyrene. *Carbon*, **19**, 27–36.

115 Al-Muhtaseb, S.A. and Ritter, J.A. (2003) Preparation and properties of resorcinol-formaldehyde organic and carbon gels. *Adv. Mater.*, **15**, 101–114.

116 Oya, A. and Otani, S. (1979) Catalytic graphitization of carbons by various metals. *Carbon*, **17**, 131–137.

117 Hasegawa, G., Kanamori, K., and Nakanishi, K. (2012) Facile preparation of macroporous graphitized carbon monoliths from iron-containing resorcinol-formaldehyde gels. *Mater. Lett.*, **76**, 1–4.

118 Paraknowitsch, J.P. and Thomas, A. (2013) Doping carbons beyond nitrogen: an overview of advanced heteroatom doped carbons with boron, sulphur and phosphorus for energy applications. *Energy Envrion. Sci.*, **6**, 2839–2855.

119 Hasegawa, G., Deguchi, T., Kanamori, K., Kobayashi, Y., Kageyama, H., Abe, T., and Nakanishi, K. (2015) High-level doping of nitrogen, phosphorus, and sulfur into activated carbon monoliths and their electrochemical capacitances. *Chem. Mater.*, **27**, 4703–4712.

120 Corriu, R.J.P. (2000) Ceramics and nanostructures from molecular

precursors. *Angew. Chem., Int. Ed.*, **39**, 1376–1398.

121 Keys, L.K. and Mulay, L.N. (1967) Magnetic susceptibility measurements of rutile and the Magnéli phases of the Ti–O system. *Phys. Rev.*, **154**, 453–456.

122 Ohkoshi, S., Tsunobuchi, Y., Matsuda, T., Hashimoto, K., Namai, A., Hakoe, F., and Tokoro, H. (2010) Synthesis of a metal oxide with a room-temperature photoreversible phase transition. *Nat. Chem.*, **2**, 539–545.

123 Walsh, F.C. and Wills, R.G.A. (2010) The continuing development of Magnéli phase titanium sub-oxides and Ebonex® electrodes. *Electrochim. Acta*, **55**, 6342–6351.

124 Kitada, A., Hasegawa, G., Kobayashi, Y., Kanamori, K., Nakanishi, K., and Kageyama, H. (2012) Selective preparation of macroporous monoliths of conductive titanium oxides Ti_nO_{2n-1} (n=2, 3, 4, 6). *J. Am. Chem. Soc.*, **134**, 10894–10898.

125 Hasegawa, G., Kitada, A., Kawasaki, S., Kanamori, K., Nakanishi, K., Kobayashi, Y., Kageyama, H., and Abe, T. (2015) Impact of electrolyte on pseudocapacitance and stability of porous titanium nitride (TiN) monolithic electrode. *J. Electrochem. Soc.*, **162**, A77–A85.

126 Hasegawa, G., Sato, T., Kanamori, K., Nakano, K., Yajima, T., Kobayashi, Y., Kageyama, H., Abe, T., and Nakanishi, K. (2013) Hierarchically porous monoliths based on N-doped reduced titanium oxides and their electric and electrochemical properties. *Chem. Mater.*, **25**, 3504–3512.

127 Hasegawa, G., Sato, T., Kanamori, K., Sun, C.J., Ren, Y., Kobayashi, Y., Kageyama, H., Abe, T., and Nakanishi, K. (2015) Effect of calcination conditions on porous reduced titanium oxides and oxynitrides via a preceramic polymer route. *Inorg. Chem.*, **54**, 2802–2808.

128 Di Valentin, C., Pacchioni, G., Selloni, A., Livraghi, S., and Giamello, E. (2005) Characterization of paramagnetic species in N-doped TiO_2 powders by EPR spectroscopy and DFT calculations. *J. Phys. Chem. B*, **109**, 11414–11419.

129 Gomathi, A., Sundaresan, A., and Rao, C.N.R. (2007) Nanoparticles of superconducting γ-Mo_2N and δ-MoN. *J. Solid State Chem.*, **180**, 291–295.

130 Giordano, C., Erpen, C., Yao, W., and Antonietti, M. (2008) Synthesis of Mo and W carbide and nitride nanoparticles via a simple "urea glass" route. *Nano Lett*, **8**, 4659–4663.

131 Schaber, P.M., Colson, J., Higgins, S., Thielen, D., Anspach, B., and Brauer, J. (2004) Thermal decomposition (pyrolysis) of urea in an open reaction vessel. *Thermochim. Acta*, **424**, 131–142.

10
Macroporous Materials Synthesized by Colloidal Crystal Templating

Jinbo Hu and Andreas Stein

University of Minnesota, Department of Chemistry, 207 Pleasant St. SE, Minneapolis, MN 55455, USA

10.1 Introduction

Colloidal crystal templating provides a way of introducing porous structure with well defined geometry and architecture into solid materials that cover a wide range of compositions. It is a so-called "hard" templating approach in which a periodic array of uniform colloidal particles acts as a mold to define the target structure. The interstitial space between particles is infiltrated with fluid precursors and the resulting composites are converted into the target solid. Concurrent or subsequent removal of the colloidal crystal template (CCT) produces a porous solid with a structure that is approximately the negative replicate structure of the original CCT (Figure 10.1).

In this chapter, we will focus on materials prepared from CCTs consisting of monodisperse spherical colloidal particles, but many of the general concepts discussed here would also apply to CCTs composed of other types of templating particles, such as cylindrical arrays that can be obtained from virus particles [2]. A cubic close-packed (ccp) array of spheres can be found in opals, and materials templated from such opaline arrays are often called inverse opals or inverted opals. They also display the opalescent appearance of the natural gems. Because the product structures are not always true inverse replicas of the original opal structure, more general terms used for these materials are three-dimensionally ordered macroporous (3DOM) or three-dimensionally ordered mesoporous (3DOm) materials, depending on the pore sizes (macropores>50 nm, mesopores between 2 and 50 nm). Hierarchical pore structure is also attainable by using a CCT together with secondary templates, such as surfactants or block-copolymers. Such materials that combine periodic macropores with mesopores are denoted as 3DOM/m materials.

Interest in 3DOM materials arises from several of their structural features, including the typically bicontinuous structure (continuous, fully interconnected

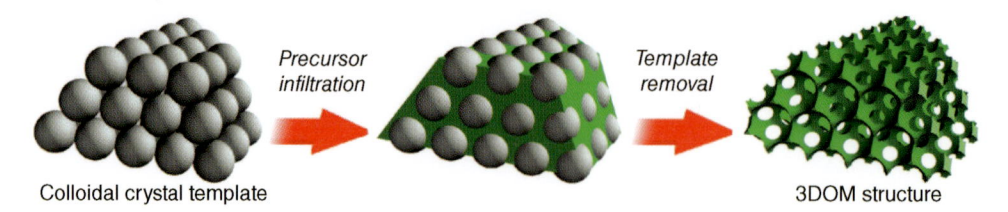

Figure 10.1 Schematic representation of a typical synthesis of a three-dimensionally ordered macroporous (3DOM) material by colloidal crystal templating. A preformed CCT is infiltrated with a precursor fluid, followed by conversion of the precursor to the target composition and finally, template removal to obtain the 3DOM material. (Adapted with permission from Ref. [1]. Copyright 2008, American Chemical Society.)

pore space and a continuous, interconnected solid skeleton), the resulting pore accessibility, enhanced surface area compared to the bulk materials, nanostructure, and periodicity in three dimensions. Such periodicity is of particular interest for optical applications when the repeat length is on the order of optical wavelengths. Other applications, especially those relying on chemical reactivity of 3DOM materials (catalysis, energy storage, sensing), do not necessarily require a periodic structure but benefit from the excellent mass transfer properties through the macropores, whose interconnectedness can be guaranteed by the periodicity. These applications also benefit from short diffusion paths through the nanostructured skeleton and large, easily accessible interfaces. In addition, the highly symmetric environment on the length scale of the pores facilitates easier modeling of materials and transport properties.

In this chapter, we will describe in some detail the varied synthetic approaches to 3DOM materials, discuss structural features, and highlight applications of these materials.

10.2
Structure

3DOM materials possess distinct structural features on multiple length scales. These are illustrated in Figure 10.2. On the macroscopic scale, the materials may be three-dimensional monoliths, two-dimensional films, one-dimensional fibers, or powdered particles (Figure 10.2a). The shape of monoliths or fibers can be defined by a macroscopic mold in which the CCT is assembled, or by shaping an already prepared CCT. Powdered particles of submillimeter size are typically obtained for 3DOM oxides, which tend to be quite brittle. 3DOM carbon materials, on the other hand, are more robust and amenable to formation of monoliths, which can be used, for example, as self-supporting electrodes without any binder. It is important to keep in mind that processing steps during the 3DOM material synthesis result in shrinkage due to loss of solvents, condensation reactions, sintering, and so on. As a result, the final pore spacing can be

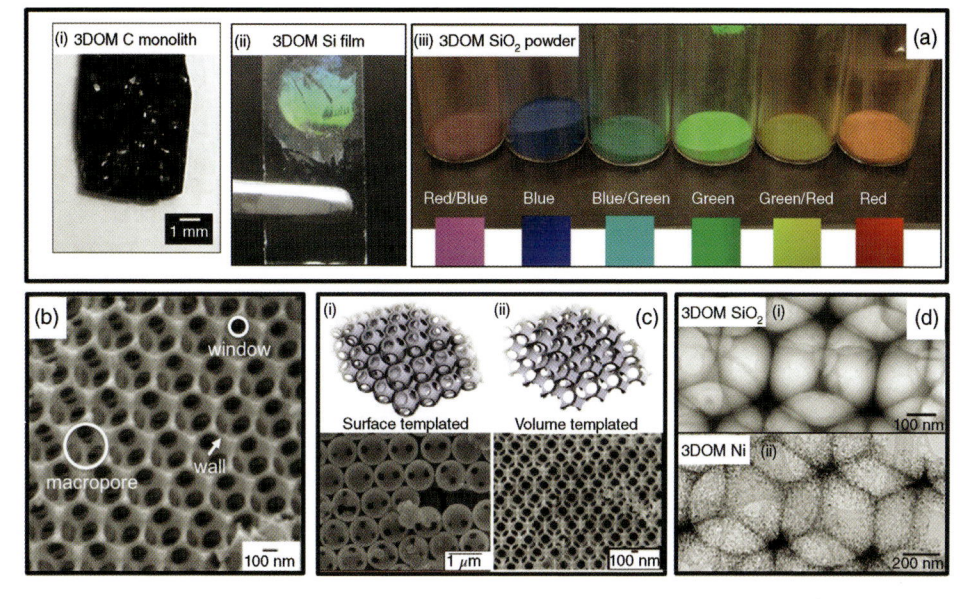

Figure 10.2 The structures of 3DOM materials on multiple length scales. (a) Photographs of a: (i) 3DOM carbon monolith. (Adapted with permission from Ref. [3]. Copyright 2006, American Chemical Society.) (ii) 3DOM Si film on a glass slide. (Adapted with permission from Ref. [4]. Copyright 2012, The Royal Society of Chemistry.) (iii) 3DOM SiO_2 powders or powder mixtures with different pore sizes leading to various colors. (Adapted with permission from Ref. [5]. Copyright 2004, Wiley-VCH Verlag GmbH.) (b) Scanning electron microscopy (SEM) image of 3DOM/m SiO_2, with white circles showing a macropore and a window, and an arrow pointing out the wall. (Adapted with permission from Ref. [6]. Copyright 2007, American Chemical Society.) (c) Schematic representation and SEM images of surface templated (i) and volume templated (ii) W inverse opals obtained from CVD and wet chemistry, respectively. (Adapted with permission from Ref. [7]. Copyright 2005, Society of Photo Optical Instrumentation Engineers.) (d) Transmission electron microscopy (TEM) images of 3DOM SiO_2 with amorphous walls (i) and 3DOM Ni that has nanocrystalline walls with small grains appearing as dark spots (ii). (Adapted with permission from Ref. [8]. Copyright 2001, Elsevier.)

significantly smaller than the repeat distance in the original template, typically between 5 and 30%. Greater shrinkage is observed with polymeric CCTs than with SiO_2 CCTs. Chemical vapor deposition (CVD), atomic layer deposition (ALD), and electrodeposition methods tend to produce much less shrinkage than solvent-based methods. These synthetic routes are discussed in Section 10.3.3.

Next, focusing on the length scale of the pores, the periodic skeletal structure (Figure 10.2b) depends on the interactions between the precursors and the template surface. Strong attractive interactions result in so-called "surface templating." As illustrated in Figure 10.2c (left), precursor material forms a shell around the sphere surface, resulting in highly curved, spheroidal pores after template removal. Additional windows are produced at the points where adjacent spheres touch (at 12 points in ccp arrays). These windows define the smallest entrances to the macropores. If precursor penetrates the templating spheres, which can be the case for polymer templates, closed shells may form, and additional small voids may be present between the shells. Weaker attractive interactions between precursors and templating spheres often result in "volume templating." All the interstitial space is filled with the precursor. If processing of the composite then results in partial loss of material as a result of condensation, sintering or solvent loss, a more strut-like skeleton forms, with much larger relative window sizes, and in the case of many oxides, less curved features (Figure 10.2c, right).

At the length scale of the pore arrays, one will also observe point, line, and planar defects. These may be introduced through existing defects in the original CCT, but also during processing as a result of sphere swelling from solvent penetration, thermal stresses, incomplete infiltration, and so on. Through process optimization, the defect density may be reduced, but it remains a challenge to avoid such defects completely.

The next length scale is that of the wall skeleton, which has cross-sections of tens of nm if macropores are a few hundred nanometers in diameter. The skeletal structure depends strongly on the material composition, phase, and processing temperature. As shown in Figure 10.2d (top), if the walls are amorphous, as is the case with 3DOM SiO_2 or 3DOM resol-based polymers, they tend to conform to the surface of the template and form a continuous, film-like structure. If, on the other hand, the walls are crystallized, they are typically composed of interconnected nanoparticles (Figure 10.2d, bottom). For some compositions, at higher processing temperatures or when crystallographic phase changes occur, excessive grain growth can result in loss of the periodic pore structure, although an interconnected pore system might still be maintained.

Interstitial spaces between nanoparticles in the wall skeleton can provide textural mesoporosity, so that the surface area of a 3DOM material may be higher than expected from geometrical considerations of the macropores only. Another approach to introduce nanopores into skeletal walls in a more controlled fashion to obtain materials with defined hierarchical porosity is through dual-templating, which is discussed in Section 10.3.3.

10.3
Synthesis

The synthesis of 3DOM materials contains several steps. As illustrated in Figure 10.1, it typically starts with the preparation of monodisperse colloidal particles, followed by assembly of colloidal particles into close-packed arrays as the CCT, infiltration of the CCT with a precursor fluid, conversion of the precursor to the target material, and finally CCT removal. Post-treatments such as surface modification may be needed, depending on the specific application. In this section, we discuss the general considerations for each step, with a focus on the CCT assembly and various templating approaches that have appeared in the literature.

10.3.1
Colloidal Spheres for CCTs

The most commonly used colloidal particles for 3DOM materials comprise spherical colloids of monodisperse SiO_2 or polymers, such as polystyrene (PS), poly(methyl methacrylate) (PMMA), copolymer poly(styrene-methyl methacrylate-acrylic acid) P(St-MMA-AA), and poly(methyl methacrylate-butyl acrylate-acrylic acid) P(MMA-BA-AA). These colloidal spheres are either commercially available, or can be readily synthesized using developed techniques. Usually, uniform SiO_2 spheres with diameters from 10 nm to several micrometers are synthesized via the hydrolysis and condensation of silicon alkoxides (i.e., the Stöber process and its modifications) [9,10], whereas monodisperse polymer spheres in the size range from 200 to 500 nm are prepared by emulsion polymerization. Standard recipes are available, such as for PS [11], PMMA [12], P(St-MMA-AA) [13], and P(MMA-BA-AA) [14]. To ensure the periodicity of the resulting CCTs, the variation of sphere sizes should be approximately 3–5% or less.

The choice and optional pretreatments of the colloidal spheres depend on the entire synthetic route. Prior to precursor infiltration, the colloidal crystal may be annealed to strengthen the contact points between spheres, leading to enhanced mechanical stability of the CCT and thus reduced defects in the final product. SiO_2 spheres can be annealed at 600 °C, while for colloidal polymers the annealing temperatures are slightly higher than their glass transition temperatures (i.e., 105–107 °C for PS or PMMA). During the infiltration process, the surface of the CCT should be sufficiently wetted by the precursor fluid. To achieve effective template–precursor interactions, the colloidal spheres may be modified with surface functional groups. When the precursor is converted to the target composition, elevated temperatures are usually employed. Therefore, the CCT should maintain its structure at least up to the temperature at which the conversion takes place. In this regard, SiO_2 spheres are suitable for synthetic routes that require high temperatures, and the products templated from SiO_2 spheres undergo less shrinkage (about 10%) than those obtained using polymer spheres (about 20–30%). The last step of the synthesis is template removal. While polymer templates can be readily removed by calcination at temperatures higher than 300 °C or by solvent extraction, SiO_2 spheres require extraction with HF or hot alkali solutions, and appropriate safety precautions must be taken.

10.3.2
Colloidal Crystal Assembly

Once the colloidal spheres are chosen, the next step is to assemble them into close-packed arrays, or the CCT. These arrays are usually hexagonally close-packed (hcp) in two dimensions and ccp in three dimensions. For a typical 3DOM structure, the theoretical volume fraction of colloidal spheres is 74.05%. Several reviews on colloidal crystal assembly processes have been published [15–17], and here we limit our discussion to the most commonly used techniques that can be readily performed in laboratories; each has its unique advantages and limitations. Before choosing a specific method, several factors should be considered, such as the quantity of the CCTs needed, the time scale of experiments, and tolerance toward defects.

Sedimentation methods, including natural sedimentation and accelerated sedimentation by centrifugation or filtration, are suitable for applications in which large quantities (i.e., several grams) of 3DOM materials are needed (Figure 10.3a). The obtained colloidal crystal is in the form of a powder, or monolith with dimensions of several millimeters. For natural sedimentation, the suspension of colloidal spheres is placed in a vessel to allow gravitational settling of spheres. Although time-consuming (it can take several weeks), it seems to be the simplest approach to form large quantities of well-ordered, three-dimensional crystalline arrays from colloid spheres (Figure 10.3b). To avoid rapid sedimentation that can result in disordered sphere aggregates, solvents with high viscosities and low evaporation rates may be needed. The

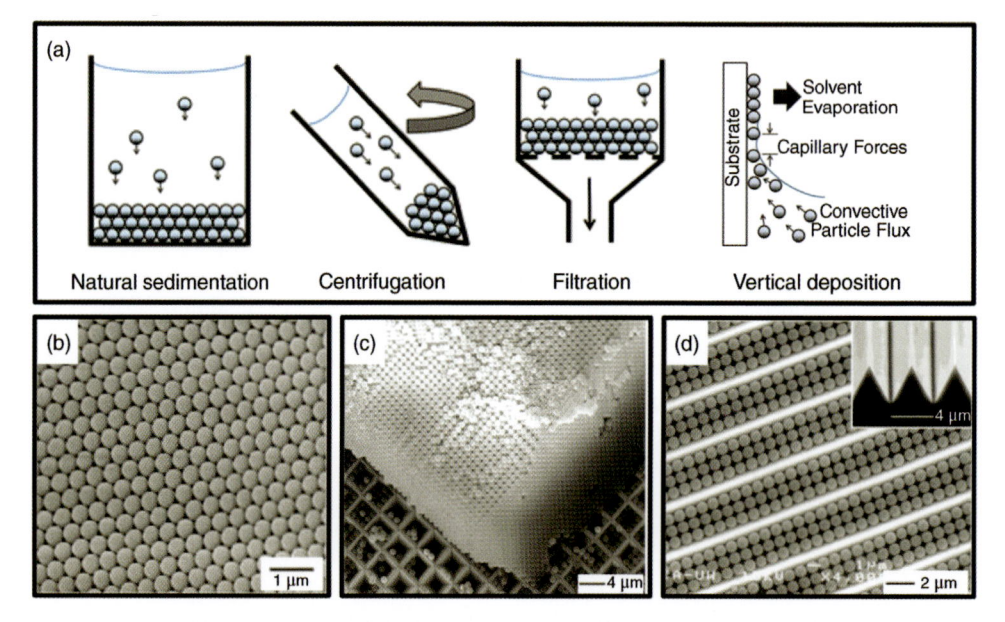

Figure 10.3 Various approaches for CCT assembly and examples of colloidal crystals. (a) Schematic representation of natural sedimentation, centrifugation, filtration, and vertical deposition methods to prepare CCT. (Adapted with permission from Ref. [20]. Copyright 2013, The Royal Society of Chemistry.) (b) SEM image of a CCT composed of fcc PMMA spheres obtained by natural sedimentation. (Adapted with permission from Ref. [6]. Copyright 2003, American Chemical Society.) (c) SEM image of a (100)-oriented PS crystal obtained from a patterned substrate with pyramidal cavities. (d) SEM image showing a (100)-oriented PS crystal grown on a V-shaped substrate. (Parts (c) and (d) adapted with permission from Ref. [18]. Copyright 2003, American Chemical Society.)

assembly time can be greatly reduced to several hours or a few days by using centrifugation. It should be noted that when polymer spheres are used, the centrifugation speed should be held below 1000 rpm, because too rapid centrifugation may cause excess densification of the resulting colloidal crystal, leaving insufficient space for the precursor to penetrate. Another accelerated sedimentation technique is through filtration, in which a membrane filter is chosen to permit the passage of solvent while retaining the colloidal spheres. Compared to natural sedimentation, colloidal crystals obtained by accelerated sedimentation methods usually contain higher levels of polycrystallinity with defects. When large and single crystal CCTs are needed with controlled orientation, a patterned substrate can be employed to induce slow crystal growth along a specific plane, such as silicon wafers with pyramidal or V-shaped cavities (Figure 10.3c and d) [18], or substrates with flowerlike or crosslike patterns [19].

Several techniques are used to prepare thin-film CCTs containing mono- or multiple layers of close-packed colloidal spheres with controlled thickness. These approaches include spin coating, evaporation-assisted assembly, and crystallization within confined spaces. Spin coating is a rapid and scalable process that is suitable for preparing CCTs on large planar substrates [21]. To ensure a high structural quality of the film, the colloidal spheres need to sufficiently wet the substrate, and a viscous solvent is usually used to avoid premature sedimentation. The spin speed also needs to be optimized, since a low spin speed produces multilayer coatings whereas a high spin speed favors monolayer coatings. When the spin speed is too high, excess voids can form within a monolayer coating. High quality thin films can also be obtained with evaporation-assisted assembly, which is also referred to as vertical deposition, convective assembly, or controlled drying [22–24]. As shown in Figure 10.3a, a planar substrate is contacted with a suspension of colloidal spheres at a controlled angle. As the solvent wets the substrate, colloidal spheres are drawn to the substrate and directed to assemble through capillary forces, thus forming the nucleus for subsequent crystal growth. This is followed by a convective particle flux induced by solvent evaporation, which draws more spheres to the already ordered array to allow crystal growth. Film thickness and structural quality are related to multiple factors, including sphere sizes, volume fractions of colloidal spheres, choice of solvent, angle of the substrate placement in the suspension, and so on [15]. When colloidal suspensions are allowed to crystallize in confined spaces, CCTs with patterned morphologies can be obtained. This technique is suitable to prepare CCTs with other designed features. Specific examples include crystallization inside parallel substrates [25], capillaries [26], and micromolds made from poly(dimethylsiloxane) [27].

10.3.3
The Templating Process and Synthetic Alternatives

The next steps involve infiltration of the assembled CCT with a precursor fluid, conversion of the precursor to the target material, and removal of the template to produce the final product. The precursor fluid can be in liquid, gas, or solid phase, with liquid-phase infiltration most commonly used and solid precursors less common. In this section, we introduce various synthetic routes to prepare 3DOM materials, focusing on general synthesis aspects. Readers interested in a specific approach are encouraged to consult the original literature for more details.

Sol–Gel Chemistry

Sol–gel chemistry is widely used for the preparation of 3DOM materials comprising metal oxides. Common examples include, but are not limited to, oxides of Si [28,29], Ti [11,26,30], Zr [11,30], Al [11,30], W [11], Fe [11], and Sn [31]. Multicomponent oxides can be obtained by doping [32], or employing precursors with mixed metal alkoxides [11,33]. In a typical sol–gel templating process, a precursor solution containing metal alkoxides, or mixtures of metal alkoxides and metal salts is infiltrated into the interstitial space between CCT spheres by natural capillary forces, or with the aid of an applied vacuum [30]. This is followed by hydrolysis and condensation reactions of the precursor to form a composite material containing the target metal oxides. These reactions can be induced by atmospheric moisture, surface functionalities on CCT spheres, or added acid or base. Eventually, the CCT is removed to render the final product.

Several criteria should be considered to ensure an efficient templating process. To start, the solvent used for the precursor solution should adequately dissolve the metal precursor, without destroying the CCT by dissolving or swelling the spheres in a polymeric CCT. The precursor solution should sufficiently wet the CCT, which may be achieved using cosolvents. Alternatively, the surface of CCT spheres could be functionalized (e.g., with a surfactant) [28]. The strength of the interactions between precursors and spheres can result in different structural features of the final 3DOM material, with "surface-templated" materials arising from strong precursor/sphere interactions and "volume-templated" materials from weak interactions [34], as discussed in Section 10.2. Dilution of precursor solutions can also help to decrease the viscosity and reactivity of moisture-sensitive metal alkoxides, which prevents extensive polymerization that can lead to incomplete filling of the CCT as well as formation of a nontemplated surface overlayer. It should be noted, however, that too dilute precursor solutions may also cause insufficient formation of solid material in the final product, because in such cases the void space in the CCT is primarily filled by the solvent, rather than the precursor. Therefore, multiple infiltration/drying cycles may be needed to increase the volume fraction of the target material. For precursors solutions with high viscosity, wetting the CCTs with solvents prior to the exposure to metal alkoxides can improve the efficacy of infiltration [35]. Besides sol–gel chemistry, these general considerations may also apply to other liquid-phase infiltration approaches, as presented below.

Salt Precipitation and Chemical Conversion

Salt precipitation and chemical conversion provides an alternative to synthesize 3DOM metal oxides, especially those with highly viscous and moisture-sensitive precursors that are difficult to process by sol–gel chemistry. Also, this approach can be used to prepare various porous metals and metal salts. To start, one infiltrates the CCT with a concentrated metal salt solution, which may be an acetate, nitrate, or oxalate salt. As solvent evaporates, the metal salt begins to precipitate within the interstitial space between CCT spheres, leading to the formation of a continuous network of the salt. By using this method, macroporous NaCl was obtained by infiltrating a PS template with a hot saturated NaCl solution [36]. If metal salt is not the desired final product, calcination can be used to chemically convert the metal salt to other compositions. To avoid phase separation caused by melting of the metal salt (e.g., certain metal acetate hydrates), oxalic acid treatment can be used to convert the low-melting-point metal salts to oxalates that undergo decomposition rather than melting at high temperatures [37]. The composition of the final product also depends on the atmosphere of calcination. Typically, calcination of the metal salt/CCT composite in air produces porous metal oxides, whereas the use of an inert or reducing atmosphere converts the composite to metals or alloys, such as Ni, Co, Fe, and $Ni_{1-x}Co_x$ [38–40].

Oxide and Salt Reduction
To obtain 3DOM materials in their reduced states, hydrogen can be used to reduce the preformed 3DOM oxides or salts (synthesized with aforementioned approaches) to the desired composition. For example, macroporous metallic Ni, Co, Fe were obtained by reducing corresponding macroporous metal oxides in an atmosphere of hydrogen. It was observed that macroporous metals obtained with this approach exhibit larger grain sizes than those prepared by direct conversion from metal salts to metals [39,40]. Similarly, macroporous Ge was synthesized from templated GeO_2 [41], whereas mesoporous Au and Pt were obtained by reducing composites of SiO_2 CCTs filled with the corresponding metal salts [42].

Polymerization
When the target 3DOM material or its precursor is a polymer, techniques involving *in situ* polymerization or infiltration with polymer solutions can be used. Using *in situ* polymerization, the CCT is first filled with liquid monomers, and then exposed to certain conditions that can induce polymerization/crosslinking of the organic monomers within the void space of the CCT. Typical treatments include thermal treatment [43], UV exposure [44], and addition of a catalyst [45]. Besides producing macroporous polymers, this method is also used to synthesize porous carbon materials. Carbon precursors comprising phenolic or resorcinol resins are allowed to infiltrate SiO_2 or PMMA CCTs and are subsequently crosslinked by thermal curing. This is followed by carbonization at high temperatures (i.e., >900 °C) to convert the resins to carbon, thus creating macroporous or mesoporous carbon frameworks [34,46]. As an alternative to *in situ* polymerization, concentrated solutions containing preformed polymers can be infiltrated into CCTs, followed by precipitation of the polymeric material within the CCT as a result of solvent evaporation. This approach has also been used to

prepare various 3DOM polymers, such as macroporous PS [47] and poly(alkylthiophenes) [48].

Nanocrystal Deposition and Sintering

Different from the aforementioned approaches where the infiltrated precursor is converted to a different chemical form to obtain the target composition, in the nanocrystal deposition method the CCT is infiltrated with preformed colloidal nanocrystals that already have the desired composition. The final product is obtained by a subsequent sintering process that merges the boundaries between nanocrystals to form the 3DOM material as a single piece. Structural shrinkage can be minimized to approximately 5–10%. The colloidal nanocrystals may enter the voids between the CCT spheres with the aid of filtration [49], solvent evaporation [50], or electrophoresis (for charged nanocrystals) [51]. Alternatively, the nanocrystals can be incorporated into the CCT through cooperative assembly, in which the ordering of colloidal spheres and formation of the nanocrystal network occur simultaneously from a mutual suspension [52,53]. The application of this approach depends on the availability of the nanocrystals for a specific composition, and the sizes of nanocrystals should be small enough to enter the voids between CCT spheres. This technique is used not only for synthesis of 3DOM materials (e.g., macroporous Au and TiO_2) [49,53], but also for loading nanocrystal catalysts into CCTs for subsequent reactions, as discussed in Section "Electroless Deposition" [54].

Electrodeposition

Electrodeposition is a versatile approach to prepare 3DOM materials with high structural qualities, especially for thin films. Mono- or multiple layers of CCT are grown on a conducting substrate (e.g., indium tin oxide-coated glass, Si wafer, Au, glassy carbon) to serve as a working electrode. Then, this working electrode is immersed into an electrolyte containing the precursor of the target material, and electrodeposition is achieved by applying a current or a voltage that can induce redox or polymerization reactions to form the desired product. The applied current/voltage should be chosen carefully to avoid undesired side-reactions, namely, oxidation or reduction of the product to a different state or decomposition of the electrolyte. Therefore, ionic liquids with wide electrochemical windows were explored to replace conventional organic solvents [4,55]. Different from other approaches, this method fills the CCT starting from the conducting substrate and builds up the product from the bottom to the top, thus avoiding pore blockage during the infiltration and allowing complete filling of the CCT. The film thickness is controlled by the quantity of applied charge, which offers opportunities for fine-tuning of the structure. Selected examples of 3DOM materials obtained by this method include chalcogenides (e.g., CdS) [56], metals (e.g., Pt) [57], metal oxides (e.g., Cu_2O) [58], semiconductors (e.g., Ge, Si) [4,55], and conducting polymers (e.g., poly(3,4-ethylenedioxythiophene)) [59].

Electrophoretic Deposition

Electrophoretic deposition is another versatile electrochemical method for the synthesis of 3DOM materials. It is used both for the assembly of colloidal spheres to prepare CCTs [60], and for the infiltration of CCTs with charged material precursors [51]. This approach utilizes a two-electrode system, in which an electric field is applied between two electrodes to induce the motion of charged colloidal particles to the working electrode bearing the opposite charge. For a specific synthesis of ZnO inverse opals, one starts with a polymer CCT that is grown on a conducting substrate. Then, this working electrode is connected to a counter electrode and placed in a suspension of preformed ZnO nanoparticles with positive charges. The application of a negative voltage between the two electrodes drives ZnO nanoparticles to the working electrode, thus allowing their infiltration into the CCT. This process is followed by calcination, which sinters the ZnO nanoparticles and removes the polymer CCT [51].

Electroless Deposition

Compared to electrodeposition and electrophoretic deposition, electroless deposition is a nongalvanic plating method that can produce a uniform and continuous coating in the CCT without an external electrical power source. It is suitable for the synthesis of 3DOM metals, because standard recipes are available for a large variety of metals as bulk materials. The reaction usually takes place with the aid of a catalyst; therefore the CCTs are preloaded with catalysts (e.g., Au nanoparticles) either by simple infiltration [54], or surface modification of the spheres [61–63]. For example, colloidal SiO_2 spheres were modified with thiol surface functionalities to anchor Au nanocrystals as catalysts. After thermal treatment to sinter the spheres and remove organic components from the Au surfaces, the catalyst-loaded CCTs were placed in electroless deposition baths to deposit various metals including Ni, Cu, Au, Pt, and Ag. Finally, the SiO_2 templates were removed by HF treatment, resulting in free-standing porous metal films [61]. Using a similar approach, bimetallic Au/Pt nanostructures with hierarchical porosity can be obtained [63].

Chemical Vapor Deposition

Besides liquid-phase infiltration, gas-phase infiltration techniques, such as CVD and ALD, are also used for preparing 3DOM materials. By avoiding the use of a solvent that may disrupt the ordering of the template, these techniques are particularly suitable for preparing photonic materials with high structural quality needed to obtain complete photonic band gaps. In a CVD process, the CCT is exposed to a flow of one or more volatile precursors, allowing infiltration and subsequent reaction of the precursors in the interstitial space between the template spheres. When polymer spheres are used as templates, the CVD temperature should be carefully optimized because a low temperature may yield inhomogeneous deposits, whereas a high temperature can soften the polymer template [64]. Selected examples of this approach include macroporous semiconductors (e.g., Si) [65], carbon [34], polymers (e.g., polyethylene) [66], metal oxides (e.g., ZnO) [64], and metal films (e.g., Au film) [67].

Atomic Layer Deposition
As an alternative to CVD, ALD allows precise conformal film deposition on the surface of CCT spheres with well-controlled film thickness. In a typical ALD process, the surface of the template is first exposed to vapor of one precursor until it is saturated to achieve full monolayer coverage. Any excess unreacted precursor vapor is removed by an inert gas flow. Then, a vapor of a second precursor is applied to the template surface to react with the previously deposited layer, forming the desired composition. Because ALD is a self-limiting process, only a certain maximum film thickness can be achieved during each cycle, and thicker films are obtained by performing multiple cycles. As an example, a TiO_2 inverse opal was synthesized with a SiO_2 CCT by sequential applications of the titanium precursor $TiCl_4$ and the oxygen precursor H_2O at 100 °C, separated by a N_2 flow. After template removal with HF, TiO_2 inverse opals with very smooth surfaces were obtained [68]. The self-limiting nature of ALD offers extra opportunities for fine-tuning of the templated structure. By optimizing the precursor vapor pressure and exposure time, the surface of the CCT is covered by chemisorbed precursor molecules only up to certain depths that are defined by the path length of Knudsen diffusion. This depth control enables a gradient in wall thickness of the obtained porous structure, leading to an inverse opal with gradient refractive index [69].

Spraying Techniques
Thin-film macroporous materials can also be prepared by various spraying techniques, such as spray pyrolysis [70], ion spraying [57], and laser spraying [57], among which spray pyrolysis is most commonly used. In a typical spray pyrolysis process, close-packed arrays of CCT spheres are deposited on a planar substrate, which is subsequently heated and sprayed over by a precursor solution. The precursor droplets can permeate the voids of CCT spheres, and undergo solvent evaporation and solute condensation. This is followed by an annealing process at a higher temperature, which removes the template and converts the precursor to the final product. For those applications in which the order of the structure is not essential (e.g., catalysis), a mutual suspension containing both of the template spheres and precursor may be sprayed directly to the heated substrate to prepare the final product in the form of porous particles [71].

Sedimentation and Aggregation
A fast and scalable approach to prepare macroporous materials with disordered pore structures is sedimentation and aggregation. In this approach, CCT spheres are allowed to settle in a precursor solution directly, thus combining CCT assembly and precursor infiltration within a single step. As an example, macroporous SiO_2 was obtained from an ethanol suspension containing tetraethyl orthosilicate and PS spheres. The PS spheres were allowed to settle, while

tetraethyl orthosilicate reacted with residual water to deposit SiO_2 on PS spheres through a homogeneous wetting process. Excess tetraethyl orthosilicate and solvent were removed by application of vacuum, and the PS spheres were decomposed by calcination [72]. Similarly, disordered macroporous metal oxides (e.g., NbO, Al_2O_3, TiO_2) were obtained from the corresponding alkoxides [73], and hierarchical SiO_2 structures were prepared in the presence of surfactants [74].

Pseudomorphic Transformations
For some 3DOM target materials that are difficult to synthesize directly by the aforementioned methods, one can use pseudomorphic transformations as an indirect approach. In this technique, a material with a composition that is easier to template is synthesized with the desired structure and then converted to the target composition while retaining its structural features. As shown in Figure 10.4a, an amorphous 3DOM SiO_2 skeleton can be converted to 3DOM

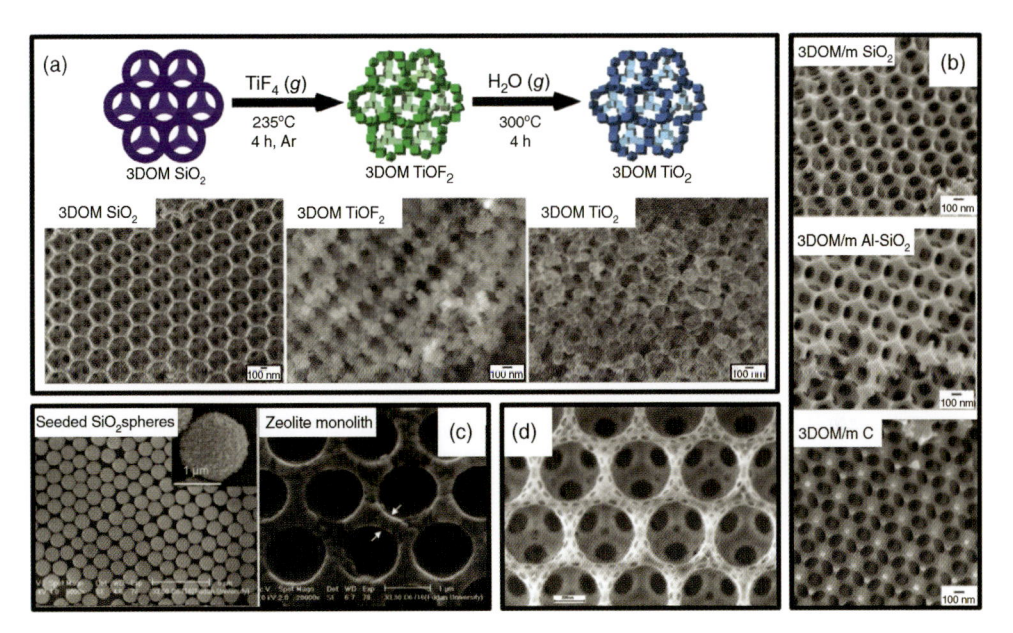

Figure 10.4 3DOM materials synthesized by various approaches. (a) Schematic overview and corresponding SEM images of pseudomorphic transformations from 3DOM SiO_2 to 3DOM $TiOF_2$ and finally to 3DOM TiO_2. (Adapted with permission from Ref. [75]. Copyright 2004, American Chemical Society.) (b) Demonstration of a double templating approach where 3DOM/m SiO_2 is first surface-modified with Al catalysts, and then infiltrated by a carbon precursor and finally converted to 3DOM/m carbon. (Adapted with permission from Ref. [6]. Copyright 2006, American Chemical Society.) (c) SEM images of core–shell SiO_2 spheres coated with silicate-1 nanoseeds and the obtained zeolite monolith, with arrows indicating adjacent, merged shells. (Adapted with permission from Ref. [78]. Copyright 2002, Wiley-VCH Verlag GmbH.) (d) SEM image of a binary SiO_2 inverse opal with hierarchical pore structure synthesized using a multimodal CCT. (Adapted with permission from Ref. [79]. Copyright 2006, American Chemical Society.)

TiOF$_2$ by reacting with gaseous TiF$_4$. During this process, the smooth wall structure of 3DOM SiO$_2$ is replaced by nanocrystalline TiOF$_2$ cubes. After reacting with water vapor, 3DOM TiOF$_2$ is further converted to crystalline 3DOM TiO$_2$ that maintains the features of the 3DOM parent structure with comparable pore spacing [75]. Using a similar approach, inverse opal WC was obtained from WO$_3$ [76], and ordered hierarchical SiC was synthesized from a SiO$_2$/C composite [77].

Double Templating
Inverse opals obtained from colloidal crystal templating can be used as templates to prepare their replicas of another composition. The final product could be a negative or positive replica of the original 3DOM material, depending on the fabrication process. When the voids within an inverse opal are completely filled by a precursor, a negative replica is obtained; whereas a thin coating of precursor on the skeleton without filling the voids leads to a positive replica. In the literature, this approach may be referred to as inverse opal templating, nanocasting, or micromolding. Monodisperse spherical colloidal crystals of various compositions (e.g., metal oxides, metals, polymers) were synthesized by infiltrating 3DOM polymer templates with the corresponding precursors. It was found that weak interactions between precursors and polymer templates led to solid colloids, whereas hollow spheres were obtained with precursors that strongly wetted the surface of a polymer template [80]. Similarly, metal sphere arrays were prepared with NiO or carbon inverse opals [81]. When a positive replica is needed, the skeleton of the inverse opal template can be modified so that the precursors of the target material can be selectively deposited and reacted on the surface of the template. To synthesize 3DOM/m carbon, a 3DOM/m SiO$_2$ template was first surface-modified with acidic catalytic Al sites, and then infiltrated with a phenolic resin carbon precursor through a gas-phase process (Figure 10.4b). These efforts ensured selective deposition of carbon on the surface of 3DOM/m SiO$_2$ template, whose structure was reproduced on multiple length scales [6]. This sequential double templating procedure is distinct from dual templating, described later, which involves multiple hard and/or soft templates in a single process to achieve structures with hierarchical porosity.

Assembly from Core–Shell Spheres
A unique approach to control the wall thickness of 3DOM materials is to use colloidal spheres with core–shell structures. Various species, such as polyelectrolytes and nanoparticles [78,82–84], can be coated on the surface of colloidal spheres one layer at a time through electrostatic attraction, thus forming the shell. In this manner, the wall thickness of the final product can be controlled by altering the shell thickness, although closed pores may be formed with thicker walls [83]. In a specific synthesis of 3DOM zeolites, colloidal mesoporous SiO$_2$ spheres with core–shell structures were obtained by sequential modification with positively charged polyelectrolytes and negatively charged zeolite nanoseeds. This was followed by natural sedimentation to form a colloidal crystal and

hydrothermal treatment in a precursor solution to produce the final 3DOM zeolite monoliths (Figure 10.4c) [78].

Dual Templating of 3DOM Materials with Hierarchical Pore Structure
Secondary templates may be used in addition to the main CCT to add meso- or micropores (<2 nm) to the skeletal walls. Secondary templates include smaller colloidal particles that also act as hard templates or surfactant or block-copolymer systems that function as soft templates. Sufficiently small secondary colloids tend to pack within the interstitial space between larger spheres, although in some cases they also decorate the complete sphere surface to form an interlayer between adjacent large spheres. After their removal, mesopores perforate the skeletal walls.

A way to combine soft templating with colloidal crystal templating is to infiltrate the CCT with a precursor mixture typically employed to produce mesoporous materials with ionic or nonionic surfactants of block-copolymers. Micellar structures then form in the interstices of the CCT, at least in regions where sufficient space for micelle generation is available. Micelles cannot form at the narrowest cusps. The mesopore architecture within the skeletal walls does not necessarily coincide with the mesostructure of bulk mesoporous materials synthesized without a CCT under otherwise identical conditions. Interactions between the surfactant/block-copolymer component, precursor, solvent, and the template surface, as well as confinement effects determine the shape and orientation of micelles within the template space [85], and such interactions can be exploited to achieve the desired pore orientations [86]. Mesopores perpendicular to the macropore surface are preferred to improve mass transport between macropores or to load mesopores with additional phases to form nanocomposites, whereas mesopores parallel to a macropore surface might be exploited to generate two mostly independent transport paths (through the skeleton and through the macropore system).

Hierarchical 3DOM materials that include micropores (3DOM/μ materials) have been prepared using either zeolitic skeletons [87], or skeletons composed of metal-organic frameworks [88]. In these cases, a structure-directing agent may or may not be needed as a secondary template for micropores, depending on the microporous material. Binary inverse opals can be obtained with trimodal colloidal crystals composed of PS, PMMA, and SiO_2 nanospheres (Figure 10.4d) [79].

Surface Modification
Surface modification is a versatile tool to optimize the properties (e.g., wettability, conductivity, catalytic activity) of 3DOM materials for specific applications. Due to their highly accessible pore structures with open and interconnected pores, 3DOM materials can be readily modified with various functionalities, such as molecular surface groups (e.g., fluoroalkylsilane) [89], polymers (e.g., polyelectrolyte) [90], and nanoparticles (e.g., TiO_2 nanoparticles) [91]. Depending on the specific 3DOM material, these functional components may directly bind to its surface, or require prior surface preparation to provide initial anchoring points. For example, fluoroalkylsilane can readily react with the

hydroxyl-rich SiO_2 surface [89], whereas 3DOM carbon is usually oxidized to bear surface functional groups (e.g., ketone, phenol, carboxylic acid) for further attachment [92]. Readers interested in surface modification of porous carbon [92], metals and metal oxides [93], and SiO_2 [94] are referred to other reviews.

10.4 Applications

Due to their porous structures and tunable properties, 3DOM materials have been used in both physical and chemical applications, such as optics, catalysis, electrochemistry, fuel cells, solar cells, bioengineering, and so on. In this section, we discuss these applications with a focus on how the unique properties of 3DOM materials can benefit a specific application.

10.4.1 Optical Applications

For both CCTs and 3DOM materials, the opalescent appearance is probably the most obvious feature one can immediately notice without a tool. Visible light interacts through scattering and diffraction with the periodic structure that has a similar length scale as the wavelength of light. This interaction gives 3DOM materials structural color along with opportunities to control and manipulate the flow of light for various optical applications.

Photonic Crystals
Photonic crystal applications are among the earliest explorations of 3DOM materials. Due to the multiple scattering of the periodic structure, light with certain wavelengths may not propagate in the 3DOM material in certain directions or in all directions, producing stop bands or a photonic band gap, respectively. To achieve such interactions, the periodicity of the 3DOM material should be on the order of a fraction of the wavelength of the light, which could be in the range from ultraviolet to visible to near-infrared (i.e., 300–2000 nm). For 3DOM materials with ccp and hcp structures, theoretical calculations indicate that a minimum refractive index contrast of 2.8 is needed to achieve a full photonic band gap [95]. Therefore, a composition with a high refractive index is usually chosen to construct the skeleton, with a ccp array of voids. In this regard, semiconductors with wide band gaps are of particular interest for photonic crystal applications in both visible (e.g., GaP) [96], and near-infrared ranges (e.g., CdSe) [97]. Readers interested in this topic are referred to several recent reviews [98–100].

Tunable Photonic Crystals
The position of the stop bands of photonic crystals depends on the d-spacing of the diffraction layers, as well as the refractive index contrast between the skeleton and void portions of the inverse opal. Therefore, a color change in the

material can be readily observed with inverse opals composed of the same material with different pore size or wall thickness, as shown in Figure 10.5a [12]. When a 3DOM photochromic hydrogel is exposed to UV irradiation or a temperature change, switching between two or multiple structural colors is observed due to the change of volume states of the photochromic material [101]. Similarly, an electrically triggered structural color change was achieved with an inverse opal of an electrochromic polyelectrolyte gel, whose lattice spacing could be tuned by an electric field [102]. Besides structure, controlled variation in refractive index contrast can also be used to tune the optical properties of 3DOM photonic crystals. This strategy is simply demonstrated by the color change upon infiltration of an inverse opal with organic solvents (Figure 10.5a) [12]. Infiltration can also be performed with responsive secondary phases that exhibit photo- or thermosensitive refractive indices, such as photochromic dyes [103], or thermosensitive liquid crystals (Figure 10.5b) [104]. Detailed discussions about tunable photonic crystals can be found in a few reviews [105,106].

Modification of Spontaneous Emission

Besides manipulating the light propagated from external sources, inverse opal photonic crystals can also alter the spontaneous emission of light sources that are embedded in the structure [107]. These light sources include dye molecules [108], rare earth ions [109], and quantum dots [110]. For example, when CdSe quantum dots were infiltrated into the pores of a TiO_2 inverse opal, either inhibited and enhanced photoluminescence decay rates were observed, depending on the emission frequency of the photoluminescence [110]. In another study to pursue color purification, the undesired blue, orange, and red emission bands

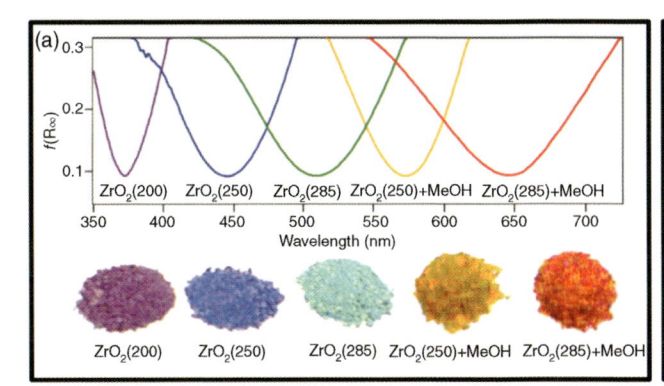

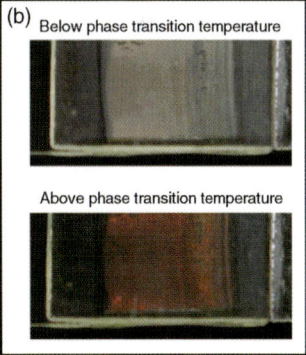

Figure 10.5 Examples of tunable photonic crystals. (a) Diffuse-reflectance UV-visible spectra (top) and photographs (bottom) of 3DOM ZrO_2 powders with different pore sizes (as indicated in the parenthesis) and samples infiltrated with methanol. (Adapted with permission from Ref. [12]. Copyright 2002, American Chemical Society.) (b) Photographs of SiO_2 inverse opal films filled with thermosensitive liquid crystals that change their refractive indices when heated above the phase transition temperature. (Adapted with permission from Ref. [104]. Copyright 2004, American Chemical Society.)

of Tb^{3+} embedded in a SiO_2 inverse opal were suppressed by the photonic stop bands, while the desired green emission was enhanced [109]. When a SiO_2 inverse opal was used as a substrate for optical sensing, the fluorescence signal was enhanced by a factor of 60, thus improving the sensitivity of the resulting sensor [111].

10.4.2
Catalytic Applications

A number of potential applications of 3DOM materials that have been investigated rely mainly on the activity or reactivity of the skeletal surface. These applications may benefit from efficient mass transport through the nontortuous macropore system, nanostructured features of the walls, and relatively large interfacial areas that are accessible even to large molecules. Such applications include catalysis, sorption, and bioactive materials.

For catalytic applications, the open 3DOM structure reduces pressure build-up compared to pressed powders. The 3DOM material can either be used as a support for the catalytic components, or serve as a catalyst itself. Advantages of the 3DOM structure are mainly seen in high-throughput, short contact-time reactions. For example, catalytic Au nanoparticles can be well dispersed within a 3DOM CeO_2 support with little aggregation, leading to enhanced activity for formaldehyde catalytic oxidation [112]. In another study, catalysts comprising hierarchical 3DOM/m $Ce_{0.6}Zr_{0.3}Y_{0.1}O_2$ loaded with Pt nanoparticles were used for methane combustion [113]. The hierarchical porosity in 3DOM/m materials combines the easy access through macropores with potential size selectivity in mesopores or micropores [114].

Other catalytic reactions, such as photocatalysis, electrocatalysis, and biocatalysis, can also benefit from the unique structure of 3DOM materials. It was reported that Pt-loaded 3DOM WO_3 and hierarchical 3DOM/m Bi_2WO_6 can enhance the photocatalytic activity of decomposition of organic compounds under light irradiation [115,116]; while 3DOM IrO_2 and $LaFeO_3$ were used as efficient electrocatalysts for oxygen evolution reactions and Li-O_2 batteries, respectively [117,118]. For biocatalytic applications, 3DOM materials can be used as a platform to load catalytic enzymes [119], or bioactive enzyme-based inverse opals can be synthesized directly using the CCT approach [120]. In many other cases though, catalytic performance is not significantly improved compared to less well-defined catalyst supports of the same composition, and, therefore, the higher cost of preparing the structure is not justified.

10.4.3
Electrochemical Energy Storage

Conductive 3DOM materials have been widely explored as electrode materials for electrochemical systems, including electrochemical energy storage devices. The nanopore structure of 3DOM materials offers several advantageous features for energy storage in rechargeable batteries and supercapacitors [121,122]. The open pore structure facilitates electrolyte transport and provides good access of the electrolyte to the electrode surface, with high contact area for rapid charge transfer. In addition, the walls of active material surrounding the pores can be very thin, reducing path lengths for ion diffusion. Together, these features are advantageous for rapid charge and discharge of an electrode, especially if the skeleton of the material is highly conductive. However, the larger surface area of a porous electrode can also enhance electrolyte decomposition and the formation of a solid/electrolyte interface layer, particularly at the anode side. Improved rate capabilities of half-cells employing 3DOM electrodes have been demonstrated for 3DOM carbon, Ni, or $LiCoO_2$. For carbon systems, adding secondary mesopores to form 3DOM/m carbon further improved rate capabilities. Although packed nanoparticles may also be considered for high rate electrodes, porous electrodes prepared by templating typically provide better interparticle contact.

The small feature sizes of the templated walls permit increased utilization of the electrode material, because in a given amount of time, a larger volume fraction is penetrated by charge carriers than in bulk electrode materials. As a result, specific capacities can be increased, especially at high charge and discharge rates. In contrast, volumetric capacities decrease with higher pore volumes. An optimal compromise between high rate capabilities and volumetric capacity may be achieved with pore size distributions in the mesopore range. A large concentration of smaller micropores limits electrical conductivity through the solid, and macropores are mainly advantageous for cells employing viscous electrolytes, otherwise limiting the volumetric energy density of the electrode. While volumetric capacities are reduced for larger pores, it is possible, although technically challenging, to utilize most pore space by coating the electrode surface with a separator membrane and filling the residual void space with a counter electrode. This concept of a 3D-interpenetrating electrochemical cell has been demonstrated in the laboratory [123] and is of interest for on-chip devices because high aerial capacities can be achieved; however, solutions for scale-up remain to be developed.

3DOM/m carbon materials with hierarchical pore structure can act as host matrices for additional active electrode materials and stabilize these during lithiation/delithiation processes. Large volume changes are typically observed during the formation of lithium alloys with high capacity anode materials such as SnO_2, Sn, Si, or Ge. These result in pulverization and loss of interparticle contact after multiple lithiation/delithiation cycles. By encapsulating nanoparticles of such alloy phases within the mesopores of 3DOM/m carbon, the nanoparticles maintain contact with the conductive host and also maintain capacity for more cycles [124].

To prepare composite electrodes consisting of active material mixed with a conductive carbon phase, the active components can be placed in specific regions of the conductive matrix to optimize electrode performance. Such site-specific placement depends on relative interactions of precursors with each other and with the template. For example, in syntheses of 3DOM TiO_2/C anode materials, discrete TiO_2 nanoparticles were placed either at the interface of the template or embedded within the templated carbon skeleton, depending on the polarity of the ligands associated with the precursor for TiO_2 [125,126]. When TiO_2 was dispersed throughout the carbon skeleton, nanoparticles were kept from sintering and remained smaller than when they formed at the template interface (Figure 10.6a). As a result, higher charge and discharge rates could be achieved. On the other hand, deep embedding of TiO_2 nanoparticles within the carbon matrix restricted access of Li^+ ions to the active material and limited the high-rate performance. In another case of 3DOM/m $LiFePO_4$/C cathode materials, the more ionic $LiFePO_4$ component occupied mainly the larger octahedral voids of the polymeric CCT, and polymer-derived carbon mainly

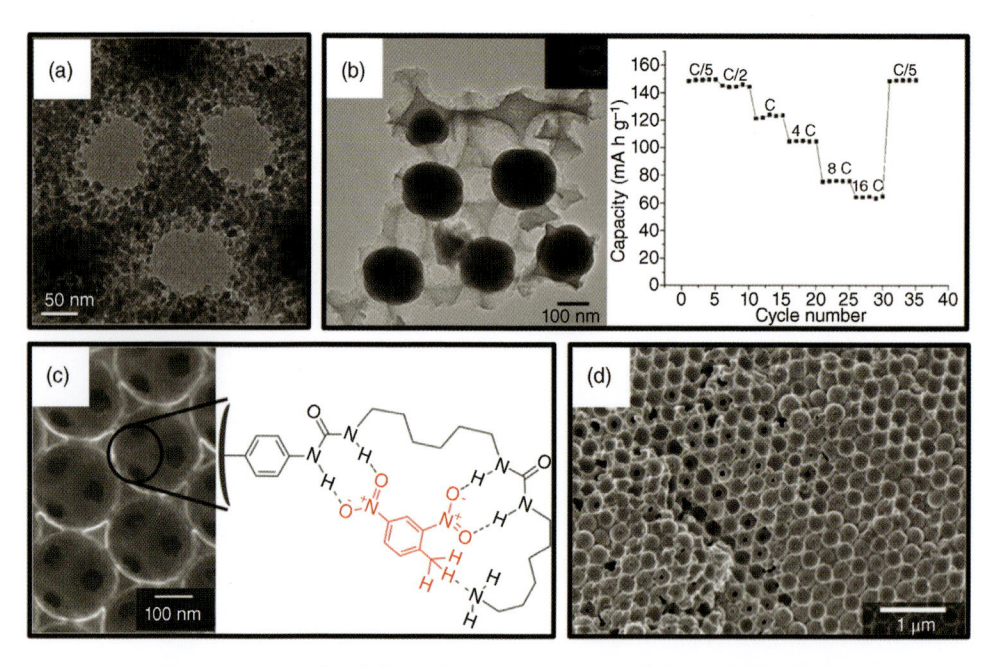

Figure 10.6 Examples of electrochemical applications of 3DOM materials. (a) TEM image of a 3DOM TiO_2/C anode material showing that the TiO_2 nanoparticles (appearing as black dots) are well-dispersed within the 3DOM carbon skeleton. (Adapted with permission from Ref. [125]. Copyright 2010, American Chemical Society.) (b) TEM image of a 3DOM/m $LiFePO_4$/C cathode material demonstrating that the $LiFePO_4$ components mainly occupy the octahedral voids of the polymer CCT during the synthesis (left), and the capacity of the resulting electrode cycled at various rates. (Adapted with permission from Ref. [127]. Copyright 2011, American Chemical Society.) (c) a schematic representation of a 3DOM carbon electrode surface-modified with a receptor (black) for the sensitive detection of 2,4-dinitrotoluene (red). (Adapted with permission from Ref. [128]. Copyright 2012, American Chemical Society.) (d) SEM image of a 3DOM carbon monolith infiltrated with a polymeric ion-sensing membrane that results in large interfacial capacitance with good signal transduction and stability. (Adapted with permission from Ref. [129]. Copyright 2014, American Chemical Society.)

occupied the tetrahedral voids which have higher surface-to-volume ratios (Figure 10.6b) [127].

For supercapacitor applications, activated carbon is frequently employed as a high surface area electrode material. Because of a wide distribution of pore sizes in activated carbon with many dead ends, not all of the surface area is readily accessibly to the electrolyte. The more uniform and open pore structure in 3DOm carbon provides better surface access, especially when viscous ionic liquids are used as an electrolyte, leading to improved capacitance at high charge/discharge rates. In a systematic study of the effects of pore size in 3DOm carbon on capacitance, an optimum pore size of 30 nm was found with an ionic liquid electrolyte system [130].

10.4.4
Electrochemical Sensing

In a similar way the open and accessible pore structure of 3DOM material benefits electrochemical energy storage applications, electrochemical sensors based on 3DOM materials exhibit high interfacial contact areas between the sensing phase and the porous skeleton, supporting charge transfer and signal transduction processes. The surface area of a specific electrode based on a 3DOM Au film is approximately 14 times higher than that of a conventional flat Au electrode [131]. This feature can lead to higher sensitivity, improved signal stability, and reduced response time for 3DOM material-based electrochemical sensors, which are combined with various electroanalytical techniques, such as voltammetry [128], amperometry [132], potentiometry [3], and electrochemical impedance spectroscopy [133].

To detect 2,4-dinitrotoluene as an analogue of explosive 2,4,6-trinitrotoluene, the surface of 3DOM carbon can be functionalized with a receptor using cyclic voltammetry and additional chemical reactions (Figure 10.6c). The resulting receptor-modified 3DOM carbon electrode can be used in square wave voltammetry, exhibiting high selectivity for 2,4-dinitrotoluene over interferents, with a detection limit of 10 uM [128]. In another study, hemoglobin was immobilized in a Au nanoparticle-doped 3DOM TiO_2 to catalyze and detect the reduction of H_2O_2. The amperometric response of the immobilized hemoglobin showed a good linear relation with the concentration of H_2O_2 and a higher sensitivity over TiO_2 nanoneedle-based electrodes [132]. When 3DOM carbon is used as a solid contact material for all-solid-state potentiometric ion sensors, outstanding signal stability can be achieved due to the large interfacial double-layer capacitance at the sensing membrane/3DOM carbon interface (Figure 10.6d) [3,129]. Improved signal stability and electrode reproducibility can be achieved using colloid-imprinted mesoporous carbon with higher interfacial contact area and lower redox-active surface functionalities [134,135]. Loading the voids of a 3DOM conducting polymer with redox mediators was also reported to enhance the electrode reproducibility of the resulting all-solid-state Ag^+ potentiometric sensors [136].

10.4.5
Fuel Cells

Fuel cells efficiently convert chemical energy from a variety of sources (e.g., H_2, CH_4, CH_3OH) into electrical energy and are therefore a promising alternative to current fossil-fuel based energy generation technologies. The structural features of 3DOM materials provide advantages for fuel cell components, and 3DOM materials have been studied as electrodes, catalysts, or membrane components in proton-exchange membrane fuel cells, solid oxide fuel cells, and biofuel cells and also in related processes to produce hydrogen fuel or remove CO from streams that could poison Pt anodes in a fuel cell.

Proton-Exchange Membrane Fuel Cells
For proton-exchange membranes that employ 3DOM materials, a 3DOM host matrix is typically infiltrated with a proton-conducting material. As a host, both 3DOM SiO_2 and 3DOM polymers have been used, and for the proton-conducting electrolyte, inorganic $Cs_{2.5}H_{0.5}PW_{12}O_{40}$ (CsHPW) cluster systems or organic gel polymers have been considered [137–142]. In all of these systems, the periodic 3DOM structure offers well-interconnected pathways for efficient proton transport. The inorganic 3DOM SiO_2/CsHPW composite membrane is of particular interest for higher temperature operation, reaching a proton conductivity of 0.25 S cm^{-1} at 170 °C at high electrolyte loading in a single cell [142]. When used to confine a proton-conducting gel polymer, a 3DOM SiO_2 or 3DOM polyimide matrix can suppress polymer swelling in the composite membrane, thereby reducing methanol permeability across the membrane and mitigating the problem of crossover [137,139]. Both of these composites with gel polymers could be prepared as self-standing membranes and improved the transfer selectivity between protons and methanol molecules by an order of magnitude compared to Nafion membranes. Methanol permeability was reduced most significantly when windows between adjacent macropores (the bottlenecks for mass transport) were smaller than 100 nm [140].

A similar configuration can also be used for the oxygen electrode in proton-exchange membrane fuel cells. For example, a solution containing CsHPW clusters and Nafion ionomer was infiltrated into a colloidal crystal to form 3DOM Nafion/CsHPW nanocomposites, whose macropores were subsequently loaded with Pt/C nanoparticles to introduce catalytic sites. The resulting porous oxygen electrodes achieved significantly higher maximum power densities than conventional Nafion-binder-based oxygen electrodes, as a result of increased proton conductivity and enhanced gas transport through the porous matrix [143].

3DOM Pt has been investigated as an electrode for the membrane electrode assembly in a proton-exchange membrane fuel cell (Figure 10.7a) [141]. Because the Pt is integrated into the membrane-electrode assembly, loss of catalyst particles is reduced. Other advantages observed for this configuration include improved mass diffusivity through the interconnected pore array with low tortuosity and more effective water management. The large interfacial area provided

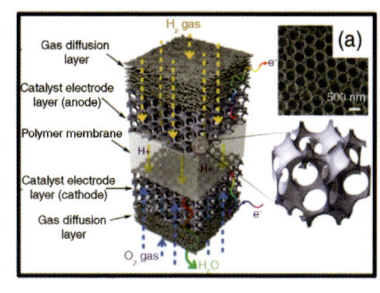

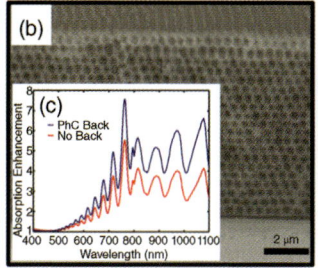

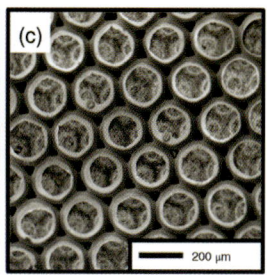

Figure 10.7 (a) A schematic overview of a proton-exchange membrane fuel cell employing 3DOM Pt electrodes for enhanced mass transfer. (Adapted with permission from Ref. [141]. Copyright 2013, Nature Publishing Group.) (b) SEM image of a 3DOM Si thin film as a back reflector for photovoltaic cells, with the inset showing the absorption enhancement of the cell. (Adapted with permission from Ref. [144]. Copyright 2013, Wiley-VCH Verlag GmbH.) (c) SEM image of 3DOM poly(D,L-lactide-coglycolide) scaffolds reinforced with hydroxyapatite nanoparticles for bone tissue engineering. (Adapted with permission from Ref. [145]. Copyright 2010, American Chemical Society.)

by the electrode surface resulted in an improvement of the apparent reaction rate.

H_2 as a fuel for proton-exchange membrane fuel cells is typically generated by catalytic reforming of hydrocarbons and contains trace amounts of CO. This CO must be removed to avoid poisoning the Pt catalysts used in the fuel cell electrodes. Catalysts that preferentially oxidize CO over H_2 are needed in H_2-rich fuel streams. In the design of such catalysts, the 3DOM architecture can also be beneficial. 3DOM CeO_2/SiO_2 with a Pt–Rh catalyst showed higher efficiency for CO removal in preferential CO oxidation than a similar wash-coated catalyst support [146]. The improvements were achieved by better dispersion of catalyst species in the templated support, enhanced gas–solid interactions and therefore, improved reaction efficiency. Other examples for preferential CO oxidation include 3DOM Au/CeO_2 [147] and more complex combinations of 3DOM Au/CeO_2 with other transition metal oxides [148]. A study of the effect of pore size (200–600 nm) on catalytic activity indicated that the highest activity was achieved with the smallest pore size material that offered the highest surface area. This trend agrees with trends observed for other 3DOM systems used in fuel cells, such as 3DOM carbon supports for Pt–Ru alloy clusters used in direct methanol fuel cells, where those supports with smaller pore sizes yielded higher surface areas and better catalytic activity [149,150].

Solid Oxide and Biofuel Cells

In solid oxide fuel cells, the open interconnected pore structure of 3DOM electrodes helps to reduce the polarization resistance, as demonstrated, for example, with 3DOM $La_{0.8}Sr_{0.2}MnO_3$/YSZ (yttria-stabilized zirconia) cathodes [151] or 3DOM NiO-YSZ anodes [152]. To maximize the surface area and number of reactive sites (which requires smaller pores), while maintaining a low polarization resistance (which benefits from larger pores), a double-layered 3DOM NiO-YSZ anode composed of a small-pore layer (1.4 μm pores) and a large-pore layer (2.5 μm pores) was fabricated. This structure improved cell performance further compared to a similar anode with only the larger macropores. Because the cells operate at high temperatures between 500 and 1000 °C, structural materials with smaller pore sizes and nanostructured walls may sinter and densify during operation, resulting in loss of surface area and potentially blocking pores. This limits the use of 3DOM materials at high temperatures.

An example of a 3DOM material in a biofuel cell is a 3DOM Au electrode coated with additional Au nanoparticles and enzymes for use in a one-compartment glucose/O_2 cell [153]. A performance improvement in the biofuel cell was achieved as a result of the larger amount of enzyme that could be supported on the macroporous electrode, compared to a planar electrode.

10.4.6
Solar Cells

Solar cells, or photovoltaic cells, are electrical devices that convert the energy of light into electricity via the photovoltaic effect. Due to the unique properties of light manipulation and enhanced charge transfer originating from the ordered and interconnected pore structures, 3DOM materials have been studied as various components for photovoltaic cells.

Photovoltaic Cells
One of the most predominant photoelectric materials for photovoltaic applications is Si. Consequently, Si inverse opal-based photonic crystals have been explored as photoelectric generation layers in photovoltaic cells, from both electrical and optical aspects. The electrical conductivity of Si inverse opals was reported to be independent of the periodicity of the structure and comparable to that of crystalline Si [154]. Improved electrical conductivity can be achieved through chemically controlled n-type and p-type doping, resulting in electrically conducting Si inverse opals acceptable for photovoltaic devices [155]. Also, 3DOM photonic crystals with tunable stop bands or complete photonic band gaps offer unique features of photon management (e.g., light trapping or suppression) over specific regions of the spectrum, which is advantageous over traditional optical manipulation techniques that treat all wavelengths equally. For 3DOM Si, slow photon enhanced photoconductivity with amplified photo-to-electron conversion efficiency was observed around their photonic band gaps [156]. When a thin-film of 3DOM Si was used as a back reflector of a commercial polycrystalline Si solar cell, an efficiency enhancement of 10% was achieved due to an enhanced reflection of the near-IR light that is most needed for the cell (Figure 10.7b) [144]. Besides Si, promising results were also obtained with other 3DOM semiconductors, such as CdS and CdSe inverse opals [157,158].

Dye-Sensitized Solar Cells

A dye-sensitized solar cell is a photoelectrochemical system that comprises a photosensitized anode (e.g., typically a porous layer of TiO_2 nanoparticles coated with a Ru-based molecular dye), a counter electrode (e.g, a conductive Pt sheet), and an electrolyte containing a redox couple (e.g., I^-/I_3^-). For a photoanode composed of 3DOM TiO_2, the nanostructured skeleton provides a highly accessible and continuous path for dye absorption, electrolyte transport, and charge transfer, thus benefiting the overall conversion efficiency of the cell [159]. A detailed study of the photoelectrochemical behavior of macroporous TiO_2 films revealed that 3DOM TiO_2 exhibits higher light absorption capability than disordered macroporous TiO_2, and the recombination rate of charge carriers seems to be less affected by the degree of order in the structure [160]. Furthermore, the reflectance of 3DOM TiO_2 can be managed to selectively enhance the light harvesting capability of the cell, whereas conventional photoanodes composed of TiO_2 nanoparticles randomly reflect light over a wide range of the spectrum [161]. In addition to photoanodes, 3DOM materials have also been tested as counter electrodes, such as 3DOM carbon [162,163], 3DOM fluorinated tin oxide [164], and 3DOM poly(3,4-ethylenedioxythiophene) [165]. For these systems, improved performance is usually observed due to the facilitated electrolyte transport and charge transfer offered by the 3DOM structure.

10.4.7
Bioactive Materials and Tissue Engineering

Bioactive glasses are materials that react with body fluids to form a biocompatible hydroxycarbonate apatite layer at the interface with bone. Colloidal-crystal templated bioglasses with typical molar compositions of 20–21% CaO: 75–80% SiO_2: 0–4% P_2O_5 have been investigated *in vitro* using simulated body fluid at body temperature [166–168]. Under these conditions the conversion to bone-like apatite was significantly faster than for a nontemplated control sample. The more accessible surface promoted the formation of amorphous calcium phosphate nodules on the surface of the 3DOM bioglass skeleton, followed by growth of spherical hydroxycarbonate apatite clusters that were eventually converted to extended crystalline hydroxycarbonate apatite aggregates. *In vitro* cell culture studies of 3DOM bioglasses with osteoblastic cells demonstrated that neither the 3DOM bioglass particles nor leachates from these particles were toxic to the cells. Cells attached themselves onto 3DOM bioglass particles, where they spread and proliferated [169,170].

The suitability of 3DOM materials as scaffolds for cell growth and tissue engineering has also been investigated and applied to tissue engineering of bone, cartilage, and neural tissue [171]. Cylindrical proteins have been accommodated in macroporous poly(vinyl alcohol) films [172], and human cell cultures in 3DOM silicate or hydrogel scaffolds templated by large polymer spheres (hundred micron range) [173,174]. Degradable chitosan inverse opal scaffolds also provide a suitable environment for cell growth [175], in particular after the macropores are further functionalized with chitosan microstructures so that cells can use the void space more efficiently [176]. Cell alignment is possible by culturing cells on stretchable polymeric inverse opal films and stretching these in appropriate directions [177]. *In vitro* mineralization by preosteoblasts has been carried out in 3DOM poly(D,L-lactide-co-glycolide) scaffolds reinforced with hydroxyapatite nanoparticles for bone tissue engineering (Figure 10.7c) [145]. Compared to scaffolds with nonuniform pores, the uniform 3DOM pore structure results in a more uniform distribution of cells and a higher degree of differentiation of preosteoblasts [178]. A study of neovascularization in these materials revealed that 3DOM scaffolds with pores <200 μm favor the formation of vascular networks with small blood vessels at high densities and poor penetration depth, whereas those with pores >200 μm favor the formation of networks with large vessels at low densities and deep penetration depth [179].

10.5
Conclusions and Outlook

Colloidal crystal templating as a method of structuring porous materials has reached a certain level of maturity at this stage, especially in regard to templating with opaline sphere arrays. It has been successfully applied to many classes of materials and compositions, producing highly symmetric, open pore architectures. Hierarchical porosity is achievable by combining colloidal crystal templates with other templating methods, and compositional complexity is accomplished either in single-step syntheses or in multiple step processes. 3DOM materials can be obtained with a variety of morphologies, including powders, fibers, thin films, and monoliths. As illustrated in this chapter, all of these features can benefit a wide range of technical applications, at least conceptually. In practice, the translation into products is still limited by scale-up of this sacrificial templating approach. Therefore, the first applications will likely involve specialized, high mark-up, small-volume applications, where the benefits derived from the 3DOM structure outweigh the cost of synthesis. The colloidal crystal templating approach can be more efficient than, for example, lithographic patterning processes, when multilayered three-dimensional and periodic structures with submicrometer features are needed, but compared to those processes it lacks flexibility in obtaining nonregular structures in 3D space. Future developments of colloidal assembly to produce colloidal crystals with lower symmetry will be helpful. The templating processes developed for opaline templates can then be transferred to other template geometries, and lessons learned from templating with ccp colloidal crystals regarding template–precursor interactions and processing conditions remain applicable. In the near future, one can expect integration of colloidal crystal templated materials into other complex systems that may utilize multiple functions of these nanostructured porous materials. Other early applications will be those that tolerate some disorder or variation in pore size and can therefore rely on materials prepared from imperfect arrays of colloids (e.g., colloid imprinting approaches) [134,135,180].

References

1 Stein, A., Li, F., and Denny, N.R. (2008) *Chem. Mater.*, **20**, 649–666.
2 Fowler, C.E., Shenton, W., Stubbs, G., and Mann, S. (2001) *Adv. Mater.*, **13**, 1266–1269.
3 Lai, C.-Z., Fierke, M.A., Stein, A., and Bühlmann, P. (2007) *Anal. Chem.*, **79**, 4621–4626.
4 Liu, X., Zhang, Y., Ge, D., Zhao, J., Li, Y., and Endres, F. (2012) *Phys. Chem. Chem. Phys.*, **14**, 5100–5105.
5 Josephson, D.P., Miller, M., and Stein, A. (2014) *Z. Anorg. Allg. Chem.*, **640**, 655–662.
6 Wang, Z., Li, F., Ergang, N.S., and Stein, A. (2006) *Chem. Mater.*, **18**, 5543–5553.
7 Denny, N.R., Han, S., Turgeon, R.T., Lytle, J.C., Norris, D.J., and Stein, A. (2005) *SPIE Proc.*, **6005**, 60050501–60050513.
8 Stein, A. and Schroden, R.C. (2001) *Curr. Opin. Solid State Mater. Sci.*, **5**, 553–564.
9 Stöber, W., Fink, A., and Bohn, E. (1968) *J. Colloid Interface Sci*, **26**, 62–69.
10 Yokoi, T., Sakamoto, Y., Terasaki, O., Kubota, Y., Okubo, T., and Tatsumi, T. (2006) *J. Am. Chem. Soc.*, **128**, 13664–13665.
11 Holland, B.T., Blanford, C.F., Do, T., and Stein, A. (1999) *Chem. Mater.*, **11**, 795–805.
12 Schroden, R.C., Al-Daous, M., Blanford, C.F., and Stein, A. (2002) *Chem. Mater.*, **14**, 3305–3315.
13 Cong, H. and Cao, W. (2004) *Langmuir*, **20**, 8049–8053.
14 You, B., Wen, N., Shi, L., Wu, L., and Zi, J. (2009) *J. Mater. Chem.*, **19**, 3594–3597.
15 Dziomkina, N.V. and Vancso, G.J. (2005) *Soft Matter*, **1**, 265–279.
16 Li, F., Josephson, D.P., and Stein, A. (2011) *Angew. Chem., Int. Ed.*, **50**, 360–388.
17 Vogel, N., Retsch, M., Fustin, C.-A., del Campo, A., and Jonas, U. (2015) *Chem. Rev.*, **115**, 6265–6311.
18 Yin, Y., Li, Z.-Y., and Xia, Y. (2003) *Langmuir*, **19**, 622–631.
19 Cai, Z., Teng, J., Xia, D., and Zhao, X.S. (2011) *J. Phys. Chem. C*, **115**, 9970–9976.
20 Stein, A., Wilson, B.E., and Rudisill, S.G. (2013) *Chem. Soc. Rev.*, **42**, 2763–2803.
21 Jiang, P. and McFarland, M.J. (2004) *J. Am. Chem. Soc.*, **126**, 13778–13786.
22 Denkov, N.D., Velev, O.D., Kralchevsky, P.A., Ivanov, I.B., Yoshimura, H., and Nagayama, K. (1993) *Nature*, **361**, 26–126.
23 Wong, S., Kitaev, V., and Ozin, G.A. (2003) *J. Am. Chem. Soc.*, **125**, 15589–15598.
24 Meng, L., Wei, H., Nagel, A., Wiley, B.J., Scriven, L.E., and Norris, D.J. (2006) *Nano Lett.*, **6**, 2249–2253.
25 Park, S.H., Qin, D., and Xia, Y. (1998) *Adv. Mater.*, **10**, 1028–1032.
26 Wijnhoven, J.E.G.J. and Vos, W.L. (1998) *Science*, **281**, 802–804.
27 Yang, P., Deng, T., Zhao, D., Feng, P., Pine, D., Chmelka, B.F., Whitesides, G.M., and Stucky, G.D. (1998) *Science*, **282**, 2244–2246.
28 Velev, O.D., Jede, T.A., Lobo, R.F., and Lenhoff, A.M. (1997) *Nature*, **389**, 447–448.
29 Velev, O.D., Jede, T.A., Lobo, R.F., and Lenhoff, A.M. (1998) *Chem. Mater.*, **10**, 3597–3602.
30 Holland, B.T., Blanford, C.F., and Stein, A. (1998) *Science*, **281**, 538–540.
31 Lytle, J.C., Yan, H., Ergang, N.S., Smyrl, W.H., and Stein, A. (2004) *J. Mater. Chem.*, **14**, 1616–1622.
32 Yin, J.S. and Wang, Z.L. (1999) *Adv. Mater.*, **11**, 469–472.
33 Petkovich, N.D., Rudisill, S.G., Venstrom, L.J., Boman, D.B., Davidson, J.H., and Stein, A. (2011) *J. Phys. Chem. C*, **115**, 21022–21033.
34 Zakhidov, A.A., Baughman, R.H., Iqbal, Z., Cui, C., Khayrullin, I., Dantas, S.O.,

Marti, J., and Ralchenko, V.G. (1998) *Science*, **282**, 897–901.

35 Chen, X., Li, Z., Ye, J., and Zou, Z. (2010) *Chem. Mater.*, **22**, 3583–3585.

36 Wijnhoven, J.E.G.J., Bechger, L., and Vos, W.L. (2001) *Chem. Mater.*, **13**, 4486–4499.

37 Yan, H.W., Blanford, C.F., Holland, B.T., Smyrl, W.H., and Stein, A. (2000) *Chem. Mater.*, **12**, 1134–1141.

38 Yan, H., Blanford, C.F., Smyrl, W.H., and Stein, A. (2000) *Chem. Commun.*, 1477–1478.

39 Yan, H., Blanford, C.F., Holland, B.T., Parent, M., Smyrl, W.H., and Stein, A. (1999) *Adv. Mater.*, **11**, 1003–1006.

40 Yan, H., Blanford, C.F., Lytle, J.C., Carter, C.B., Smyrl, W.H., and Stein, A. (2001) *Chem. Mater.*, **13**, 4314–4321.

41 Míguez, H., Meseguer, F., López, C., Holgado, M., Andreasen, G., Mifsud, A., and Fornés, V. (2000) *Langmuir*, **16**, 4405–4408.

42 Egan, G.L., Yu, J.S., Kim, C.H., Lee, S.J., Schaak, R.E., and Mallouk, T.E. (2000) *Adv. Mater.*, **12**, 1040–1042.

43 Jiang, P., Hwang, K.S., Mittleman, D.M., Bertone, J.F., and Colvin, V.L. (1999) *J. Am. Chem. Soc.*, **121**, 11630–11637.

44 Park, S.H. and Xia, Y. (1998) *Chem. Mater.*, **10**, 1745–1747.

45 Míguez, H., Meseguer, F., López, C., López-Tejeira, F., and Sánchez-Dehesa, J. (2001) *Adv. Mater.*, **13**, 393–396.

46 Fan, W., Snyder, M.A., Kumar, S., Lee, P.-S., Yoo, W.C., McCormick, A.V., Lee Penn, R., Stein, A., and Tsapatsis, M. (2008) *Nat. Mater.*, **7**, 984–991.

47 Qian, W., Gu, Z.-Z., Fujishima, A., and Sato, O. (2002) *Langmuir*, **18**, 4526–4529.

48 Yoshino, K., Kawagishi, Y., Tatsuhara, S., Kajii, H., Lee, S., Fujii, A., Ozaki, M., Zakhidov, A.A., Vardeny, Z.V., and Ishikawa, M. (1999) *Microelectron. Eng.*, **47**, 49–53.

49 Velev, O.D., Tessier, P.M., Lenhoff, A.M., and Kaler, E.W. (1999) *Nature*, **401**, 548–1548.

50 Vlasov, Y.A., Yao, N., and Norris, D.J. (1999) *Adv. Mater.*, **11**, 165–169.

51 Chung, Y.-W., Leu, I.-C., Lee, J.-H., and Hon, M.-H. (2009) *Electrochim. Acta*, **54**, 3677–3682.

52 Subramania, G., Constant, K., Biswas, R., Sigalas, M.M., and Ho, K.-M. (1999) *Appl. Phys. Lett.*, **74**, 3933–3935.

53 Meng, Q.B., Fu, C.H., Einaga, Y., Gu, Z.Z., Fujishima, A., and Sato, O. (2002) *Chem. Mater.*, **14**, 83–88.

54 Cong, H. and Cao, W. (2004) *J. Colloid Interface Sci.*, **278**, 423–427.

55 Meng, X., Al-Salman, R., Zhao, J., Borissenko, N., Li, Y., and Endres, F. (2009) *Angew. Chem., Int. Ed.*, **48**, 2703–2707.

56 Braun, P.V. and Wiltzius, P. (1999) *Nature*, **402**, 603–604.

57 Luo, Q., Liu, Z., Li, L., Xie, S., Kong, J., and Zhao, D. (2001) *Adv. Mater.*, **13**, 286–289.

58 Kim, J., Kim, H.S., Choi, J.H., Jeon, H., Yoon, Y., Liu, J., Park, J.-G., and Braun, P.V. (2014) *Chem. Mater.*, **26**, 7051–7058.

59 Bognár, J., Szűcs, J., Dorkó, Z., Horváth, V., and Gyurcsányi, R.E. (2013) *Adv. Funct. Mater.*, **23**, 4703–4709.

60 Rogach, A.L., Kotov, N.A., Koktysh, D.S., Ostrander, J.W., and Ragoisha, G.A. (2000) *Chem. Mater.*, **12**, 2721–2726.

61 Jiang, P., Cizeron, J., Bertone, J.F., and Colvin, V.L. (1999) *J. Am. Chem. Soc.*, **121**, 7957–7958.

62 Lu, L., Randjelovic, I., Capek, R., Gaponik, N., Yang, J., Zhang, H., and Eychmüller, A. (2005) *Chem. Mater.*, **17**, 5731–5736.

63 Lu, L., Capek, R., Kornowski, A., Gaponik, N., and Eychmüller, A. (2005) *Angew. Chem., Int. Ed.*, **44**, 5997–6001.

64 Juarez, B.H., Garcia, P.D., Golmayo, D., Blanco, A., and Lopez, C. (2005) *Adv. Mater.*, **17**, 2761–2765.

65 Míguez, H., Chomski, E., García-Santamaría, F., Ibisate, M., John, S., López, C., Meseguer, F., Mondia, J.P., Ozin, G.A., Toader, O., and van Driel, H.M. (2001) *Adv. Mater.*, **13**, 1634–1637.

66 Zhang, X., Yan, W., Yang, H., Liu, B., and Li, H. (2008) *Polymer*, **49**, 5446–5451.

67 Au, R.H.W. and Puddephatt, R.J. (2007) *Chem. Vapor. Depos.*, **13**, 20–22.
68 King, J.S., Graugnard, E., and Summers, C.J. (2005) *Adv. Mater.*, **17**, 1010–1013.
69 Karuturi, S.K., Liu, L., Su, L.T., Chutinan, A., Kherani, N.P., Chan, T.K., Osipowicz, T., and Tok, A.I.Y. (2011) *Nanoscale*, **3**, 4951–4954.
70 Lee, S., Teshima, K., Fujisawa, M., Fujii, S., Endo, M., and Oishi, S. (2009) *Phys. Chem. Chem. Phys.*, **11**, 3628–3633.
71 Balgis, R., Sago, S., Anilkumar, G.M., Ogi, T., and Okuyama, K. (2013) *ACS Appl. Mater. Interfaces*, **5**, 11944–11950.
72 Vaudreuil, S., Bousmina, M., Kaliaguine, S., and Bonneviot, L. (2001) *Adv. Mater.*, **13**, 1310–1312.
73 Vaudreuil, S., Bousmina, M., Kaliaguine, S., and Bonneviot, L. (2001) *Microporous Mesoporous Mater.*, **44–45**, 249–258.
74 Danumah, C., Vaudreuil, S., Bonneviot, L., Bousmina, M., Giasson, S., and Kaliaguine, S. (2001) *Microporous Mesoporous Mater.*, **44–45**, 241–247.
75 Lytle, J.C., Yan, H., Turgeon, R.T., and Stein, A. (2004) *Chem. Mater.*, **16**, 3829–3837.
76 Lytle, J.C., Denny, N.R., Turgeon, R.T., and Stein, A. (2007) *Adv. Mater.*, **19**, 3682–3686.
77 Shi, Y., Zhang, F., Hu, Y.-S., Sun, X., Zhang, Y., Lee, H.I., Chen, L., and Stucky, G.D. (2010) *J. Am. Chem. Soc.*, **132**, 5552–5553.
78 Dong, A., Wang, Y., Tang, Y., Zhang, Y., Ren, N., and Gao, Z. (2002) *Adv. Mater.*, **14**, 1506–1510.
79 Wang, J., Li, Q., Knoll, W., and Jonas, U. (2006) *J. Am. Chem. Soc.*, **128**, 15606–15607.
80 Jiang, P., Bertone, J.F., and Colvin, V.L. (2001) *Science*, **291**, 453–457.
81 Xu, L., Zhou, W., Kozlov, M.E., Khayrullin, I.I., Udod, I., Zakhidov, A.A., Baughman, R.H., and Wiley, J.B. (2001) *J. Am. Chem. Soc.*, **123**, 763–764.
82 Caruso, F., Caruso, R.A., and Möhwald, H. (1998) *Science*, **282**, 1111–1114.
83 Wang, D., Caruso, R.A., and Caruso, F. (2001) *Chem. Mater.*, **13**, 364–371.
84 Hotta, Y., Jia, Y., Kawamura, M., Omura, N., Tsunekawa, K., Sato, K., and Watari, K. (2006) *J. Mater. Sci.*, **41**, 2779–2786.
85 Li, F., Wang, Z., Ergang, N.S., Fyfe, C.A., and Stein, A. (2007) *Langmuir*, **23**, 3996–4004.
86 Li, F., Wilker, M.B., and Stein, A. (2012) *Langmuir*, **28**, 7484–7491.
87 Holland, B.T., Abrams, L., and Stein, A. (1999) *J. Am. Chem. Soc.*, **121**, 4308–4309.
88 Wu, Y.-n., Li, F., Zhu, W., Cui, J., Tao, C.-A., Lin, C., Hannam, P.M., and Li, G. (2011) *Angew. Chem., Int. Ed.*, **50**, 12518–12522.
89 Yue, L., Weiping, C., Bingqiang, C., Guotao, D., Fengqiang, S., Cuncheng, L., and Lichao, J. (2006) *Nanotechnology*, **17**, 238.
90 Yeo, S.J., Kang, H., Kim, Y.H., Han, S., and Yoo, P.J. (2012) *ACS Appl. Mater. Interfaces*, **4**, 2107–2115.
91 Wang, Z., Ergang, N.S., Al-Daous, M.A., and Stein, A. (2005) *Chem. Mater.*, **17**, 6805–6813.
92 Stein, A., Wang, Z., and Fierke, M.A. (2009) *Adv. Mater.*, **21**, 265–293.
93 Dumee, L.F., He, L., Lin, B., Ailloux, F.-M., Lemoine, J.-B., Velleman, L., She, F., Duke, M.C., Orbell, J.D., Erskine, G., Hodgson, P.D., Gray, S., and Kong, L. (2013) *J. Mater. Chem. A*, **1**, 15185–15206.
94 Linares, N., Serrano, E., Rico, M., Mariana Balu, A., Losada, E., Luque, R., and Garcia-Martinez, J. (2011) *Chem. Commun.*, **47**, 9024–9035.
95 Busch, K. and John, S. (1998) *Phys. Rev. E*, **58**, 3896–3908.
96 Norris, D.J. and Vlasov, Y.A. (2001) *Adv. Mater.*, **13**, 371–376.
97 Neale, N.R., Lee, B.G., Kang, S.H., and Frank, A.J. (2011) *J. Phys. Chem. C*, **115**, 14341–14346.
98 lvaro, B. and Cefe, L. (2006) *Annual Review of Nano Research*, World Scientific, Singapore, pp. 81–152.
99 von Freymann, G., Kitaev, V., Lotsch, B.V., and Ozin, G.A. (2013) *Chem. Soc. Rev.*, **42**, 2528–2554.
100 Zhao, Y., Shang, L., Cheng, Y., and Gu, Z. (2014) *Acc. Chem. Res.*, **47**, 3632–3642.

101 Matsubara, K., Watanabe, M., and Takeoka, Y. (2007) *Angew. Chem., Int. Ed.*, **46**, 1688–1692.

102 Ueno, K., Sakamoto, J., Takeoka, Y., and Watanabe, M. (2009) *J. Mater. Chem.*, **19**, 4778–4783.

103 Kubo, S., Gu, Z.-Z., Takahashi, K., Ohko, Y., Sato, O., and Fujishima, A. (2002) *J. Am. Chem. Soc.*, **124**, 10950–10951.

104 Kubo, S., Gu, Z.-Z., Takahashi, K., Fujishima, A., Segawa, H., and Sato, O. (2004) *J. Am. Chem. Soc.*, **126**, 8314–8319.

105 Aguirre, C.I., Reguera, E., and Stein, A. (2010) *Adv. Funct. Mater.*, **20**, 2565–2578.

106 Josephson, D. and Stein, A. (2013) *Responsive Photonic Nanostructures: Smart Nanoscale Optical Materials*, The Royal Society of Chemistry, pp. 63–90.

107 Schroden, R.C., Al-Daous, M., and Stein, A. (2001) *Chem. Mater.*, **13**, 2945–2950.

108 Bechger, L., Lodahl, P., and Vos, W.L. (2005) *J. Phys. Chem. B*, **109**, 9980–9988.

109 Shrivastava, V.P., Sivakumar, S., and Kumar, J. (2015) *ACS Appl. Mater. Interfaces*, **7**, 11890–11899.

110 Lodahl, P., Floris van Driel, A., Nikolaev, I.S., Irman, A., Overgaag, K., Vanmaekelbergh, D., and Vos, W.L. (2004) *Nature*, **430**, 654–657.

111 Li, H., Wang, J., Pan, Z., Cui, L., Xu, L., Wang, R., Song, Y., and Jiang, L. (2011) *J. Mater. Chem.*, **21**, 1730–1735.

112 Zhang, J., Jin, Y., Li, C., Shen, Y., Han, L., Hu, Z., Di, X., and Liu, Z. (2009) *Appl. Catal. B*, **91**, 11–20.

113 Arandiyan, H., Dai, H., Ji, K., Sun, H., Zhao, Y., and Li, J. (2015) *Small*, **11**, 2366–2371.

114 Zhu, W., Tao, S., Tao, C.-A., Li, W., Lin, C., Li, M., Wen, Y., and Li, G. (2011) *Langmuir*, **27**, 8451–8457.

115 Sadakane, M., Sasaki, K., Kunioku, H., Ohtani, B., Ueda, W., and Abe, R. (2008) *Chem. Commun.*, 6552–6554.

116 Sun, S., Wang, W., and Zhang, L. (2012) *J. Mater. Chem.*, **22**, 19244–19249.

117 Hu, W., Wang, Y., Hu, X., Zhou, Y., and Chen, S. (2012) *J. Mater. Chem.*, **22**, 6010–6016.

118 Xu, J.-J., Wang, Z.-L., Xu, D., Meng, F.-Z., and Zhang, X.-B. (2014) *Energy Environ. Sci.*, **7**, 2213–2219.

119 Jiang, Y., Shi, L., Huang, Y., Gao, J., Zhang, X., and Zhou, L. (2014) *ACS Appl. Mater. Interfaces*, **6**, 2622–2628.

120 Jiang, Y., Cui, C., Huang, Y., Zhang, X., and Gao, J. (2014) *Chem. Commun.*, **50**, 5490–5493.

121 Vu, A., Qian, Y.Q., and Stein, A. (2012) *Adv. Energy Mater.*, **2**, 1056–1085.

122 Li, Y., Fu, Z.-Y., and Su, B.-L. (2012) *Adv. Funct. Mater.*, **22**, 4634–4667.

123 Ergang, N.S., Fierke, M.A., Wang, Z., Smyrl, W.H., and Stein, A. (2007) *J. Electrochem. Soc.*, **154**, A1135–A1139.

124 Wang, Z., Fierke, M.A., and Stein, A. (2008) *J. Electrochem. Soc.*, **155**, A658–A663.

125 Petkovich, N.D., Rudisill, S.G., Wilson, B.E., Mukherjee, A., and Stein, A. (2014) *Inorg. Chem.*, **53**, 1100–1112.

126 Petkovich, N.D., Wilson, B.E., Rudisill, S.G., and Stein, A. (2014) *ACS Appl. Mater. Interfaces*, **6**, 18215–18227.

127 Vu, A. and Stein, A. (2011) *Chem. Mater.*, **23**, 3237–3245.

128 Fierke, M.A., Olson, E.J., Bühlmann, P., and Stein, A. (2012) *ACS Appl. Mater. Interfaces*, **4**, 4731–4739.

129 Fierke, M.A., Lai, C.-Z., Bühlmann, P., and Stein, A. (2010) *Anal. Chem.*, **82**, 680–688.

130 Vu, A., Li, X., Phillips, J., Han, A., Smyrl, W.H., Bühlmann, P., and Stein, A. (2013) *Chem. Mater.*, **25**, 4137–4148.

131 Chen, X., Wang, Y., Zhou, J., Yan, W., Li, X., and Zhu, J.-J. (2008) *Anal. Chem.*, **80**, 2133–2140.

132 Wei, N., Xin, X., Du, J., and Li, J. (2011) *Biosens. Bioelectron.*, **26**, 3602–3607.

133 Li, X.-H., Dai, L., Liu, Y., Chen, X.-J., Yan, W., Jiang, L.-P., and Zhu, J.-J. (2009) *Adv. Funct. Mater.*, **19**, 3120–3128.

134 Hu, J., Zou, X.U., Stein, A., and Bühlmann, P. (2014) *Anal. Chem.*, **86**, 7111–7118.

135 Hu, J., Ho, K.T., Zou, X.U., Smyrl, W.H., Stein, A., and Bühlmann, P. (2015) *Anal. Chem.*, **87**, 2981–2987.

136 Szűcs, J., Lindfors, T., Bobacka, J., and Gyurcsányi, R.E. (2016) *Electroanalysis*, **28**, 778–786.

137 Kanamura, K., Mitsui, T., and Munakata, H. (2005) *Chem. Mater.*, **17**, 4845–4851.

138 Chen, S.-L., Xu, K.-Q., and Dong, P. (2005) *Chem. Mater.*, **17**, 5880–5883.

139 Yamamoto, D., Munakata, H., and Kanamura, K. (2008) *J. Electrochem. Soc.*, **155**, B303–B308.

140 Munakata, H., Yamamoto, D., and Kanamura, K. (2008) *J. Power Sources*, **178**, 596–602.

141 Kim, O.-H., Cho, Y.-H., Kang, S.H., Park, H.-Y., Kim, M., Lim, J.W., Chung, D.Y., Lee, M.J., Choe, H., and Sung, Y.-E. (2013) *Nat. Commun.*, **4**, 1–9.

142 Liang, C., Li, J., Tang, H., Zhang, H., Zhang, H., and Mu, P. (2014) *J. Mater. Chem. A*, **2**, 753–760.

143 Li, J., Tang, H., Chen, R., Liu, D., Xie, Z., Pan, M., and Jiang, S.P. (2015) *J. Mater. Chem. A*, **3**, 15001–15007.

144 Varghese, L.T., Xuan, Y., Niu, B., Fan, L., Bermel, P., and Qi, M. (2013) *Adv. Opt. Mater.*, **1**, 692–698.

145 Choi, S.-W., Zhang, Y., Thomopoulos, S., and Xia, Y. (2010) *Langmuir*, **26**, 12126–12131.

146 Guan, G., Zapf, R., Kolb, G., Hessel, V., Löwe, H., Ye, J., and Zentel, R. (2008) *Int. J. Hydrogen Energy*, **33**, 797–801.

147 Liu, Y., Liu, B., Wang, Q., Liu, Y., Li, C., Hu, W., Jing, P., Zhao, W., and Zhang, J. (2014) *RSC Adv.*, **4**, 5975–5985.

148 Liu, Y., Liu, B., Wang, Q., Li, C., Hu, W., Liu, Y., Jing, P., Zhao, W., and Zhang, J. (2012) *J. Catal.*, **296**, 65–76.

149 Chai, G., Yoon, S.B., Kang, S., Choi, J.H., Sung, Y.E., Ahn, Y.S., Kim, H.S., and Yu, J.S. (2004) *Electrochim. Acta*, **50**, 823–826.

150 Chai, G.S., Yoon, S.B., Yu, J.-S., Choi, J.-H., and Sung, Y.-E. (2004) *J. Phys. Chem. B*, **108**, 7074–7079.

151 Zhang, N., Li, J., Li, W., Ni, D., and Sun, K. (2012) *RSC Adv.*, **2**, 802–804.

152 Munakata, H., Otani, M., Katsuki, Y., and Kanamura, K. (2009) *ECS Trans.*, **25**, 1855–1860.

153 Deng, L., Wang, F., Chen, H., Shang, L., Wang, L., Wang, T., and Dong, S. (2008) *Biosens. Bioelectron.*, **24**, 329–333.

154 Suezaki, T., O'Brien, P.G., Chen, J.I., Loso, E., Kherani, N.P., and Ozin, G.A. (2009) *Adv. Mater.*, **21**, 559–563.

155 Suezaki, T., Chen, J.I.L., Hatayama, T., Fuyuki, T., and Ozin, G.A. (2010) *Appl. Phys. Lett.*, **96**, 242102.

156 Suezaki, T., Yano, H., Hatayama, T., Ozin, G.A., and Fuyuki, T. (2011) *Appl. Phys. Lett.*, **98**, 072106.

157 Ling, T., Kulinich, S.A., Zhu, Z.-L., Qiao, S.-Z., and Du, X.-W. (2014) *Adv. Funct. Mater.*, **24**, 707–715.

158 Zheng, X.-L., Qin, W.-J., Ling, T., Pan, C.-F., and Du, X.-W. (2015) *Adv. Mater. Interfaces*, **2**, 1400464.

159 Zhao, Z., Liu, G., Li, B., Guo, L., Fei, C., Wang, Y., Lv, L., Liu, X., Tian, J., and Cao, G. (2015) *J. Mater. Chem. A*, **3**, 11320–11329.

160 Sordello, F., Maurino, V., and Minero, C. (2011) *J. Mater. Chem.*, **21**, 19144–19152.

161 Liu, L., Karuturi, S.K., Su, L.T., and Tok, A.I.Y. (2011) *Energy Environ. Sci.*, **4**, 209–215.

162 Kang, D.-Y., Lee, Y., Cho, C.-Y., and Moon, J.H. (2012) *Langmuir*, **28**, 7033–7038.

163 Chen, Y., Zhu, Y., and Chen, Z. (2013) *Thin Solid Films*, **539**, 122–126.

164 Yang, Z., Gao, S., Li, W., Vlasko-Vlasov, V., Welp, U., Kwok, W.-K., and Xu, T. (2011) *ACS Appl. Mater. Interfaces*, **3**, 1101–1108.

165 Park, S.H., Kim, O.-H., Kang, J.S., Lee, K.J., Choi, J.-W., Cho, Y.-H., and Sung, Y.-E. (2014) *Electrochim. Acta*, **137**, 661–667.

166 Yan, H., Zhang, K., Blanford, C.F., Francis, L.F., and Stein, A. (2001) *Chem. Mater.*, **13**, 1374–1382.

167 Zhang, K., Yan, H., Bell, D.C., Stein, A., and Francis, L.F. (2003) *J. Biomed. Mater. Res. A*, **66**, 860–869.

168 Zhang, K., Washburn, N.R., Antonucci, J.M., and Simon, C.G. (2005) *Key Eng. Mater.*, **284**, 655–658.

169 Zhang, K., Washburn, N.R., and Simon, C.G., Jr (2005) *Biomaterials*, **26**, 4532–4539.

170 Zhang, K., Francis, L.F., Yan, H., and Stein, A. (2005) *J. Am. Ceram. Soc.*, **88**, 587–592.

171 Zhang, Y.S., Choi, S.-W., and Xia, Y. (2013) *Soft Matter*, **9**, 9747–9754.
172 Batra, D., Vogt, S., Laible, P.D., and Firestone, M.A. (2005) *Langmuir*, **21**, 10301–10306.
173 Liu, Y., Wang, S., Lee, J.W., and Kotov, N.A. (2005) *Chem. Mater.*, **17**, 4918–4924.
174 Zhang, Y., Wang, S., Eghtedari, M., Motamedi, M., and Kotov, N.A. (2005) *Adv. Funct. Mater.*, **15**, 725–731.
175 Choi, S.-W., Xie, J., and Xia, Y. (2009) *Adv. Mater.*, **21**, 2997–3001.
176 Zhang, Y., Choi, S.-W., and Xia, Y. (2012) *Macromol. Rapid. Commun.*, **33**, 296–301.
177 Lu, J., Zou, X., Zhao, Z., Mu, Z., Zhao, Y., and Gu, Z. (2015) *ACS Appl. Mater. Interfaces*, **7**, 10091–10095.
178 Choi, S.-W., Zhang, Y., and Xia, Y. (2010) *Langmuir*, **26**, 19001–19006.
179 Choi, S.-W., Zhang, Y., MacEwan, M.R., and Xia, Y. (2013) *Adv. Healthcare Mater.*, **2**, 145–154.
180 Li, Z. and Jaroniec, M. (2001) *J. Am. Chem. Soc.*, **123**, 9208–9209.

11
Optical Properties of Hybrid Organic–Inorganic Materials and their Applications – Part I: Luminescence and Photochromism

Stephane Parola,[1] Beatriz Julián-López,[2] Luís D. Carlos,[3,4] and Clément Sanchez[5]

[1]*Université de Lyon, Ecole Normale Supérieure de Lyon, Laboratoire de Chimie ENS Lyon, Université Lyon 1, CNRS UMR 5182, 46 allée d'Italie, 69364 Lyon, France*
[2]*University Jaume I, Department of Inorganic and Organic Chemistry - INAM, Av. Vicent Sos Baynat, Castellón de La Plana 12071, Spain*
[3]*University of Aveiro, Department of Physics, Campus Universitário de Santiago, 3810193 Aveiro, Portugal*
[4]*University of Aveiro, CICECO – Aveiro Institute of Materials, Campus Universitário de Santiago, 3810193 Aveiro, Portugal*
[5]*Collège de France, Laboratoire Chimie de la Matière Condensée Paris, UMR 7574, 75005 Paris, France*

11.1
Introduction

The field of hybrid organic–inorganic materials has been growing intensively during the past 20 years and is certainly nowadays one of the major fields of research and unambiguously one of the most exciting. The association of organic and inorganic systems has been developed in order to optimize the final properties of the materials. However, the difficulty to associate materials that possess low compatibility for each other's has led the chemists and physicists to imagine numerous possible types of architectures. The rise of new characterization techniques and knowledge at the nanometer scale and even the molecular scale has been a driving force toward imagination of hybrid architectures in many major fields such as sensors, electronics, optics, lighting, medicine, catalysis, energy storage, energy conversion. Moreover, the need for better materials for complex devices with improved properties and multifunctional responses poses a perpetual demand that often can be fulfilled by the use of hybrid materials. In particular, regarding the broad field of optical materials, ranging from lighting to energy, bioimaging, screen design, optoelectronics, and many others, the design of hybrid materials has been particularly productive (Figure 11.1). One reason is that combining unique optical responses of organic or organometallic molecular species with either mechanical properties or optical properties of inorganic counterparts provides a unique way to elaborate

Figure 11.1 Hybrid materials and optical properties, from ancient time and art design to highly innovative functional devices.

highly innovative optical systems at the macro- or the nanoscale. Considering the usually high sensitivity of optical responses to the environment, the interfaces between the organics and inorganics is a crucial parameter in most systems. Of course, large number of books and reviews were previously published in the different fields related to optics, by chemists, physicists, and biologists, showing the large community concerned with this topic. This chapter aims to review the latest reported hybrid optical materials, with a strong focus on the 10 last years, and tries to provide an overview of the relationship between the structures and the measured properties, as well as ways to control the interfaces in the hybrids. Considering the number of work in the field, this chapter cannot be exhaustive, but can give an insight into the actual state of the art in the field of hybrid materials devoted to optics and, in particular, the relationship between the different types of structure, the organic–inorganic interfaces and the final properties of the systems.

11.2
Light-Emitting Hybrid Materials

11.2.1
Introduction to Luminescence

Luminescence is a general term that describes any nonthermal processes in which energy is emitted in the ultraviolet, visible, or infrared spectral regions from an electronically excited species. Generally, the emission occurs at a higher wavelength than that at which light is absorbed (a process termed as downshifting). The term broadly includes the commonly used categories of fluorescence and phosphorescence, depending upon the electronic configuration of the excited state and the emission pathway [1,2]. The distinction between the various types of luminescence is usually made according to the mode of excitation. For instance, bioluminescence is related to light emission from live animals and plants, cathodoluminescence results from excitation by electron beams, chemiluminescence is the emission occurring during a chemical reaction, radioluminescence is produced by ionizing radiation (α, β, γ, and X-rays), triboluminescence is ascribed to rubbing, mechanical action, and fracture, electroluminescence is the conversion of electrical energy into light, and photoluminescence results from excitation by photons.

In a first-order absorption process, when the photon energy of the incident radiation is lower than the energy difference between two electronic states, the photons are not absorbed and the material is transparent to such radiation energy. For higher photon energies, absorption occurs (typically in 10^{-15} s) and the valence electrons will make a transition between two electronic energy levels. The excess energy will be dissipated through vibrational processes that occur throughout the near infrared (NIR) spectral region. If this transition does not involve spin inversion, the excited state is also a singlet (S), that is, it has the same state multiplicity as the ground level (S_0). However, if there is spin inversion, the two electrons have the same spin, $S=1$ and $2S+1=3$, and the excited state is called a triplet (T). Then, the excited atoms may return to the original level through radiative (fluorescence or phosphorescence) and nonradiative transitions (internal conversion) (Figure 11.2). It should be noted that an absorption involving a triplet state is forbidden by the spin selection rule: allowed transitions must involve the promotion of electrons without a change in their spin ($\Delta S = 0$). The relaxation of the spin selection rule can occur though strong spin–orbit coupling, which is, for instance, what happens in the case of Ln^{3+} ions.

Following are the examples of nonradiative processes:

- *Internal conversion*: an electron close to a ground state vibrational energy level, relaxes to the ground state via transitions between vibrational energy levels giving off the excess energy to other molecules as heat (vibrational energy). The time scale of the internal conversation and vibrational relaxation processes is 10^{-14}–10^{-11} s.
- *Intersystem crossing*: the electron transition in an upper S_1 excited state to a lower energy level, such as T_1.
- *Delayed fluorescence*: after a fast intersystem crossing to T_1 and a thermally popped back into S_1. The lifetime of S_1 increases and, in the limit, would be nearly equal to the lifetime of T_1 [3]. Delayed fluorescence can also occur when two excited states interact and annihilate forming an emitting state, for example, in triplet–triplet annihilation [4,5]. Triplet–triplet annihilation is an encouraging up-conversion approach due to its low excitation power density

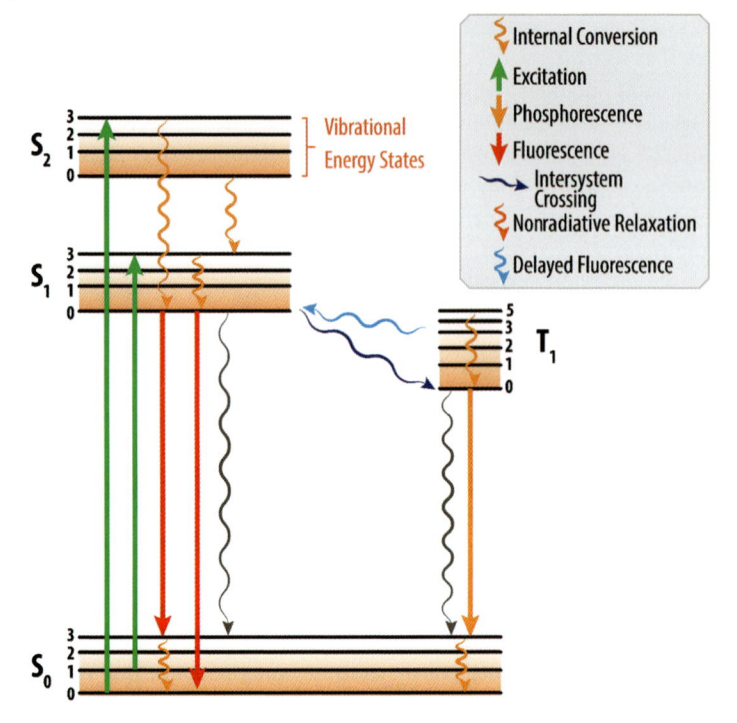

Figure 11.2 Jablonski diagram summarizing the typical radiative and nonradiative transitions within an electronically excited specie.

(solar radiation is enough), high quantum yield, tunable excitation/emission wavelength, and strong absorption [6].

Following are the examples of radiative processes:

- *Fluorescence*: emission of a photon from S_1 to the vibrational states of S_0 occurring on a time scale of 10^{-9}–10^{-7} s.
- *Phosphorescence*: emission of a photon from T_1 to the vibrational sates of S_0. This process is much slower than fluorescence (10^{-3}–10^2 s) because it involves two states of distinct multiplicity. For very long luminescence time decays (minutes and even hours), the emission is called persistent luminescence or afterglow.

Due to the nonradiative transitions, fluorescence and phosphorescence will occur at lower energy (longer wavelengths) than that of the absorbed photons. The energetic difference between the maximum of the emission and absorption spectra ascribed to the same electronic transition is known as Stokes shift [7].

Electroluminescence results from the radiative recombination of electrons and holes injected into an inorganic, organic, or Organic–inorganic hybrid (OIH) semiconductor material. Electroluminescent devices include light emitting diodes (LEDs), which produce light when a current is applied to a doped p–n junction of a semiconductor, and matrix-addressed displays [7]. In the past decades, LEDs and organic light-emitting diodes (OLEDs) induced a deep revolution in the lighting industry. The first examples of the use of hybrid materials in lighting (more specifically in solid-state lighting, SSL) appeared in 2001 with layered crystalline organic–inorganic perovskites [8]. Although the interest in these hybrid perovskites as single-phase white light emitters continues [9,10], the wide range of materials with potential application in SSL would include dye-bridged [11–17], dye-doped [18], and quantum dot-doped [19] siloxane-based organic–inorganic hybrids, and metal organic frameworks [20]. Despite the interest of organic–inorganic hybrid perovskites in lighting, the most exciting application of these materials is in solar cells. Perovskite solar cells have shown remarkable progress in the last decade, with rapid increases in conversion efficiency, from 3.8% in 2009 [21] to 20% in 2015 [22], demonstrating the potential competitiveness to traditional commercial solar cells and offering the prospective for an earth-abundant and low-energy-production solution to large-scale manufacturing of photovoltaic modules [23].

The potential of light-emitting hybrid materials relies on the possibility of fully exploiting the synergy between the optical features of the emitting centers and the intrinsic characteristics of the sol–gel derived hosts. Hybrid materials present several advantages for photonic and optical applications, such as: (i) versatile shaping and patterning, depending on the foreseen application; (ii) optimization of composition and processing conditions, yielding excellent optical quality, high transmission, low processing temperature (<200 °C), and easy control of the refractive index by changing the relative proportion of the different precursors; and (iii) photosensitivity, mechanical integrity, corrosion protection, and suitable adhesion properties [24–30]. In general, the embedding of organic dyes, quantum dots (QDs), and trivalent lanthanide (Ln^{3+}) complexes into hybrid hosts, with the corresponding formation of covalent (Class II) or noncovalent (Class I) [31] host–guest interactions, improves the thermal stability, the mechanical resistance, and the aging and environmental stability of the guest emitting centers, relatively to what is observed for isolated centers [32,33]. Moreover, the dispersion of these centers within the hybrid framework allows their incorporation in larger amounts, isolated from each other and protected by the hybrid host, improving the emission quantum yield and preventing the emission degradation [34–38]. An illustrative example is the control of the molecular aggregation of chromophores in the solid state, namely, the formation of fluorescent J-aggregates instead of nonfluorescent H-aggregates, using fragments of hydrolysable triethoxysilane (TEOS) [12].

In this section, we cover recent developments of luminescent and electroluminescent hybrid materials, with particular emphasis to specific applications, such as LEDs, random and feedback lasers, luminescent solar concentrators (LSCs), and luminescent thermometers. Examples will address organic–inorganic hybrids (essentially siloxane-based ones) embedding organic dyes, with the formation of covalent (dye-bridged hybrids) or noncovalent (dye-doped hybrids) dye–matrix interactions, QDs, inorganic nanoparticles, and Ln^{3+} complexes.

Hybrids with photochromic features will be discussed in Section 11.3, whereas hybrids for phosphors [39] and for luminescent coatings (based on organic dyes [40], fluorine polymers [41], cooper iodide clusters [42], and on Ln^{3+} ions [43–45]) were not reviewed in detail, being only addressed in the context of LEDs and LSCs, respectively. General and comprehensive reviews on luminescent organic–inorganic hybrid materials embedding organic dyes and Ln^{3+} ions were published by Carlos et al. [25,28], Sanchez et al. [27,46–48], Escribano and coworkers [49], Binnemans [50], Zhang and coworkers [51], and Ribeiro et al. [52]. Concerning electroluminescence, there is only one recent review addressing the application of dye-doped hybrids on LEDs [53]. In view of the current trends of the subject, and balancing the literature published since these reviews, we decide not to discuss the applications of luminescent hybrid materials in integrated optics and optical telecommunications, in biomedicine, and in solar cells. While in integrated optics and optical telecommunications the review by Ferreira et al. [26] is relatively updated, in solar cells and in biomedicine the amount of work published in the past 5 years justifies independent review publications. In fact, perovskite solar cells largely dominate the current research trend of hybrid materials in photovoltaics, while in biomedicine ligand-decorated QDs and Ln^{3+}-based inorganic NPs, Ln^{3+} chelates embedded within inorganic matrices and more complex core–shell and core–corona hybrid architectures have been designed as biosensing platforms for *in vivo* imaging, diagnostics, targeting, and therapy. The subject underwent an enormous expansion during the last decade that can be tracked, for instance, in the recent reviews of Prasad and coworkers [54,55], Prodi and coworkers [56], and Bünzli [57].

11.2.2
White Light Emission and LEDs

The interest in organic–inorganic hybrids for LEDs has grown considerably during the last three decades, since the seminal work of Tang and VanSlyke on dye-based OLEDs operating at low driving voltages [58] and the first reports demonstrated the possibility of applying these materials in solid-state lasers [59–61]. The first examples of white LEDs (WLEDs) based on organic–inorganic hybrids date back to the end of the last century comprising dye-modified silanes incorporating hole- or electron-transporting units and light-emitting species in the orange [62] and green [11] spectral regions. Later on, more efficient WLEDs were reported involving silsesquioxane hybrid matrices, as, for example, that based on the phenylenevinylenediimide precursor, with a luminance value of 10 cd/m^2 for voltages lower than 30 V [12], and that based on polyhedral oligomeric silsesquioxanes bearing in the structure a dye molecule from the cyanine family, with a threshold operating voltage of 4 V [63]. An intriguing example is the fabrication of a WLEDs by coating a commercial UV LED (390 nm) with a periodic mesoporous organosilica (PMO) film doped with rhodamine 6G (Rh6G) and synthesized by surfactant-templated sol–gel polycondensation using a 1,3,6,8-tetraphenylpyrene (TPPy)-containing organosilane precursor [13]. The blue emission of the films, with an emission quantum yield of 0.70, overlaps the absorption spectra of the dye and thus efficient energy transfer occurred and the white-light is achieved by combining the blue emission of the host with the yellow light of the guest. UV-pumped WLEDs have huge potential because they are easy to manufacture, the white light emission is only due to the down-converting phosphors, they exhibit a low color point variation as a function of the forward-bias currents, and they have superior temperature stability. Moreover, as the human eyes are insensitive to UV radiation, the white color is independent of the pumping LED and of the thickness of the phosphor layer.

Poly(2-hydroxyethyl methacylate-silica hybrids doped with organoboron dyes emitting in the blue, green, and red spectral regions [16] and dye-bridged epoxy functional oligosiloxanes emitting in the green and red spectral regions with absolute emission quantum yields of 0.85 and 0.41, respectively [17], were used to produce efficient multicolor light-emitting hybrids. In this latter example, WLEDs were produced using a commercial blue LED as the excitation source and by controlling the dyes concentration and the ratio between the red and green emitting species, (Figure 11.3). The best device has a correlated color temperature (CCT) of 4810 K, a color rendering index (CRI) of 85 and a luminous efficacy of 23.7 lm/W, being thermally stable at 120 °C for 1200 h [17].

Another example comprises dye molecules containing trialkoxysilyl terminal groups embedded into silica nanoparticles emitting in the red and yellowish-green spectral regions [15]. In this case, a WLED was fabricated by combining such fluorescent organosilica nanoparticles with a commercial blue InGaN LED. The resulting three-color RGB LED exhibited a CRI of 86.7, a CCT of 5452.6 K, and CIE (x, y) coordinates of (0.3334, 0.3360) [15].

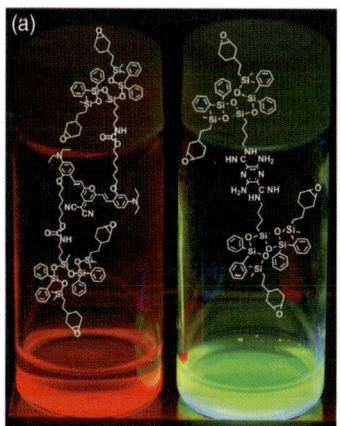

Figure 11.3 Photographs of (a) red and green dye-bridged oligosiloxanes under UV excitation at 365 nm and (b) dye-bridged nanohybrid-based white LED, "Commission Internationale de l'Eclairage" (CIE) color coordinates of (0.348, 0.334), fabricated encapsulating a blue LED (445 nm) with a blend of red and green dyes. (Taken with permission from Ref. [17]. Copyright 2011, Wiley-VCH Verlag GmbH.)

The possible use of organically doped layered phyllosilicate clays was introduced in the late 1990s [64,65]. Although not fabricated as WLEDs, white light-emitting soft hybrids were reported based on the supramolecular coassembly of organoclays and ionic chromophores [66–68]. Layered magnesium phyllo (organo)silicate with aminopropyl pendants (AC in Figure 11.4a) was used as the inorganic counterpart, while coronene tetra-carboxylate (CS), sulforhodamine G (SRG), and tetraphenylethylene derivatives (TPTS) were employed as

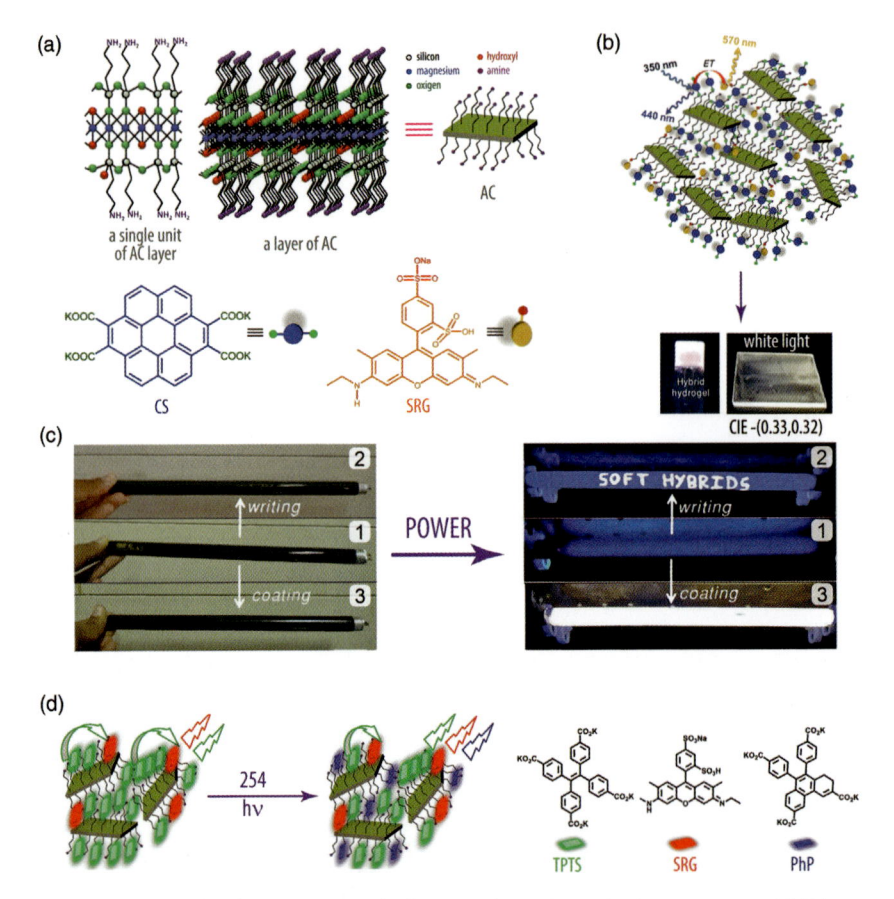

Figure 11.4 (a) Molecular structures and schematic representation of the organofunctionalized clay and organic chromophores. (b) Proposed schematic representation of the coassembled AC-CS clay-chromophore soft-hybrids and the energy-transfer process to the acceptor red-emitting SRG molecules. Photographs of the white-light-emitting hybrids in gel and film phases are also shown. (c) Schematic showing the strategy for photomodulation of the clay hybrids. (d) The hybrids were used to paint and write on commercial UV-lamps (365 nm) (1) uncoated lamp, (2) written as "SOFT HYBRIDS" on the surface of the lamp and (3) lamp fully coated with the soft-hybrids. Hybrid-coated lamps are exposed to the UV irradiation by glowing lamps, which showed bright white light for both written letters and the fully-coated lamp. (Taken with permission from Refs [66,68]. Copyright 2013, Wiley-VCH Verlag GmbH.)

donor–acceptor organic chromophores (Figure 11.4a). White light emission was attained via partial excitation energy transfer from the AC–CS hybrid to the coassembled acceptor SRG dye molecules [66,67] (Figure 11.4b) or by *in situ* photogeneration of a blue component modifying the clay hybrid containing green (TPTS) and red (SRG) emitting components (Figure 11.4c) [68]. The solution processability, high transmittance, color tunability, and the environmentally benign solvent medium of these "soft hybrids" were exploited for large-area device fabrication. An example is the coating of a commercial 365 nm UV-lamp (surface area of 125 cm^2) with the soft-hybrids that glowed with a bright white color when the lamp was connected to the electrical power (Figure 11.4d).

QDs are also used as light-emitting centers in WLEDs, although typically the host framework is a polymer (e.g., polyfluorene, PFO, or poly(phenylene vinylene)-PFO copolymers) and not an organic–inorganic hybrid [69,70]. A recent illustrative example is the embedding of yellow-, orange-, or red-emitting QDs into a polyfluorene composite (called Green B). WLEDs with luminous efficiency >17.21 m/W, correlated color temperatures of 3500 and 5500 K and high-color-rendering index (CRI up to 90 at 3500 K) were fabricated [71]. Nanocomposites of CdS QDs and silica-based carbon dots (CDs) were also proposed for WLEDs. An illustrative example are the hybrids emitting bright white light under UV excitation formed by the cohydrolysis of blue fluorescent CDs functionalized by (3-aminopropyl) triethoxysilane (APTES) with yellow-emitting CdS nanocrystals modified by (3-mercaptopropyl) trimethoxysilane (MPTMS). A WLED was fabricated by the as-prepared CD/CdS QDs hybrids as converters, which exhibits white light with a color coordinate of (0.27, 0.32) [72].

Highly efficient WLEDs can also be fabricated with inorganic or organic–inorganic hybrid nanoparticles. A very intriguing example was reported by Ferreira and coworkers [73] in which WLEDs were produced by combining a commercial UV-LED chip (InGaAsN, 390 nm) and boehmite(γ-AlOOH)-based organic–inorganic hybrid material as white down-converting phosphor (Figure 11.5a). The hybrid organic–inorganic nanoparticles consist of few nanometer thick boehmite nanoplates capped with *in situ* formed benzoate ligands prepared by a one-pot nonaqueous reaction (Figure 11.5b). The efficient white light emission results from a synergic energy transfer between the triplet level of the organic phase (benzoate ligands, T_1) and the triplet state of the inorganic component (bohemite F-centers, ^{3}P). The efficient energy transfer results from two main aspects: (i) the near-resonance between the T_1 and ^{3}P states and the (ii) large spin–orbit effect that induce a high triplet radiative rate at room temperature due to the presence of Al atoms coordinated to the benzoate groups. As a direct result of these two effects, the overall quantum yields of the boehmite hybrid nanoparticles are the highest reported so far for ultraviolet-pumped white phosphors. The WLEDs are able to emit white light with "Commission Internationale de l'Éclairage" coordinates, color-rendering index, and correlated color temperature values of (0.32, 0.33), 85.5 and 6111 K, respectively; overwhelming state-of-the-art single-phase UV-pumped WLEDs phosphors. We note that an important advantage of these boehmite hybrid phosphors lies in the fact that they are made of nontoxic, abundant, and

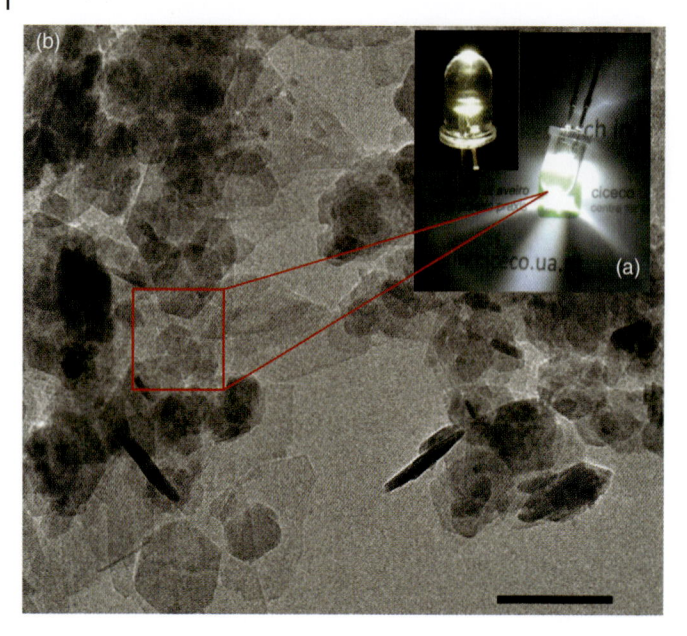

Figure 11.5 (a) Photographs of the as-fabricated WLED using boehmite-based organic–inorganic hybrid nanoparticles operating at 3.0 V. (b) Representative TEM image of the boehmite hybrid nanoparticles (scale bar, 100 nm) [73].

low-cost materials that is desirable from an industrial and environmental viewpoint.

Lanthanide-bearing organic–inorganic hybrids were also proposed for WLEDs. One of the first examples reported the fabrication of WLEDs by combining near-UV LED emission (390–420 nm) with a hybrid phosphor comprising two strontium aluminates, $SrAl_2O_4$: Eu^{2+} (green emission) and $Sr_4Al_{14}O_{25}$: Eu^{2+} (blue emission), and $Eu(btfa)_3phen$ (red emission), where $btfa^- = $ 4,4,4-trifluoro-1-phenyl-1,3-butanedionate and phen is 1,10-phenanthroline [74].

A distinct approach involves the deposition of blue-, green-, and red-emitting organic–inorganic hybrid materials as homogeneous and transparent thin films (about 50 nm). In the example reported by Huang et al., a sol–gel-derived hybrid showing emission from the blue to the yellow-green in a wide range of excitation wavelengths (254–380 nm) was synthesized with poly(9,9-dihexylfluorene-alt-9,9-dioctylfluorene) embedded into a silica matrix (blue-emitting component) and Tb^{3+} and Eu^{3+} ions coordinated in the matrix (green- and red-emitting components, respectively). Although white light was obtained for excitation at 340 nm, only the Eu^{3+} material was investigated as a potential phosphor coated on an UV LED [75].

WLEDs were also fabricated using lanthanide-bearing metal organic frameworks (MOFs) through a remarkable methodology. Eu^{3+} ions were first encapsulated into MOF-253 using a postsynthetic method. The uncoordinated bipyridyl group of MOF-253 is ideal for chelating and sensitizing the Eu^{3+} ions. The

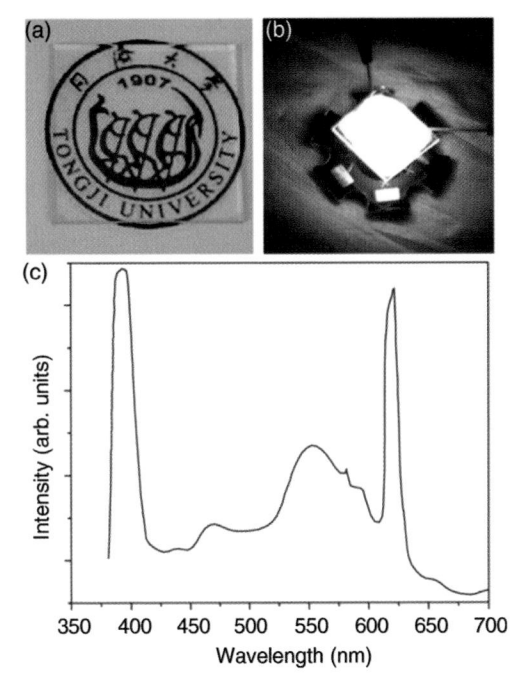

Figure 11.6 (a) A photograph of MOF–PEMA-3.5 hybrid materials. (b) The hybrids bright white light emission under excitation by a 395 nm GaN chip and (c) the corresponding LED electroluminescent spectrum. (Taken with permission from Ref. [76]. Copyright 2014, RSC.)

resulting MOF was further modified with tta (tta = 2-thenoyltrifluoroacetone) and functionalized with ethyl methacrylate to achieve transparent polymer-MOF hybrid materials (MOF–PEMA-3.5) through radical polymerization. Finally, the polymer-MOF hybrid is assembled on a near-UV GaN chip to fabricate near-UV WLED (Figure 11.6) operating at 350 mA with a CCT of 3742 K and a CRI of 87.34 [76].

Monochromatic LEDs were also fabricated based on dye-bridged hybrids [14–17] and Ln^{3+} complexes [77]. Concerning the former, an illustrative example is the functionalization of silsesquioxane cores with pyrene and 4-heptylbenzene that produced green emitting LEDs [14]. The resulting amorphous materials offer numerous advantages for OLEDs, for example, high glass transition temperatures, low polydispersity, solubility in common solvents, and high purity via column chromatography. The devices are among the highest efficient OLEDs fabricated up to now with fluorescent silsesquioxane, characterized by a high-external quantum efficiency of 3.64% and a luminous efficiency of 9.56 cd/A [14].

With Ln^{3+} complexes, an interesting example involves the Eu^{3+}-bearing poly (MMA-MA-co-Eu(tta)$_2$phen) copolymer, where MA and MMA stand, respectively, for maleic acid and methyl methacrylate. A red-emitting LED was fabricated by combining the Eu-copolymer (quantum yield of 24% under near-UV light excitation) with a 395 nm-emitting InGaN chip [77].

11.2.3
Random and Feedback Lasers

Interesting applications of dye-doped organic–inorganic hybrids are random and distributed feedback lasers. A random laser is an open source of stimulated emission comprising a number of phenomena related to the emission of light by spatially inhomogeneous disordered materials (not bounded by any artificial mirrors) [78,79]. Laser emission is produced by multiple scattering processes that increase the dwell time of photons inside the material allowing amplification and creating gain saturation [80]. Random lasers were engineered to provide low spatial coherence and to generate images with superior quality than images generated with spatially coherent illumination [81,82]. By providing intense laser illumination without the drawback of coherent artifacts (as those produced by lasers and superluminescent diodes that corrupt image formation), random lasers are well suited for full-field imaging applications, such as full-field microscopy and digital light projector systems [82].

With organic–inorganic hybrid materials, examples comprise ZnO nanoparticles dispersed into a polymer matrix [83,84], Rh6G-doped SiO_2 nanoparticles [78,85], Rh6G-bearing diureasils [37,80], polymer films embedded with silver nanoparticles [86], and poly(2-hydroxyethyl methacrylate) (pHEMA) incorporating silsesquioxane nanoparticles (POSS) doped with the LDS722 and LDS730 red-emitting dyes [87,88].

Focusing on the example of Rh6G-bearing di-ureasils, the emission features of the ground powders were compared with those of a silica gel containing Rh6G-doped SiO_2 nanoparticles revealing a slightly larger slope efficiency and a lower threshold for laser-like emission in the later case (Figure 11.7) [37,78]. However,

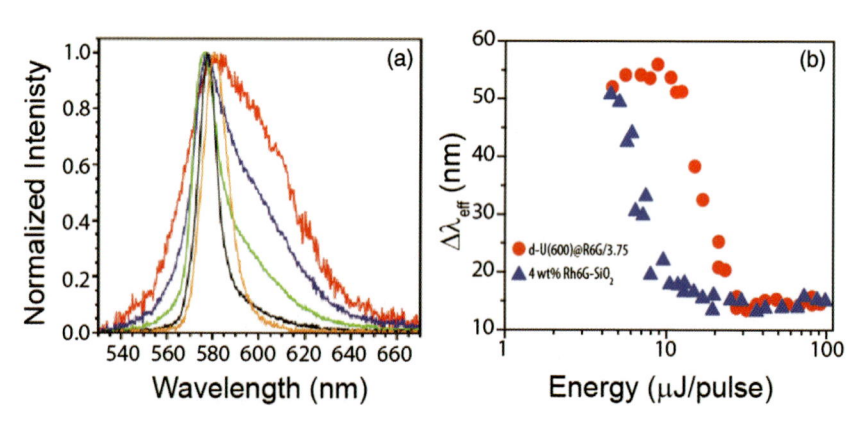

Figure 11.7 (a) Normalized emission spectra of the ground powder of Rh6G-bearing diureasils obtained at 11 μJ/pulse (red), 18 μJ/pulse (blue), 20 μJ/pulse (green), 24 μJ/pulse (black), and 100 μJ/pulse (orange). (b) Spectral narrowing of the ground powders of the diureasil hybrid (red dots) and the bulk silica gel (blue triangles). (Taken with permission from Ref. [37]. Copyright 2010, OSA.)

it is worth noting that these random laser performances (threshold and efficiency) in the diureasils have been obtained with a Rh6G concentration four orders of magnitude lower than the one used in the silica gel that makes the diureasil hybrids far more attractive for applications [37].

A distributed feedback (DFB) laser effect was also reported with dye-doped organic–inorganic hybrid materials, comprising polymers [89,90], biopolymers (such as silk fibroin [91]) and di-ureasils [92]. DFB lasers are devices operating in longitudinal single-mode oscillation due to a grating structure existing throughout the gain medium, providing the feedback for lasing, with potential applications in medical diagnosis and communications [91]. The nanofabrication capability of hybrid materials together with their superior physical, thermal, and optical properties, compared with those of isolated dyes, open up the possibility of using them as an alternative source for photonic devices, including random and DFB lasers.

11.2.4
Luminescent Solar Concentrators

Luminescent solar concentrators (LSCs) are cost-effective components easily integrated in photovoltaics. Despite the first reports dating back to 1976 [93,94], LSCs reappeared in the last decade as an effective approach to collect and concentrate sunlight in an economic way, enhancing solar cells' performance and promoting the integration of photovoltaics architectural elements into buildings, with unprecedented possibilities for energy harvesting in façade design, urban furnishings, and wearable fabrics [95–100]. Conventional LSCs are optical plastic waveguides doped with phosphors (e.g., organic dyes, QDs, metal halide nanoclusters, or Ln^{3+} ions), or glass transparent (or semitransparent) substrates coated with optically active layers embedding those phosphors. When exposed to direct and diffuse sunlight, part of the absorbing radiation is re-emitted at longer (downshifting or down-conversion) or shorter (up-conversion) wavelengths. Part of the emitted light is lost at the surface and the rest is trapped within the waveguide, or the substrate, and guided, through total internal reflection, to the edges, where it is collected and converted into electricity by conventional photovoltaic cells (Figure 11.8a–c).

The research in LSCs has increased substantially over the past three decades, with the major advances of the field highlighted in several recent reviews [95–100]. The development of LSCs faces numerous challenges, many of which are related to the materials used, particularly related to the loss mechanisms that limit conversion efficiency (e.g., emission quantum yield, reabsorption losses, incomplete utilization of the solar spectrum, and escape cone losses) and long-term photostability [102]. Various authors (even since the very beginning of the field [103]) concluded that it is unlikely that a single (organic or inorganic) material can overcome these issues, and organic–inorganic hybrids should play a key role on the LSC design optimization [99,102,104]. Moreover, despite the quite limited use of hybrid materials in the fabrication of LSC, their efficiency

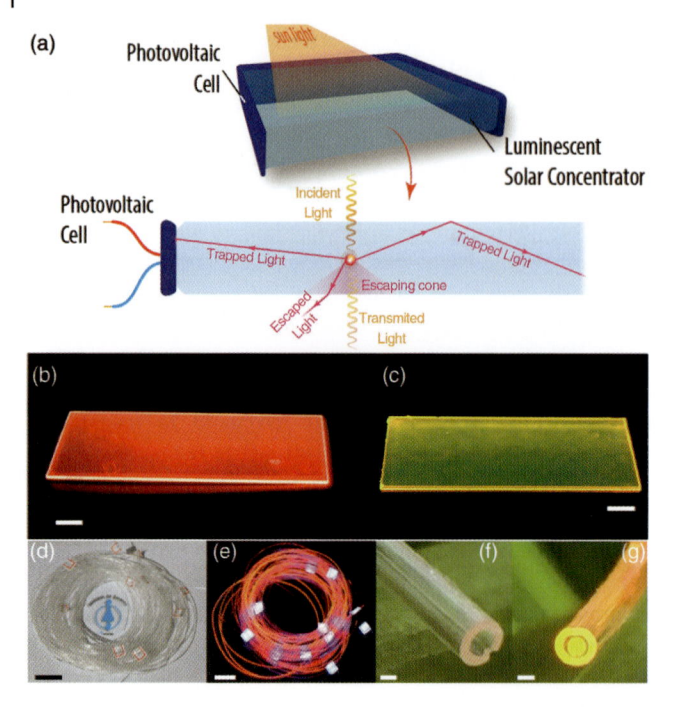

Figure 11.8 (a) Schematic representation of the working principle of a LSC. Photographs of two LSCs under UV irradiation (365 nm) based on (b) a diureasil hybrid doped with [Eu(btfa)$_3$–(MeOH)$_2$]$_2$bpta$_2$, (bpta$^-$ = trans-1,2-bis (4-pyridil) ethane and MeOH = methanol) and (c) a triureasil hybrid doped with PTMS (phenyltrimethoxysilane) and Rh6G. Photographs of meter-length Eu-based diureasil LSCs under (d) daylight conditions, and (e) UV irradiation (scale bars of 10^{-2} m). Detailed view of the Eu- (f) and Rh6G-based (g) LSCs extremities under outdoor illumination highlighting the light concentration (scale bars of 10^{-3} m). (Taken with permission from Ref. [101]. Copyright 2016, RSC.)

values are of the same order of magnitude as those of pure organic LSCs [100]. Recent examples of hybrid materials used in LSCs comprise polymers doped with QDs [105,106], metal halide nanocluster blends [107] and organic dyes [104], Eu^{3+}-based bridged silsesquioxanes [108–110], and di- and triureasils doped with organic dyes and Eu^{3+} β-diketonate complexes [101,111–113].

An intriguing example of lightweight and mechanically flexible high-performance waveguiding photovoltaics is the fabrication of cylindrical LSCs of plastic optical fibers (POFs) coated (bulk fibers) or filled (hollow-core fibers) with Rh6G- or Eu^{3+}-doped organic–inorganic hybrids (Figure 11.8d–g) [101,112]. Cylindrical LSCs have a large potential compared with that of planar ones, despite the very small number of examples involving short length (10^{-2} m) bulk or hollow-core POFs that can be found in the literature [114–116]. First, the concentration factor F (that dominates the devices' performance) of a cylindrical LSC can be twice higher than that of a square-planar one of equivalent collection area and volume [117]. Second, the cylindrical geometry allows for easier coupling with optical fibers that could transport light to a remote place for lighting or power production and renders easier photovoltaic urban integration [101]. In the example illustrated in Figure 11.8, a drawing optical fiber facility is used to scale up the area of the devices demonstrating the possibility of obtaining large area LSCs (length up to 2.5 m) based on bulk-coated and hollow-filled POFs with unprecedented concentration factors, up to 11.75.

11.2.5
Luminescent Thermometers

The already mentioned unique characteristics of organic–inorganic hybrids, in particular their capability to incarcerate luminescence organic and inorganic thermometric probes (e.g., organic dyes, QDs, and Ln^{3+} ions), preventing the aggregation and the emission degradation, makes them suitable for the design of thermometric systems. Moreover, the combination of organic and inorganic counterparts can also induce synergic effects resulting in an enhancement of the thermometric efficiency, relative to that of the organic or inorganic thermometric probes alone, and multiplying the design possibilities of luminescent micro- and nanothermometers [118]. Luminescent thermometry exploits the temperature dependence of the light emission features of the thermometric probes, namely, emission intensity [119–121], peak position [122], and excited-states lifetime [123,124] or risetime [125,126], possessing the unique advantage of high-resolution contactless measurement, even in harsh environments and under strongly electromagnetic fields [127–130]. Although relatively recent (luminescent thermometry exploded over the past 5 years), the technique appears to be beneficial to many technological applications in a great variety of areas, such as microelectronics, microfluidics, bio-, and nanomedicine [131].

Examples of luminescent thermometers based on organic–inorganic hybrids include metal-organic molecular compounds [132], layered double hydroxides [133], metal-organic frameworks [134], polymer nanocomposites [135], QDs in polymers [136], inorganic NPs coated with an organic (or hybrid) layer [137], and di-ureasil films codoped with Eu^{3+} and Tb^{3+} β-diketonate complexes [119,138]. These later films were used as self-referenced and efficient luminescent probes to map temperature in microelectronic circuits [119,138,139] and optoelectronic devices [140], demonstrating an intriguing application of hybrid materials in microelectronics. The thermal gradients generated at submicrometer scale by the millions of transistors contained in integrated circuits are becoming the key limiting factor for device integration in micro- and nanoelectronics and, then, noncontact thermometric techniques with high-spatial resolution (such as luminescent thermometry) are essential for noninvasive off-chip characterization and heat management [141]. The diureasil films incorporating [Eu(btfa)$_3$(MeOH)(bpeta)] and [Tb(btfa)$_3$(MeOH)(bpeta)] complexes allowed temperature mapping in wired-board circuits and in a Mach–Zehnder interferometer using commercial detectors and excitation sources [118,138,140]. For

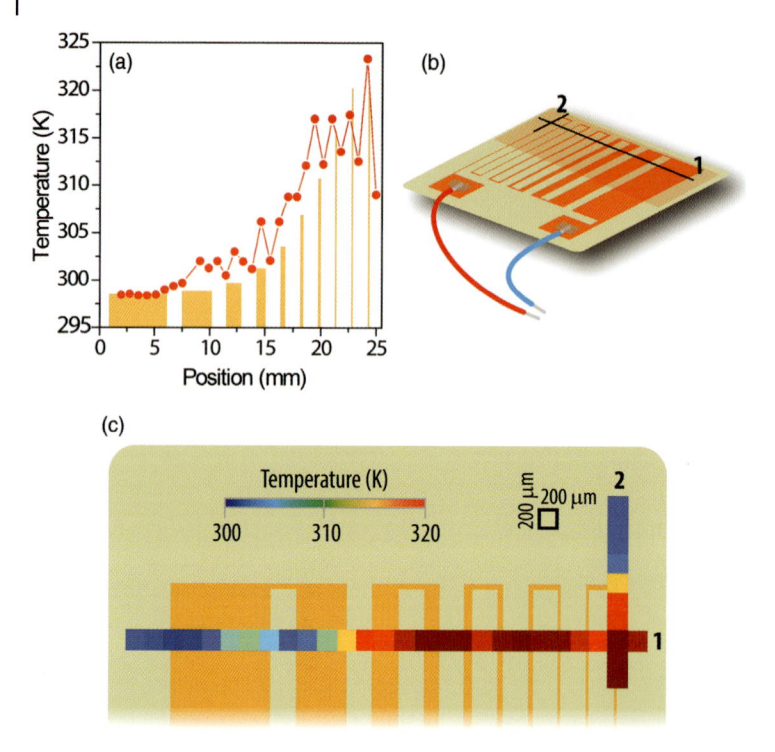

Figure 11.9 (a) Temperature profile obtained with a Eu^{3+}/Tb^{3+} codoped diureasil film of the FR-4 wire-board depicted in (b). The emission of the film (excited at 365 nm) is collected with a 200 μm core diameter fiber along the direction denoted by 1 using a scanning step of 200 μm. The temperature uncertainly is 0.5 K. The orange shadowed areas correspond to the distinct copper tracks. (c) Pseudocolor temperature maps reconstructed from the emission of the diureasil along the two perpendicular directions denoted by 1 and 2 of the FR4 printed wiring board.

instance, Figure 11.9 shows temperature profiles of a FR4 printed wiring board reconstructed from the emission spectra of the Eu^{3+}/Tb^{3+}-containing di-ureasil. The higher spatial resolution obtained, 0.42 μm, is 4.5 times lower than the Rayleigh limit of diffraction (1.89 μm) in the experimental conditions used, and much lower than that recorded with a state-of-the-art commercial IR thermal camera (~130 μm). As the temperature readout results from a spectroscopic measurement, it is not limited by the Rayleigh criterion and, thus, the spatial resolution is only limited by the experimental setup used that produces a field-of-view averaged temperature change above the sensitivity of the detector [138]. The measured temporal resolution is on the order of the integration time of the detectors used (5–100 ms). Although there is not a single technique able to combine submicron and submillisecond resolutions, up to now, the examples that have the best performance are those based on luminescent Ln^{3+}-doped organic–inorganic hybrid thermometers.

11.3
Photochromic Hybrid Materials

11.3.1
Introduction to Photochromism

According to the International Union of Pure and Applied Chemistry (IUPAC) [142] definition, photochromism is "a reversible transformation of a chemical specie induced in one or both directions by absorption of electromagnetic radiation between two forms, A and B, having different absorption spectra." The thermodynamically stable form A is transformed by irradiation into form B, and the back reaction can occur thermally (T-type photochromism) or photochemically (P-type photochromism). In most cases, a change in color toward longer wavelengths absorption takes place (positive photochromism). However, when $\lambda_{max}(A) > \lambda_{max}(B)$, photochromism is called negative or inverse. Figure 11.10 summarizes the concept of reversible photochromism.

This interconversion between two states is accompanied by the change in color but also change in refractive index, dielectric constant, redox potentials, solubility, viscosity, surface wettability, magnetism, luminescence, or a mechanical effect. Therefore, besides the well-established use of photochromic materials in ophthalmics (lenses for sunglasses), there is a growing list of other real or potential application areas, including cosmetics, security, displays and filters, optical memories, photooptical switches, photography, photometry, and so on.

Many types of chemical species exhibit photochromism. Most of them are organic molecules, which are able to absorb one photon (one-photon mechanism) to form B from the singlet ($1A^*$) or triplet ($3A^*$) excited states, or to absorb two photons (two-photon mechanism) where B is formed from the population of an upper excited state. The main families of organic photochromic compounds are spiropyrans (SP), spirooxazines, chromenes, fulgides, diarylethenes, dithienylethene, spirodihydroindolizines, azo compounds, polycyclic aromatic compounds, anils, polycyclic quinones, viologens, triarylmethanes, and biological receptors as retinal proteins and phytochrome. The chemical

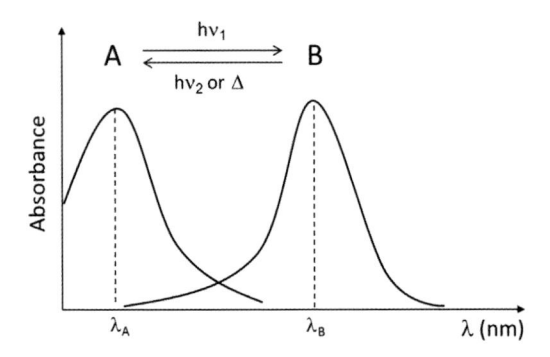

Figure 11.10 Reversible photochromism. (Adapted with permission from Ref. [142].)

processes usually involved in the organic photochromism are pericyclic reactions, cis-trans (E/Z) isomerizations, intramolecular hydrogen/group transfers, dissociation processes, and electron transfers.

In contrast to organic molecules, photochromic inorganic compounds are comparatively few. They include metal halides of group IB metals (such as silver halides), oxides, polyoxometalates, and carbonyls of VIB transition metals (i.e., WO_3, MoO_3), oxides of group IVB and VB metals (TiO_2, V_2O_5, Nb_2O_5, etc.), zinc sulfide, alkali metal azides, sodalite, mercury salts, Ni–Al layered double hydroxides, valence-tautomeric Prussian blue analogs, and so on. The origin of the color change is the ability of these compounds to change stoichiometry, and their performance is related to electron/hole pairs interacting with their environment (air, liquid solution, adsorbed water or solvent molecules, rigid matrices, etc.). Thus, the photochromic properties show a critical dependence on the surrounding medium, which is often hard to predict and systematize.

For a more detailed accounts on the classical photochromic phenomenon, materials and applications we invite the readers to consult the books edited by Crano and Guglielmetti [143,144], Durr and Bouas-Laurent [145], or Bamfield and Hutchings [146].

The research on photochromic systems from the individual organic or inorganic fields has been very difficult. OIH photochromic materials can be extremely versatile in terms of physical and chemical properties, compositions, and processing techniques. Thus, they offer a wide range of possibilities to fabricate tailor-made photochromic materials in which the final response will depend not only on the chemical nature of each component but also on the interface and synergy between both organic and inorganic counterparts. The adequate design of OIH allows: (i) to improve photophysical and photochemical properties (especially via molecular modification in organic chromophores), (ii) to offer easy-to-shape materials for a real commercial deployment (via integration of the active species into functional matrices), and (iii) to explore new photoresponsive multicomponent systems. Excellent reviews have been written by He and Yao [147], Guo and coworkers [148], and Levy and coworkers [149,150].

But their economic potential is still not being fully realized due to the difficulty to simultaneously fulfill some relevant technical requirements for application. Some of them are as follows:

1) The material must develop a strong color rapidly upon UV/vis irradiation.
2) The fade rate back to the colorless state must be controllable.
3) The response must be constant through many coloration cycles.

In the last decade, great efforts have been devoted to accomplish these issues but for some applications the achievements are still far from a real implementation. Also, the choice of the system and the specific requirements depend on the targeted application. In sunglasses or protective coatings, for instance, the consumer needs robust systems with an immediate reaction to a change of external conditions. Also, a change from white (or colorless) to gray or even black is preferred, especially for large-volume displays (display panels, front and rear windows, and mirrors for cars and trucks, etc.). For other applications, multicolored systems can be desirable. Indeed, most of the photochromic compounds do change between two or more different colors, and that is useful for sensors or markers. For memory devices, data storage or switches, for instance, a good contrast between the two states is probably the most valuable property. Here, we give a brief overview of the most recent progress in this area, focusing on photochromic hybrids where photochromism comes mainly from organic molecules, from inorganic species or combination of both. Their applications will be progressively presented and discussed in terms of their structural and physicochemical features.

11.3.2
Organic Photochromism in Hybrids

Organic photochromism offers several tens of thousands of molecules with which one can tailor their applicability by the shade of color change, the direction of the photochromism, the rate of change in color intensity, reversibility, and so on. To achieve robust and functional photochromic-based devices that could be easily handled and integrated in solid materials, the main approach followed is the micro-/nanostructuring of photochromic molecules into organic–inorganic hybrid materials (classical sol–gel glasses, polymers, nanoparticles, etc.) via weak (Class I hybrid) [31] or strong (Class II hybrid) chemical interactions [151–165]. Among the photochromic organic molecules, spirooxazines, azobenzenes, diarylethenes, and fulgides have been the most extensively studied [144]. These molecules can easily be entrapped in ORMOSIL (organically modified silanes) hybrid matrices [166]. A huge work was done on this area in the late 1990's and early 2000's with the boom of soft chemistry [167]. The synthesis of these matrices involves inorganic polymerizations at room temperature from metal-organic precursors, so under these mild conditions, the active molecules can be stored preserving or improving their photochemical and photophysical properties. The main advantages of silane matrices over pure organic polymers (PMMA, PAA, etc.) are the higher thermal stability and the possibility of chemical modification that can be of interest for a particular application (for instance, including specific functional groups or changing properties like refractive index, the hydrophilic/phobic character, etc.). This approach is also more convenient than obtaining photochromic single crystals [168–170]. Few organic molecules preserve their photochromic properties in the solid state [171], but even so the shaping and conformation into practical devices is really difficult. In the last years, extended literature can still be found on photochromic dyes randomly embedded (covalently bonded or just doped) into amorphous hybrid silica derivates (nanoparticles [172], films [173], fibers [174]) and also in layered materials [175]. As a difference to classical works, the more recent articles deal with the combination of photochromic properties to other functionality to get multifunctional systems. For instance, Fölling et al. [172] have designed a new functional (amino reactive) highly efficient fluorescent molecular switch (FMS) with a photochromic diarylethene and a rhodamine fluorescent dye on silica

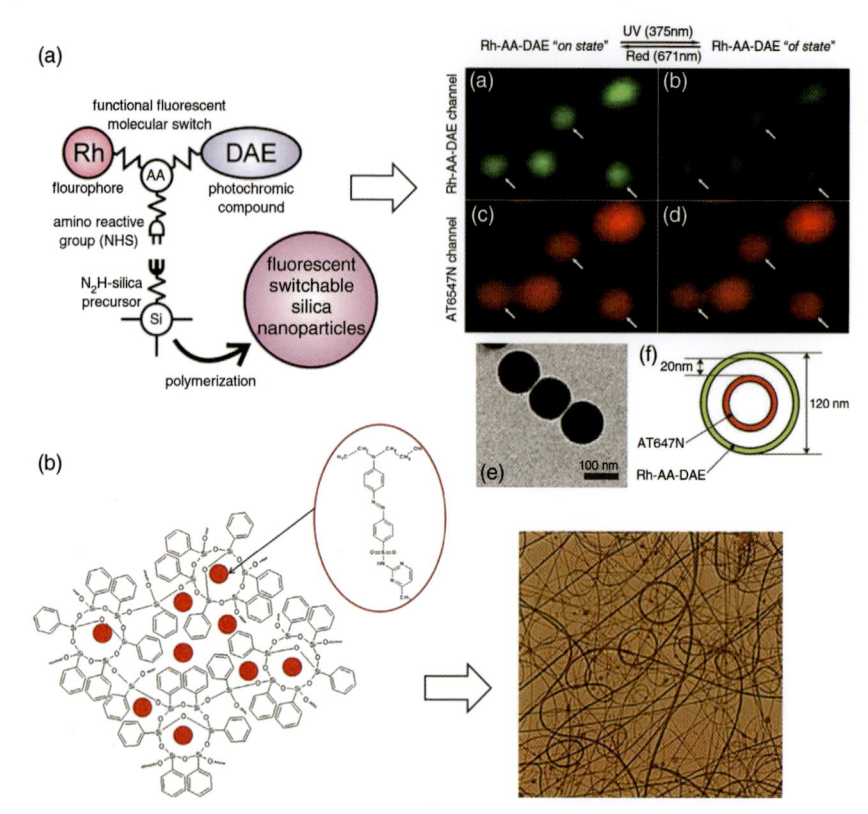

Figure 11.11 (a) Chemical structure of the fluorescent molecular switch Rh-AA-DAE and the photochromic reaction responsible for the fluorescence modulation. The fluorophore moiety is excited with green light: red light is emitted in the on state, while resonant energy transfer prevents this emission in the off state. Confocal images of fluorescent NPs (120 nm) doubly stained with Rh-AA-DAE and AT647N, recorded in two channels. The lower panel shows a TEM image of the particles (E) and a scheme with the distribution of the dyes inside each particle (F). (Adapted with permission from Ref. [172].) (b) The guest–host system containing photoactive dye of SMERe incorporated into triethoxyphenylsilane matrix. SEM pictures showing the fibrous structure of the hybrid materials. (Adapted with permission from Ref. [174].)

nanoparticles, with application in fluorescent microscopy (Figure 11.11a). Other works are devoted to develop new procedures of shaping and conformation, such as the fibers shown in Figure 11.11b.

An interesting idea to facilitate photochromic transformations is to include the organic dye into mesostructured materials with inorganic and organic domains separated at the nanometric scale. Stucky and coworkers [176] reported a block-copolymer/silica nanocomposite used as host for two photochromic dyes, a spirooxazine, and a spiropyran, where the dyes are incorporated predominantly

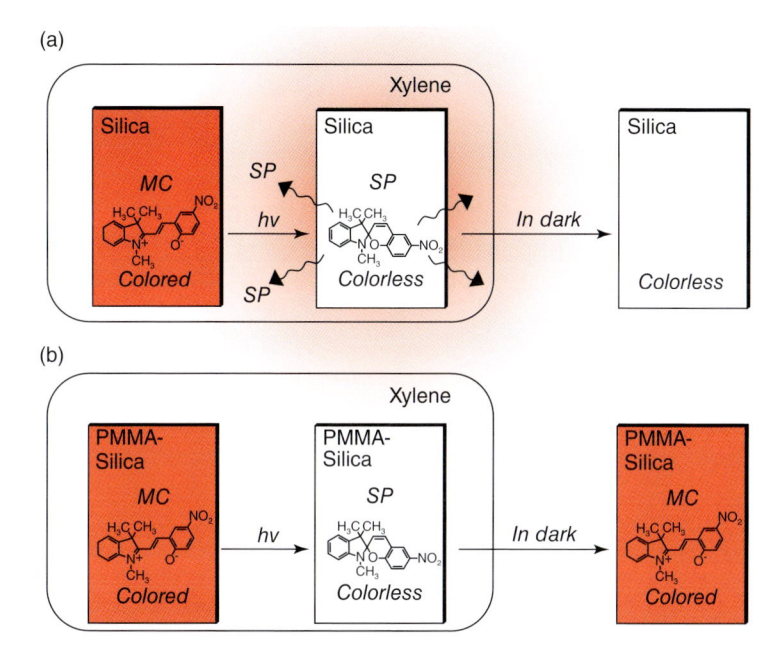

Figure 11.12 Schematic illustration on the changes in the state of spiropyran molecules during soaking in xylene for pure silica (a) and PMMA-silica (b) systems under visible light and in the dark. (Adapted with permission from Ref. [177].)

within the hydrophobic occlusions. The materials exhibit direct photochromism with faster response times, being at that time in the range of the best values reported so far for solid-state composites. These silica/block-copolymers can also be processed easily in any desired shape, including fibers, thin films, monoliths, waveguide structures, and optical coatings. Similar systems based on spiropyrane doped PMMA-silica have been reported, in which the presence of PMMA reduces the polar character of the matrix, facilitates the solubility of the dyes, modify the stability of particular isomers, and increases the chemical durability of the hybrid (Figure 11.12) [177].

The rational design of the hybrid nanocomposite is of great importance because the dye–matrix interaction is responsible for a good chemical compatibility. Indeed, the main limitations in these materials are usually the low content of photochromic species, low photochromic conversions in the solid state, or alterations in the chemical structure of the isomers during the polymerization process. Ribot et al. demonstrated in spirooxazine-doped tin-based nanobuilding blocks (NBBs) embedded in PEG copolymers that the interfaces can be tuned to afford a fast photochromic response together with a high-dye content [178]. Some publications have addressed this dye–matrix interaction to fully exploit the properties and tuneability of the photochromic hybrid systems. Evans et al. [179] reveal that

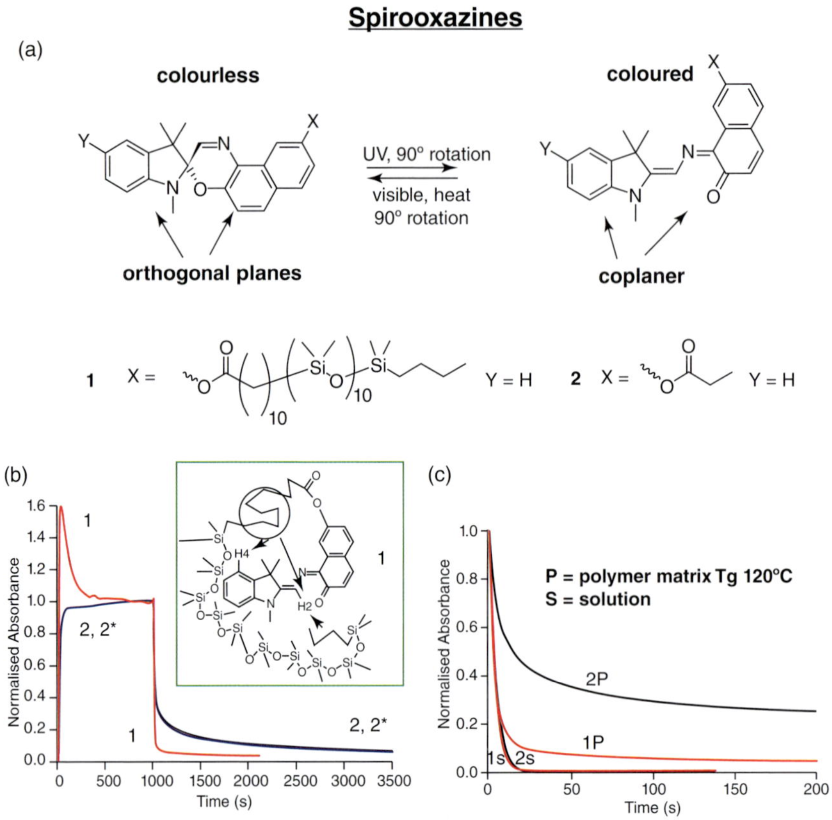

Spirooxazines

Figure 11.13 (a) Structural changes of spirooxazine compounds during photochromic switching. (b) Positive effect of conjugating a spirooxazine (compound 2) with a short PDMS chain (compound 1) on coloration and decoloration speeds. The covalent bonding between dye and silane oligomer is crucial for increasing the transformation speed (compare compound 2 with 2*, where 2* is the spirooxazine with PDMS oligomers added to the matrix, without chemical bonding). The intramolecular interaction of PDMS units with dye is shown in a green square.(c): The solution-like fade performance of compound 1 in the rigid silane matrix (curve 1P) as compared to 1 and 2 in toluene solution (curves 1S and 2S) and compound 2 in the identical rigid silane matrix (curve 2P). (Adapted with permission from Ref. [179].)

the intramolecular interactions of flexible PDMS (poly-dimethylsiloxane) oligomers with the photochromic dyes make possible faster chemical processes, greatly increasing the dye switching speed in a rigid hybrid system such as PDMS ophthalmic lenses. Both coloration and fade behavior indicate that the spirooxazine 1 (Figure 11.13) is in a highly mobile, near solution-like environment within the rigid matrix. The greatest impact was observed in the thermal fade parameters $T_{1/2}$ and $T_{3/4}$ – the times it takes for the optical density to reduce by half and three quarters of the initial optical density of the colored state – which were reduced by 75% and 94% for this dye in a host polymer with a glass-transition temperature of 120 °C. This is a quite surprising result because bonding photochromic dyes to polymers in low concentrations usually slows down the switching process [180]. The key difference in the Evans' work is that a low T_g oligomer provides a localized favorable switching environment where it is needed, that is, near/around the dye. Thus, the compatibility between the covalently bonded oligomer to the dye and the matrix allowed a greater molecular mobility for the switching process, which can find utility in data recording or optical switching.

The rigid environment of a dye molecule within the hybrid matrix can be beneficial, for instance, to avoid a nondesired photochromic transformation. This fact is detailed in amine-alcohol-silicate hybrid materials doped with naphthopyran [181], where the isomerization of one specific colored isomer (which is a space demanding reaction) is hindered, and films present fast and fully reversible coloration/transparent cycles upon irradiation/dark (Figure 11.14). This behavior contrasts with the classical performance of naphthopyrans that generate two-colored species, associated with two isomeric species, with different kinetics. In this case, the structural design prevents the formation of the long-lived colored specie, leading to faster and reversible photoswitchable materials.

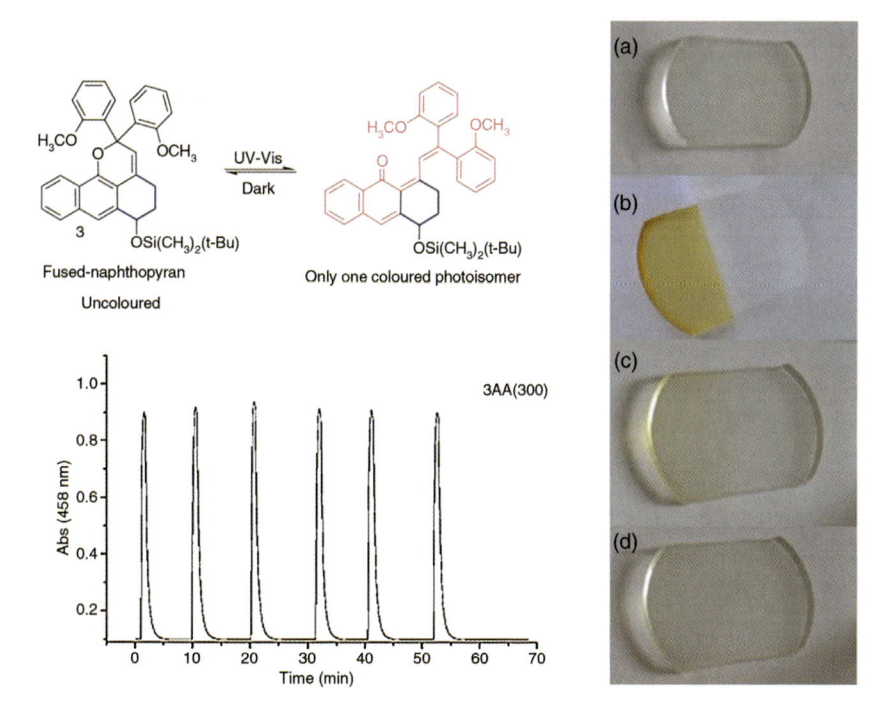

Figure 11.14 Photochromic equilibrium for the fused-naphthol[1,2-b]pyran. UV-vis irradiation/dark cycles of gels doped with naphthopyran 3 measured at 458 nm. Samples 3AA(300) in the dark (a), exposure of half of the sample to sunlight for 1 min (b), sample after 60 s in the dark (c) and after 120 s in the dark (d). (Adapted with permission from Ref. [181].)

Another interesting approach is to exploit the dye–matrix interaction to get periodically organized nanostructures through a template effect. Numerous studies about hybrids including viologen cations (1,1′-disubstituted-4,4′-bipyridinium, V^{2+} cations) use this effect. Viologens are electron-acceptor species that, in the solid state and upon irradiation, can accept one electron from a donor entity such as halides or pseudohalides X^- or a neutral molecule D, to afford stable separated charge state systems that find applications in electrochromic displays, molecular electronics, solar energy conversion, and so on. Mercier published an excellent review showing peculiar interactions between viologen dications and anions of hybrid structures [182]. One pioneering result was found in (MV)[MX$_{5-x}$X′$_x$] ((M = Bi^{III}, Sb^{III}; X = Cl, Br, I; $x=0-5$) compounds that exhibited ferroelectric properties thanks to the electron–donor interaction of a methylviologen dication with polar MX$_5$ chains of *trans*-connected octahedra (Figure 11.15) [183].

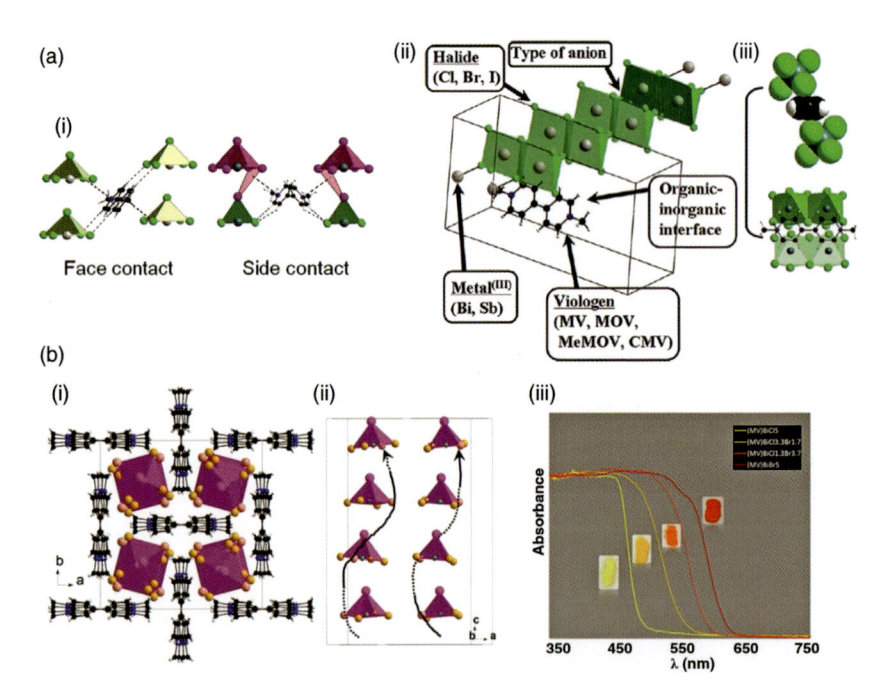

Figure 11.15 (a) Interactions at the organic–inorganic interface between one methylviologen cation and two anionic chains showing face and side contacts (part i). Part of the structure of (MV)[Bi$_2$Cl$_8$] showing one methylviologen entity and the inorganic network (part ii) and interactions between the organic cation and two inorganic anions (part iii: side view in space-filling representation and viewed along the direction perpendicular to the viologen plane), and the different parameters that can influence the photo-induced charge-transfer process in viologen halometalate salts. (Adapted with permission from Ref. [182].) (b) Structure of (MV)[SbBr$_{3.8}$I$_{1.2}$] viewed along the chain axis showing the tetragonal arrangement (part i), and syn-coupling of two neighboring chains that displays opposite chiralities (part ii). UV-vis spectra and pictures of (MV)[Bi$_2$Cl$_{5-x}$Br$_x$] ($x=0$, 3.7, 5) hybrids (part iii). (Adapted with permission from Ref. [183].)

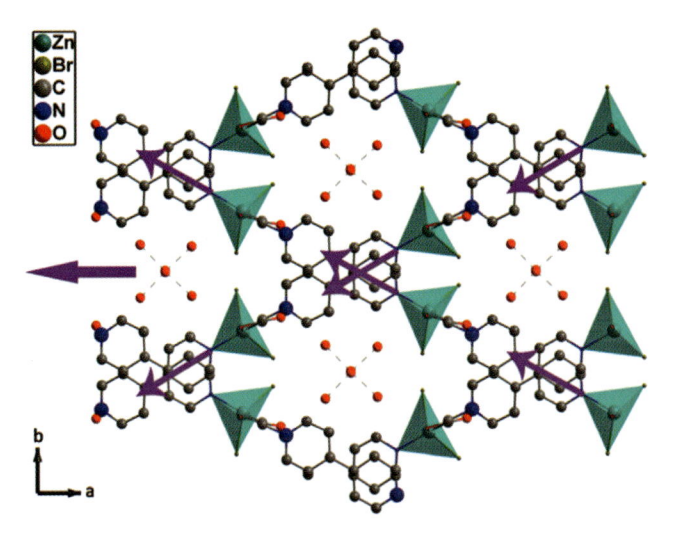

Figure 11.16 Molecular packing in a crystal of [ZnBr$_2$(μ-CEbpy).3H$_2$O] view along the c-axis. The violet arrows indicate the whole remnant polarity. (Adapted with permission from Ref. [184].)

Asymmetric viologen ligands can also be coordinated to Zn(II) centers through a template effect, leading to the first example of bulk electron-transfer photochromic compound with intrinsic second-order nonlinear optical (NLO) photoswitching properties [184]. The electron transfer is possible, thanks to the synergetic interaction between the ligand and a metal center with acentric coordination geometry. Briefly, the structure exhibits a 1D polar chain infinite structure with continuous head-to-tail linking of metallorganic units (one zinc atom linked to an acceptor viologen moiety and two electron donor species) that cross-stack according to the acentric space group (Cc), giving rise to the macroscopic polarization in the bulk (Figure 11.16).

The use of light as an external trigger to switch the second-order NLO activity has been increasingly addressed for nondestructive data storage or opto-optical switching in the emerging field of photonic devices [185]. However, there is a long way to use them in practical applications because of the low reaction speeds and the limited contrast for the photoswitching of NLO properties. To overcome these problems, Schulze et al. [186] propose the anchoring of photochromic fulgimide molecules into a self-assembly monolayer (SAM) on a Si(111) surface. The use of silicon, one of the most relevant materials for semiconductor devices, as a substrate for SAM formation and the further anchoring and aligning of molecular switches (photochromic molecules) paves the way for an efficient switching of the NLO response with higher contrast (Figure 11.17).

Periodically organized systems have demonstrated to enhance the photochemical processes over amorphous photoresponsive materials. Thus, the covalent grafting of organic molecules (flavilium, spyropyrans, etc.) in different types of mesoporous matrices (MCM-41, SBA-15, polymers, etc.) has also been

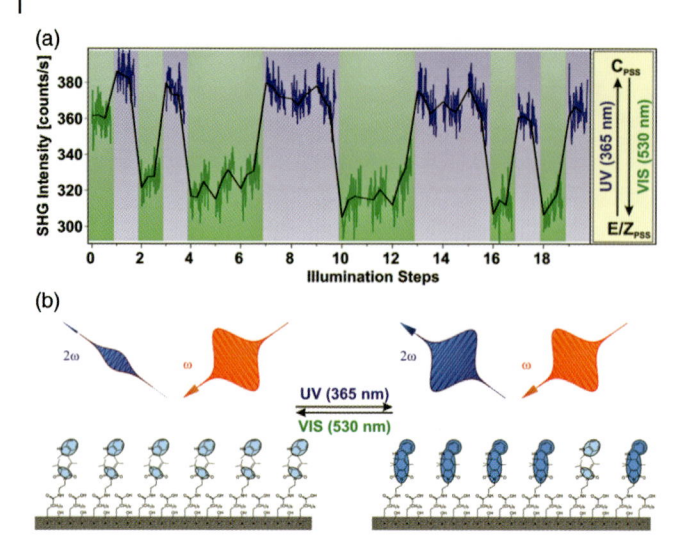

Figure 11.17 (a) Changes in the second harmonic generation (SHG) signal amplitude as a function of illumination with different wavelengths (365 and 530 nm), demonstrating the light-induced reversible changes in the NLO interfacial response due to the ring-opening/closure reaction. (b) Scheme of the reversible photoinduced switching probed with SHG. (Adapted with permission from Ref. [185].)

investigated. The ordered nanostructure and the strong dye–matrix bonding enhance their photostability and avoid leaching of the dyes. To mention some of the most interesting achievements, Gago et al. [187] prepared a solid state pH-dependent photochromic material with fast kinetics in color change thanks to the highly ordered hexagonal arrays of channels in mesoporous silica.

More sophisticated designs, such as core–shell or hollow spheres [188] are also of great interest. An example can be the smart nanocapsules reported by Allouche et al. [189] that are made of a dense silica core with a mesoporous photochromic (SP) silica shell in a dual templating sol–gel method. A different strategy is explored by Hernando and coworkers [190] in which photochromic molecules are encapsulated into liquid-filled polyamide capsules to achieve high switching speeds in solid materials. The authors state that its strategy is universal and does not require synthetic modification of commercially available photochromes, what is of great interest for industry. Furthermore, the hollow structures provide photochromic solid materials with solution-like color fading kinetics, which are about one order of magnitude faster than those measured upon direct dispersion of the photosensitive molecules into rigid polymer thin films and solid particles.

Regarding nonsilica systems, Andersson et al. [191] synthesized ordered honeycomb porous films made of a spiropyran functional PAA polymer that shows a rapid and intense color changes upon irradiation with UV and visible light. An attractive point is that these nanostructures can also entrap metal ions (Pt, Pd,

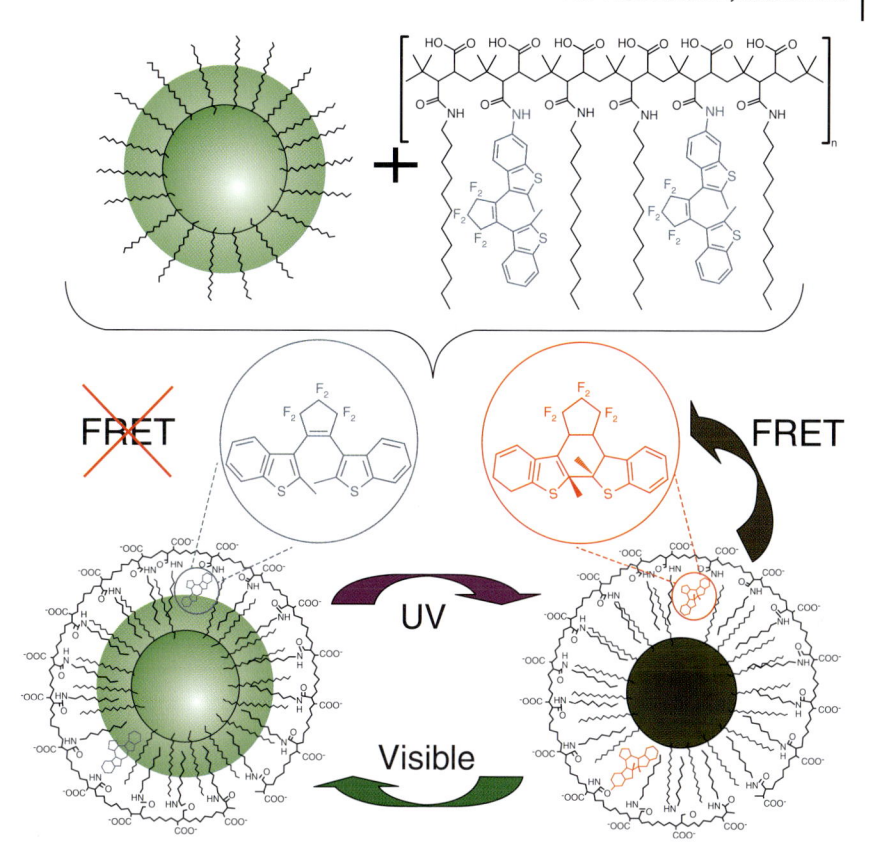

Figure 11.18 Photoswitching QDs coated with an amphiphilic photochromic polymer. The fluorescence of the psQD is toggled with FRET by modulating the absorbance of the photochromic polymer with UV and visible light. (Adapted from Ref. [199].)

etc.) or nanoparticles that shift the absorption and the chromic response, acting as reversible metal ion sensors [192]. Complex architectures can be designed to make possible energy transfers between photochromic dyes and other luminescent species such as quantum dots [193], up-converting nanocrystals [194–196], or plasmonic metal nanoparticles [197,198]. One nice example is the decoration of CdSe/ZnS QDs with an amphiphilic photochromic polymer coating [199]. The core–shell architecture permits the accommodation of hydrophobic acceptor dyes in a close proximity to the QD donor, thereby providing an effective fluorescence resonance energy transfer (FRET) probe in a small ($\sim$7 nm diameter) water-soluble package (Figure 11.18) that opens up numerous applications in biological and live cell studies.

Hybrid nanomaterials combining organic photochromes with plasmonic nanoparticles have also gained considerable interest in the fields of data storage, photovoltaics, biosensing, and so on [200,201]. The mutual interplay between organic molecules and metal properties can be used to finely tune the photochrome reactivity as well as the energy of the localized surface plasmon resonance (LSPR) of the metallic nanoparticles. In some cases, the light-induced photochromic process (for instance, photoisomerization between the *trans* and *cis* states in azobenzene molecules) results in significant changes in other properties like refractive index, with interest for optical switching and filtering. Most of the studies reported so far deal with grafted diarylethene [202] and spiropyrane [203] derivatives, characterized by an initial colorless state and a colored photoreacted state that causes an energy shift of the LSPR band in a privileged manner. Ledin *et al.* [200] reports the fabrication of silver nanocubes coated with photochromic azosilsesquioxane hybrid derivatives and deposited onto quartz substrates (Figure 11.19 *top*). The photochromic transformation induces variations in the refractive-index medium and a reversible tuning of the plasmonic modes of noble-metal nanostructures. Also, gold nanorods can be coated with silica and functionalized with grafted fluorescent and photochromic derivatives (Figure 11.19 *bottom*) [204]. Spectroscopic investigations demonstrated that cross-coupled interactions between plasmonic, photochromic, and fluorescence properties play a major role in such nanosystems, depending on the thickness of the silica spacer, leading to multisignal photoswitchability.

Multicolored systems have been developed by embedding two or more active species in matrices. These materials can be used in multifrequency photochromic recording, multicolor displays, inks, barcodes, and other applications, and are able to benefit from the control offered by a single material with several interconverted states. One example is reported by Wigglesworth and Branda [205], where three dithienylethene (DTE) derivatives providing the three primary colors are covalently linked to norbonene-based water-soluble monomers. Thus, a new family of multiaddressable photoresponsive copolymers is prepared by ring-opening metathesis polymerization (ROMP). Since DTE compounds exhibit fast response times, good thermal stability of both isomers (colorless ring-open and colored ring-closed), high fatigue resistance and their structures can be easily modified, they are some of the most promising candidates for use in devices. For instance, a three-component hybrid system (CP4, Figure 11.20) has been successfully tested as multicolor barcode. Readers with interest on DTE-including materials and their application in memories, switches, and actuators can consult the review of Irie *et al.* [206].

Multicolor photochromism can also be provided by supramolecular coordination polymers accommodating guest water molecules and anions (perchlorates, halides, pseudohalides, etc.) [207]. The soft metal organic hybrid material interacts with the environment, leading to a solvent- and anion-controlled photochromism via different charge transfers with the electron-accepting bipyridinium (viologen) moieties.

Multicolor systems combining photochromic dyes with graphene have recently been published. Sharker *et al.* developed a stimuli-responsive material based on graphene oxide coupled with a polymer conjugated with photochromic

11.3 Photochromic Hybrid Materials | **303**

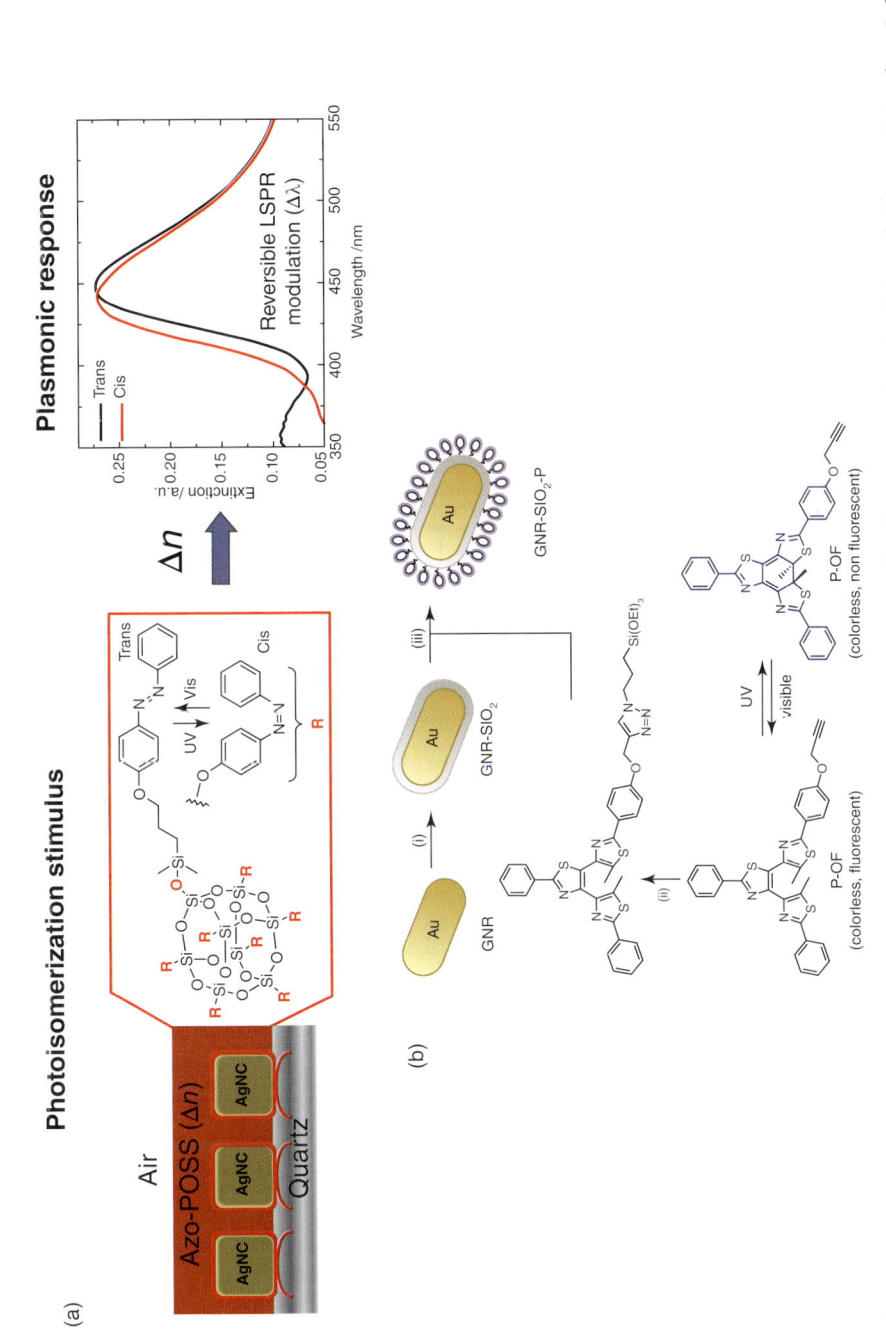

Figure 11.19 (a) Branched Azo-POSS conjugates as a variable-refractive-index matrix for plasmonic nanoparticles. (Adapted with permission from Ref. [200].) (b) Synthesis of the silica-coated gold nanorods with surface grafted photochromic and fluorescent derivatives. (Adapted with permission from Ref. [204].)

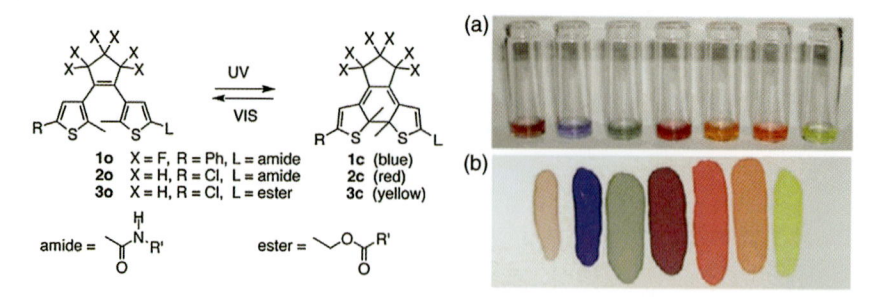

Figure 11.20 DTE derivatives whose ring-closed isomers exhibit primary colors. (a) Colors mixing of DTE homopolymers P1–P3 after THF solutions are irradiated with 313-nm light. The solutions contain (from left to right) [P1 + P2 + P3], [P1], [P1 + P3], [P1 + P2], [P2], [P2 + P3], and [P3]. (b) Samples of homo- and copolymers (from left to right) CP4, P1, CP2, CP1, P2, CP3, and P3 painted onto a silica plate followed by irradiation with 313-nm light. (Adapted with permission from Ref. [205].)

spiropyran dye and hydrophobic boron dipyrromethane dye, for application in triggered target multicolor bioimaging [208]. The different color of the functionalized graphene oxide is induced by both irradiation with UV light and by changing the pH from acidic to neutral. The stability, biocompatibility, and quenching efficacy of this nanocomposite open a different perspective for cell imaging in different independent colors, sequentially and simultaneously. Despite the efforts of the scientists on developing multicolor photochromic systems, most of them are based on polymers [39], single crystals [207], or organogels [209,210] containing organic moieties (no organic–inorganic hybrids), and they still present strong limitations for a real implementation in commercial devices.

11.3.3
Inorganic and Organometallic Photochromism in Hybrids

The inorganic photochromism is mainly provided by silver nanoparticles, transition metal oxides such as WO_3, MoO_3 and Nb_2O_5, and polyoxometalates (POMs). The color changes are completely different to those of organic dyes, and the mechanisms usually involve the generation of electron–hole pairs and valence state variations in oxygen defect structures. Some composites and hybrid systems including inorganic photochromes have already been fabricated. Despite all the efforts of scientists, their use in technical applications such as erasable optical storage media, large-area displays, chemical sensors, control of radiation intensity, or self-developing photography is still incipient due to the necessity of improving the variety of colors and the kinetics of coloration and fading.

A brief review of photochromic hybrids of metal halides, cyanides and chalcogenides, polyoxometalates, and metal–organic complexes was reported in 2010 [148]. It is worth noting that there are references in the literature employing the term hybrid in the sense of composite material, and no organic moiety is included in the final compound. In most of the cases, organic molecules are employed in the synthetic procedure as structure-directing agents or templates. The readers will find below some of the major advances in inorganic and organometallic photochromism.

The photochromic activity of silver halides relies on its dissociation into colloidal silver upon UV radiation. Its use in ophthalmic lenses is well-known since the 1960s, but the development of new materials with a wider palette of colors, faster bleaching kinetics, lower density, and so on is still a subject of extensive investigation. Numerous papers reporting smart photochromic properties of Ag [211–215] species deposited onto TiO_2 substrates have appeared in the last years. Interesting properties are developed in titanate nanotubes loaded with AgCl or Ag NPs [216]. For AgCl–titanate nanotubes (TNTs), the photochromic behavior consists of the red colouration on the material after irradiation by red light due to the photoreduction of silver halide to Ag nanoparticles. In the case of Ag NPs loaded TNTs, the materials exhibit multicolor photochromism corresponding to that of incident light associated particle-plasmon-assisted electron transfer from Ag nanoparticles to TiO_2. This example reveals how important is to control the active species and the photochromic mechanism involved in the photo- and electro-optical processes.

One interesting advance in terms of industrial application is reported by Tricot et al. [214] who prepared a flexible Ag: TiO_2 photochromic material using ink-jet and flexography printing processes compatible with industrial scale production (Figure 11.21). For that, a titanium precursor solution was printed on a plastic substrate thanks to adaptation of printing processes. After incorporating silver, the coating showed reversible photochromic behavior with a good contrast between the colored and bleached state. This breakthrough technology

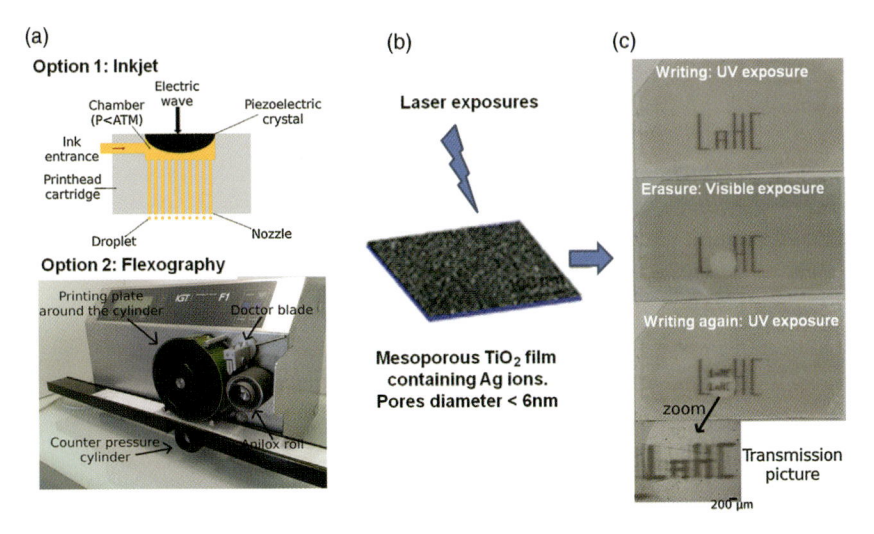

Figure 11.21 Principle of inkjet and flexography process (a), description of the Ag: TiO_2 material (b) and pictures of UV exposure inscription and visible exposure erasing on a film coated by flexography (c).

offers a new means to store updatable data and to secure products in smart cards or goods packaging areas.

Cheaper Cu NPs exhibiting unique photochromic properties onto TiO_2 substrates have also appeared. The special feature is that copper can exhibit three oxidation states, Cu^{2+}, Cu^+, and Cu^0. The redox process associated with photochromism can be controlled by varying the light source and exposure time. For instance, Tobaldi et al. [217] reports $Cu-TiO_2$ hybrid nanoparticles with tunable photochromism under both UVA and visible-light exposure. The material is 2 nm Cu NPs decorating the surface of ~10 nm TiO_2 NPs. Under UVA, Cu^{2+} is completely reduced to Cu^0 in few minutes (the d–d absorption band lowers to 95% after 10 min). Under visible-light, Cu^{2+} reduces only to Cu^+ and in a lesser extent (80% in 1 h under visible-light). This rapid and sensitive effect can potentially be used to modify, tune, or monitor the progress of photoactivated behavior in a new generation of smart/active multifunctional materials and photoactive devices or sensors.

Regarding photochromic WO_3 and MoO_3 materials, significant advances have been made in the past decades, such as the response to visible light, and the improved photochromism by proton donors [218,219]. However, organic–inorganic hybrid derivatives could solve other critical questions: (i) to improve reversibility; (ii) to fabricate systems exhibiting a wide variety of colors; and (iii) to enhance the sensitivity of the photochromic effect in the near-ultraviolet range. Some groups have investigated the intercalation of organic molecules such as diaminoalkane, phenethylamine, pyridine, and poly(ethylene glycol) in the interlayer space of both structures, WO_3 and MoO_3 [220–223]. A nice example is the preparation of tungsten oxide layers intercalated with acetic acid and hexamethylenetetramine (HMT) molecules with a 3D flowerlike morphology [224]. The most important result of the work is the good reversibility (as high as 98.4%, which is much higher than conventional WO_3 materials) and the good coloration response expanded to the visible-light region, which is a critical challenge for practical applications using solar energy and other visible laser sources. A second approach is to develop anisotropic organic–inorganic nanohybrids [225], in which the 1D organization of MoO_x nanoclusters enhances the photochromic sensitivity, and tunable components endow them with tailored performance.

Numerous studies have reported the preparation of photochromic WO_3 and MoO_3 shaped as thin films [226–228], nanorods [229], nanowires [229], or nanoflakes [230] in the last decade, but very few are devoted to real organic–inorganic hybrids. In those reports, organic molecules (EDTA [229], citric acid [231], etc.) are used to prevent the uncontrolled precipitation of the inorganic oxides and also act as directing agents, giving to tunable morphologies and photochromism. Only one paper [232] reports that tungsten oxide included in a fluoropolymer matrix yields a modest photochromic response. Only a 40% of transparency is obtained, so further improvements are necessary for a real application.

The photochromic properties of tungsten and molybdenum oxides can be modified by interaction with different inorganic species such as CdS NPs [233],

Ag [234]/Cu [227] species, TiO$_2$ [235,236], silica [237], or cellulose [238]. These nanocomposite materials are not strictly hybrid organic–inorganic materials, so these issues are out of the scope of this manuscript.

Photochromic polyoxometalates (POMs) are by far the most studied systems in the last years. A complete report about photochromism in composites and hybrid materials based on polyoxometalates was reported in 2006 [147]. POMs can be organically modified inorganic metal-oxygen cluster anions (i.e., polyoxomolybdates and polyoxotungstates), and they are promising candidates for photochromic applications due to their highly versatile and tunable structural, chemical, and redox properties. POMs, upon UV or VIS radiation, are able to accept electrons and/or protons from organic donor counter-cations in a reversibly exchange without decomposing or undergoing changes to its structural arrangement, to become mixed valence colored species (*heteropolyblues* or *heteropolybrowns*). In these complex structures, the organic moieties play an important role in the optical processes, since purely inorganic POMs have no reversible photochromism and the color change is not attractive. In general, the photogenerated colors of these hybrids depend upon the chemical composition and topology of the POMs, while the coloration and fading kinetics of these compounds are related to the nature of the organic cations and their interactions at the organic–inorganic interface. Therefore, much attention has been dedicated to the effective modification of photochromic properties using different organic moieties. Among them, we can cite an interesting work in which photochromic organic–inorganic materials are constructed via the coupling of liquid-crystalline nonionic surfactants and polyoxometalates (POMs) [239]. The chemical interaction of the complex nanostructures with organic molecules from the environment (solvent, air, etc.) has a clear effect on the photochromic response, providing new sensors and smart catalysts. An interesting work was reported in 2009 [240], metal-organic frameworks constructed from titanium-oxo clusters and dicarboxylate linkers, exhibit a reversible photochromic behavior induced by alcohol adsorption.

The combination of POMs with organic dyes to enhance the activity or to provide new multichromic hybrid organic–inorganic supramolecular assemblies has also been explored. For instance, spiropyran molecules, with high photochromic performance, have been recently integrated in POM structures via covalent bonds [241–243] and noncovalent interactions [244,245] with different performances.

The systematic study of Hakouk *et al.* [242] is especially interesting since they analyze the photochromic behavior taking into account several physical parameters (Spiropyran (SP) structural characteristics, SP/POM and SP/solvent solid-state interactions, molar volume, and so on). The study reveals that the coloration of the materials before UV exposure is governed by a low-energy intermolecular charge-transfer (CT) transition between SP donor and POM acceptor (see Figure 11.22). The CT transition energy can be tailored by tuning the intrinsic ligand-to-metal charge-transfer (LMCT) of the POM unit, which allows drastic improvement of the photocoloration contrasts.

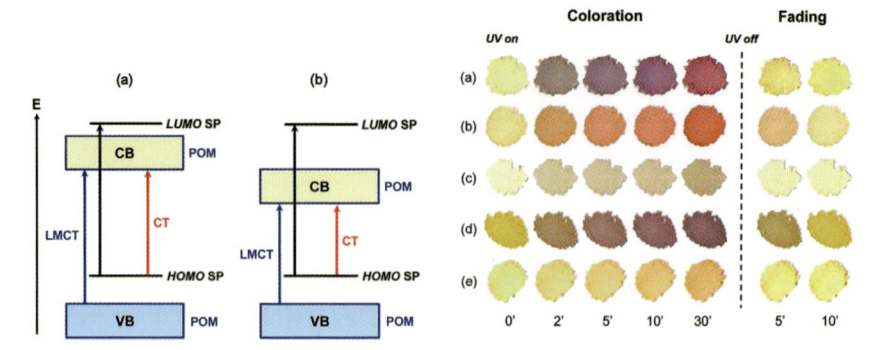

Figure 11.22 Schematic energetic diagrams displaying the three absorption phenomena predictable in SP-POM self-assembled structures when the SP cation is assembled with a POM unit having (a) a high-energy LMCT transition and (b) a low-energy LMCT transition. Photographs of powders of (a) SP$_4$Mo$_8$·CH$_3$CN, (b) SP$_4$Mo$_8$·DMSO, (c) SP$_4$Mo$_8$·DMF, (d) SP$_3$Mo$_8$, and (e) SP$_2$AlMo$_6$ at different time during the coloration process under UV irradiation (365 nm – 6 W), and the fading process under ambient light at room temperature [242].

Furthermore, most of these systems exhibit electrochromism (Figure 11.23). Also cucurbituril-POMs dyads [246] exhibit reversible photochromic properties as well as excellent photocatalytic activities toward the degradation of methyl orange (MO) and rhodamine-B (RB) under visible light irradiation. Interconversion pathways and chemical factors affecting the stabilization of the different species are highlighted and discussed in hybrids based on DABCO [247], piperazine, and molybdate [248].

Some exotic systems such as 3D iodoplumbate open-framework material exhibited interesting wavelength-dependent photochromic properties [249].

However, to apply them in useful devices, the challenge is to encapsulate or integrate POMs into organic, polymeric, or inorganic matrices or substrates to find new materials with adequate optical, mechanical, and chemical properties and applications such as catalysis, energy storage, or biomedicine [250]. The first

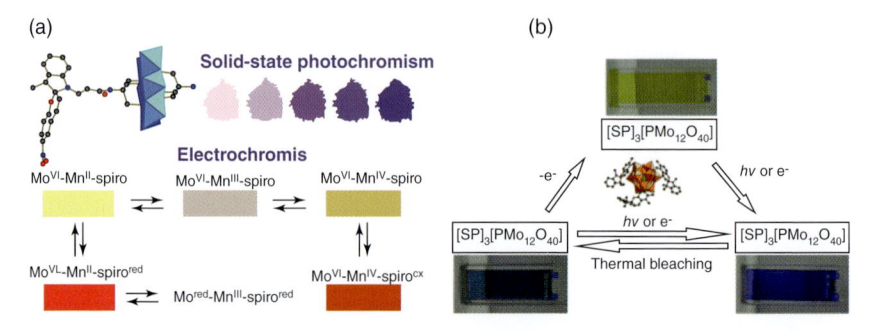

Figure 11.23 Structures of dual photochromic/electrochromic compounds based on spiropyrans and polyoxometalates with "covalent bonding" (a) [241] and "electrostatic interaction" (b) [244].

attempt to process photoactive porous POMs as thin films with high optical quality have already been done [251]. The work presents a simple route for the preparation of colloids of a flexible porous iron carboxylate (MIL-89) with tailored sorption properties that can be extended to other polyoxometallates and metal-organic frameworks.

Mesoporous bulk silicas have been used as supports for the immobilization of photochromic POMs by cocondensation and direct post-grafting, both based on covalent bonding and by impregnation [245,252,253]. Photochromic POMs have also been introduced in bulk silica matrices by sol–gel to produce POM/silica hybrid films [254,255]. Other works explore noncovalent interactions between the photochromic species and the matrix, such as the lanthano phosphomolybdates anions immobilized through electrostatic forces onto positively-charged silica nanoparticles [256]. Ormosils like triureasils have also been used to embed POMs providing transparent, flexible and rubbery photochromic materials [245,257].

References

1 Lakowicz, J.R. (2006) *Principles of Fluorescence Spectroscopy*, Springer, NY, USA.
2 Valeur, B. and Berberan-Santos, M.N. (2012) *Molecular Fluorescence: Principles and Applications*, Wiley-VCH Verlag GmbH, Weinheim.
3 Turro, N.J. (1991) *Modern Molecular Photochemistry*, University Science Books, Sausalito, CA.
4 Kohler, A., Wilson, J.S., and Friend, R.H. (2002) *Adv. Mater.*, **14**, 701.
5 Ng, K.K. and Zheng, G. (2015) *Chem. Rev.*, **115**, 11012.
6 Zhao, J.Z., Ji, S.M., and Guo, H.M. (2011) *RSC Adv.*, **1**, 937.
7 Blasse, G. and Grabmaier, B.C. (1994) *Luminescent Materials*, Springer-Verlag, Berlin.
8 Mitzi, D.B., Chondroudis, K., and Kagan, C.R. (2001) *IBM J. Res. Dev.*, **45**, 29.
9 Dohner, E.R., Hoke, E.T., and Karunadasa, H.I. (2014) *J. Am. Chem. Soc.*, **136**, 1718.
10 Kim, Y.H., Cho, H., Heo, J.H., Kim, T.S., Myoung, N., Lee, C.L., Im, S.H., and Lee, T.W. (2015) *Adv. Mater.*, **27**, 1248.
11 Dantas de Morais, T., Chaput, F., Boilot, J.P., Lahlil, K., Darracq, B., and Levy, Y. (2000) *C.R. Acad. Sci. IV-Phys.*, **1**, 479.
12 Dautel, O.J., Wantz, G., Almairac, R., Flot, D., Hirsch, L., Lere-Porte, J.P., Parneix, J.P., Serein-Spirau, F., Vignau, L., and Moreau, J.J.E. (2006) *J. Am. Chem. Soc.*, **128**, 4892.
13 Mizoshita, N., Goto, Y., Maegawa, Y., Tani, T., and Inagaki, S. (2010) *Chem. Mater.*, **22**, 2548.
14 Yang, X.H., Giovenzana, T., Feild, B., Jabbour, G.E., and Sellinger, A. (2012) *J. Mater. Chem.*, **22**, 12689.
15 Jung, H.S., Kim, Y.J., Ha, S.W., and Lee, J.K. (2013) *J. Mater. Chem. C*, **1**, 5879.
16 Kajiwara, Y., Nagai, A., Tanaka, K., and Chujo, Y. (2013) *J. Mater. Chem. C*, **1**, 4437.
17 Kwak, S.Y., Yang, S., Kim, N.R., Kim, J.H., and Bae, B.S. (2011) *Adv. Mater.*, **23**, 5767.
18 Carregal-Romero, E., Llobera, A., Cadarso, V.J., Darder, M., Aranda, P., Dominguez, C., Ruiz-Hitzky, E., and Fernandez-Sanchez, C. (2012) *ACS Appl. Mater. Interfaces*, **4**, 5029.
19 Yuan, Y. and Kruger, M. (2012) *Polymers*, **4**, 1.
20 Furman, J.D., Warner, A.Y., Teat, S.J., Mikhailovsky, A.A., and Cheetham, A.K. (2010) *Chem. Mater.*, **22**, 2255.
21 Kojima, A., Teshima, K., Shirai, Y., and Miyasaka, T. (2009) *J. Am. Chem. Soc.*, **131**, 6050.

22 Yang, W.S., Noh, J.H., Jeon, N.J., Kim, Y.C., Ryu, S., Seo, J., and Seok, S.I. (2015) *Science*, **348**, 1234.

23 Zhao, Y.X. and Zhu, K. (2016) *Chem. Soc. Rev.*, **45**, 655.

24 Houbertz, R., Domann, G., Cronauer, C., Schmitt, A., Martin, H., Park, J.U., Frohlich, L., Buestrich, R., Popall, M., Streppel, U., Dannberg, P., Wachter, C., and Brauer, A. (2003) *Thin Solid Films*, **442**, 194.

25 Carlos, L.D., Ferreira, R.A.S., de Zea Bermudez, V., and Ribeiro, S.J.L. (2009) *Adv. Mater.*, **21**, 509.

26 Ferreira, R.A.S., Andre, P.S., and Carlos, L.D. (2010) *Opt. Mater.*, **32**, 1397.

27 Sanchez, C., Belleville, P., Popall, M., and Nicole, L. (2011) *Chem. Soc. Rev.*, **40**, 696.

28 Carlos, L.D., Ferreira, R.A.S., de Zea Bermudez, V., Julián-López, B., and Escribano, P. (2011) *Chem. Soc. Rev.*, **40**, 536.

29 Lebeau, B. and Innocenzi, P. (2011) *Chem. Soc. Rev.*, **40**, 886.

30 Zayat, M., Pardo, R., Castellón, E., Torres, L., Almendro, D., Parejo, P.G., Álvarez, A., Belenguer, T., García-Revilla, S., Balda, R., Fernández, J., and Levy, D. (2011) *Adv. Mater.*, **23**, 5318.

31 Sanchez, C. and Ribot, F. (1994) *New J. Chem.*, **18**, 1007.

32 Reisfeld, R., Shamrakov, D., and Jorgensen, C. (1994) *Sol. Energy Mater. Sol. Cells*, **33**, 417.

33 Yang, Y., Wang, M.Q., Qian, G.D., Wang, Z.Y., and Fan, X.P. (2004) *Opt. Mater.*, **24**, 621.

34 Lebeau, B., Herlet, N., Livage, J., and Sanchez, C. (1993) *Chem. Phys. Lett.*, **206**, 15.

35 Nhung, T.H., Canva, M., Chaput, F., Goudket, H., Roger, G., Brun, A., Manh, D.D., Hung, N.D., and Boilot, J.P. (2004) *Opt. Commun.*, **232**, 343.

36 Julián-López, B., Corberan, R., Cordoncillo, E., Escribano, P., Viana, B., and Sanchez, C. (2004) *J. Mater. Chem.*, **14**, 3337.

37 Pecoraro, E., García-Revilla, S., Ferreira, R.A.S., Balda, R., Carlos, L.D., and Fernandez, J. (2010) *Opt. Express*, **18**, 7470.

38 Kajiwara, Y., Nagai, A., and Chujo, Y. (2010) *J. Mater. Chem.*, **20**, 2985.

39 Zhang, X., Liu, W., Wei, G.Z., Banerjee, D., Hu, Z.C., and Li, J. (2014) *J. Am. Chem. Soc.*, **136**, 14230.

40 Dire, S., Babonneau, F., Sanchez, C., and Livage, J. (1992) *J. Mater. Chem.*, **2**, 239.

41 de Francisco, R., Hoyos, M., Garcia, N., and Tiemblo, P. (2015) *Langmuir*, **31**, 3718.

42 Roppolo, I., Messori, M., Perruchas, S., Gacoin, T., Boilot, J.P., and Sangermano, M. (2012) *Macromol. Mater. Eng.*, **297**, 680.

43 Viana, B., Koslova, N., Aschehoug, P., and Sanchez, C. (1995) *J. Mater. Chem.*, **5**, 719.

44 Pradal, N., Boyer, D., Chadeyron, G., Therias, S., Chapel, A., Santilli, C.V., and Mahiou, R. (2014) *J. Mater. Chem. C*, **2**, 6301.

45 Potdevin, A., Chadeyron, G., Therias, S., and Mahiou, R. (2012) *Langmuir*, **28**, 13526.

46 Sanchez, C., Lebeau, B., Chaput, F., and Boilot, J.P. (2003) *Adv. Mater.*, **15**, 1969.

47 Sanchez, C., Julian, B., Belleville, P., and Popall, M. (2005) *J. Mater. Chem.*, **15**, 3559.

48 Nicole, L., Rozes, L., and Sanchez, C. (2010) *Adv. Mater.*, **22**, 3208.

49 Escribano, P., Julián-López, B., Planelles-Aragó, J., Cordoncillo, E., Viana, B., and Sanchez, C. (2008) *J. Mater. Chem.*, **18**, 23.

50 Binnemans, K. (2009) *Chem. Rev.*, **109**, 4283.

51 Feng, J. and Zhang, H. (2013) *Chem. Soc. Rev.*, **42**, 387.

52 Ribeiro, S.J.L., dos Santos, M.V., Silva, R.R., Pecoraro, E., Gonçalves, R.R., and Caiut, J.M.A. (2015) *The Sol-Gel Handbook*, vol. **3** (eds D. Levy and M. Zayat), Wiley-VCH Verlag GmbH, Weinheim, p. 929.

53 Freitas, V.T., Ferreira, R.A.S., and Carlos, L.D. (2015) *The Sol-Gel Handbook*, vol. **3** (eds D. Levy and M. Zayat), Wiley-VCH Verlag GmbH, Weinheim, p. 883.

54 Chen, G.Y., Qju, H.L., Prasad, P.N., and Chen, X.Y. (2014) *Chem. Rev.*, **114**, 5161.

55 Chen, G.Y., Roy, I., Yang, C.H., and Prasad, P.N. (2016) *Chem. Rev.*, **116**, 2826.

56 Montalti, M., Prodi, L., Rampazzo, E., and Zaccheroni, N. (2014) *Chem. Soc. Rev.*, **43**, 4243.
57 Bunzli, J.C.G. (2016) *J. Lumin.*, **170**, 866.
58 Tang, C.W. and Vanslyke, S.A. (1987) *Appl. Phys. Lett.*, **51**, 913.
59 Reisfeld, R., Brusilovsky, D., Eyal, M., Miron, E., Burstein, Z., and Ivri, J. (1989) *Chem. Phys. Lett.*, **160**, 43.
60 Altman, J.C., Stone, R.E., Dunn, B., and Nishida, F. (1991) *IEEE Photonic. Tech. Lett.*, **3**, 189.
61 He, X.X., Duan, J.H., Wang, K.M., Tan, W.H., Lin, X., and He, C.M. (2004) *J. Nanosci. Nanotechnol.*, **4**, 585.
62 Dantas de Morais, T., Chaput, F., Lahlil, K., and Boilot, J.P. (1999) *Adv. Mater.*, **11**, 107.
63 Olivero, F., Carniato, F., Bisio, C., and Marchese, L. (2012) *J. Mater. Chem.*, **22**, 25254.
64 Burkett, S.L., Press, A., and Mann, S. (1997) *Chem. Mater.*, **9**, 1071.
65 Patil, A.J. and Mann, S. (2008) *J. Mater. Chem.*, **18**, 4605.
66 Rao, K.V., Datta, K.K.R., Eswaramoorthy, M., and George, S.J. (2013) *Adv. Mater.*, **25**, 1713.
67 Rao, K.V., Jain, A., and George, S.J. (2014) *J. Mater. Chem. C*, **2**, 3055.
68 Jain, A., Achari, A., Eswaramoorthy, M., and George, S.J. (2016) *J. Mater. Chem. C*, **4**, 2748.
69 Huang, C.Y., Su, Y.K., Wen, T.C., Guo, T.F., and Tu, M.L. (2008) *IEEE Photonic Tech. Lett.*, **20**, 282.
70 Huang, C.Y., Huang, T.S., Cheng, C.Y., Chen, Y.C., Wan, C.T., Rao, M.V.M., and Su, Y.K. (2010) *IEEE Photonic Tech. Lett.*, **22**, 305.
71 Lin, H.Y., Wang, S.W., Lin, C.C., Chen, K.J., Han, H.V., Tu, Z.Y., Tu, H.H., Chen, T.M., Shih, M.H., Lee, P.T., Chen, H.M.P., and Kuo, H.C. (2016) *IEEE J. Sel. Top. Quant. Electron.*, **22**, 35–41.
72 Chen, J., Liu, W., Mao, L.-H., Yin, Y.-J., Wang, C.-F., and Chen, S. (2014) *J. Mater. Sci.*, **49**, 7391.
73 Bai, X., Caputo, G., Hao, Z.D., Freitas, V.T., Zhang, J.H., Longo, R.L., Malta, O.L., Ferreira, R.A.S., and Pinna, N. (2014) *Nat. Commun.*, **5**, 5702.
74 Lee, K., Cheah, K., An, B., Gong, M., and Liu, Y.L. (2005) *Appl. Phys. A*, **80**, 337.
75 Huang, X.G., Zucchi, G., Tran, J., Pansu, R.B., Brosseau, A., Geffroy, B., and Nief, F. (2014) *New J. Chem.*, **38**, 5793.
76 Lu, Y. and Yan, B. (2014) *Chem. Commun.*, **50**, 15443.
77 Yan, H.G., Wang, H.H., He, P., Shi, J.X., and Gong, M.L. (2011) *Synth. Met.*, **161**, 748.
78 García-Revilla, S., Fernandez, J., Illarramendi, M.A., García-Ramiro, B., Balda, R., Cui, H., Zayat, M., and Levy, D. (2008) *Opt. Express*, **16**, 12251.
79 Wiersma, D.S. (2008) *Nat. Phys.*, **4**, 359.
80 García-Revilla, S., Fernandez, J., Barredo-Zuriarrain, M., Carlos, L.D., Pecoraro, E., Iparraguirre, I., Azkargorta, J., and Balda, R. (2015) *Opt. Express*, **23**, 1456.
81 Redding, B., Choma, M.A., and Cao, H. (2011) *Opt. Lett.*, **36**, 3404.
82 Redding, B., Choma, M.A., and Cao, H. (2012) *Nat. Photonics*, **6**, 355.
83 Giannelis, E.P., Stasinopoulos, A., Psyllaki, M., Zacharakis, G., Das, R.N., Anglos, D., Anastasiadis, S.H., and Vaia, R.A. (2002) *Mater. Res. Soc. Symp. Proc.*, **726**, 11.
84 Anglos, D., Stassinopoulos, A., Das, R.N., Zacharakis, G., Psyllaki, M., Jakubiak, R., Vaia, R.A., Giannelis, E.P., and Anastasiadis, S.H. (2004) *J. Opt. Soc. Am. B*, **21**, 208.
85 García-Revilla, S., Zayat, M., Balda, R., Al-Saleh, M., Levy, D., and Fernandez, J. (2009) *Opt. Express*, **17**, 13202.
86 Meng, X.G., Fujita, K., Zong, Y.H., Murai, S., and Tanaka, K. (2008) *Appl. Phys. Lett.*, **92**, 201112.
87 Cerdan, L., Costela, A., and Garcia-Moreno, I. (2012) *Org. Electron.*, **13**, 1463.
88 Costela, A., Garcia-Moreno, I., Cerdan, L., Martin, V., Garcia, O., and Sastre, R. (2009) *Adv. Mater.*, **21**, 4163.
89 Gorrn, P., Lehnhardt, M., Kowalsky, W., Riedl, T., and Wagner, S. (2011) *Adv. Mater.*, **23**, 869.
90 Sakhno, O.V., Stumpe, J., and Smirnova, T.N. (2011) *Appl. Phys. B-Lasers O*, **103**, 907.
91 da Silva, R.R., Dominguez, C.T., dos Santos, M.V., Barbosa-Silva, R., Cavicchioli, M., Christovan, L.M., de Melo, L.S.A., Gomes, A.S.L., de Araujo,

C.B., and Ribeiro, S.J.L. (2013) *J. Mater. Chem. C*, **1**, 7181.

92 Oliveira, D.C., Messaddeq, Y., Dahmouche, K., Ribeiro, S.J.L., Goncalves, R.R., Vesperini, A., Gindre, D., and Nunzi, J.M. (2006) *J. Sol–Gel Sci. Technol.*, **40**, 359.

93 Reisfeld, R. and Neuman, S. (1978) *Nature*, **274**, 144.

94 Reisfeld, R. and Kalisky, Y. (1980) *Nature*, **283**, 281.

95 van der Ende, B.M., Aarts, L., and Meijerink, A. (2009) *Phys. Chem. Chem. Phys.*, **11**, 11081.

96 Giebink, N.C., Wiederrecht, G.P., and Wasielewski, M.R. (2011) *Nat. Photonics*, **5**, 695.

97 Huang, X., Han, S., Huang, W., and Liu, X. (2013) *Chem. Soc. Rev.*, **42**, 173.

98 Chou, C.H., Chuang, J.K., and Chen, F.C. (2013) *Sci. Rep.*, **3**, 2244.

99 Bünzli, J.-C.G. and Chauvin, A.-S. (2014) *Handbook on the Physics and Chemistry of Rare-Earths*, vol. 44 (eds J.-C.G. Bünzli and V.K. Pecharsky), Elsevier B. V., Amsterdam, p. 169.

100 Correia, S.F.H., de Zea Bermudez, V., Ribeiro, S.J.L., André, P.S., Ferreira, R.A.S., and Carlos, L.D. (2014) *J. Mater. Chem. A*, **2**, 5580.

101 Correia, S.F.H., Lima, P.P., Carlos, L.D., André, P.S., and Ferreira, R.A.S. (2016) *Prog. Photovolt: Res. Appl.* doi: 10.1002/pip.2772

102 Rowan, B.C., Wilson, L.R., and Richards, B.S. (2008) *IEEE J. Sel. Top. Quant.*, **14**, 1312.

103 Reisfeld, R. and Jorgensen, C.K. (1982) *Struct. Bond.*, **49**, 1.

104 Reisfeld, R. (2010) *Opt. Mater.*, **32**, 850.

105 Meinardi, F., Colombo, A., Velizhanin, K.A., Simonutti, R., Lorenzon, M., Beverina, L., Viswanatha, R., Klimov, V.I., and Brovelli, S. (2014) *Nat. Photonics*, **8**, 392.

106 Meinardi, F., McDaniel, H., Carulli, F., Colombo, A., Velizhanin, K.A., Makarov, N.S., Simonutti, R., Klimov, V.I., and Brovellii, S. (2015) *Nat. Nanotechnol.*, **10**, 878.

107 Zhao, Y. and Lunt, R.R. (2013) *Adv. Energy Mater.*, **3**, 1143.

108 Graffion, J., Cattoën, X., Wong Chi Man, M., Fernandes, V.R., André, P.S., Ferreira, R.A.S., and Carlos, L.D. (2011) *Chem. Mater.*, **23**, 4773.

109 Graffion, J., Cojocariu, A.M., Cattoën, X., Ferreira, R.A.S., Fernandes, V.R., André, P.S., Carlos, L.D., Wong Chi Man, M., and Bartlett, J.R. (2012) *J. Mater. Chem.*, **22**, 13279.

110 Freitas, V.T., Fu, L.S., Cojocariu, A.M., Cattoën, X., Bartlett, J.R., Parc, R.Le., Bantignies, J.L., Wong Chi Man, M., André, P.S., Ferreira, R.A.S., and Carlos, L.D. (2015) *ACS Appl. Mater. Interfaces*, **7**, 8770.

111 Nolasco, M.M., Vaz, P.M., Freitas, V.T., Lima, P.P., André, P.S., Ferreira, R.A.S., Vaz, P.D., Ribeiro-Claro, P., and Carlos, L.D. (2013) *J. Mater. Chem. A*, **1**, 7339.

112 Correia, S.F.H., Lima, P.P., André, P.S., Ferreira, R.A.S., and Carlos, L.D. (2015) *Sol. Energy Mater. Sol. Cells*, **138**, 51.

113 Kaniyoor, A., McKenna, B., Comby, S., and Evans, R.C. (2016) *Adv. Opt. Mater.*, **4**, 444.

114 Wu, W., Wang, T., Wang, X., Wu, S., Luo, Y., Tian, X., and Zhang, Q. (2010) *Sol. Energy*, **84**, 2140.

115 Inman, R.H., Shcherbatyuk, G.V., Medvedko, D., Gopinathan, A., and Ghosh, S. (2011) *Opt. Express*, **19**, 24308.

116 Banaei, E.H. and Abouraddy, A.F. (2013) *Proc. SPIE*, **8821**, 882102.

117 McIntosh, K.R., Yamada, N., and Richards, B.S. (2007) *Appl. Phys. B*, **88**, 285.

118 Millán, A., Carlos, L.D., Brites, C.D.S., Silva, N.J.O., Piñol, R., and Palacio, F. (2016) *Thermometry at the Nanoscale: Techniques and Selected Applications* (eds L.D. Carlos and F. Palacio), Royal Society of Chemistry, Oxfordshire, p. 237.

119 Brites, C.D.S., Lima, P.P., Silva, N.J.O., Millán, A., Amaral, V.S., Palacio, F., and Carlos, L.D. (2010) *Adv. Mater.*, **22**, 4499.

120 Jung, W., Kim, Y.W., Yim, D., and Yoo, J.Y. (2011) *Sens. Actuator A-Phys.*, **171**, 228.

121 Wawrzynczyk, D., Bednarkiewicz, A., Nyk, M., Strek, W., and Samoc, M. (2012) *Nanoscale*, **4**, 6959.

122 Maestro, L.M., Rodriguez, E.M., Rodriguez, F.S., la Cruz, M.C.I., Juarranz, A., Naccache, R., Vetrone, F., Jaque, D., Capobianco, J.A., and Sole, J.G. (2010) *Nano Lett.*, **10**, 5109.

123 Okabe, K., Inada, N., Gota, C., Harada, Y., Funatsu, T., and Uchiyama, S. (2012) *Nat. Commun.*, **3**, 705.
124 Savchuk, O.A., Haro-Gonzalez, P., Carvajal, J.J., Jaque, D., Massons, J., Aguilo, M., and Diaz, F. (2014) *Nanoscale*, **6**, 9727.
125 Khalid, A.H. and Kontis, K. (2009) *Meas. Sci. Technol.*, **20**, 025305.
126 Lojpur, V., Antic, Z., and Dramicanin, M.D. (2014) *Phys. Chem. Chem. Phys.*, **16**, 25636.
127 Brites, C.D.S., Lima, P.P., Silva, N.J.O., Millán, A., Amaral, V.S., Palacio, F., and Carlos, L.D. (2012) *Nanoscale*, **4**, 4799.
128 Jaque, D. and Vetrone, F. (2012) *Nanoscale*, **4**, 4301.
129 Wang, X.D., Wolfbeis, O.S., and Meier, R.J. (2013) *Chem. Soc. Rev.*, **42**, 7834.
130 Zhou, H.Y., Sharma, M., Berezin, O., Zuckerman, D., and Berezin, M.Y. (2016) *Chem. Phys. Chem.*, **17**, 27.
131 Carlos, L.D. and Palacio, F. (2016) *Thermometry at the Nanoscale: Techniques and Selected Applications*, Royal Society of Chemistry, Oxfordshire.
132 Lupton, J.M. (2002) *Appl. Phys. Lett.*, **81**, 2478.
133 Yan, D.P., Lu, J., Ma, J., Wei, M., Evans, D.G., and Duan, X. (2011) *Angew. Chem., Int. Ed.*, **50**, 720.
134 Wang, Z.P., Ananias, D., Carne-Sanchez, A., Brites, C.D.S., Imaz, I., Maspoch, D., Rocha, J., and Carlos, L.D. (2015) *Adv. Funct. Mater.*, **25**, 2824.
135 Lee, J., Govorov, A.O., and Kotov, N.A. (2005) *Angew. Chem., Int. Ed.*, **44**, 7439.
136 Li, J., Hong, X., Liu, Y., Li, D., Wang, Y.W., Li, J.H., Bai, Y.B., and Li, T.J. (2005) *Adv. Mater.*, **17**, 163.
137 Piñol, R., Brites, C.D., Bustamante, R., Martínez, A., Silva, N.J., Murillo, J.L., Cases, R., Carrey, J., Estepa, C., Sosa, C., Palacio, F., Carlos, L.D., and Millán, A. (2015) *ACS Nano*, **9**, 3134.
138 Brites, C.D.S., Lima, P.P., Silva, N.J.O., Millán, A., Amaral, V.S., Palacio, F., and Carlos, L.D. (2013) *Front. Chem.*, **1**, 9.
139 Brites, C.D.S., Lima, P.P., Silva, N.J.O., Millán, A., Amaral, V.S., Palacio, F., and Carlos, L.D. (2013) *J. Lumin.*, **133**, 230.
140 Ferreira, R.A.S., Brites, C.D.S., Vicente, C.M.S., Lima, P.P., Bastos, A.R.N., Marques, P.G., Hiltunen, M., Carlos, L.D., and André, P.S. (2013) *Laser Photonics Rev.*, **7**, 1027.
141 Rodrigues, M., Piñol, R., Antorrena, G., Brites, C.D.S., Silva, N.J.O., Murillo, J.L., Cases, R., Díez, I., Palacio, F., Torras, N., Plaza, J.A., Pérez-García, L., Carlos, L.D., and Millán, A. (2016) *Adv. Funct. Mater.*, **26** 200.
142 Bouas-Laurent, H. and Durr, H. (2001) *Pure Appl. Chem.*, **73**, 639.
143 Crano, J.C. and Guglielmetti, R.J. (eds) (1999) *Organic Photochromic and Thermochromic Compounds*, vol. **1**, Plenum Press, New York.
144 Crano, J.C. and Guglielmetti, R.J. (eds) (1999) *Organic Photochromic and Thermochromic Compounds*, vol. **2**, Plenum Press, New York.
145 Dürr, H. and Bouas-Laurent, H. (eds) (2003) *Photochromism: Molecules and Systems*, Elsevier, Amsterdam.
146 Bamfield, P. and Hutchings, M.G. (eds) (2010) *Chromic Phenomena: Technological Applications of Colour Chemistry*, RSC Publishing, Cambridge.
147 He, T. and Yao, J. (2006) *Prog. Mater. Sci.*, **51**, 810.
148 Wang, M.-S., Xu, G., Zhang, Z.-J., and Guo, G.-C. (2010) *Chem. Commun.*, **46**, 361.
149 Levy, D. (1997) *Chem. Mater.*, **9**, 2666.
150 Pardo, R., Zayat, M., and Levy, D. (2011) *Chem. Soc. Rev.*, **40**, 672.
151 Avnir, D., Levy, D., and Reisfeld, R. (1984) *J. Phys. Chem.*, **88**, 5956.
152 Behar-Levy, H. and Avnir, D. (2002) *Chem. Mater.*, **14**, 1736.
153 Levy, D. and Avnir, D. (1991) *J. Photochem. Photobiol. A*, **57**, 41.
154 Levy, D., Einhorn, S., and Avnir, D. (1989) *J. Non Cryst. Solids*, **113**, 137.
155 Bentivegna, F., Canva, M., Brun, A., Chaput, F., and Boilot, J.P. (1996) *J. Appl. Phys.*, **80**, 4655.
156 Bentivegna, F., Canva, M., Brun, A., Chaput, F., and Boilot, J.P. (1997) *J. Sol–Gel Sci. Technol.*, **9**, 33.
157 Bentivegna, F., Canva, M., Georges, P., Brun, A., Chaput, F., Malier, L., and Boilot, J.P. (1993) *Appl. Phys. Lett.*, **62**, 1721.

158 Boilot, J.P., Biteau, J., Chaput, F., Gacoin, T., Brun, A., Darracq, B., Georges, P., and Levy, Y. (1998) *Pure Appl. Opt.*, **7**, 169.

159 Canva, M., Roger, G., Cassagne, F., Levy, Y., Brun, A., Chaput, F., Boilot, J.P., Rapaport, A., Heerdt, C., and Bass, M. (2002) *Opt. Mater.*, **18**, 391.

160 Dubois, A., Canva, M., Brun, A., Chaput, F., and Boilot, J.P. (1996) *Synth. Met.*, **81**, 305.

161 Dubois, A., Canva, M., Brun, A., Chaput, F., and Boilot, J.P. (1996) *Appl. Opt.*, **35**, 3193.

162 Fournier, T., Tranthi, T.H., Herlet, N., and Sanchez, C. (1993) *Chem. Phys. Lett.*, **208**, 101.

163 Lebeau, B., Brasselet, S., Zyss, J., and Sanchez, C. (1997) *Chem. Mater.*, **9**, 1012.

164 Lebeau, B., Maquet, J., Sanchez, C., Toussaere, E., Hierle, R., and Zyss, J. (1994) *J. Mater. Chem.*, **4**, 1855.

165 Lebeau, B., Sanchez, C., Brasselet, S., Zyss, J., Froc, G., and Dumont, M. (1996) *New J. Chem.*, **20**, 13.

166 Schaudel, B., Guermeur, C., Sanchez, C., Nakatani, K., and Delaire, J.A. (1997) *J. Mater. Chem.*, **7**, 61.

167 Lafuma, A., Chodorowski-Kimmes, S., Quinn, F.X., and Sanchez, C. (2003) *Eur. J. Inorg. Chem.*, **2**, 331.

168 Morimoto, M., Kobatake, S., and Irie, M. (2003) *Photochem. Photobiol. Sci.*, **2**, 1088.

169 Morimoto, M., Kobatake, S., and Irie, M. (2003) *J. Am. Chem. Soc.*, **125**, 11080.

170 Irie, M., Lifka, T., Kobatake, S., and Kato, N. (2000) *J. Am. Chem. Soc.*, **122**, 4871.

171 Amimoto, K. and Kawato, T. (2005) *J. Photochem. Photobiol. C-Photochem. Rev.*, **6**, 207.

172 Folling, J., Polyakova, S., Belov, V., van Blaaderen, A., Bossi, M.L., and Hell, S.W. (2008) *Small*, **4**, 134.

173 Kinashi, K., Harada, Y., and Ueda, Y. (2008) *Thin Solid Films*, **516**, 2532.

174 Bucko, A., Zielinska, S., Ortyl, E., Larkowska, M., and Barille, R. (2014) *Opt. Mater.*, **38**, 179.

175 Okada, T., Sohmiya, M., and Ogawa, M. (2015) *Photofunctional Layered Materials*, vol. 166 (eds D. Yan and M. Wei), Springer International Publishing Ag, Cham, p. 177.

176 Wirnsberger, G., Scott, B.J., Chmelka, B.F., and Stucky, G.D. (2000) *Adv. Mater.*, **12**, 1450.

177 Yamano, A. and Kozuka, H. (2011) *Thin Solid Films*, **519**, 1772.

178 Ribot, F., Lafuma, A., Eychenne-Baron, C., and Sanchez, C. (2002) *Adv. Mater.*, **14**, 1496.

179 Evans, R.A., Hanley, T.L., Skidmore, M.A., Davis, T.P., Such, G.K., Yee, L.H., Ball, G.E., and Lewis, D.A. (2005) *Nat. Mater.*, **4**, 249.

180 Such, G., Evans, R.A., Yee, L.H., and Davis, T.P. (2003) *J. Macromol. Sci. Polymer Rev.*, **C43**, 547.

181 Coelho, P.J., Silva, C.J.R., Sousa, C., and Moreira, S.D.F.C. (2013) *J. Mater. Chem. C*, **1**, 5387.

182 Mercier, N. (2013) *Eur. J. Inorg. Chem.*, **2013**, 19.

183 Leblanc, N., Mercier, N., Allain, M., Toma, O., Auban-Senzier, P., and Pasquier, C. (2012) *J. Solid State Chem.*, **195**, 140.

184 Li, P.-X., Wang, M.-S., Zhang, M.-J., Lin, C.-S., Cai, L.-Z., Guo, S.-P., and Guo, G.-C. (2014) *Angew. Chem., Int. Ed.*, **53**, 11529.

185 Coe, B.J. (2006) *Acc. Chem. Res.*, **39**, 383.

186 Schulze, M., Utecht, M., Hebert, A., Rueck-Braun, K., Saalfrank, P., and Tegeder, P. (2015) *J. Phys. Chem. Lett.*, **6**, 505.

187 Gago, S., Fonseca, I.M., and Parola, A.J. (2013) *Microporous Mesoporous Mater.*, **180**, 40.

188 Wu, Y., Qu, X.Z., Huang, L.Y., Qiu, D., Zhang, C.L., Liu, Z.P., Yang, Z.Z., and Feng, L. (2010) *J. Colloid Interface Sci.*, **343**, 155.

189 Allouche, J., Beulze, A.Le., Dupin, J.-C., Ledeuil, J.-B., Blanc, S., and Gonbeau, D. (2010) *J. Mater. Chem.*, **20**, 9370.

190 Vazquez-Mera, N., Roscini, C., Hernando, J., and Ruiz-Molina, D. (2013) *Adv. Opt. Mater.*, **1**, 631.

191 Andersson, N., Alberius, P., Ortegren, J., Lindgren, M., and Bergstrom, L. (2005) *J. Mater. Chem.*, **15**, 3507.

192 Connal, L.A., Franks, G.V., and Qiao, G.G. (2010) *Langmuir*, **26**, 10397.

193 Díaz, S.A., Giordano, L., Jovin, T.M., and Jares-Erijman, E.A. (2012) *Nano Lett.*, **12**, 3537.
194 Carling, C.-J., Boyer, J.-C., and Branda, N.R. (2009) *J. Am. Chem. Soc.*, **131**, 10838.
195 Boyer, J.-C., Carling, C.-J., Gates, B.D., and Branda, N.R. (2010) *J. Am. Chem. Soc.*, **132**, 15766.
196 Carling, C.-J., Boyer, J.-C., and Branda, N.R. (2012) *Org. Biomol. Chem.*, **10**, 6159.
197 Klajn, R., Stoddart, J.F., and Grzybowski, B.A. (2010) *Chem. Soc. Rev.*, **39**, 2203.
198 Nishi, H., Asahi, T., and Kobatake, S. (2009) *J. Phys. Chem. C*, **113**, 17359.
199 Díaz, S.A., Menéndez, G.O., Etchehon, M.H., Giordano, L., Jovin, T.M., and Jares-Erijman, E.A. (2011) *ACS Nano*, **5**, 2795.
200 Ledin, P.A., Russell, M., Geldmeier, J.A., Tkachenko, I.M., Mahmoud, A.M., Shevchenko, V., El-Sayed, M.A., and Tsukruk, V.V. (2015) *ACS Appl. Mater. Interfaces*, **7**, 4902.
201 Snell, K.E., Mevellec, J.-Y., Humbert, B., Lagugne-Labarthet, F., and Ishow, E. (2015) *ACS Appl. Mater. Interfaces*, **7**, 1932.
202 Yamaguchi, H., Matsuda, K., and Irie, M. (2007) *J. Phys. Chem. C*, **111**, 3853.
203 Shiraishi, Y., Tanaka, K., Shirakawa, E., Sugano, Y., Ichikawa, S., Tanaka, S., and Hirai, T. (2013) *Angew. Chem., Int. Ed.*, **52**, 8304.
204 Ouhenia-Ouadahi, K., Yasukuni, R., Yu, P., Laurent, G., Pavageau, C., Grand, J., Guerin, J., Leaustic, A., Felidj, N., Aubard, J., Nakatani, K., and Metivier, R. (2014) *Chem. Commun.*, **50**, 7299.
205 Wigglesworth, T.J. and Branda, N.R. (2005) *Chem. Mat.*, **17**, 5473.
206 Irie, M., Fulcaminato, T., Matsuda, K., and Kobatake, S. (2014) *Chem. Rev.*, **114**, 12174.
207 Sun, J.-K., Wang, P., Yao, Q.-X., Chen, Y.-J., Li, Z.-H., Zhang, Y.-F., Wu, L.-M., and Zhang, J. (2012) *J. Mater. Chem.*, **22**, 12212.
208 Sharker, S.M., Jeong, C.J., Kim, S.M., Lee, J.-E., Jeong, J.H., In, I., Lee, H., and Park, S.Y. (2014) *Chem. Asian J.*, **9**, 2921.
209 Chen, Q., Zhang, D., Zhang, G., Yang, X., Feng, Y., Fan, Q., and Zhu, D. (2010) *Adv. Funct. Mater.*, **20**, 3244.
210 Matsubara, K., Watanabe, M., and Takeoka, Y. (2007) *Angew. Chem., Int. Ed.*, **46**, 1688.
211 Crespo-Monteiro, N., Destouches, N., Epicier, T., Balan, L., Vocanson, F., Lefkir, Y., and Michalon, J.-Y. (2014) *J. Phys. Chem. C*, **118**, 24055.
212 Diop, D.K., Simonot, L., Destouches, N., Abadias, G., Pailloux, F., Guerin, P., and Babonneau, D. (2015) *Adv. Mater. Interfaces*, **2**, 1500134.
213 Nakato, T., Ishida, S., Kaneda, J.-Y., and Mouri, E. (2015) *J. Ceram. Soc. Jpn.*, **123**, 809.
214 Tricot, F., Vocanson, F., Chaussy, D., Beneventi, D., Party, M., and Destouches, N. (2015) *RSC Adv.*, **5**, 84560.
215 Bois, L., Chassagneux, F., Battie, Y., Bessueille, F., Mollet, L., Parola, S., Destouches, N., Toulhoat, N., and Moncoffre, N. (2010) *Langmuir*, **26**, 1199.
216 Miao, L., Ina, Y., Tanemura, S., Jiang, T., Tanemura, M., Kaneko, K., Toh, S., and Mori, Y. (2007) *Surf. Sci.*, **601**, 2792.
217 Tobaldi, D.M., Rozman, N., Leoni, M., Seabra, M.P., Škapin, A.S., Pullar, R.C., and Labrincha, J.A. (2015) *J. Phys. Chem. C*, **119**, 23658.
218 He, T. and Yao, J.N. (2003) *J. Photochem. Photobiol. C*, **4**, 125.
219 He, T. and Yao, J. (2007) *J. Mater. Chem.*, **17**, 4547.
220 Ikake, H., Fukuda, Y., Shimizu, S., Kurita, K., and Yano, S. (2002) *Kobunshi Ronbunshu*, **59**, 608.
221 R. F., deFarias. (2005) *Mater. Chem. Phys.*, **90**, 302.
222 Ingham, B., Chong, S.V., and Tallon, J.L. (2005) *J. Phys. Chem. B*, **109**, 4936.
223 Polleux, J., Pinna, N., Antonietti, M., and Niederberger, M. (2005) *J. Am. Chem. Soc.*, **127**, 15595.
224 Zhao, Z.G. and Miyauchi, M. (2009) *Chem. Commun.*, 2204.
225 Gao, Q., Wang, S., Fang, H., Weng, J., Zhang, Y., Mao, J., and Tang, Y. (2012) *J. Mater. Chem.*, **22**, 4709.
226 Liu, X.X., Bian, L.J., Zhang, L., and Zhang, L.J. (2007) *J. Solid State Electrochem.*, **11**, 1279.
227 Gavrilyuk, A.I. (2010) *Sol. Energy Mater. Sol. Cells*, **94**, 515.

228 Rouhani, M., Foo, Y.L., Hobley, J., Pan, J.S., Subramanian, G.S., Yu, X.J., Rusydi, A., and Gorelik, S. (2013) *Appl. Surf. Sci.*, **273**, 150.

229 Ha, J.H., Muralidharan, P., and Kim, D.K. (2009) *J. Alloy. Compd.*, **475**, 446.

230 He, Y.P. and Zhao, Y.P. (2008) *J. Phys. Chem. C*, **112**, 61.

231 Chen, W., Shen, H., Zhu, X., Xing, Z., and Zhang, S. (2015) *Ceram. Int.*, **41**, 12638.

232 DeJournett, T.J. and Spicer, J.B. (2014) *Sol. Energy Mater. Sol. Cells*, **120**, 102.

233 Zhao, Z.-G., Liu, Z.-F., and Miyauchi, M. (2010) *Adv. Funct. Mater.*, **20**, 4162.

234 Gavrilyuk, A.I. (2009) *Sol. Energy Mater. Sol. Cells*, **93**, 1885.

235 Yang, J.K., Zhang, X.T., Liu, H., Wang, C.H., Liu, S.P., Sun, P.P., Wang, L.L., and Liu, Y.C. (2013) *Catal. Today*, **201**, 195.

236 Zhang, J., He, T., Wang, C., Zhang, X., and Zeng, Y. (2011) *Opt. Laser Technol.*, **43**, 974.

237 Luo, Z.K., Yang, J.J., Cai, H.H., Li, H.Y., Ren, X.Z., Liu, J.H., and Liang, X. (2008) *Thin Solid Films*, **516**, 5541.

238 Yamazaki, S., Ishida, H., Shimizu, D., and Adachi, K. (2015) *ACS Appl. Mater. Interfaces*, **7**, 26326.

239 Poulos, A.S., Constantin, D., Davidson, P., Imperor, M., Pansu, B., Panine, P., Nicole, L., and Sanchez, C. (2008) *Langmuir*, **24**, 6285.

240 Dan-Hardi, M., Serre, C., Frot, T., Rozes, L., Maurin, G., Sanchez, C., and Ferey, G. (2009) *J. Am. Chem. Soc.*, **131**, 10857.

241 Oms, O., Hakouk, K., Dessapt, R., Deniard, P., Jobic, S., Dolbecq, A., Palacin, T., Nadjo, L., Keita, B., Marrot, J., and Mialane, P. (2012) *Chem. Commun.*, **48**, 12103.

242 Hakouk, K., Oms, O., Dolbecq, A., Marrot, J., Saad, A., Mialane, P., Bekkachi, H.El., Jobic, S., Deniard, P., and Dessapt, R. (2014) *J. Mater. Chem. C*, **2**, 1628.

243 Parrot, A., Izzet, G., Chamoreau, L.-M., Proust, A., Oms, O., Dolbecq, A., Hakouk, K., El Bekkachi, H., Deniard, P., Dessapt, R., and Mialane, P. (2013) *Inorg. Chem.*, **52**, 11156.

244 Mialane, P., Zhang, G., Mbomekalle, I.M., Yu, P., Compain, J.-D., Dolbecq, A., Marrot, J., Secheresse, F., Keita, B., and Nadjo, L. (2010) *Chem. Eur. J.*, **16**, 5572.

245 Ferreira-Neto, E.P., Ullah, S., de Carvalho, F.L.S., de Souza, A.L., Oliveira J Jr., M.de., Schneider, J.F., Mascarenhas, Y.P., Jorge, A.M. Jr., and Rodrigues-Filho, U.P. (2015) *Mater. Chem. Phys.*, **153**, 410.

246 Lu, J., Lin, J.-X., Zhao, X.-L., and Cao, R. (2012) *Chem. Commun.*, **48**, 669.

247 Dessapt, R., Gabard, M., Bujoli-Doeuff, M., Deniard, P., and Jobic, S. (2011) *Inorg. Chem.*, **50**, 8790.

248 Coue, V., Dessapt, R., Bujoli-Doeuff, M., Evain, M., and Jobic, S. (2007) *Inorg. Chem.*, **46**, 2824.

249 Zhang, Z.-J., Xiang, S.-C., Guo, G.-C., Xu, G., Wang, M.-S., Zou, J.-P., Guo, S.-P., and Huang, J.-S. (2008) *Angew. Chem., Int. Ed.*, **47**, 4149.

250 Casan-Pastor, N. and Gomez-Romero, P. (2004) *Front. Biosci.*, **9**, 1759.

251 Horcajada, P., Serre, C., Grosso, D., Boissiere, C., Perruchas, S., Sanchez, C., and Ferey, G. (2009) *Adv. Mater.*, **21**, 1931.

252 Luo, X. and Yang, C. (2011) *Phys. Chem. Chem. Phys.*, **13**, 7892.

253 de Oliveira, M. Jr., Lopes de Souza, A., Schneider, J., and Rodrigues-Filho, U.P. (2011) *Chem. Mater.*, **23**, 953.

254 Qi, W., Li, H., and Wu, L. (2008) *J. Phys. Chem. B*, **112**, 8257.

255 Huang, Y., Pan, Q.Y., Dong, X.W., and Cheng, Z.X. (2006) *Mater. Chem. Phys.*, **97**, 431.

256 Pinto, T.V., Fernandes, D.M., Pereira, C., Guedes, A., Blanco, G., Pintado, J.M., Pereira, M.F.R., and Freire, C. (2015) *Dalton Trans.*, **44**, 4582.

257 Molina, E.F., Marcal, L., de Carvalho, H.W.P., Nassar, E.J., and Ciuffi, K.J. (2013) *Polym. Chem.*, **4**, 1575.

12
Optical Properties of Hybrid Organic–inorganic Materials and their Applications – Part II: Nonlinear Optics and Plasmonics

Stephane Parola,[1] Beatriz Julián-López,[2] Luís D. Carlos,[3,4] and Clément Sanchez[5]

[1]Université de Lyon, Ecole Normale Supérieure de Lyon, Laboratoire de Chimie ENS Lyon, Université Lyon 1, CNRS UMR 5182, 46 allée d'Italie, 69364 Lyon, France
[2]University Jaume I, Institute of Advanced Materials – INAM, 12071 Castellón de La Plana, Spain
[3]University of Aveiro, Department of Physics, P-3810193 Aveiro, Portugal
[4]University of Aveiro, CICECO – Aveiro Institute of Materials, P-3810193 Aveiro, Portugal
[5]Collège de France, UPMC Univ Paris 06, Laboratoire Chimie de la Matière Condensée Paris, UMR 7574, F-75005 Paris, France

12.1
Hybrid Materials for Nonlinear Optics

12.1.1
Introduction to Nonlinear Optics

Nonlinear optics phenomena appear when light meets nonlinear media giving rise to a nonlinear response of the dielectric polarization versus the incident electric field. Even though the phenomena were observed before the discovery of lasers [1–4], they occur essentially at high light intensity. Thus, the strong development of nonlinear optics is directly related to the development of lasers in the 1960s [5]. The polarizability represents the ability of the charges in a material to be displaced by an electric field (E). When the electric field is lower than the internal fields, the overall macroscopic polarization P can be expressed as a function of the electrical susceptibility (χ):

$$P = P_0 + \chi^{(1)}E. \tag{12.1}$$

In Eq. (12.1), P_0 is the permanent polarization; P is the total macroscopic polarization; $\chi^{(1)}$ is the linear susceptibility of the first order, and $\chi^{(1)}E$ is the term for the induced linear polarization. However, when increasing the intensity of the

Handbook of Solid State Chemistry, First Edition. Edited by Richard Dronskowski, Shinichi Kikkawa, and Andreas Stein.
© 2017 Wiley-VCH Verlag GmbH & Co. KGaA. Published 2017 by Wiley-VCH Verlag GmbH & Co. KGaA.

light such as in the case of lasers, the induced polarization is not linear any more. It can be expressed as in Eq. (12.2):

$$P = P_0 + \chi^{(1)}E + \chi^{(2)}E^2 + \chi^{(3)}E^3 + \cdots. \tag{12.2}$$

In Eq. (12.2), P is the total induced macroscopic polarization, $\chi^{(1)}$ is the linear polarizability, $\chi^{(2)}$ and $\chi^{(3)}$ are the nonlinear susceptibility tensors that correspond respectively to the hyperpolarizabilities for the quadratic and cubic terms (β and γ) at the molecular level.

The response to an external field E for the induced molecule is shown by Eq. (12.3):

$$\mu = \mu_0 + \alpha E + \beta E^2 + \gamma E^3 + \cdots, \tag{12.3}$$

where μ is the total dipole moment, $\mu_{(0)}$ is the permanent dipole moment, α is the molecular polarizability, and β and γ are the second- and third-order hyperpolarizabilities.

For large field intensities with a frequency ω, $\chi^{(2)}$ is responsible for the generation of an oscillating field with double frequency 2ω. It can be used for instance for optical storage, electrooptical modulation, optical switchers, and wavelength conversion. This phenomenon is only observed in molecules or materials possessing a noncentrosymmetrical structure, which represents a specific requirement in the design of materials. The $\chi^{(3)}$ is the term that generates the nonlinear absorption activity or multiphoton absorption (MPA). It can be used for instance in optical limiting systems (laser protection), two-photon imaging, 3D-data storage, and microfabrication [6]. In this case, there is no symmetry requirement. The nonlinear absorption is proportional to the imaginary part of the third-order susceptibility. In the same way, the refractive index is proportional to the real part of the third-order susceptibility. In this chapter, the most recent developments are discussed in terms of hybrid materials design, properties, and applications for second- and third-order nonlinearities.

12.1.2
Second-Order Nonlinear Materials

12.1.2.1 Dye-Doped Inorganic Matrices

The second harmonic generation of dye-doped hybrid materials, as well as for pure molecular systems, is directly dependent on the orientation of the molecular dipoles. Most common matrices for dye dispersion are polymers or inorganic sol–gel networks, since they do not require harsh treatment for their final stabilization. In all cases, the poling process, which induces the overall noncentrosymmetry, and the final material structure stability, which prevents molecular relaxation and loss of noncentrosymmetry are critical. The orientation can be achieved for instance through common corona poling [7–10] or optical poling [11–13]. Previous reviews emphasized the potential strong impact of the interface between the dyes and the matrix on the nonlinear optic (NLO) response and the stability of the material [14,15]. In the sol–gel process, typically two approaches are commonly proposed, simple dispersion with low-energy coupling (Class I) or covalent bonding with strong coupling (Class II), the latter being often mentioned as the most stable one. Since the very precursor works in the early 1990s when efficient doping and orientation of dyes in inorganic matrices was demonstrated [16], a large number of reports explored the capacity of sol–gel matrices to stabilize the orientation of chromophores. Large second-order optical nonlinearities were observed using covalent grafting of the dyes to the silica backbone [17,18]. The optical response was evidently correlated to the matrix condensation level and rigidity [19]. Since then numerous work reported on the stabilization of the hybrid system and optimization of the optical and nonlinear optical responses [14,16–33]. It is now well established that the strong binding of the molecular guest to the hosting matrix induces an improvement of the stability of the final nonlinear optical material in particular after the orientation process as it was already reported [14,15].

Comparisons between polymer matrices (often polymethylmethacrylate PMMA) and sol–gel matrices evaluated for both second- and third-order nonlinearities are often in favor of the siloxane chains that usually show a higher stability combined with good optical properties [34–37]. The grafting on a sol–gel matrix can easily be achieved using silylated chromophores, and numerous work reported the influence of the grafting and the matrix composition on the nonlinear optical performance. Even though a large amount of important work has been reported since the early 90 s, there is still a tendency to improve the properties and evaluate the process parameters and the impact on the NLO, without any real breakthrough in terms of second-harmonic generation (SHG) values. For instance Qian and coworkers investigated the thermal behavior and possibility to enhance the second-harmonic generation and improve the stability of the optical properties by using covalent grafting to the matrix [34,37–48]. Several types of chromophores bearing silane groups were developed (Figure 12.1). Interestingly, the studies showed that thermal stability of the nonlinear optical properties was improved by combining covalent bonding of the dyes and hydrogen bonding or by playing with the sterical hindrance, which impact the temporal stability of the oriented dipole [39,40,44]. Such interactions were for example obtained by incorporation of aniline–silane in the matrix [44]. Moreover, it was shown that the topological localization of the silane groups along the structure and the type of spacer used impacts the nonlinear optical properties [48]. This can be anticipated since it probably influences the mobility of the dyes, as well as the molecular structure and conformation (and consequently the electron transfers through the structure) and interchromophores interactions.

Another important parameter to consider is the heating conditions during the whole process (precuring and poling). It was shown that the precuring and poling temperature considerably impacts the nonlinear optical responses due to modification of the cross-linking [26,28,49]. The precuring temperature should not be too high to prevent from densification of the network. Such excessive cross-linking would decrease the mobility of the dyes and consequently lower

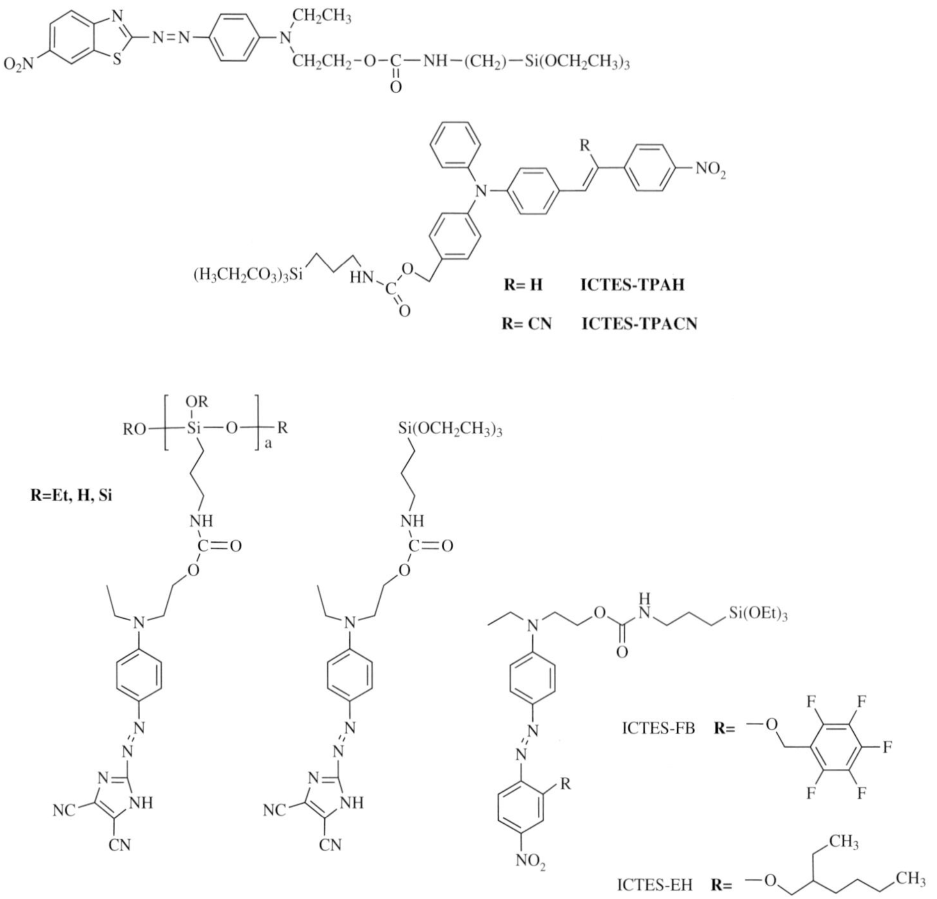

Figure 12.1 Structures of sol–gel NLO dyes. (Adapted with permission from Ref. [40], copyright 2007 American Chemical Society; and from Refs. [34,38–48].)

the poling efficiency. This depends also on the rotational capacity of the dyes. For instance a two-dimensional alkoxysilane chromophore based on a carbazole core incorporated in a tetramethoxysilane-based matrix was compared to the well-known alkoxysilane-modified DR19. A two-dimensional system exhibited improved thermal stability compared to the one-dimensional one [49,50]. The number of silane groups per chromophore units must also be considered since the molecular structure can be impacted by the rigidity of the overall system. For instance chromophores with two or more silane groups were prepared for nonlinear optics and showed impact on the final properties either for second- or third-order nonlinearities [51,52]. These materials generally exhibited excellent stability, in particular thermal stability. The grafting was achieved through photo-cross-linking [53] or polymerization reaction, for instance free-radical polymerization in an interpenetrating methacrylate/silica hybrid matrix [54]. In

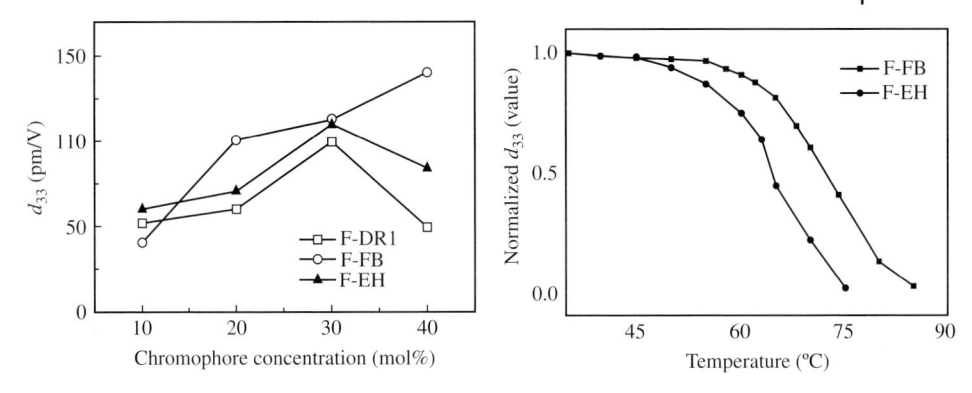

Figure 12.2 SHG and thermal stability of DR1 hybrid compared to fluorinated species (FB and EH in Figure 12.1). (Adapted with permission from Ref. [47].)

this case, a chromophore bearing allyl glycidyl ether was reacted with (γ-methacrylpropyl)-silsesquioxane. This strategy allowed improvement of the cross-linking, higher T_g for the final material and better stability. The poling is achieved at 200 °C. The highest T_g, d_{33}, and stability are obtained for a film prepared with 15 wt.% polyhedral oligomeric silsesquioxane (POSS) [54]. In all cases, particular attention should be given to the structure of the dyes considering that each system (molecule structure + matrix structure) is different. For instance, it was shown that the use of fluorinated groups on dyes induced a better nonlinear optical activity than the basic chromophore (DR1 in this example, Figure 12.2) [47]. The thermal stability was also improved using bulky fluorinated substituting groups.

Precise control of the heating conditions seems to be crucial but needs to be optimized for each molecular structure. This has to be considered together with the cross-linking process parameters (hydrolysis–condensation conditions), which will impact the densification, orientation capacity, structure of the guest molecule and final stability, which was also shown to strongly impact NLO properties [36,51,55–57].

Besides using corona or optical poling, the orientation can also be achieved through the control of the deposition process. For instance, a layer-by-layer approach combined with the sol–gel process allows easy access to self-oriented materials [58,59]. In this case, the orientation can be induced by the strong repulsive forces between the negatively charged ZrO_2 and the sulfonate groups of the chromophore (Figure 12.3) [58]. Such an approach can be extended to a wide range of optical materials and more generally light-activated systems combining for instance nanoparticles, dyes, and polyoxometallates with polyelectrolytes [60].

Another way of controlling the orientation is to intercalate the chromophore in the interlayer space of layered inorganic materials. This strategy was used in the early 1990s by Lacroix *et al.* who managed to intercalate

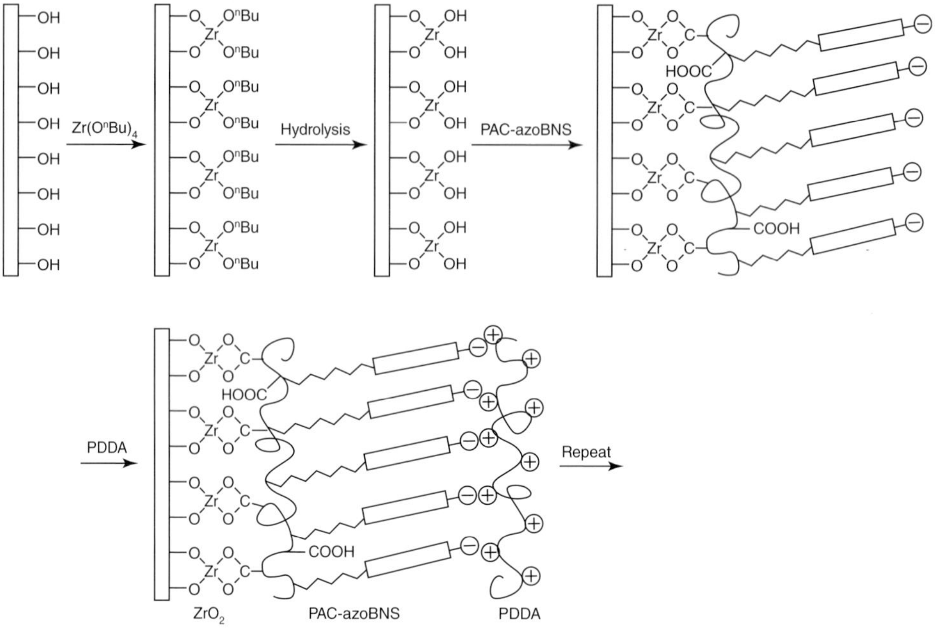

Figure 12.3 Self-orientation of azobenzene chromophore using a layer-by-layer deposition. (Adapted with permission from Ref. [58], copyright 2007 American Chemical Society.)

stilbazolium dyes with spontaneous poling into manganese-based inorganic-layered material (MPS$_3$) rising to efficient second-harmonic generation combined with permanent magnetization below 40 K due to the presence of Mn^{2+} ions [61]. Kawamata and coworkers reviewed the strategies based on incorporation in clay minerals [62]. Using this approach, very regular orientation of the dyes can be obtained and thus enhancement of the optical nonlinearities. For instance, Kuroda *et al.* reported the inclusion of 4-nitroaniline into kaolinite (Al$_2$Si$_2$O$_5$(OH)$_4$) interlayers [63]. The 4-nitroaniline spontaneously self-orient during intercalation thanks to hydrogen bonding with the silicate nanosheets present on one side of the interlayers (Figure 12.4), thus giving rise to a SHG signal.

Finally the use of crystals of dyes grown in an inorganic composite layer can be an alternative to the dispersion of molecular entities in sol–gel matrices. Ibanez and coworkers advantageously developed this approach in particular for third-order nonlinearities [64–67]. However, the poling process is reported to be hard to achieve due to the ionic conductivity at high temperature and also disorientation during the cooling down [67]. On the other hand, the use of noncentrosymmetric crystal organization can be a way to achieve strong SHG at the molecular level. This has been recently developed through the use of the wide possibilities offered by the coordination chemistry and supramolecular hybrid approaches and is developed in the next part.

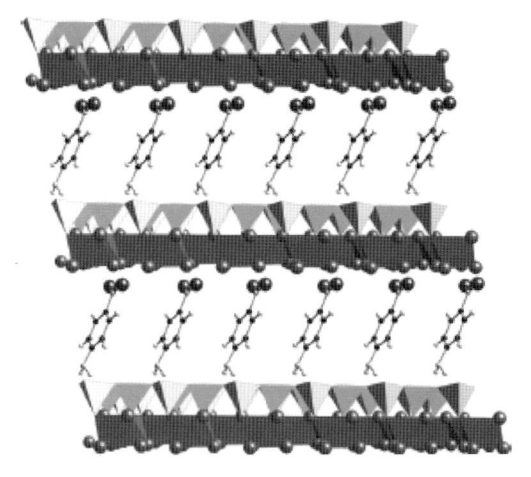

Figure 12.4 Oriented intercalation of 4-nitroaniline into kaolinite interlayers by Kuroda and coworkers [62,63].

12.1.2.2 Coordination and Organometallic Compounds Based Hybrid Systems

Coordination chemistry offers a large range of possible molecular and supramolecular architectures by association of metallic centers and organic ligands. The introduction of metal centers has shown interesting impact on the NLO response of coordinating organic dyes in a push–pull configuration [68]. The charge transfer is usually less efficient when introducing a metallic center between the push and pull systems. In the case of 3D self-assembled materials, typically crystalline materials, the noncentrosymmetry can be achieved by controlling the organization of the complexes in the crystal structure during the crystallization process [68,69]. A high degree of organization and thus SHG response can be expected in such molecular arrangement. During the past decade, numerous works were devoted to the use of such molecular-based self-assembly in crystals for preparation of hybrid materials for NLO applications. Metal–organic frameworks (MOFs) or polyoxometalates (POMs) were particularly investigated for nonlinear optical applications with different proposed mechanism for the optical nonlinearities [70,71].

Crystallization in noncentrosymmetric space groups being hard to achieve, a possible way of controlling the noncentrosymmetry during the crystallization process is to use chiral ligands. For instance, Dolbecq and coworkers used a chiral stilbazolium cation (CHIDAMS$^+$) that was crystallized with $PW_{12}O_{40}^{3-}$ counterions [72]. In this system, a charge transfer between the electron-donor organic part and the electron-acceptor inorganic counterpart, the POMs, was observed. The presence of three independent organic molecules on one POM prevents the formation of head-to-tail dimeric units, which would reduce the noncentrosymmetry and thus the NLO activity. This complex exhibited high SHG response, which was evaluated to be 30 times higher than the one measured in a KDP crystal. Theoretical investigations on charge transfer and NLO response of Lindqvist-type POMs (hexamolybdate/molybdate anions) functionalized with side chains organic ligands showed that the charge transfer from molybdate central core to the ligands has a central role in the enhancement of the SHG and significant values of second-order polarizability can be obtained in such configuration [73,74].

In the same way, metal polycarboxylates [69,75–78] based crystals have shown promising potential in terms of second-harmonic generation. Cariati and coworkers reported several complexes using *trans*-4-(4-dimethylaminostyryl)-1-methylpyridinium (DAMS$^+$) cationic ligands with general formula [DAMS]$_4$[M$_2$M'(C$_2$O$_4$)$_6$]·2DAMBA·2H$_2$O (where M = Rh, Fe, Cr; M' = Mn, Zn) giving rise to noncentrosymmetric layered structures and thus strong SHG [77]. In this case, the very large SHG is mostly explained by the important quadratic polarizability of the ligands DAMS$^+$ (much more important than the SHG from the metal-to-ligand charge-transfers contribution) combined with their self-organization in a layered structure with an optimized J-type aggregate configuration (Figure 12.5). Considering their high potential in terms of SHG and since such molecular self-assembled crystalline materials can easily be incorporated in composite materials (inclusion in polymers or sol–gel matrices, layer-by-layer assembly on surfaces . . .) [79,80], it can certainly be expected some future development of systems for applications in nonlinear optics.

12.1.3
Third-Order Nonlinear Materials

12.1.3.1 Dispersion of Dyes in Sol–Gel or Organic Materials

The general design of hybrid organic–inorganic materials, which exhibit third-order nonlinearities, is often very similar to the materials for SHG with

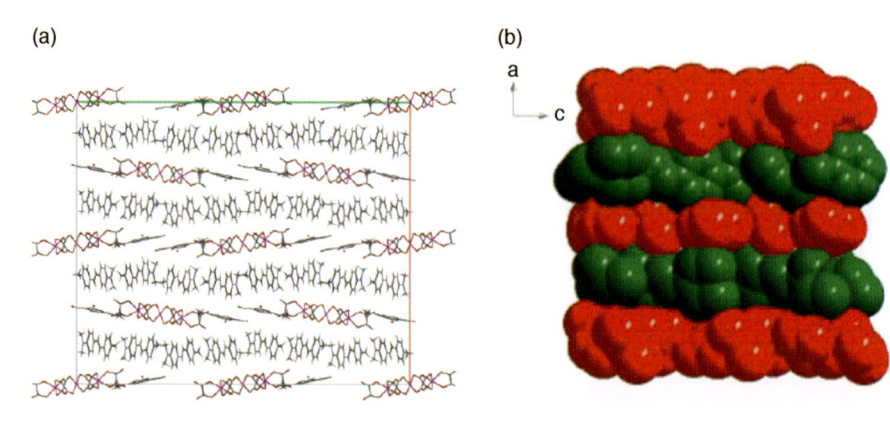

Figure 12.5 General structure showing the packing of alternate anionic organic/inorganic and cationic organic layers in the complexes [DAMS]$_4$[M$_2$M'(C$_2$O$_4$)$_6$]·2DAMBA·2H$_2$O, along the c-axis (a) and b-axis (b). (Adapted with permission from Refs. [77,78], copyright 2007, 2010 American Chemical Society.)

association of optical dyes with nonlinear activity to inorganic or hybrid matrices. However, one important difference is the symmetry rules. In the case of third-harmonic generation (THG), no symmetry rule is required. The preparation of the materials can thus focus on the interface control, concentration optimization and structure preservation of the dyes during the process. The main and almost unique application to this family of materials concerns the nonlinear absorption phenomena and the capacity of the systems to spontaneously block incident intense light radiation. It means that these materials find important use as filters to protect all types of optical sensors, sights, or cameras against lasers damage. Such materials are called optical power limiting (OPL) materials [81,82]. The two main mechanisms are the reverse saturable absorption (RSA) and the two-photon absorption (TPA) or multi-photon absorption (MPA) [6,56,83–87]. The reverse saturable absorption (RSA) appears for molecular systems possessing an intersystem crossing (ISC) with a lifetime of the triplet excited state (T_1), at least comparable or longer than the duration of the laser pulse, and an absorption coefficient of the T_1 compared to that of the ground state and can thus provide an efficient limiting of pulsed laser irradiance [88,89].

The molecular design for third-order nonlinear applications and in particular OPL (optimization of two-photon cross section through the molecular structure optimization . . .), using either RSA or MPA mechanisms, has been intensively reported and optical measurements are mostly reported on molecules in solution [6]. However, the exploitation of such nonlinear phenomena in real applications requires that the optically active molecules are introduced in a solid optical host material that enables post-processing such as cutting, polishing, and gluing. Solid-state materials are easily incorporated in optical devices and propose several advantages compared to liquid cells:

- Shape design for easy optical device construction
- Higher chemical, physical, and mechanical stability
- Eco-friendly devices (no solvent)

The preparation of solid materials performing efficient nonlinear absorption appears to be the most appropriate route toward an efficient protection in any environmental conditions. One should also mention the risk that a liquid container is susceptible to breakage, during a laser attack or following a harsh treatment, leading to leakage of hazardous liquid material. The main drawback of solid materials is that the laser damage threshold (an important parameter to consider) is generally lower compared to liquids. This is explained by the heat diffusion mechanisms and molecules mobility, which are much higher in liquids than in solids. The liquid samples can efficiently dissipate the heat through the evaporation and formation of bubbles. The typical requirements for a material suitable for efficient optical protection using nonlinear absorption mechanisms were previously reported and discussed [90–92]. The material has to be processed preferentially as monolith for optical protection or in some cases can be used as thin films. Hybrid materials appear to be an interesting alternative to provide efficient nonlinear absorption, optical quality, processability, and stability. Thus dispersion of chromophores or nanostructures in optically transparent matrices (sol–gel, polymers) such as presented for second-order nonlinear materials is a consistent approach to efficient devices. The requirements for these third-order nonlinear materials are (i) to preserve the molecular structure of the dyes in the final solid in order to optimize the final nonlinear properties, (ii) to control the concentration as high as possible with homogeneous dispersion preventing from molecule-to-molecule interactions, (iii) to keep the linear transmission at maximum value, (iv) the capacity to achieve codoping with several molecules and/or nanostructures to ensure broad band covering of the final system. It is important to note that no symmetry and orientation requirement is necessary at this stage in contrary to the second-order systems. The control of organization and orientation might affect the nonlinear response but will not be discussed here due to lack of reports and only proposed as future possible implementation of on-going technologies. The ideal matrix possesses a high optical quality and a high damage threshold under the laser exposure, and a good compatibility with the chromophores. The main possible matrices are polymers or silica-based materials. Polymers are interesting because they can exhibit high compatibility with the guest chromophores, good optical quality, and are easy to prepare. Inorganic materials, such as silica-based systems used in optical glasses, are also promising candidates because of high optical quality and high damage thresholds. Even if few articles report the use of thin films [93], the preparation of monolithic materials represents the best potential approach since they can be designed and shaped conveniently to be adapted to the final OPL setup. Moreover, such 3D monolithic materials present higher loading capacity in chromophores than films and a better focusing efficiency of the beam.

The first reports on solid-state materials for nonlinear absorption applications were launched in the early 1990s. The first systems were not optimized neither on the active molecular entities nor on the interface with the surrounding matrices. The dispersion of metallophtalocyanine dyes with interesting RSA into SiO_2 sol–gel matrices was first reported [24,94]. The reverse saturable absorption of the dyes was also observed in the solid materials demonstrating the capacity of the strategy. However, the spectroscopy showed that the overall systems needed optimization. The presence of dimers showed that the dyes were not totally homogeneously dispersed, and protonated species that showed that the process should be adapted to the dyes sensitivity (Control of pH . . .) [94]. The final nonlinear properties are comparable to the one in buffer solution. In the same period, investigations on the use of polymeric matrices were reported [84,85,95]. Although those precursor works were launching the research area, the important developments and breakthrough on materials for nonlinear absorption started from the late 1990s, and several important works appeared during the past 15 years. Innocenzi and coworkers reported the entrapment of functionalized fullerenes in inorganic matrices for optical limiting applications [15,96–98]. Fullerenes were selected because they present interesting RSA in the visible wavelengths [97–100]. One drawback with these systems is their poor solubility and their poor compatibility with the silica-based matrix. Improved compatibility was achieved by adding silane or dendrimer pending units on the fullerenes structures [97–99]. In most of these materials, the nonlinear performances of the chromophores in solution were almost totally recovered in the solid. The nonlinear response of fullerenes was similar in solution and in the solid state, exhibiting RSA mechanisms with nonlinear scattering and nonlinear refraction contributions at high fluence [15]. The main problem regarding the final application was that the clamping levels in these materials were still too high due to the low concentration of chromophores. Moreover, these systems absorb strongly in the visible.

Another original strategy adopted in order to increase the concentration of nonlinear active species and try to reach the requirements was to grow nanocrystals of dyes in inorganic glasses [64–67]. Stilbene 3 nanocrystals were grown in sol–gel glasses [101]. Interestingly, the final composite materials showed strong nonlinear absorption at visible wavelengths with a three-photon absorption (two-photon absorption mechanism followed by an excited-state absorption) [66]. Impregnation of a previously prepared porous matrix was also used for aggregation of dyes (naphthalocyanine for instance) into the cavities of the xerogel, which exhibited then interesting RSA response and optical limiting properties [102].

Another family of dyes, based on a platinum central atom and acetylides type ligands, was widely investigated due to its advantage of presenting very low absorption in the visible wavelengths and strong nonlinear behavior with intersystem crossing (ISC) and triplet excited-state absorption. The heavy atom facilitates ISC for instance in the platinum(II) square-planar complex $trans$-$[P(n$-$Bu)_3]_2Pt[(C\equiv C$-p-C_6H_4-$C\equiv C$-p-$C_6H_5]_2$ [83,103,104]. This complex was reported for its extremely interesting nonlinear absorption properties in solution [83,103–105] and series of platinum-based chromophores was prepared with improved nonlinear properties [51,106–115]. The use of PMMA monolithic material hosting platinum acetylide derivatives was reported to be an efficient route toward OPL materials [35,116]. The chromophores were either simply dispersed in the matrix (host–guest system) or covalently bonded to the PMMA network using methacrylate-substituted dyes. The measured linear transmissions of the PMMA materials were high, nearly 90% in the visible for wavelengths >550 nm. Nonlinear absorption properties of the chromophores were maintained in the solid and clamping levels in the range 3–4 μJ were reached (incident intensity 115 μJ). Interestingly the type of interaction (covalent bond or dispersion) between the chromophore and the matrix was observed to play a role in the final optical performance. The structural rigidity of the grafted dyes limited possibilities of structural changes and relaxation compared to host–guest systems, leading to a loss of OPL efficiency. Moreover, the damage threshold was impacted and decreased when using grafting processes. It is important to note that in contrast to what was observed for second-order NLO, grafting of the dyes affected the NLO efficiency of the materials negatively in this case. Similar investigations were carried out in parallel on sol–gel hybrid hosting matrices [36,55]. Functional chromophores were designed in order to improve the compatibility with the silica-based matrices and thus the optimum concentration of dyes in the inorganic phase [36,51]. The grafted and dispersed systems were compared. In the case of grafted systems, it was demonstrated that the hydrolysis step procedure could be used to control the way the dyes were dispersed (aggregated in the case of separate hydrolysis or homogeneously dispersed in the case of simultaneous *in-situ* hydrolysis). The material structure and properties could be optimized using silane functionalized dyes precursors for improving the concentration and *in situ* hydrolysis for homogeneous dispersion (Figure 12.6) [36]. This material exhibited efficient nonlinear absorption in the visible wavelengths and was effectively used to protect cameras from lasers.

The pH control during the process was also reported to be a critical parameter. Many organic dyes are known to be pH sensitive. In this sense, it becomes often important to have a fine control over the pH during the addition of the dyes to the precursors. The use of a sonogel was mentioned by Morales-Saavedra *et al.* as an alternative to avoid the use of acidic or basic catalysts and thus preserve the integrity of the dyes in particular for THG [117,118]. In the case of the use of catalysts, neutralization of the sol after hydrolysis allowed easy insertion of sensitive dyes into the silica system [36,55,57]. In connection with the pH control, the loading capacity of the glass with the dyes (extremely important in the case of optical applications) was efficiently tuned by controlling the condensation step kinetics without any need for covalent grafting (which requires often complicated chemistry on the dyes) [55,57]. A stable neutral sol was mixed with the dyes. Provided that the condensation ratio of the sol was sufficiently high to remain liquid and have only few remaining -OH groups to condense (typically in

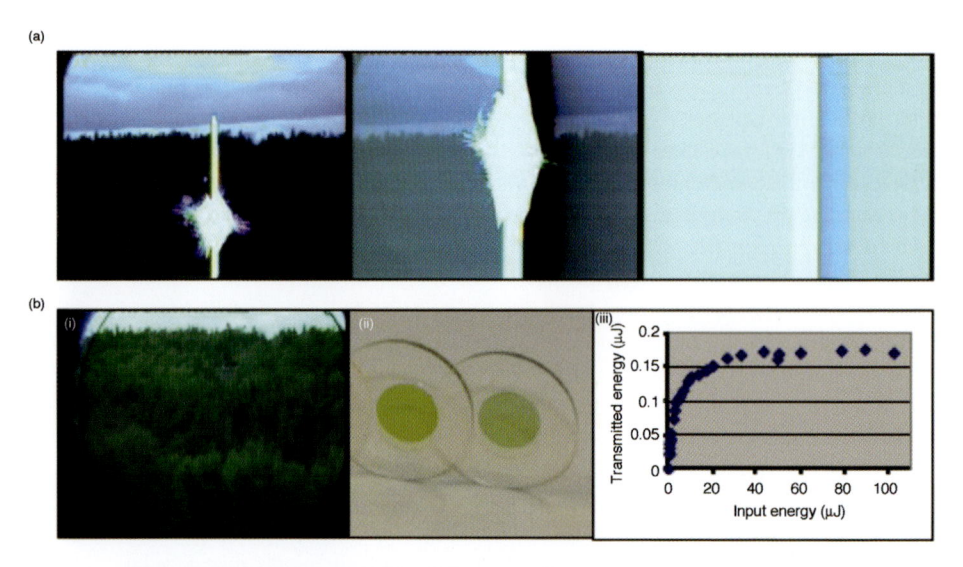

Figure 12.6 (a) Visualization of the destruction of a camera under laser irradiation. (b) similar view from the same camera protected by highly transparent hybrid materials (i), the filters (ii) and nonlinear absorption response of the filter at 532 nm in the nanosecond regime allowing the efficient protection (iii) [36].

the range 70–90%), the liquid to solid transition during the condensation was efficiently achieved within a very short time (which can occur from a few seconds to several minutes depending on the conditions) by adding a small amount of base (3-aminopropyltriethoxysilane (APTES)). The strong advantage of the method is that the concentration of dyes can be extremely high (typically 40–50%) without any aggregation and high optical quality of the final materials. The drawback in the case of grafted systems in terms of optical response is that the structure of the inserted dyes can be affected by the strong interaction with the matrix and mechanical stresses. Thus, this approach, which allows high loading without grafting can be a real alternative for third-order nonlinear optics (obviously extendable to other optical systems). It was also efficiently used recently for preparing the first materials activated in the near-infrared wavelengths [57]. The study evidenced the important role of the microstructure of the matrix on the final properties. Optimization of the porosity allowed control of oxygen diffusion and thus optimization of the triplet excited states lifetime, the intersystem crossing and consequently the nonlinear absorption. Figure 12.7

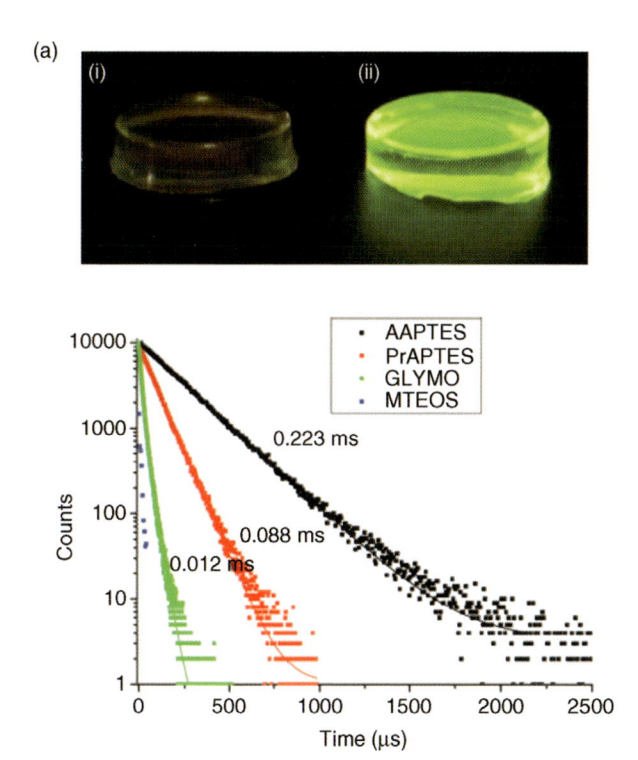

Figure 12.7 Impact of the porosity on the optical properties of platinum acetylides doped hybrid silica xerogels. (a) Irradiation under UV of hybrid materials prepared with MTEOS (i) and MTEOS/GLYMO (ii) precursors. (b) Lifetime changes versus the matrix composition. (Adapted with permission from reference [55], copyright 2012 American Chemical Society.)

shows two observable consequences, on the phosphorescence emission (quenched by oxygen diffusion in pure methyltriethoxysilane (MTEOS) matrix) and the lifetime, which was strongly improved by using functional precursors reducing the microporosity of the system.

A comparison between polymer and the sol–gel approach in this specific case showed few advantages for the sol–gel method. The T_g of the polymer was a limitation for use in harsh environments. The nonlinear response control appeared to be easier in the inorganic matrix. Finally the damage thresholds under intense laser irradiation are often higher in inorganic than in organic systems. Comparison between PMMA and sol–gel silica doped with Rhodamine 6G gave the same conclusions [119]. A silica host matrix appeared to be more efficient than PMMA and the nonlinear parameters (n_2, β, $Re[\chi^{(3)}]$, $Im[\chi^{(3)}]$) were better in the hybrid system.

12.1.3.2 Polysilsesquioxanes Hybrids for Nonlinear Absorption

Silsesquioxanes (POSS) were functionalized with optically active dyes and can be considered as intermediates between the molecular level and a nanocomposite material. Su and coworkers reported the synthesis of POSS modified with azobenzene, acetylene, or stilbene derivatives exhibiting nonlinear absorption [120–122]. These systems exhibited large $\chi^{(3)}$ susceptibility and efficient reverse saturable absorption with enhanced thermal stability. They were processed as thin films. Carbon nanotubes, well known for their optical limiting response due to nonlinear scattering of *in-situ* formed microbubbles [123,124], were also associated to POSS units (covalent or ionic bonding) in order to improve their solubility and optical properties (Figure 12.8) [125,126]. The prepared nanocomposites with multiwall carbon nanotubes (MWCNTs) and POSS showed much better stability in suspension and a better processability. These hybrids exhibited efficient optical limiting performances for nanosecond pulses at 532 nm, attributed to the nonlinear scattering thermally induced [125,126]. It also revealed a good photostability under the laser irradiation. However, even if the hybrid composites materials based on polysilsesquioxane bricks appeared to improve noticeably the nonlinear optical responses and stabilities, the processability in films or monoliths and their optical characterizations have not been reported so far.

12.1.3.3 Graphene Based Hybrid Materials for Nonlinear Absorption

Graphene was found to present efficient nonlinear absorption with consequently a strong potential for optical protection applications [127–129]. Graphene oxide (GO) suspensions in DMF exhibited both two-photon absorption in the picosecond regime and excited state absorption in the nanosecond regime and differed from the nonlinear absorption mechanisms observed in carbon nanotubes based essentially on nonlinear scattering [128]. The graphene dispersion in DMA showed important nonlinear scattering in the nanosecond regime, due to the formation of solvent bubbles, and broadband optical limiting (532 and 1064 nm) [129]. Also, association between dyes and graphene was successfully

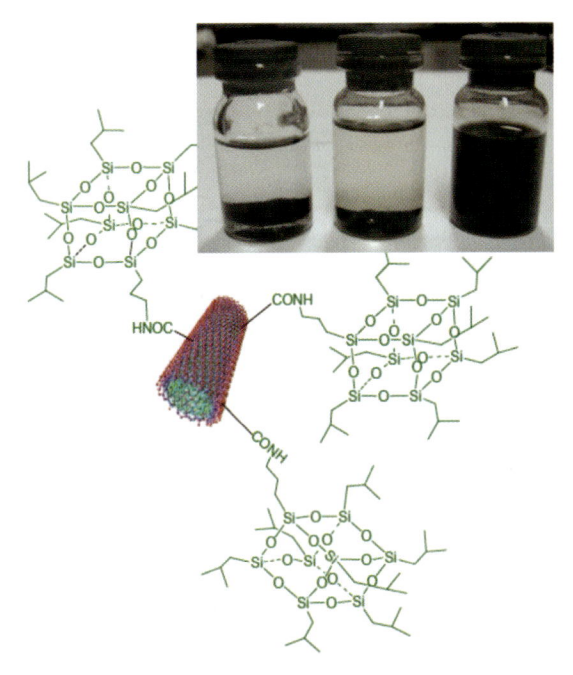

Figure 12.8 POSS functionalized multiwall carbon nanotubes and photo showing the suspension stability improvement of the MWCNTs/POSS (right) compared to MWCNTs (left) and MWCNTs-COOH (middle). (Adapted from Ref. [125].)

attempted in order to improve the nonlinear absorption properties, both experimentally and theoretically [130,131]. Consequently, numerous works recently reported the potential use of graphene-based materials for nonlinear absorption. In the case of hybrid nanomaterials based on the association between graphene oxide bearing –COOH pending groups and iron oxide Fe_3O_4, enhancement of optical limiting response due to both contributions of nonlinear absorption (TPA) and nonlinear scattering was observed [132]. Bulk materials were prepared by dispersion or codispersion of graphene derivatives in either organic or inorganic matrices. Zhan and coworkers reported the efficient synthesis of graphene embedded in sol–gel ormosil glasses [133,134]. They used both nanosheets (GONSs) and nanoribbons (GONRs) (Figure 12.9). The prepared materials showed high linear transmission in the visible, good thermostability, and good optical limiting response attributed to both nonlinear scattering and nonlinear absorption. Similarly, Xu and coworkers reported the dispersion of GO in sol–gel matrices using modified GO for covalent grafting to the silica network [135]. The GO–COOH was modified using 3-aminopropyltriethoxysilane (APTES) providing -CO-NH- bond between the COOH and the amino group of APTES. The GO–APTES was then cohydrolyzed and cocondensed with the silicon precursors (TEOS). The functionalization of GO with APTES and covalent bonding to the silica network allowed better dispersion and homogeneity of the

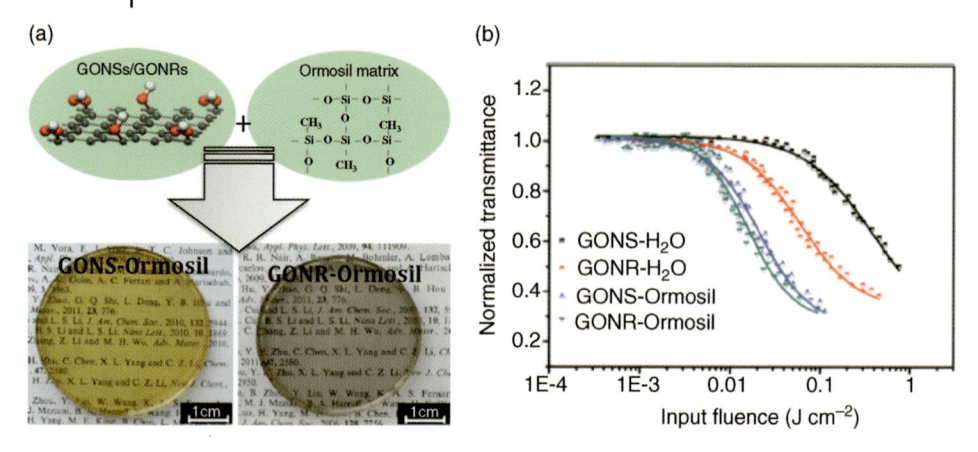

Figure 12.9 The GONS–Ormosil and GONR–Ormosil hybrid materials (a) and normalized transmittance versus the input fluence (b) [133].

prepared sol–gel materials while GO tend to aggregate when simply dispersed in the matrix.

The NLO properties were evaluated in the solid and attributed to a TPA process since no bubble can be formed in the solid environment in contrast to suspensions in solvents. The performances were improved in the solid compared to the solution. This was explained by the difference in mobility between the liquid and the solid phases. In the liquid, thermal effects under the laser beam can induce an ejection of GO out of the beam due to a temperature gradient and diffusion. The consequence was a decrease in the GO concentration at the focal point followed by a decrease in nonlinear absorption efficiency. Such a phenomenon could not occur in the solid, preserving the local concentration in GO and thus the nonlinear efficiency. Similarly, composite materials with graphene and metal nanoparticles in sol–gel glasses were also recently proposed [136]. Hybrid composites with graphene and ZnO or CdS nanoparticles dispersed in PMMA glasses were proposed by Chen and coworkers [137,138]. They showed that the interfacial charge transfer between ZnO and graphene induced enhanced nonlinear absorption and nonlinear scattering properties. Graphene-based hybrid materials appear clearly as extremely promising composites for THG applications both in the nanosecond and the picosecond regimes. A combination of graphene with other type of dyes or nanostructures in hybrid composites could provide an important breakthrough in the field of THG materials in the near future.

Hybrid sol–gel monolithic materials appear to be versatile systems allowing easy dispersion of chromophores or nanostructures with relatively high concentrations and good stability in particular under laser irradiation. The growing knowledge regarding the relationships between the process, the microstructure, the dyes localizations, and the optical properties open a wide access to highly improved new optical materials with third-order nonlinearities. Hybrid materials exhibiting synergetic effects between the guest systems (molecules and nanostructures) will certainly bring strong improvements in the optical responses and important innovations in the field of nonlinear optical devices.

12.2 Plasmonic Hybrid Materials

12.2.1 Optical Properties of Metal Nanoparticles

Plasmonic materials are among the most studied systems in the field of light-matter interactions, and numerous articles and reviews were centered on this topic for the past decade [139–155]. Association of plasmonic nanostructure with organic entities, in particular molecules with optical properties, has been intensively investigated for several purposes, such as stabilization of the metal nanostructures or metal-to-dye interactions. Indeed, the specific optical response of such metallic nanostructures gives rise to unexpected interactions with optically active molecules. The methods for controlling these interactions in hybrid systems are reviewed in this part as well as the impact on the optical responses with respect to theoretical predictions.

When nanoparticles, which possess a sufficiently high free carrier concentration in their structure (typically noble metals or semiconductors), are excited by an incoming light source with wavelengths much larger than their size, they express a specific optical response due to collective oscillation of the free-electrons in their structure [156–165]. This phenomenon is commonly called localized surface plasmon resonance (LSPR) and can be experimentally easily observed as a modification of the absorption spectra. It provides also a large electromagnetic field enhancement at the metallic surface vicinity and local thermal effects and in some cases important scattering of light. Gold- and silver-based nanostructures were the most reported due to their easy accessibility in terms of synthesis and structure versatility and stability. The absorption bands relative to the LSPRs are essentially dependent on the metal, the particle size and shape, the potential collaborative effects between particles, and the refractive index of the surrounding media [149,154,160,163,166–191]. Portehault and coworkers demonstrated recently the impact of the ligand (aromatic thiolates)-to-gold particle charge transfer on the plasmon resonance position using an original approach by spectroscopic ellipsometry [192]. In the field of optics, these nanoobjects are extremely relevant since they provide, under light excitation, a way to strongly affect the optical responses both of the metallic nanostructures (with plasmon resonances tunable from the visible to the IR wavelengths) and of optical systems in close interaction with the metallic surfaces (such as molecular dyes or semiconductors). The control of these interactions in hybrid systems has been an important field of scientific contributions for the past 10 years. The first important reports concerned the enhancement of sensitivity for molecular detection using Raman spectroscopy. A large amount of work was then launched related to surface-enhanced Raman spectroscopy (SERS) in the field of sensing and detection with control of metallic surface interactions with various molecular entities. Such plasmon-molecule or plasmon-particle interactions opened a broad field of research related to enhancement of luminescence and absorption properties as well as nonlinear optical properties for applications in medical imaging, photodynamic therapy, lightning, LEDs, solar cells, sensors, and photocatalysis.

12.2.2
Hybrids with Dyes and Plasmonic Nanostructures: Luminescence and Nonlinear Optical Properties

Interactions between a dye and a metallic surface can be achieved and controlled through different architectural approaches for the material design, typically either on a surface, on nanoparticles, in nanoshells or associated in a composite material, depending essentially on the application mode (Figure 12.10). In all cases the impact of the interaction on the properties strongly depends on the molecule-to-surface distance but also on the respective orientations of the dipoles. Theoretical reports have shown that the chromophore optical response

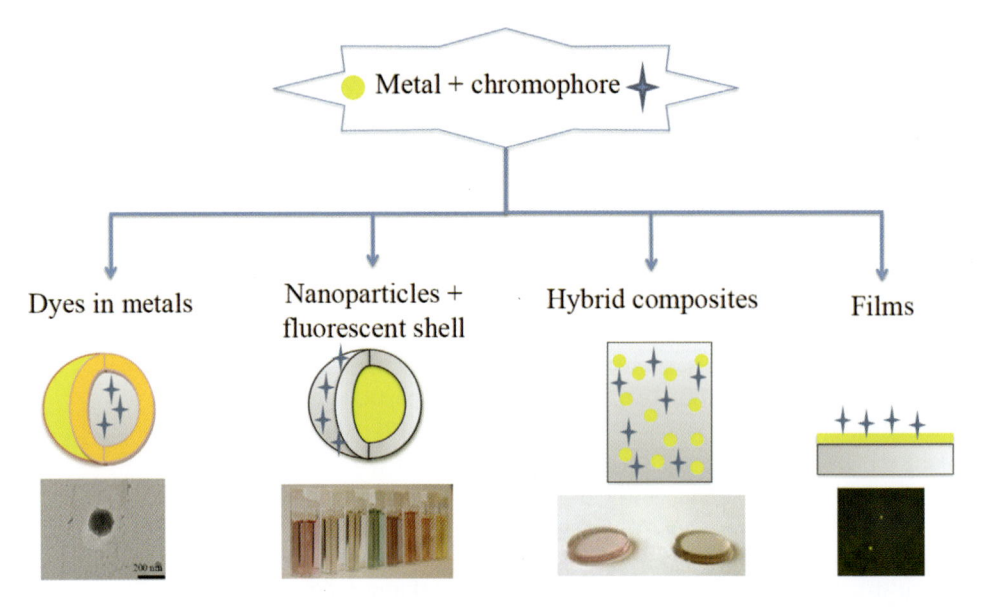

Figure 12.10 Principal strategies to control interactions between chromophores and metal nanoparticles, respectively from left to right: Incorporation of dyes into nanoshells, surface modification of nanoparticles with dyes, codoping into solid matrix (polymer or sol–gel), and multilayered thin films. The two first approaches on colloidal systems are often proposed for biomedical imaging or therapy (injectable nanoplatforms) and the composites/films for optical devices design (optical filters, photovoltaic, sensors, etc.).

can be strongly affected by the presence of the local enhanced electromagnetic field (provided by either single particles or dimers, aggregates . . .), in particular the emission, depending on the distance, dipole orientations and spectral overlaps between the LSPR band and the emission band [155,183,193–195]. Experimentally, the control of the dipole orientation remains extremely complicated and poorly documented while a large amount of work was published on different ways of controlling the dye-to-nanoparticle distance and evaluating the photophysical properties in each case. The different strategies were often correlated with modeling approaches [155]. The following part reports thus on the common strategies used to combine plasmonic effect with optical responses of molecules (emission, absorption, nonlinear reponses) based essentially on the distance control between the systems.

12.2.2.1 Surface Functionalization of Metal Nanoparticles

The functionalization of metallic surfaces remains the easiest way to induce interactions between the two systems since surface chemistry is well controlled on metals. Regarding the synthesis of the metallic cores, this is extremely well documented and previously reviewed. It is important to note that the chemistry and growth mechanisms of metallic nanostructures, especially silver and gold, are nowadays well mastered. This allows a fine control over the size and morphologies (spheres, rods, cubes, stars, bipyramids . . .) and thus a precise tuning of the SPR band from the visible to the NIR [142,148,166,196]. The surface state of the native nanostructures depends on the synthesis route, and the particles are usually stabilized through the presence of an organic ligand (i.e., citrate) or surfactants (i.e., CTAB) at their surface. The total or partial replacement of the stabilizing molecules can be easily achieved through ligand exchanges reactions with functional systems bearing thiol or thioctic acid pending groups [147]. These groups possess a strong affinity for the metal surface and ensure efficient binding. The use of electrostatic interactions (i.e., with polyelectrolytes) can also be a way for surface modification [147]. Chen et al. used direct interactions between a sulfonated aluminum phtalocyanine and gold nanorods or nanocubes to enhance their fluorescence [197]. Nanorods showed limited fluorescence enhancement due to the overlap between the SPR and the emission band of the chromophore. The use of nanocubes, which combines stronger SPR than spheres and no overlapping, instead of rods, successfully lead to better enhancement in the emission. Gold nanocubes were used for fluorescence imaging of cancer cells using two-photon excitation in the NIR (Figure 12.11).

Dyes bearing thiol groups are easily attached to metal surfaces. Numerous examples can be recorded from the literature. Russel and coworkers used gold nanoparticles modified by zinc phtalocyanines and later on by antibody to achieve efficient oxygen singlet generation and photodynamic therapy of cancer cells (Figure 12.12) [198,199]. In this case, the distance between the dyes and the surface is mostly determined by the C11 chain between the molecule core and the thiol groups or the disulfides.

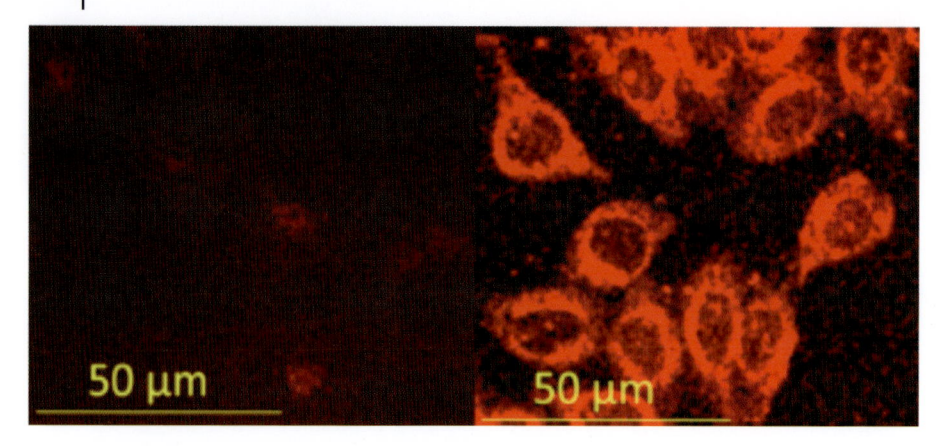

Figure 12.11 Two-photon fluorescence imaging of cancer cell lines using pure aluminum phthalocyanine (a) and aluminum phthalocyanine combined with gold nanocubes (b) (Exc. 800 nm/fs laser). (Adapted with permission from Ref. [197], copyright 2014 American Chemical Society.)

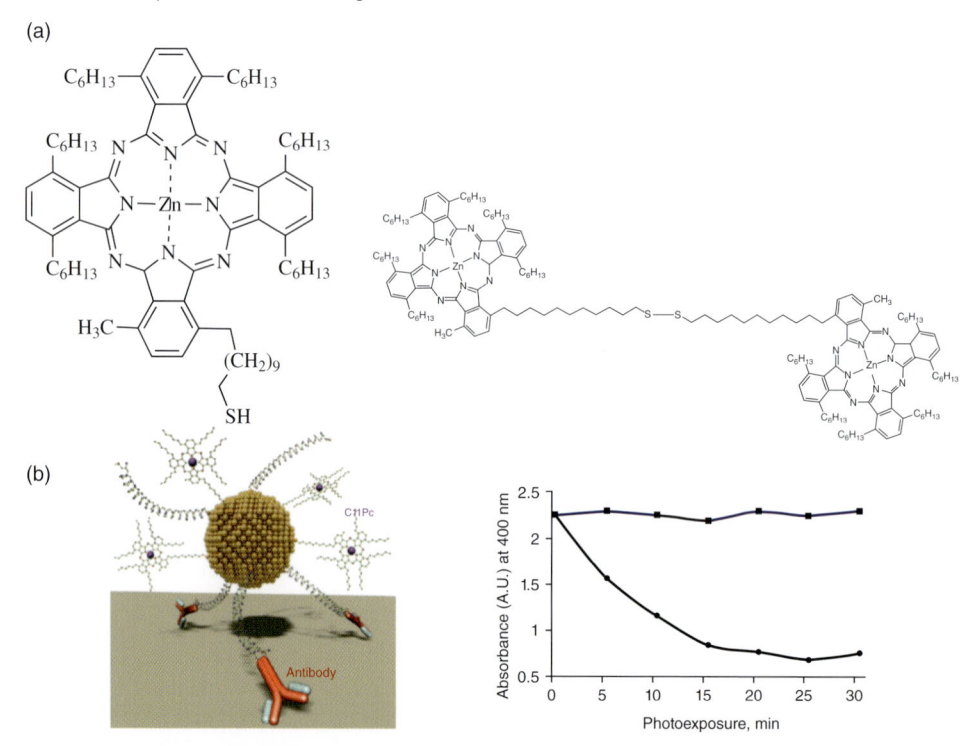

Figure 12.12 Two structures of phthalocyanine-based photosensitizers (a) and preparation of the conjugate with Au-NPs (b). Bottom right shows the generation of oxygen singlet evidenced by the decreasing in the absorption spectrum of the disodium 9,10-anthracenedipropionic acid (ADPA) probe (dots) versus the control sample without the photosensitizer (squares) [198,199]. (Adapted from Ref. [198], copyright 2002 American Chemical Society.)

Since the chromophore to particle distance is a crucial parameter in the control of the charge or energy transfers between the two systems, several spacing strategies were investigated. The control of the distance can be achieved by using molecular spacers such as oligomers between the thiol-anchoring group and the chromophores. The synthesis of block copolymers bearing chromophores along the chains was proposed and showed efficient control over the distance by altering the polymer structure, and the electrostatic repulsion and sterical hindrance between the chains, which contributed to the linear structuration of the polymers [200,201]. For instance, a copolymer functionalized with Lucifer yellow (LY) allowed preparation of luminescent gold hybrid nanostructures by manipulating the polymer structural arrangement at the surface of the particle. It was shown that the size of the particle, and thus the curvature angle, impacted the surface organization and structure of the polymer going from mushroom type to an extended structure (Figure 12.13) [200]. This was confirmed by measurements of the fluorescence intensity as a function of the average size of the particles. The optimized nanostructures were successfully used for cell imaging and photodynamic therapy of cancer cells using a similar copolymer with a dibromobenzene derivative as photosensitizer [200]. Interestingly, it was also shown that the use of anisotropic gold cores, such as nanobipyramids, allowed enhancement of the fluorescence of a similar luminescent polymer, while this was not observed in the case of spherical nanoparticles [201].

The distances of interactions can be also tuned by using charged spacers such as polyelectrolytes for surface modifications of nanoparticles through layer-by-layer self-assembly stabilized by electrostatic interactions (Figure 12.14) [202–205], which was initially introduced to immobilize particles on surfaces [206] or for biocompatibilization [207] to achieve bioimaging. Up to 11 layers of spacer were introduced in the structure leading to shell thicknesses in the range 1–20 nm [203]. The presence of the layers can easily be monitored using zeta potential measurements.

In the case of nanostars coated with polyelectrolytes, it was shown that the fluorescence emission of the dyes could be recovered with a distance of about 10 nm [203]. However in this case no enhancement could be observed on the

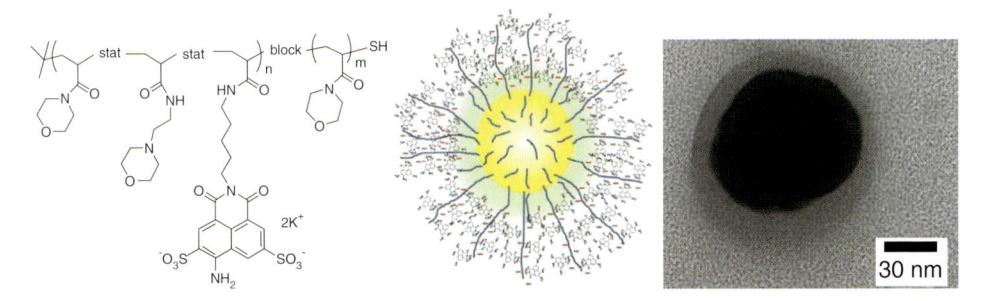

Figure 12.13 Structure of the diblock copolymer functionalized with LY dyes and hybrid gold nanoparticles bearing the luminescent polymers [200].

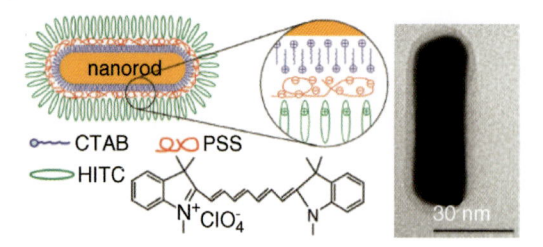

Figure 12.14 Alternate deposition of polyelectrolytes in order to control the shell thickness on gold nanorods. (Adapted from Ref. [204], copyright 2008 American Chemical Society. The dyes were deposited on the outer layer using electrostatic interactions.)

emission. Interestingly, Murphy and coworkers showed that the use of nanorods with similar strategy allowed an enhancement of the two photon absorption cross section of an organic chromophore (AF348-3A) nearby the surface with a decrease in the enhancement when going from 3 to 12 nm from the surface [202].

Halas et al. have used human serum albumin as a spacer of about 8 nm between NIR fluorophores (IR800) and gold nanoshells or nanorods, attached to the surface through electrostatic interactions [208]. They were able to observe large increases in the quantum yield compared to isolated chromophores. Nanoshells appeared to be more efficient for emission enhancement (40-fold enhancement, 86% QY), and this is mostly attributed to their important scattering cross section at the emission wavelength of the chromophores compared to nanorods. The use of DNA is also an interesting alternative to control optical responses of hybrid nanosystems. For instance, thiol-terminated DNA can easily bind to the metal surface, and DNA hybridization can thus be used for fluorescence enhancement [209]. Thiol-modified DNA was also used to control plasmon coupling between Au nanoparticle aggregates and the induced two-photon luminescence from the metal, which could be enhanced by a factor 265 in the best case (2 nm distance for 41 nm aggregates) and used for DNA sequences detection [210]. Willner and coworkers used DNA tweezers modified with a 10 nm Au NP and a fluorophore (Cy3) to build switchable closure and opening with controlled fluorescence enhancement or quenching depending on the state [211].

Among the most investigated strategies is the use of core shell nanostructures with metal core and silica-based shell, which also enables spacing control between the dyes and the metallic surface [212–222]. The dye can be either directly incorporated in the silica shell during the synthesis of post-grafted at the surface of the silica. Noginov used this configuration to generate core-shell structure functionalized at the surface by Oregon Green 488 fluorophore encapsulated in the outer silica shell [223]. This structure, with a spherical gold core of 14 nm in diameter and a silica shell of 15 nm thickness, allowed the demonstration of a spaser-based nanolaser in the visible range [223]. Murphy and coworkers showed that it was possible to fine tune the distance and carry out

12.2 Plasmonic Hybrid Materials | 339

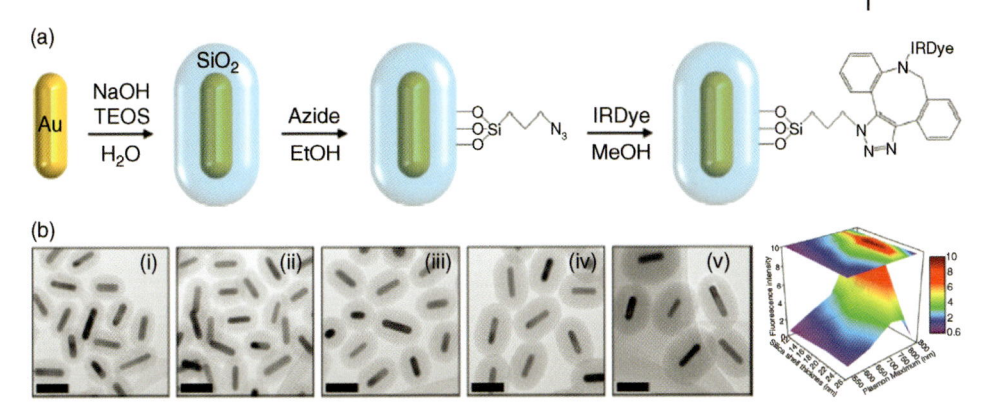

Figure 12.15 Strategy from Murphy et al. using surface click chemistry on the silica layer to bind the chromophore and control the distance to the metal surface (a). (Adapted from Ref. [215], copyright 2014 American Chemical Society. The TEM images show the increase of silica shell thickness on one type of nanorods (i). The 3D two-tier contour of emission intensity versus both the shell thickness and SPR band shows the hotspot fluorescence in the best configuration (ii).

surface grafting using click-chemistry reactions (Figure 12.15) [215]. Interestingly, they observed an enhancement of fluorescence of an IR dye (800CW DBCO) in their systems with hotspots (10-fold enhancement) in the range 14–22 nm thicknesses for the silica shell and SPR maxima at 750–800 nm overlapping the emission spectrum of the dye. The main mechanism was strong coupling between the dye and the metal, and almost no contribution of the scattering effects was detected. A strong decrease in the lifetime was also observed when the shell thickness was decreased.

A similar strategy but following a different chemical route is to ensure grafting through amide bonds for instance by reacting between an amine NH_2 supported on the silica and a carboxylic acid group (Figure 12.16) [213,214]. Xu et al. reported a two-photon enhancement of fluorescence maximum of 11.8 at a 20 nm distance due to the electric field amplification.

Durand and coworkers introduced grafted chromophores directly in the silica nanoparticles and either incorporated Au NPs in the core or grafting them at the surface of the fluorescent silica (Figure 12.17) [222]. Comparison between the different systems showed that the introduction of the Au NPs provided strong enhancement of the emission in both configurations and these hybrid particles could efficiently be internalized in cells and used for two-photon imaging with a much better efficiency than without the Au NPs. Fluorescent pH sensors can also be achieved through similar approach. For instance Ag@SiO_2 particles on which a pH-sensitive dye was adsorbed exhibited enhanced sensitivity to pH sensing [217].

Mesoporous silica can also be used to coat metal nanoparticles. These mesoporous structures can efficiently be loaded with photosensitizers [212,221] or upconversion nanoparticles [219]. Such a strategy was for instance effectively

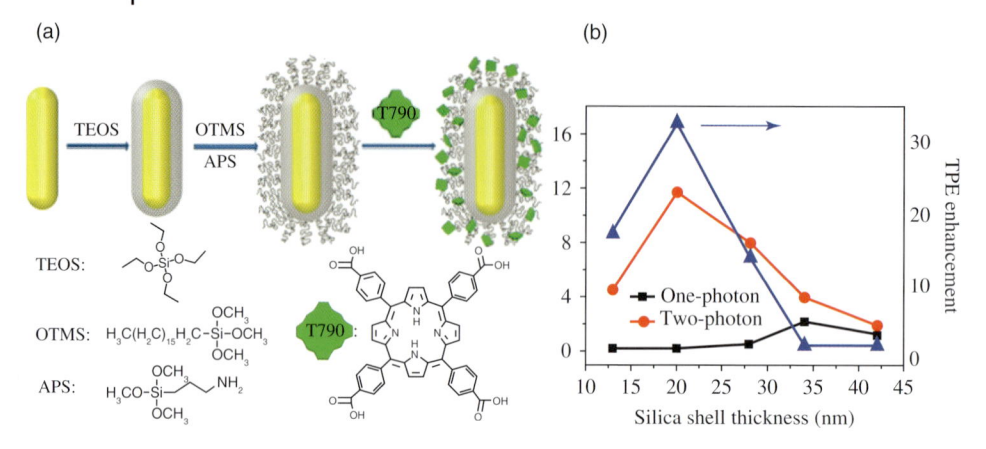

Figure 12.16 Synthesis of a photosensitizer doped silica shell on gold nanorods. (Adapted from Ref. [214], copyright 2014 American Chemical Society (a) and two-photon excitation fluorescence enhancement factors versus the silica shell thickness (b).

used by Khlebtsov and coworkers to treat tumors *in vivo* using combined photodynamic and photothermal therapy [220]. This approach with porous silica shell can also be used to control the diffusion of molecules in the silica membrane and achieve further efficient SERS detection [224].

Regarding the perspectives on innovative metal-based nanostructures, researchers have started to investigate smaller size metal nanoparticles such as gold quantum dots (AuQDs). Such systems show interesting absorption/emission properties in the NIR corresponding to the commonly reported biological window. The main drawback of such small AuQDs remains their cytotoxicity. It was recently demonstrated that incorporation of such AuQDs into hollow mesoporous silica nanoparticles (Quantum rattles QRs) is an alternative to stabilize the metal while preserving its specific optical and magnetic properties [225]. The QRs were successfully used for multimodal imaging (fluorescence, photoacoustic), drug delivery, and photothermal therapy of tumors (Figure 12.18). Hybrid materials using specific responses of stabilized small metal NPs, which have not been fully explored yet, represent certainly a wide range of possible future innovations routes in the fields of nanomedicine or catalysis.

Consequently, the use of hybrid fluorescent metal nanoparticles in different configurations showed several important advantages for applications, in particular in biomedical fields. These strategies usually allow high chromophore loading capacity, good stability, enhanced cellular uptake efficiency, enhanced fluorescence and enhanced photodynamic therapy using the SPR effects, subcellular characterization through possible combined fluorescence imaging (fluorescence of dyes and of metal nanoparticles), photoacoustic, scattering of metal (dark field imaging), and photo-theranostic by using photoinduced therapies (photothermal, photodynamic). Moreover, the enhanced permeability and retention (EPR)

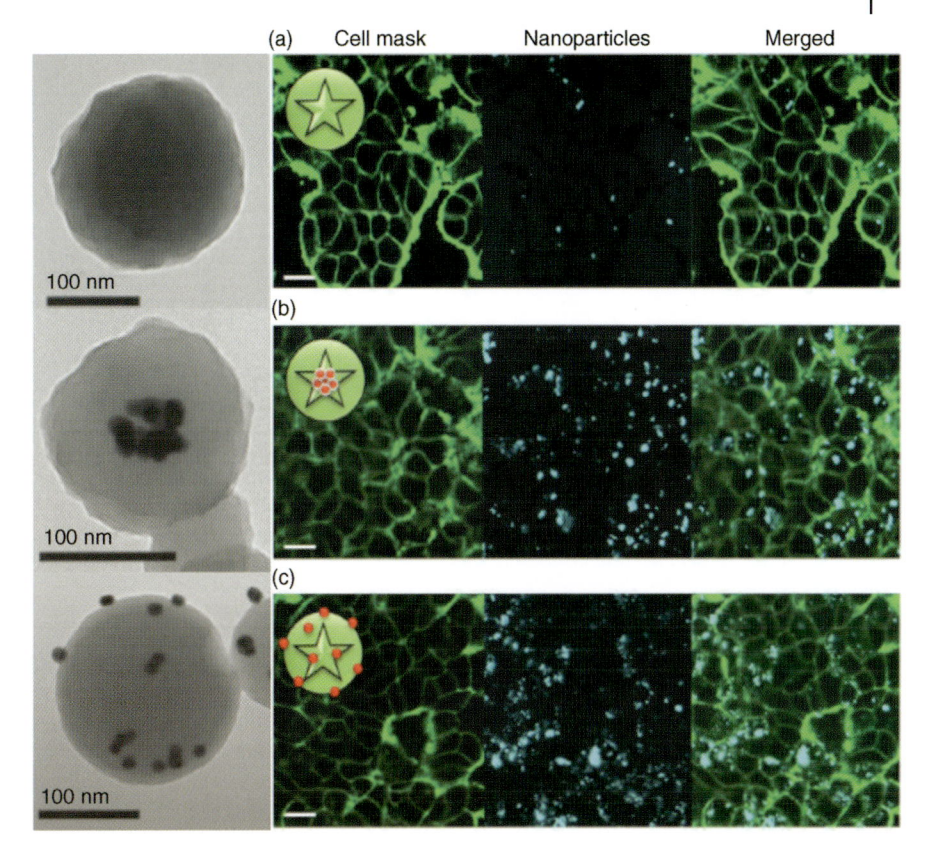

Figure 12.17 Two-photon imaging of MCF-7 cancer cells using three architectures and respective TEM images: (a) a reference two-photon photosensitizer-doped silica nanoparticle, (b) introduction of a gold core in the particle, and (c) attaching gold NPs at the surface of the hybrid fluorescent silica core. (Adapted from Ref. [222].)

effect can be used to better accumulate the hybrid particles in tumoral tissues and opens important opportunities in terms of targeting.

12.2.2.2 Encapsulation of Dyes in Metals

Encapsulation of dyes in the metal can be a way to achieve strong coupling between the molecules and the metallic surface and to investigate unusual interactions. Moreover this can provide a method to efficiently protect the organic chromophores against photobleaching. In this sense, the use of metallic nanoshells can be a route to built core-shell structures entrapping dyes in the core. Both core-shell structures and hollow nanospheres were reported in the literature since they were supposed to exhibit specific optical resonances in particular in the NIR [226–228]. Moreover nanoshells showed important capacity to produce photothermal effect and were consequently effectively used in photothermal therapy of cancers or in laser-tissue welding [229–234]. Thus

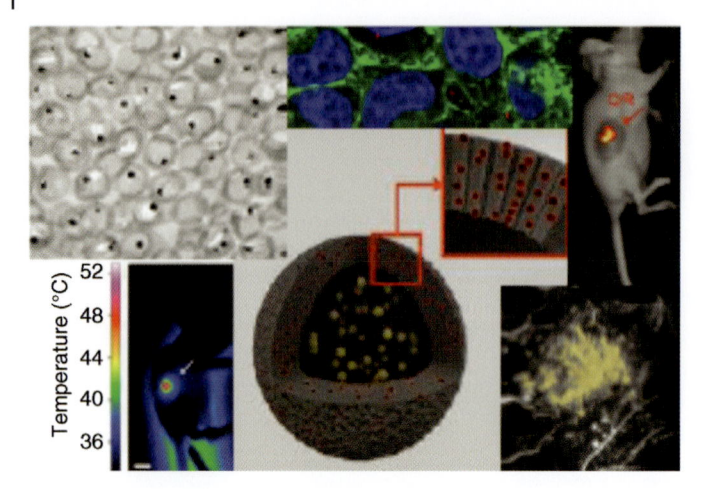

Figure 12.18 Structure of the Au@SiO$_2$ quantum rattles (QRs) with small AuNPs grown in the pores and their use in multimodal imaging and photothermal therapy. (Adapted from Ref. [225], courtesy of the National Academy of Science.)

several chemical strategies were adapted to build such nanostructures [226,228,230,235–240]. Using the proposed chemical routes, it was possible to introduce organic entities inside the core of the nanoparticles and investigate the properties and in particular the optical responses in the confined media. For instance, Rhodamine 610 was included in hollow gold nanospheres (Figure 12.19) [183,241]. Interestingly this confinement showed a strong decrease in the emission lifetime from 1.7 ns in water to less than 150 ps in the shell together with a strong improvement of the photostability of the dyes (an order of magnitude). The object also appeared a little brighter than the pure fluorophore [183].

Recently, strong exciton-plasmon coupling was demonstrated in a Rhodamine doped silica core surrounded by a gold shell in a very similar configuration [242].

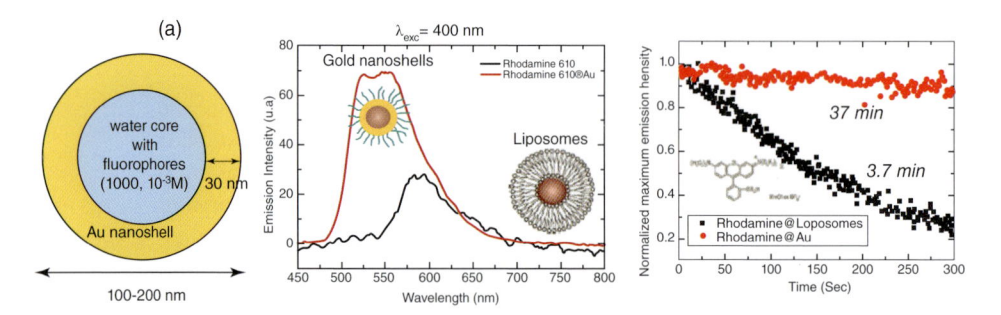

Figure 12.19 Structure of hybrid luminescent gold nanoshell (a), emission spectrum of the hybrid nanoshells and liposomes with the same fluorophore concentration in the core (b), fluorescence photoresistance of the hybrid and the free fluorophore (c). (Adapted from Ref. [183], copyright 2011 American Chemical Society.)

Silica-doped core combined with gold shells were also used to achieve SERS inside the nanoshells and exhibited a giant enhancement due to the cavity effect that concentrates the electromagnetic radiation [243]. Nanovesicles built from self-assembly of primary gold spherical nanoparticles bearing semifluorinated ligands can be used to encapsulate dyes or drugs [244]. The encapsulated molecules can then be released efficiently upon laser excitation and photothermal conversion of the laser electromagnetic energy.

Avnir and coworkers have introduced original ways to entrap molecular dyes into different metals (Ag, Au, Cu, Fe) [245–253]. Indeed, they generate the metal nanoparticles in the presence of the molecular guest and induce nanoaggregations entrapping the guests (Figure 12.20). They also developed an electroless deposition process to grow hybrid films of metal aggregates entrapping molecular guests [252]. Such structures can lead to rather homogeneous hybrid films with very interesting properties such as enhanced Raman response. They also showed that they were able to recover part of the emission of the Rhodamine [252]. Such optical systems are promising hybrid materials for enhanced detection and highly sensitive sensors.

12.2.2.3 Composite Materials and Thin Films

Preparation of solid shapeable materials is extremely important on the way to the design of efficient devices. Regarding optical systems, most of the materials are processed as thin films or bulk monoliths. In both case, the use of plasmonic nanostructures has been investigated in different configurations, either introduced directly in the matrix as composite material or deposited on a surface. In the case of composite materials, basically two approaches can be found, either *in situ* synthesis of metal nanostructures in the matrix using chemical, thermal, or photoinduced reductions; or incorporation of colloidal dispersions during the

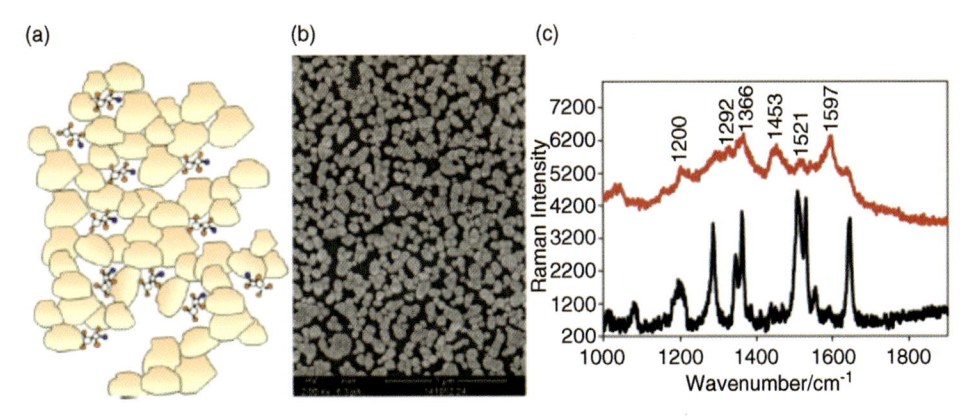

Figure 12.20 Suggested structure of molecularly doped metals from reference [253], copyright 2014 American Chemical Society (a). SEM photo of a gold film doped with Rhodamine B (b). Enhanced Raman spectrum of Rhodamine B in gold aggregates (red) versus the spectrum of pure Rhodamine B powder (black) (c) [252].

matrix preparation. The matrix can typically be inorganic [254–257], sol–gel-based matrix [258–269] or polymers [270–273]. Regarding the *in situ* growth, the main drawbacks are the limitations in terms of control of the shape of the particles, which are often spherical or wire-type structures and the homogeneity in size and dispersion. The control of the growth using laser and in particular two-photon-induced fabrication is limited in resolution to the micrometer size. In this sense, the stabilization of colloidal suspensions of metal nanoparticle and further dispersion in a solid matrix appear as a good alternative to controlled dispersion and also codoping of optically active molecule and plasmonic nanostructures. For instance, recent work has strongly focused on the introduction of plasmonic nanostructures in the field of nonlinear optics to provide either direct nonlinear absorption or enhancement of nonlinear properties of chromophores. The first works were reported on the use of gold nanoparticles either precipitated in glass [254,255], polymers [274,275], or as colloidal suspension and aggregates [37,276] and showed nonlinear absorption response essentially due to the surface plasmon resonance and RSA or nonlinear scattering processes. Regarding the possibility of using hybrid systems doped with metal nanoparticles for enhancement of optical responses, composites using luminescent polymers with dispersion of metal were mostly investigated hitherto. For instance, Halas and coworkers reported the use of core-shell silica–gold nanoparticles incorporated into semiconducting luminescent polymers to induce a quenching of the polymer triplet exciton, which consequently reduces capacity to transfer to oxygen (singlet oxygen formation) and thus decreases the photooxydation and enhance the active lifetime of the polymer [277]. Similarly, core-shell Au@SiO$_2$ NPs were incorporated into the emitting layer of a LED to successfully enhance the luminous efficiency [278]. The design of thiolated silicon polymers allowed surface modification of metal nanoparticles of different shapes and efficient dispersion in sol–gel matrices [279]. Such an approach was used to prepare Co-doped sol–gel monoliths, which unexpectedly showed unusually long-distance enhancement of nonlinear absorption processes [280].

Multilayers systems can be an alternative to control interactions between metal NPs and luminescent polymers [272,281–283]. Mitsuishi *et al.* also used a hybrid polymer approach, combining polymer nanosheets nanoassemblies through Langmuir–Blodgett technology and introduced metal nanoparticles layers [281,282]. The polymer layer part included chromophores (for luminescence or NLO), and the distance to the nanoparticles deposited on the films or between two films was controlled by the length and the thickness of the polymers. These researchers were able to demonstrate SHG and luminescence enhancement on free-standing ultrathin films (Figures 12.21 and 12.22). The SHG enhancement was dominated by the dipole LSPR coupling (confirmed by finite-difference time-domain (FDTD) calculations) and controlled at the nanometer scale [281].

Scattering of nanoparticles can be efficiently used to enhance the absorption of devices, in particular in the field of photovoltaics [284]. Future applications of

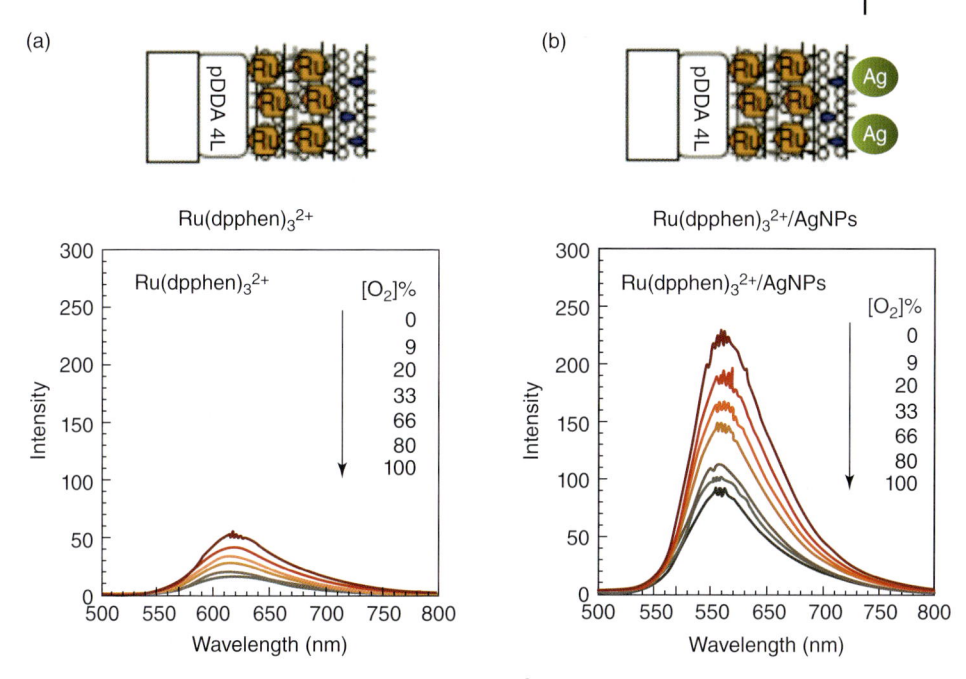

Figure 12.21 Hybrid polymer thin films with Ru(ddphen)$_3^{2+}$ and Ag NPs and the corresponding luminescence spectra showing the enhancement of the emission [282].

polymer–metal NPs interaction in solar cells are thus extremely promising, and can be used for instance to increase drastically the light absorption of the organic photovoltaic device through the local field enhancement [285,286]. Indeed, in the work from Wu *et al.* the gold nanoparticles scattering effect

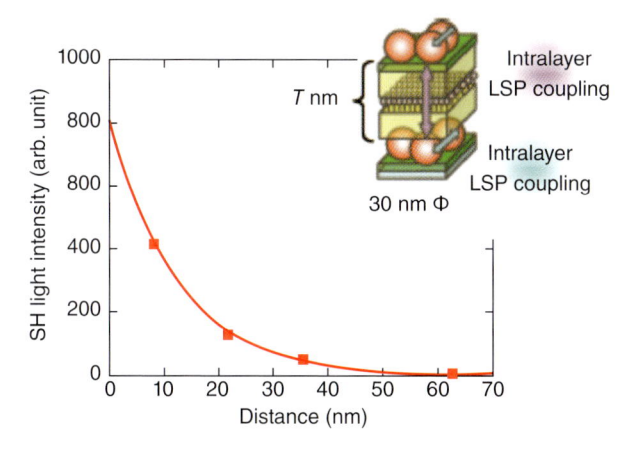

Figure 12.22 Impact of the distance between the nanoparticles layers surrounding the SHG polymeric layer in the hybrid configuration on the second harmonic response. (Adapted from Ref. [281], copyright 2009 American Chemical Society.)

increased the optical path in the layer improving the absorption. Moreover, the rate of generation of the excitons was improved by the presence of the LSPR, which reduced the recombinations and thus the exciton loss [285].

All the reported systems tend to show that the control of optical properties of chromophore using plasmonic interactions is not trivial and each case is specific in the interpretation of the optical response. The optimization of the hybrid structures needs to take into consideration a multiparameter approach and in particular the molecular absorption/emission mechanisms. In particular, recent results tend to show that interesting innovations might be expected in the future, with for instance short- or long-distances effects, which were not at all considered experimentally at this point.

12.3
General Conclusion and Perspectives (Parts I and II)

The field of hybrid materials with optical properties is growing intensively nowadays with breakthroughs related to the fields of imaging, sensing, protection, therapy, energy, and many other applications. This interest is mostly based on the fact that light is a natural source of energy (photovoltaics, lasers), and is a critical environmental issue (lighting) that has become crucial for humanity. Moreover, optical systems are often related to human eyes (color changes, protection, cameras . . .), which make them of great concern for human beings. The strong interest of researchers in the development of devices activated by light, generating light or modulating the optical signal is thus a logical consequence, and is boosted by the technological developments and implementation of the scientific knowledge. This review shows clearly the high level of control reached by scientists in the architectural aspects of the hybrid systems, in particular in the mastering of the interactions between the organic entities and their inorganic counterparts. This induces consequently a precise tuning of the optical responses of the materials, an increased efficiency and the possibility to integrate them in final operating devices such as light-emitting diodes (LEDs) for screens or lighting, photovoltaics for energy conversion, optical filters in cameras, or optical detection for biological entities. Even more interesting for the future, the use of interactive optically active organic and inorganic systems remains to be fully explored, and, for instance, the controlled coupling between plasmonics and luminescence, intensively investigated, is opening a land of opportunities in the field of optical systems.

Finally, hybrid materials are not only a domain of interfacing for chemists (organic, inorganic, analytic), but it has become a strong node of fruitful interactions for very different communities such as chemists, physicists, physicians, radiologists, biochemists, theoreticians, computer scientists, textile developers, and many more, in particular when related to optics. Important breakthroughs should be expected in terms of developments at these crucial interfaces in the future years.

References

1 Wawilov, W.L.L.S.I. (1926) *Z. Phys.*, **35**, 920.
2 Vavilov, S.I. (ed.) (1950) *Microstructure of Light*, USSR Acad. Sci., Moscow.
3 Boyd, R.W. (2003) *Nonlinear Optics*, 2nd edn, Academic, San Diego.
4 Träger, F. (ed.) (2007) *Springer Handbook of Lasers and Optics*, Springer.
5 Maiman, T.H. (1960) *Nature*, **187**, 493.
6 He, G.S., Tan, L.-S., Zheng, Q., and Prasad, P.N. (2008) *Chem. Rev.*, **108**, 1245–1330.
7 Pantelis, P., Hill, J.R., and Davies, G.J. (1988) *Nonlinear Optical and Electroactive Polymers* (eds P.N. Prasad and D.R. Ulrich), Plenum, New York, p. 229.
8 Mortazavi, M.A., Knoesen, A., Kowel, S.T., Higgins, B.G., and Dienes, A. (1989) *J. Opt. Soc. Am. B*, **6**, 733.
9 Singer, K.D., Sohn, J.E., and Lalama, S.J. (1986) *Appl. Phys. Lett.*, **49**, 248.
10 Singer, K.D., Kuzyk, M.G., Holland, W.R., Sohn, J.E., Lalama, S.J., Comizzoli, R.B., Katz, H.E., and Schilling, M.L. (1988) *Appl. Phys. Lett.*, **53**, 1800.
11 Charra, F., Devaux, F., Nunzi, J.M., and Raimond, P. (1992) *Phys. Rev. Lett.*, **68**, 2440.
12 Charra, F., Kajzar, F., Nunzi, J.M., Raimond, P., and Idiart, E. (1993) *Opt. Lett.*, **18**, 941.
13 Fiorini, C., Charra, F., Nunzi, J.M., Samuel, I.D., and Zyss, J. (1995) *Opt. Lett.*, **20**, 2469.
14 Sanchez, C., Lebeau, B., Chaput, F., and Boilot, J.P. (2003) *Adv. Mater.*, **15**, 1969.
15 Innocenzi, P. and Lebeau, B. (2005) *J. Mater. Chem.*, **15**, 3821.
16 Griesmar, P., Sanchez, C., Puccetti, G., Ledoux, I., and Zyss, J. (1991) *Mol. Eng.*, **1**, 205.
17 Lebeau, B., Sanchez, C., Brasselet, S., Zyss, J., Froc, G., and Dumont, M. (1996) *New J. Chem.*, **20**, 13.
18 Lebeau, B., Brasselet, S., Zyss, J., and Sanchez, C. (1997) *Chem. Mater.*, **9**, 1012.
19 Lebeau, B., Maquet, J., Sanchez, C., Toussaere, E., Hierle, R., and Zyss, J. (1994) *J. Mater. Chem.*, **4**, 1855.
20 Avnir, D., Levy, D., and Reisfeld, R. (1984) *J. Phys. Chem.*, **88**, 5956.
21 Levy, D. and Avnir, D. (1991) *J. Photochem. Photobiol. A*, **57**, 41.
22 Levy, D., Einhorn, S., and Avnir, D. (1989) *J. Non-Cryst. Solids*, **113**, 137.
23 Schaudel, B., Guermeur, C., Sanchez, C., Nakatani, K., and Delaire, J.A. (1997) *J. Mater. Chem.*, **7**, 61.
24 Bentivegna, F., Canva, M., Georges, P., Brun, A., Chaput, F., Malier, L., and Boilot, J.P. (1993) *Appl. Phys. Lett.*, **62**, 1721.
25 Riehl, D., Chaput, F., Levy, Y., Boilot, J.P., Kajzar, F., and Chollet, P.A. (1995) *Chem. Phys. Lett.*, **245**, 36.
26 Bentivegna, F., Canva, M., Brun, A., Chaput, F., and Boilot, J.P. (1996) *J. Appl. Phys.*, **80**, 4655.
27 Dubois, A., Canva, M., Brun, A., Chaput, F., and Boilot, J.P. (1996) *Appl. Opt.*, **35**, 3193.
28 Bentivegna, F., Canva, M., Brun, A., Chaput, F., and Boilot, J.P. (1997) *J. Sol-Gel Sci. Technol.*, **9**, 33.
29 Boilot, J.P., Biteau, J., Chaput, F., Gacoin, T., Brun, A., Darracq, B., Georges, P., and Levy, Y. (1998) *Pure Appl. Opt.*, **7**, 169.
30 Goudket, H., Canva, M., Levy, Y., Chaput, F., and Boilot, J.P. (2001) *J. Appl. Phys.*, **90**, 6044.
31 Le Duff, A.C., Canva, M., Levy, Y., Brun, A., Ricci, V., Pliska, T., Meier, J., Stegeman, G.I., Chaput, F., Boilot, J.P., and Toussaere, E. (2001) *J. Opt. Soc. Am. B*, **18**, 1827.
32 Canva, M., Roger, G., Cassagne, F., Levy, Y., Brun, A., Chaput, F., Boilot, J.P., Rapaport, A., Heerdt, C., and Bass, M. (2002) *Opt. Mater.*, **18**, 391.
33 Nhung, T.H., Canva, M., Dao, T.T.A., Chaput, F., Brun, A., Hung, N.D., and Boilot, J.P. (2003) *Appl. Opt.*, **42**, 2213.
34 Chen, L., Jin, X., Cui, Y., Gao, J., Qian, G., and Wang, M. (2007) *J. Sol-Gel Sci. Technol.*, **43**, 329.
35 Westlund, R., Malmström, E., Lopes, C., Öhgren, J., Rodgers, T., Saito, Y., Kawata, S., Glimsdal, E., and Lindgren, M. (2008) *Adv. Funct. Mater.*, **18**, 1939.

36 Zieba, R., Desroches, C., Chaput, F., Carlsson, M., Eliasson, B., Lopes, C., Lindgren, M., and Parola, S. (2009) *Adv. Funct. Mater.*, **19**, 235.

37 Cui, Y., Chen, L., Wang, M., and Qian, G. (2006) *J. Phys. Chem. Solids*, **67**, 1590.

38 Chen, L., Cui, Y., Qian, G., and Wang, M. (2007) *Dyes Pigm.*, **73**, 338.

39 Chen, L., Qian, G., Cui, Y., Jin, X., Wang, Z., and Wang, M. (2006) *J. Phys. Chem. B*, **110**, 19176.

40 Chen, L., Qian, G., Jin, X., Cui, Y., Gao, J., Wang, Z., and Wang, M. (2007) *J. Phys. Chem. B*, **111**, 3115.

41 Chen, L., Zhong, Q., Cui, Y., Qian, G., and Wang, M. (2008) *Dyes Pigm.*, **76**, 195.

42 Cui, Y., Chen, L., Qian, G., and Wang, M. (2008) *J. Non-Cryst. Solids*, **354**, 1211.

43 Cui, Y., Li, B., Yu, C., Yu, J., Gao, J., Yan, M., Chen, G., Wang, Z., and Qian, G. (2009) *Thin Solid Films*, **517**, 5075.

44 Cui, Y., Qian, G., Chen, L., Gao, J., and Wang, M. (2007) *Opt. Commun.*, **270**, 414.

45 Cui, Y., Qian, G., Chen, L., Wang, Z., and Wang, M. (2007) *Macromol. Rapid Commun.*, **28**, 2019.

46 Cui, Y., Qian, G., Gao, J., Chen, L., Wang, Z., and Wang, M. (2005) *J. Phys. Chem. B*, **109**, 13295.

47 Liang, T., Cui, Y., Yu, J., Lin, W., Yang, Y., and Qian, G. (2013) *Thin Solid Films*, **544**, 407.

48 Liang, T., Cui, Y., Yu, J., Lin, W., Yang, Y., and Qian, G. (2013) *Dyes Pigm.*, **98**, 377.

49 Chang, P.-H., Tsai, H.-C., Chen, Y.-R., Chen, J.-Y., and Hsiue, G.-H. (2008) *Langmuir*, **24**, 11921.

50 Wang, S., Zhao, L., Sun, J., Cui, Z., and Zhang, D. (2011) *Polym. Adv. Technol.*, **22**, 759.

51 Desroches, C., Lopes, C., Kessler, V., and Parola, S. (2003) *Dalton Trans.*, 2085.

52 Zhang, X., Li, M., Shi, Z., and Cui, Z. (2011) *Mater. Lett.*, **65**, 1404.

53 Wang, S., Zhao, L., Yang, S., Pang, S., and Cui, Z. (2009) *Mater. Lett.*, **63**, 292.

54 Wang, D., Chen, X., Zhang, X., Wang, W., Liu, Y., and Hu, L. (2009) *Curr. Appl. Phys.*, **9**, S170.

55 Chateau, D., Chaput, F., Lopes, C., Lindgren, M., Brannlund, C., Ohgren, J., Djourelov, N., Nedelec, P., Desroches, C., Eliasson, B., Kindahl, T., Lerouge, F., Andraud, C., and Parola, S. (2012) *ACS Appl. Mater. Interfaces*, **4** 2369.

56 Bouit, P.-A., Maury, O., Feneyrou, P., Parola, S., Kajzar, F., and Andraud, C. (2011) *Multiphoton Processes in Organic Materials and their Applications* (eds I. Rau and F. Kajzar), Archives Contemporaines & Old City Publishing, Paris– Philadelphia, p. 275.

57 Château, D., Bellier, Q., Chaput, F., Feneyrou, P., Berginc, G., Maury, O., Andraud, C., and Parola, S. (2014) *J. Mater. Chem. C*, **2**, 5105.

58 Kang, E.-H., Bu, T., Jin, P., Sun, J., Yanqiang, Y., and Shen, J. (2007) *Langmuir*, **23**, 7594.

59 Heflin, J.R., Guzy, M.T., Neyman, P.J., Gaskins, K.J., Brands, C., Wang, Z., Gibson, H.W., Davis, R.M., and Van Cott, K.E. (2006) *Langmuir*, **22**, 5723.

60 Borges, J., Rodrigues, L.C., Reis, R.L., and Mano, J.F. (2014) *Adv. Funct. Mater.*, **24**, 5624.

61 Lacroix, P.G., Clement, R., Nakatani, K., Zyss, J., and Ledoux, I. (1994) *Science*, **263**, 658.

62 Suzuki, Y., Tenma, Y., Nishioka, Y., and Kawamata, J. (2012) *Chem. Asian J.*, **7**, 1170.

63 Takenawa, R., Komori, Y., Hayashi, S., Kawamata, J., and Kuroda, K. (2001) *Chem. Mater.*, **13**, 3741.

64 Monnier, V., Sanz, N., Botzung-Appert, E., Bacia, M., and Ibanez, A. (2006) *J. Mater. Chem.*, **16**, 1401.

65 Sanz, N., Baldeck, P.L., Nicoud, J.-F., Le Fur, Y., and Ibanez, A. (2001) *Solid State Sci.*, **3**, 867.

66 Sanz, N., Ibanez, A., Morel, Y., and Baldeck, P.L. (2001) *Appl. Phys. Lett.*, **78**, 2569.

67 Wang, I., Baldeck, P.L., Botzung, E., Sanz, N., and Ibanez, A. (2002) *Opt. Mater.*, **21**, 569.

68 Cariati, E., Pizzotti, M., Roberto, D., Tessore, F., and Ugo, R. (2006) *Coord. Chem. Rev.*, **250**, 1210.

69 Ramakrishna Matte, H.S.S., Cheetham, A.K., and Rao, C.N.R. (2009) *Solid State Commun.*, **149**, 908.

70 Wang, C., Zhang, T., and Lin, W. (2012) *Chem. Rev.*, **112**, 1084.
71 Lopez, X., Carbo, J.J., Bo, C., and Poblet, J.M. (2012) *Chem. Soc. Rev.*, **41**, 7537.
72 Compain, J.D., Mialane, P., Dolbecq, A., Marrot, J., Proust, A., Nakatani, K., Yu, P., and Secheresse, F. (2009) *Inorg. Chem.*, **48**, 6222.
73 Song, Y., Janjua, M.R.S.A., Jamil, S., Haroon, M., Nasir, S., Nisar, Z., Zafar, A., Nawaz, N., Batool, A., and Abdul, A. (2014) *Synth. Met.*, **198**, 277.
74 Janjua, M.R.S.A., Liu, C.-G., Guan, W., Zhuang, J., Muhammad, S., Yan, L.-K., and Su, Z.-M. (2009) *J. Phys. Chem. A*, **113**, 3576.
75 Zhang, X., Huang, Y.-Y., Zhang, M.-J., Zhang, J., and Yao, Y.-G. (2012) *Cryst. Growth Des.*, **12**, 3231.
76 Wang, F., Ke, X., Zhao, J., Deng, K., Leng, X., Tian, Z., Wen, L., and Li, D. (2011) *Dalton Trans.*, **40**, 11856.
77 Cariati, E., Macchi, R., Roberto, D., Ugo, R., Galli, S., Casati, N., Macchi, P., Sironi, A., Bogani, L., Caneschi, A., and Gatteschi, D. (2007) *J. Am. Chem. Soc.*, **129**, 9410.
78 Cariati, E., Ugo, R., Santoro, G., Tordin, E., Sorace, L., Caneschi, A., Sironi, A., Macchi, P., and Casati, N. (2010) *Inorg. Chem.*, **49**, 10894.
79 Zhu, Q.-L. and Xu, Q. (2014) *Chem. Soc. Rev.*, **43**, 5468.
80 Genovese, M. and Lian, K. (2015) *Curr. Opin. Solid State Mater. Sci.*, **19**, 126.
81 Hollins, R.C. (1999) *Curr. Opi. Solid State Mater. Sci.*, **4**, 189.
82 Spangler, C.W. (1999) *J. Mater. Chem. A*, **9**, 2013.
83 McKay, T.J., Bolger, J.A., Staromlynska, J., and Davy, J.R. (1998) *J. Chem. Phys.*, **108**, 5537.
84 He, G.S., Bhawalkar, J.D., Zhao, C.F., and Prasad, P.N. (1995) *Appl. Phys. Lett.*, **67**, 2433.
85 He, G.S., Gvishi, R., Prasad, P.N., and Reinhardt, B.A. (1995) *Opt. Commun.*, **117**, 133.
86 Ehrlich, J.E., Wu, X.L., Lee, I.Y.S., Hu, Z.Y., Rockel, H., Marder, S.R., and Perry, J.W. (1997) *Opt. Lett.*, **22**, 1843.
87 Reinhardt, B.A., Brott, L.L., Clarson, S.J., Dillard, A.G., Bhatt, J.C., Kannan, R., Yuan, L.X., He, G.S., and Prasad, P.N. (1998) *Chem. Mater.*, **10**, 1863.
88 Li, C.F., Zhang, L., Yang, M., Wang, H., and Wang, Y.X. (1994) *Phys. Rev. A*, **49**, 1149.
89 Perry, J.W., Mansour, K., Lee, I.Y.S., Wu, X.L., Bedworth, P.V., Chen, C.T., Marder, D.N.S.R., Miles, P., Wada, T., Tian, M., and Sasabe, H. (1996) *Science*, **273**, 1533.
90 Miller, M.J., Mott, A.G., and Ketchel, B.P. (1998) *Proc. SPIE*, **24**, 3472.
91 Miles, P.A. (1994) *Appl. Opt.*, **33**, 6965.
92 Justus, B.L., Huston, A.L., and Campillo, A.J. (1993) *Appl. Phys. Lett.*, **63**, 1483.
93 Doyle, J.J., Wang, J., O'Flaherty, S.M., Chen, Y., Slodek, A., Hegarty, T., Carpenter Ii, L.E., Wöhrle, D., Hanack, M., and Blau, W.J. (2008) *J. Opt.*, **10**, 075101.
94 Fuqua, P.D., Mansour, K., Alvarez, J.D., Marder, S.R., Perry, J.W., and Dunn, B.S. (1992) *Proc. SPIE*, **1758**, 499.
95 Jiang, H., DeRosa, M., Su, W., Brant, M., McLean, D., and Bunning, T. (1998) *Proc. SPIE*, **3472**, 157.
96 Innocenzi, P., Brusatin, G., Guglielmi, M., Signorini, R., Bozio, R., and Maggini, M. (2000) *J. Non-Cryst. Solids*, **265**, 68.
97 Signorini, R., Meneghetti, M., Bozio, R., Maggini, M., Scorrano, G., Prato, M., Brusatin, G., Innocenzi, P., and Guglielmi, M. (2000) *Carbon*, **38**, 1653.
98 Brusatin, G., Guglielmi, M., Innocenzi, P., Martucci, A., and Scarinci, G. (2000) *J. Electroceram.*, **4**, 151.
99 Kopitkovas, G., Chugreev, A., Nierengarten, J.F., Rio, Y., Rehspringer, J.L., and Hönerlage, B. (2004) *Opt. Mater.*, **27**, 285.
100 Innocenzi, P. and Brusatin, G. (2001) *Chem. Mater.*, **13**, 3126.
101 Sanz, N., Baldeck, P.L., and Ibanez, A. (2000) *Synth. Met.*, **115**, 229.
102 Kuznetsova, R.T., Savenkova, N.S., Maier, G.V., Arabei, S.M., Pavich, T.A., and Solov'ev, K.N. (2007) *J. Appl. Spectrosc.*, **74**, 485.
103 Staromlynska, J., McKay, T.J., Bolger, J.A., and Davy, J.R. (1998) *J. Opt. Soc. Am. B*, **15**, 1731.
104 Cooper, T.M., McLean, D.G., and Rogers, J.E. (2001) *Chem. Phys. Lett.*, **349**, 31.

105 Rogers, J.E., Cooper, T.M., Fleitz, P.A., Glass, D.J., and McLean, D.G. (2002) *J. Phys. Chem. A*, **106**, 10108.

106 Vestberg, R., Westlund, R., Eriksson, A., Lopes, C., Carlsson, M., Eliasson, B., Glimsdal, E., Lindgren, M., and Malmström, E. (2006) *Macromolecules*, **39**, 2238.

107 Lindgren, M., Minaev, B., Glimsdal, E., Vestberg, R., Westlund, R., and Malmstrom, E. (2007) *J. Lumin.*, **124**, 302.

108 Glimsdal, E., Carlsson, M., Eliasson, B., Minaev, B., and Lindgren, M. (2007) *J. Phys. Chem. A*, **111**, 244.

109 Yeates, A.T., Glimsdal, E., Eriksson, A., Vestberg, R., Malmstrom, E., and Lindgren, M. (2005) *Proc. SPIE*, **5934**, 59340N.

110 Vicente, J., Chicote, M.-T., Alvarez-Falcon, M.M., and Jones, P.G. (2005) *Organometallics*, **24**, 2764.

111 Tao, C.H., Zhu, N., and Yam, V.W. (2005) *Chem. Eur. J.*, **11**, 1647.

112 Zhou, G.J., Wong, W.Y., Ye, C., and Lin, Z. (2007) *Adv. Funct. Mater.*, **17**, 963.

113 Vestberg, R., Westlund, R., Carlsson, M., Eliasson, B., Glimsdal, E., Örtengren, J., Lindgren, M., and Malmström, E. (2005) *Polym. Mater. Sci. Eng.*, **92**, 622.

114 Zhou, G.J. and Wong, W.Y. (2011) *Chem. Soc. Rev.*, **40**, 2541.

115 Guo, F., Sun, W., Liu, Y., and Schanze, K. (2005) *Inorg. Chem.*, **44**, 4055.

116 Price, R.S., Dubinina, G., Wicks, G., Drobizhev, M., Rebane, A., and Schanze, K.S. (2015) *ACS Appl. Mater. Interfaces*, **7**, 10795.

117 Morales-Saavedra, O.G., Huerta, G., Ortega-Martínez, R., and Fomina, L. (2007) *J. Non-Cryst. Solids*, **353**, 2557.

118 Morales-Saavedra, O.G. and Rivera, E. (2006) *Polymer*, **47**, 5330.

119 Sharma, S., Mohan, D., and Ghoshal, S.K. (2008) *Opt. Commun.*, **281**, 2923.

120 Su, X., Xu, H., Deng, Y., Li, J., Zhang, W., and Wang, P. (2008) *Mater. Lett.*, **62**, 3818.

121 Su, X., Guang, S., Xu, H., Yang, J., and Song, Y. (2010) *Dyes Pigm.*, **87**, 69.

122 Su, X., Guang, S., Li, C., Xu, H., Liu, X., Wang, X., and Song, Y. (2010) *Macromolecules*, **43**, 2840.

123 Sun, X., Xiong, Y., Chen, P., Lin, J., Ji, W., Lim, J.H., Yang, S.S., Hagan, D.J., and Van Stryland, E.W. (2000) *Appl. Opt.*, **39**, 1998.

124 Vivien, L., Riehl, D., Lancon, P., Hache, F., and Anglaret, E. (2001) *Opt. Lett.*, **26**, 223.

125 Zhang, B., Chen, Y., Wang, J., Blau, W.J., Zhuang, X., and He, N. (2010) *Carbon*, **48**, 1738.

126 Cardiano, P., Fazio, E., Lazzara, G., Manickam, S., Milioto, S., Neri, F., Mineo, P.G., Piperno, A., and Lo Schiavo, S. (2015) *Carbon*, **86**, 325.

127 Lim, G.K., Chen, Z.L., Clark, J., Goh, R.G.S., Ng, W.H., Tan, H.W., Friend, R.H., Ho, P.K.H., and Chua, L.L. (2011) *Nat. Photonics*, **5**, 554.

128 Liu, Z.B., Wang, Y., Zhang, X.L., Xu, Y.F., Chen, Y.S., and Tian, J.G. (2009) *Appl. Phys. Lett.*, **94**, 021902.

129 Wang, J., Hernandez, Y., Lotya, M., Coleman, J.N., and Blau, W.J. (2009) *Adv. Mater.*, **21**, 2430.

130 Zhang, L., Zou, L.-Y., Guo, J.-F., Ren, A.-M., Wang, D., and Feng, J.-K. (2014) *New J. Chem.*, **38**, 5391.

131 Song, W., He, C., Zhang, W., Gao, Y., Yang, Y., Wu, Y., Chen, Z., Li, X., and Dong, Y. (2014) *Carbon*, **77**, 1020.

132 Zhang, X.-L., Zhao, X., Liu, Z.-B., Shi, S., Zhou, W.-Y., Tian, J.-G., Xu, Y.-F., and Chen, Y.-S. (2011) *J. Opt.*, **13**, 075202.

133 Zheng, X., Feng, M., and Zhan, H. (2013) *J. Mater. Chem. C*, **1**, 6759.

134 Zheng, X., Feng, M., Li, Z., Song, Y., and Zhan, H. (2014) *J. Mater. Chem. C*, **2**, 4121.

135 Tao, L., Zhou, B., Bai, G., Wang, Y., Yu, S.F., Lau, S.P., Tsang, Y.H., Yao, J., and Xu, D. (2013) *J. Phys. Chem. C*, **117**, 23108.

136 Zheng, C., Zheng, Y., Chen, W., and Wei, L. (2015) *Opt. Laser Technol.*, **68**, 52.

137 Ouyang, Q., Xu, Z., Lei, Z., Dong, H., Yu, H., Qi, L., Li, C., and Chen, Y. (2014) *Carbon*, **67**, 214.

138 Ouyang, Q., Yu, H., Xu, Z., Zhang, Y., Li, C., Qi, L., and Chen, Y. (2013) *Appl. Phys. Lett.*, **102**, 031912.

139 de Aberasturi, D.J., Serrano-Montes, A.B., and Liz-Marzan, L.M. (2015) *Adv. Opt. Mater.*, **3**, 602.

140 Lakowicz, J.R. (2006) *Plasmonics*, **1**, 5.
141 Lauchner, A., Schlather, A.E., Manjavacas, A., Cui, Y., McClain, M.J., Stec, G.J., de Abajo, F.J.G., Nordlander, P., and Halas, N.J. (2015) *Nano Lett.*, **15**, 6208.
142 Zheng, Y.B., Kiraly, B., Weiss, P.S., and Huang, T.J. (2012) *Nanomedicine*, 7, 751.
143 Guerrero-Martinez, A., Grzelczak, M., and Liz-Marzan, L.M. (2012) *ACS Nano*, **6**, 3655.
144 Liz-Marzan, L.M. (2013) *J. Phys. Chem. Lett.*, **4**, 1197.
145 Kauranen, M. and Zayats, A.V. (2012) *Nat. Photonics*, **6**, 737.
146 Dreaden, E.C., Alkilany, A.M., Huang, X.H., Murphy, C.J., and El-Sayed, M.A. (2012) *Chem. Soc. Rev.*, **41**, 2740.
147 Daniel, M.C. and Astruc, D. (2004) *Chem. Rev.*, **104**, 293.
148 Zhao, P.X., Li, N., and Astruc, D. (2013) *Coord. Chem. Rev.*, **257**, 638.
149 Jain, P.K., Huang, X.H., El-Sayed, I.H., and El-Sayed, M.A. (2008) *Acc. Chem. Res.*, **41**, 1578.
150 Zeng, S.W., Baillargeat, D., Ho, H.P., and Yong, K.T. (2014) *Chem. Soc. Rev.*, **43**, 3426.
151 Giljohann, D.A., Seferos, D.S., Daniel, W.L., Massich, M.D., Patel, P.C., and Mirkin, C.A. (2010) *Angew. Chem., Int. Ed.*, **49**, 3280.
152 Bauch, M., Toma, K., Toma, M., Zhang, Q., and Dostalek, J. (2013) *Plasmonics*, **9**, 781.
153 Webb, J.A. and Bardhan, R. (2014) *Nanoscale*, **6**, 2502.
154 Sau, T.K., Rogach, A.L., Jackel, F., Klar, T.A., and Feldmann, J. (2010) *Adv. Mater.*, **22**, 1805.
155 Lakowicz, J.R., Ray, K., Chowdhury, M., Szmacinski, H., Fu, Y., Zhang, J., and Nowaczyk, K. (2008) *Analyst*, **133**, 1308.
156 Kriebig U. and Vollmer M. (eds) (1995) *Optical Properties of Metal Clusters*, Springer-Verlag, Berlin Heidelberg, Germany.
157 Bohren C.F. and Huffman D.R. (eds) (1983) *Absorption and Scattering of Light by Small Particles*, Wiley-Interscience, New York.
158 Henglein, A. (1989) *Chem. Rev.*, **89**, 1861.
159 Kreibig, U. and Genzel, L. (1985) *Surf. Sci.*, **156**, 678.
160 Link, S. and El-Sayed, M.A. (1999) *J. Phys. Chem. B*, **103**, 4212.
161 Link, S., Wang, Z.L., and El-Sayed, M.A. (1999) *J. Phys. Chem. B*, **103**, 3529.
162 Hutter, E. and Fendler, J.H. (2004) *Adv. Mater.*, **16**, 1685.
163 Eustis, S. and El-Sayed, M.A. (2006) *Chem. Soc. Rev.*, **35**, 209.
164 Liu, N., Hentschel, M., Weiss, T., Alivisatos, A.P., and Giessen, H. (2011) *Science*, **332**, 1407.
165 Luther, J.M., Jain, P.K., Ewers, T., and Alivisatos, A.P. (2011) *Nat. Mater.*, **10**, 361.
166 Jain, P.K., Lee, K.S., El-Sayed, I.H., and El-Sayed, M.A. (2006) *J. Phys. Chem. B*, **110**, 7238.
167 Link, S., Mohamed, M.B., and El-Sayed, M.A. (1999) *J. Phys. Chem. B*, **103**, 3073.
168 Grzelczak, M., Perez-Juste, J., Mulvaney, P., and Liz-Marzan, L.M. (2008) *Chem. Soc. Rev.*, **37**, 1783.
169 Chateau, D., Liotta, A., Vadcard, F., Navarro, J.R., Chaput, F., Lerme, J., Lerouge, F., and Parola, S. (2015) *Nanoscale*, **7**, 1934.
170 Yin, Y., Erdonmez, C., Aloni, S., and Alivisatos, A.P. (2006) *J. Am. Chem. Soc.*, **128**, 12671.
171 Metraux, G.S., Cao, Y.C., Jin, R.C., and Mirkin, C.A. (2003) *Nano Lett.*, **3**, 519.
172 Sherry, L.J., Jin, R.C., Mirkin, C.A., Schatz, G.C., and Van Duyne, R.P. (2006) *Nano Lett.*, **6**, 2060.
173 Millstone, J.E., Hurst, S.J., Metraux, G.S., Cutler, J.I., and Mirkin, C.A. (2009) *Small*, **5**, 646.
174 Zhang, J., Li, S.Z., Wu, J.S., Schatz, G.C., and Mirkin, C.A. (2009) *Angew. Chem., Int. Ed.*, **48**, 7787.
175 Zhang, J.A., Langille, M.R., Personick, M.L., Zhang, K., Li, S.Y., and Mirkin, C.A. (2010) *J. Am. Chem. Soc.*, **132**, 14012.
176 Personick, M.L., Langille, M.R., Zhang, J., and Mirkin, C.A. (2011) *Nano Lett.*, **11**, 3394.
177 Langille, M.R., Personick, M.L., Zhang, J., and Mirkin, C.A. (2012) *J. Am. Chem. Soc.*, **134**, 14542.
178 Ringe, E., Zhang, J., Langille, M.R., Mirkin, C.A., Marks, L.D., and Van

Duyne, R.P. (2012) *Nanotechnology*, **23**, 444005.

179 Personick, M.L., Langille, M.R., Wu, J.S., and Mirkin, C.A. (2013) *J. Am. Chem. Soc.*, **135**, 3800.

180 Personick, M.L., Langille, M.R., Zhang, J., Wu, J.S., Li, S.Y., and Mirkin, C.A. (2013) *Small*, **9**, 1947.

181 Padmos, J.D., Personick, M.L., Tang, Q., Duchesne, P.N., Jiang, D.E., Mirkin, C.A., and Zhang, P. (2015) *Nat. Commun.*, **6**, 7664.

182 Lux, F., Lerouge, F., Bosson, J., Lemercier, G., Andraud, C., Vitrant, G., Baldeck, P.L., Chassagneux, F., and Parola, S. (2009) *Nanotechnology*, **20**, 355603.

183 Zaiba, S., Lerouge, F., Gabudean, A.M., Focsan, M., Lerme, J., Gallavardin, T., Maury, O., Andraud, C., Parola, S., and Baldeck, P.L. (2011) *Nano Lett.*, **11**, 2043.

184 Navarro, J.R., Manchon, D., Lerouge, F., Blanchard, N.P., Marotte, S., Leverrier, Y., Marvel, J., Chaput, F., Micouin, G., Gabudean, A.M., Mosset, A., Cottancin, E., Baldeck, P.L., Kamada, K., and Parola, S. (2012) *Nanotechnology*, **23** 465602.

185 Navarro, J.R., Manchon, D., Lerouge, F., Cottancin, E., Lerme, J., Bonnet, C., Chaput, F., Mosset, A., Pellarin, M., and Parola, S. (2012) *Nanotechnology*, **23**, 145707.

186 Johnson, C.J., Dujardin, E., Davis, S.A., Murphy, C.J., and Mann, S. (2002) *J. Mater. Chem.*, **12**, 1765.

187 Busbee, B.D., Obare, S.O., and Murphy, C.J. (2003) *Adv. Mater.*, **15**, 414.

188 Orendorff, C.J., Sau, T.K., and Murphy, C.J. (2006) *Small*, **2**, 636.

189 Murphy, C.J., Thompson, L.B., Chernak, D.J., Yang, J.A., Sivapalan, S.T., Boulos, S.P., Huang, J.Y., Alkilany, A.M., and Sisco, P.N. (2011) *Curr. Opin. Colloid Interface Sci.*, **16**, 128.

190 Lohse, S.E. and Murphy, C.J. (2013) *Chem. Mater.*, **25**, 1250.

191 Grzelczak, M., Sánchez-Iglesias, A., Rodríguez-González, B., Alvarez-Puebla, R., Pérez-Juste, J., and Liz-Marzán, L.M. (2008) *Adv. Funct. Mater.*, **18**, 3780.

192 Goldmann, C., Lazzari, R., Paquez, X., Boissiere, C., Ribot, F., Sanchez, C., Chaneac, C., and Portehault, D. (2015) *ACS Nano*, **9**, 7572.

193 Thomas, M., Greffet, J.-J., Carminati, R., and Arias-Gonzalez, J.R. (2004) *Appl. Phys. Lett.*, **85**, 3863.

194 Tam, F., Goodrich, G.P., Johnson, B.R., and Halas, N.J. (2007) *Nano Lett.*, 7, 496.

195 Meng, X., Grote, R.R., Dadap, J.I., Panoiu, N.C., and Osgood, R.M. (2014) *Opt. Express*, **22**, 22018.

196 Lal, S., Link, S., and Halas, N.J. (2007) *Nat. Photonics*, **1**, 641.

197 Xu, Y.K., Hwang, S., Kim, S., and Chen, J.Y. (2014) *ACS Appl. Mater. Interfaces*, **6**, 5619.

198 Hone, D.C., Walker, P.I., Evans-Gowing, R., FitzGerald, S., Beeby, A., Chambrier, I., Cook, M.J., and Russell, D.A. (2002) *Langmuir*, **18**, 2985.

199 Stuchinskaya, T., Moreno, M., Cook, M.J., Edwards, D.R., and Russell, D.A. (2011) *Photochem. Photobiol. Sci.*, **10**, 822.

200 Navarro, J.R., Lerouge, F., Cepraga, C., Micouin, G., Favier, A., Chateau, D., Charreyre, M.T., Lanoe, P.H., Monnereau, C., Chaput, F., Marotte, S., Leverrier, Y., Marvel, J., Kamada, K., Andraud, C., Baldeck, P.L., and Parola, S. (2013) *Biomaterials*, **34** 8344.

201 Navarro, J.R., Lerouge, F., Micouin, G., Cepraga, C., Favier, A., Charreyre, M.T., Blanchard, N.P., Lerme, J., Chaput, F., Focsan, M., Kamada, K., Baldeck, P.L., and Parola, S. (2014) *Nanoscale*, **6** 5138.

202 Sivapalan, S.T., Vella, J.H., Yang, T.K., Dalton, M.J., Swiger, R.N., Haley, J.E., Cooper, T.M., Urbas, A.M., Tan, L.S., and Murphy, C.J. (2012) *Langmuir*, **28**, 9147.

203 Navarro, J.R., Liotta, A., Faure, A.C., Lerouge, F., Chaput, F., Micouin, G., Baldeck, P.L., and Parola, S. (2013) *Langmuir*, **29**, 10915.

204 Ni, W., Yang, Z., Chen, H., Li, L., and Wang, J. (2008) *J. Am. Chem. Soc.*, **130**, 6692.

205 Schneider, G., Decher, G., Nerambourg, N., Praho, R., Werts, M.H., and Blanchard-Desce, M. (2006) *Nano Lett.*, **6**, 530.

206 Gole, A. and Murphy, C.J. (2005) *Chem. Mater.*, **17**, 1325.

207 Ding, H., Yong, K.-T., Roy, I., Pudavar, H.E., Law, W.C., Bergey, E.J., and Prasad, P.N. (2007) *J. Phys. Chem. C*, **111**, 2552.

208 Bardhan, R., Grady, N.K., Cole, J.R., Joshi, A., and Halas, N.J. (2009) *ACS Nano*, **3**, 744.
209 Gu, X., Wu, Y., Zhang, L., Liu, Y., Li, Y., Yan, Y., and Wu, D. (2014) *Nanoscale*, **6**, 8681.
210 Yuan, P., Ma, R., Guan, Z., Gao, N., and Xu, Q.H. (2014) *ACS Appl. Mater. Interfaces*, **6**, 13149.
211 Shimron, S., Cecconello, A., Lu, C.H., and Willner, I. (2013) *Nano Lett.*, **13**, 3791.
212 Jiang, Z., Dong, B., Chen, B., Wang, J., Xu, L., Zhang, S., and Song, H. (2013) *Small*, **9**, 604.
213 Huang, P., Lin, J., Wang, S., Zhou, Z., Li, Z., Wang, Z., Zhang, C., Yue, X., Niu, G., Yang, M., Cui, D., and Chen, X. (2013) *Biomaterials*, **34**, 4643.
214 Zhao, T., Yu, K., Li, L., Zhang, T., Guan, Z., Gao, N., Yuan, P., Li, S., Yao, S.Q., Xu, Q.H., and Xu, G.Q. (2014) *ACS Appl. Mater. Interfaces*, **6**, 2700.
215 Abadeer, N.S., Brennan, M.R., Wilson, W.L., and Murphy, C.J. (2014) *ACS Nano*, **8**, 8392.
216 Ke, X., Wang, D., Chen, C., Yang, A., Han, Y., Ren, L., Li, D., and Wang, H. (2014) *Nanoscale Res. Lett.*, **9**, 2492.
217 Bai, Z., Chen, R., Si, P., Huang, Y., Sun, H., and Kim, D.H. (2013) *ACS Appl. Mater. Interfaces*, **5**, 5856.
218 Pang, Y., Rong, Z., Wang, J., Xiao, R., and Wang, S. (2015) *Biosens. Bioelectron.*, **66**, 527.
219 Niu, N., He, F., Ma, P., Gai, S., Yang, G., Qu, F., Wang, Y., Xu, J., and Yang, P. (2014) *ACS Appl. Mater. Interfaces*, **6**, 3250.
220 Terentyuk, G., Panfilova, E., Khanadeev, V., Chumakov, D., Genina, E., Bashkatov, A., Tuchin, V., Bucharskaya, A., Maslyakova, G., Khlebtsov, N., and Khlebtsov, B. (2014) *Nano Res.*, **7**, 325.
221 Li, Y., Wen, T., Zhao, R., Liu, X., Ji, T., Wang, H., Shi, X., Shi, J., Wei, J., Zhao, Y., Wu, X., and Nie, G. (2014) *ACS Nano*, **8**, 11529.
222 Croissant, J., Maynadier, M., Mongin, O., Hugues, V., Blanchard-Desce, M., Chaix, A., Cattoen, X., Man, M.W.C., Gallud, A., Gary-Bobo, M., Garcia, M., Raehm, L., and Durand, J.O. (2015) *Small*, **11** 295.
223 Noginov, M.A., Zhu, G., Belgrave, A.M., Bakker, R., Shalaev, V.M., Narimanov, E.E., Stout, S., Herz, E., Suteewong, T., and Wiesner, U. (2009) *Nature*, **460**, 1110.
224 Maryuri Roca, A.J.H. (2008) *J. Am. Chem. Soc.*, **130**, 14273.
225 Hembury, M., Chiappini, C., Bertazzo, S., Kalber, T.L., Drisko, G.L., Ogunlade, O., Walker-Samuel, S., Krishna, K.S., Jumeaux, C., Beard, P., Kumar, C.S., Porter, A.E., Lythgoe, M.F., Boissiere, C., Sanchez, C., and Stevens, M.M. (2015) *Proc. Natl. Acad. Sci. USA*, **112** 1959.
226 Oldenburg, R.D.A.S.J., Westcott, S.L., and Halas, N.J. (1998) *Chem. Phys. Lett.*, **288**, 243.
227 Oldenburg, S.J., Jackson, J.B., Westcott, S.L., and Halas, N.J. (1999) *Appl. Phys. Lett.*, **75**, 2897.
228 Han-Pu Liang, L.-J.W., Bai, Chun-Li, and Jiang, Li (2005) *J. Phys. Chem. B*, **109**, 7795.
229 Gobin, M.H.L.A.M., Halas, N.J., James, W.D., Drezek, R.A., and West, J.L. (2007) *Nano Lett.*, **7**, 1929.
230 Bardhan, R., Chen, W., Perez-Torres, C., Bartels, M., Huschka, R.M., Zhao, L.L., Morosan, E., Pautler, R.G., Joshi, A., and Halas, N.J. (2009) *Adv. Funct. Mater.*, **19**, 3901.
231 Gobin, A.M., O'Neal, D.P., Watkins, D.M., Halas, N.J., Drezek, R.A., and West, J.L. (2005) *Lasers Surg. Med.*, **37**, 123.
232 Lal, S., Clare, S.E., and Halas, N.J. (2008) *Acc. Chem. Res.*, **41**, 1842.
233 O'Neal, D.P., Hirsch, L.R., Halas, N.J., Payne, J.D., and West, J.L. (2004) *Cancer Lett.*, **209**, 171.
234 Bernardi, R.J., Lowery, A.R., Thompson, P.A., Blaney, S.M., and West, J.L. (2008) *J. Neurooncol.*, **86**, 165.
235 Westcott, S.L., Oldenburg, S.J., Lee, T.R., and Halas, N.J. (1998) *Langmuir*, **14**, 5396.
236 Li, X., Li, Y., Yang, C., and Li, Y. (2004) *Langmuir*, **20**, 3734.
237 Shi, W., Sahoo, Y., Swihart, M.T., and Prasad, P.N. (2005) *Langmuir*, **21**, 1610.
238 Yugang, S., Mayers, B.T.M., and Xia, Y. (2002) *Nano Lett.*, **2**, 481.
239 Yang, M., Yang, X., and Huai, L. (2008) *Appl. Phys. A*, **92**, 367.

240 Jin, Y. and Gao, X. (2009) *J. Am. Chem. Soc.*, **131**, 17774.
241 Gabudean, A.-M., Lerouge, F., Gallavardin, T., Iosin, M., Zaiba, S., Maury, O., Baldeck, P.L., Andraud, C., and Parola, S. (2011) *Opt. Mater.*, **33**, 1377.
242 De Luca, A., Dhama, R., Rashed, A.R., Coutant, C., Ravaine, S., Barois, P., Infusino, M., and Strangi, G. (2014) *Appl. Phys. Lett.*, **104**, 103103.
243 Zhang, P. and Guo, Y. (2009) *J. Am. Chem. Soc.*, **131**, 3808.
244 Niikura, K., Iyo, N., Matsuo, Y., Mitomo, H., and Ijiro, K. (2013) *ACS Appl. Mater. Interfaces*, **5**, 3900.
245 Behar-Levy, H. and Avnir, D. (2002) *Chem. Mater.*, **14**, 1736.
246 Yosef, I. and Avnir, D. (2006) *Chem. Mater.*, **18**, 5890.
247 Yosef, I., Abu-Reziq, R., and Avnir, D. (2008) *J. Am. Chem. Soc.*, **130**, 11880.
248 Ben-Knaz, R. and Avnir, D. (2009) *Biomaterials*, **30**, 1263.
249 Ben-Efraim, Y. and Avnir, D. (2012) *J. Mater. Chem.*, **22**, 17595.
250 Aouat, Y., Marom, G., Avnir, D., Gelman, V., Shter, G.E., and Grader, G.S. (2013) *J. Phys. Chem. C*, **117**, 22325.
251 Aouat, Y., Azan, Y., Marom, G., and Avnir, D. (2013) *Sci. Adv. Mater.*, **5**, 598.
252 Naor, H. and Avnir, D. (2014) *J. Mater. Chem. C*, **2**, 7768.
253 Avnir, D. (2014) *Acc. Chem. Res.*, **47**, 579.
254 Qu, S., Gao, Y., Jiang, X., Zeng, H., Song, Y., Qiu, J., Zhu, C., and Hirao, K. (2003) *Opt. Commun.*, **224**, 321.
255 Qu, S.L., Zhao, C.J., Jiang, X.W., Fang, G.Y., Gao, Y.C., Zeng, H.D., Song, Y.L., Qui, J.R., Zhu, C.S., and Hirao, K. (2003) *Chem. Phys. Lett.*, **368**, 352.
256 Mangelson, B.F., Jones, M.R., Park, D.J., Shade, C.M., Schatz, G.C., and Mirkin, C.A. (2014) *Chem. Mater.*, **26**, 3818.
257 Wang, D., Zhou, Z.H., Yang, H., Shen, K.B., Huang, Y., and Shen, S. (2012) *J. Mater. Chem.*, **22**, 16306.
258 Besson, S., Gacoin, T., Ricolleau, C., and Boilot, J.P. (2003) *Chem. Commun.*, 360.
259 Gacoin, T., Besson, S., and Boilot, J.P. (2006) *J. Phys. Condens. Matter*, **18**, S85.
260 Wu, P.-W., Cheng, W., Martini, I.B., Dunn, B., Schwartz, B.J., and Yablonovitch, E. (2000) *Adv. Mater.*, **12**, 1438.
261 Battie, Y., Destouches, N., Bois, L., Chassagneux, F., Moncoffre, N., Toulhoat, N., Jamon, D., Ouerdane, Y., Parola, S., and Boukenter, A. (2009) *J. Nanopart. Res.*, **12**, 1073.
262 Bois, L., Bessueille, F., Chassagneux, F., Battie, Y., Destouches, N., Hubert, C., Boukenter, A., and Parola, S. (2008) *Colloids Surf. A*, **325**, 86.
263 Bois, L., Chassagneux, F., Battie, Y., Bessueille, F., Mollet, L., Parola, S., Destouches, N., Toulhoat, N., and Moncoffre, N. (2010) *Langmuir*, **26**, 1199.
264 Bois, L., Chassagneux, F., Desroches, C., Battie, Y., Destouches, N., Gilon, N., Parola, S., and Stephan, O. (2010) *Langmuir*, **26**, 8729.
265 Bois, L., Chassagneux, F., Parola, S., Bessueille, F., Battie, Y., Destouches, N., Boukenter, A., Moncoffre, N., and Toulhoat, N. (2009) *J. Solid State Chem.*, **182**, 1700.
266 De, S. and De, G. (2008) *J. Phys. Chem. C*, **112**, 10378.
267 Ferrara, M.C., Mirenghi, L., Mevoli, A., and Tapfer, L. (2008) *Nanotechnology*, **19**, 365706.
268 Pal, S. and De, G. (2008) *Phys. Chem. Chem. Phys.*, **10**, 4062.
269 Martinez, E.D., Boissiere, C., Grosso, D., Sanchez, C., Troiani, H., and Soler-Illia, G.J.A.A. (2014) *J. Phys. Chem. C*, **118**, 13137.
270 Formanek, F., Takeyasu, N., Tanaka, T., Chiyoda, K., Ishikawa, A., and Kawata, S. (2006) *Opt. Express*, **14**, 800.
271 Kaneko, K., Sun, H.-B., Duan, X.-M., and Kawata, S. (2003) *Appl. Phys. Lett.*, **83**, 1426.
272 Kalfagiannis, N., Karagiannidis, P.G., Pitsalidis, C., Hastas, N., Panagiotopoulos, N.T., Patsalas, P., and Logothetidis, S. (2014) *Thin Solid Films*, **560**, 27.
273 Stellacci, F., Bauer, C.A., Meyer-Friedrichsen, T., Wenseleers, W., Alain, V., Kuebler, S.M., Pond, S.J.K., Zhang, Y., Marder, S.R., and Perry, J.W. (2002) *Adv. Mater.*, **14**, 194.

274 Porel, S., Venkatram, N., Rao, D.N., and Radhakrishnan, T.P. (2007) *J. Appl. Phys.*, **102**, 033107-1.

275 Porel, S., Venkatram, N., Rao, D.N., and Radhakrishnan, T.P. (2007) *J Nanosci. Nanotechnol.*, **7**, 1887.

276 Sun, W., Dai, Q., Worden, J.G., and Huo, Q. (2005) *J. Phys. Chem. B*, **109**, 20854.

277 Hale, G.D., Jackson, J.B., Shmakova, O.E., Lee, T.R., and Halas, N.J. (2001) *Appl. Phys. Lett.*, **78**, 1502.

278 Peng, J., Xu, X., Tian, Y., Wang, J., Tang, F., and Li, L. (2014) *Appl. Phys. Lett.*, **105**, 173301.

279 Lundén, H., Liotta, A., Chateau, D., Lerouge, F., Chaput, F., Parola, S., Brännlund, C., Ghadyani, Z., Kildemo, M., Lindgren, M., and Lopes, C. (2015) *J. Mater. Chem. C*, **3**, 1026.

280 Chateau, D., Liotta, A., Lundén, H., Lerouge, F., Chaput, F., Krein, D., Cooper, T., Lopes, C., El-Amay, A.A.G., Lindgren, M., and Parola, S. (2016) *Adv. Funct. Mater.*, **26** (33), 6005.

281 Ishifuji, M., Mitsuishi, M., and Miyashita, T. (2009) *J. Am. Chem. Soc.*, **131**, 4418.

282 Mitsuishi, M., Ishifuji, M., Endo, H., Tanaka, H., and Miyashita, T. (2007) *Polym. J.*, **39**, 411.

283 Zhang, W., Chen, Y., Gan, L., Qing, J., Zhou, X., Huang, Y., Yang, Y., Zhang, Y., Ou, J., Chen, X., and Zhang, M. Qiu (2014) *J. Phys. Chem. Solids*, **75**, 1340.

284 Ferry, V.E., Munday, J.N., and Atwater, H.A. (2010) *Adv. Mater.*, **22**, 4794.

285 Wu, J.-L., Chen, F.-C., Hsiao, Y.-S., Chien, F.-C., Chen, P., Kuo, C.-H., Huang, M.H., and Hsu, C.-S. (2011) *ACS Nano*, **5**, 959.

286 Standridge, S.D., Schatz, G.C., and Hupp, J.T. (2009) *J. Am. Chem. Soc.*, **131**, 8407.

13
Bioactive Glasses

Hirotaka Maeda and Toshihiro Kasuga

Division of Advanced Ceramics, Nagoya Institute of Technology, Gokiso-cho, Showa-ku, Nagoya 466-8555, Japan

13.1
Introduction

Biological grafts or synthetic materials are used to replace or supplement the functions of damaged or diseased living bones. Although the gold standard treatment is to use autografts, their wider application is limited because of restricted supply and the need for secondary surgery. The clinical needs for synthetic bone substitutes for replacing bone defects have been growing recently in direct relation to the increase in human population [1]. The biomaterials employed in bone substitutes should meet certain criteria to ensure biocompatibility and bioactivity. For a material to display bioactive features, instigation of a specific biological response at the interface of the material, resulting in the formation of a bond between the living bone and the material is required [2]. Since Hench first discovered bioactive glasses (45S5 S, Bioglass®) in 1969 [3], the use of bioactive glasses in the biomedical field has generated increasing interest.

Glass consists of an amorphous structure without a long-range order. Specially, the glass structure consists of three components, that is, network former, network modifier, and intermediate oxides. SiO_2, P_2O_5, and B_2O_3 act as linking or polymerizing anionic complexes as a network former, which can form the glass structure without additional components. Network modifiers convert bridging oxygen (e.g., Si—O—Si) into nonbridging oxygen (e.g., Si—O$^-$M$^+$ $^+$M$^-$O—Si, where M$^+$ is the cation of the network modifier), thereby affording the preparation of different glass structure. Examples of typical modifiers are alkali or alkali-earth metal oxides. Intermediate oxides can depolymerize the glass network or act as a network former or a network modifier depending on the coordination number.

Many research studies have shown that several types of inorganic ions, such as silicate and calcium ions, released from bioactive glass can enhance the bone-forming properties of osteoblasts owing to gene activation [4]. The effect of inorganic ions was extensively reviewed by Boccaccini and his coworkers, as summarized in Table 13.1 [5]. A new approach to designing biomaterials involves activating the interaction between host tissue and the material in addition to ensuring bioactive properties; such as types of materials are known as third-generation biomaterials. More attention has been paid to glass as a novel bioactive material owing to its great advantage, whereby its chemical, physical, and biological properties can be manipulated by adjusting its chemical compositions. Thus, bioactive glass can be classified into three groups, that is, silicate-, phosphate-, and borate-based glass systems depending on the network former oxide. These types of bioactive glasses are reviewed in this chapter.

13.2
Silicate Glasses

13.2.1
Melt-Quenched Derived Glasses

Silicate-based bioactive glass (45S5) was discovered in 1969 and has been used in clinical applications since 1985, as approved by US Food and Drug Administration (FDA) via the 510 (k) process. Silicate-based bioactive glass is typically composed of 46.1 mol% SiO_2, 24.4 mol% Na_2O, 26.9 mol% CaO, and 2.6 mol% P_2O_5. Since the discovery of 45S5, many research studies have reported different types of silicate-based bioactive glasses with different chemical compositions for use in various medical applications. The 10 most important publication milestones from the concept to the clinical applications of Bioglass® have been reviewed therein [6]. Examples of other typical glass compositions (in mol%) are 53.9SiO_2–22.7Na_2O–21.8CaO–1.7P_2O_5 (S53P4) [7], 49.46SiO_2–26.38Na_2O–23.08CaO–1.07P_2O_5 (ICIE1) [8], 54.6SiO_2–6Na_2O–22.1CaO–1.7P_2O_5–7.9K_2O–7.7MgO (13–93) [9], 49.4SiO_2–14.9Na_2O–16.6CaO–2.5P_2O_5–3.5K_2O–13.2MgO (6P50) [10].

The use of bioactive glasses results in energetic responses after implantation in the bone defects as shown in Figure 13.1 [11]. Chemical interactions at the interface of the glass lead to formation of hydroxycarbonate apatite (HCA) on the surface. HCA formation is a key factor to generating strong chemical bonding between the glass and living bone, leading to *in vivo* bioactivity. Bioactivity in Bioglass is thought to arise from interfacial reactions as follows [12]. The first step involves rapid exchange of sodium ions in glass with hydrogen ions from solution. In the second step, silanol groups form at the surface upon rupture of Si—O—Si bonds in the glass network structure, and soluble silica is released to the solution. A SiO_2-rich layer is formed on the surface by condensation and repolymerization as the third step. Subsequently, calcium and phosphate ions from the solution migrate to the surface of the SiO_2-rich layer, resulting in the formation of an amorphous calcium phosphate layer in the fourth step. Finally, crystallization of the amorphous phase to HCA occurs upon incorporation of

13.2 Silicate Glasses

Table 13.1 Biological responses to inorganic ions released from silicate-based bioactive glasses.

Glass composition by mol%	Ions released	Concentration (ppm)	Result	Remarks, conclusions
45 S5 46.1 SiO$_2$–24.3 Na$_2$O–27.0CaO–2.6P$_2$O$_5$	Si	16.5	• Ionic dissolution products directly stimulated the up-regulation of most genes in HOC (up to fivefold) known for their relevant role in bone metabolism	Which released ion responsible is unclear
	Ca	88.3		
	P	30.4		
	Na	2937.0		
	Si	16.58	• Osteoblast number increased to 150%	Ionic products of 45 S5 Bioglass may increase proliferation rate
	Ca	88.35	• Increase in osteoblast number implies increase in cell proliferation	
	P	30.45		
	Na	2938.0		
	Si	55.48	• Si concentration shortens osteoblast growth cycle and promotes the proliferation	Exact molecular mechanism should be clarified
	Ca	72.70		
	P	23.45		
	Na	3341.66		
	—	—	• Upregulation of *BSP* and *ALP* in FOB cells	Ionic concentration in culture medium is not known
	—	—	• Increased cell proliferation (MG 63) on BG pellets (2D)compared to control (Thermanox®) among which noncrystallized BG shows best results	Better proliferation on amorphous BG is due to slower degradation of sintered partially crystallization BG
			• Significantly higher cell proliferation on BG-based scaffolds (3D) compared to control	
	Si	14.5–19.4	• BG extracts create an extracellular environment for support of osteoblast phenotype expression	Osteogenic supplements effect on gene expression profiling and mineralization
	Ca	47.5–49.9	• Increased osteoblast differentiation and enhanced ECM deposition and mineralization	
	P	24.6–23.6	• Increased osteocalcin (OCN) and collagen I synthesis	
	Na	4983.8–4895.8		

(*continued*)

Table 13.1 (Continued)

Glass composition by mol%	Ions released	Concentration (ppm)	Result	Remarks, conclusions
MBG 85 85 SiO_2– 10CaO–5P_2O_5	Si Ca P	59.2 64.0 16.9	• High viability of osteoblasts (Saos-2), murine fibroblasts (1929) and murine lymphocytes (SR.D10 T) treated with BG extracts • But decreased osteoblast and fibroblast proliferation	—
BG60 S 59.9 SiO_2– 38.4CaO– 1.7P_2O_5	—	—	• 50% increase in osteoblast proliferation • Increased collagen production	No glass dissolution studies on the BG were performed; ion concentration. In culture medium is unknown
	Si Ca P	58.06 70.07 20.02	• Extracellular Ca^{2+} concentration, via bioglass dissolution, increases the mobilization of intracellular Ca^{2+} in osteoblast cells	—
S520 52 SiO_2– 21 Na_2O–	Si Ca P	30 90 10	• Good attachment, proliferation, and nodule formation of osteoblast	—
7 K_2O–18CaO– 2P_2O_5	Si Ca P	47.7 0 67.63	• Good osteoblast attachment on BG scaffold • Nodules formation and mineralization after 10 days of seeding (without supplements of mineralization agents) • Higher Si concentrations (>120 ppm) of ions lead to apoptosis	Si concentration is likely the key factor for mediating mineralization and nodule formation and cell death
58 S 60 SiO_2– 36CaO–4P_2O_5	Si Ca P Si	50.02 95.8 26.4 50.2	• No statistically significant effect on proliferation of HOC	Author suggests higher Si content in medium to achieve stimulating effects

	Ca	95.8	• Upregulation of several genes in HOC, for example, *gp130*, *MAPK3/ERK1*, *MAPKAPK2*, and *IGF-I*	First study to confirm gene expression of sol–gel derived bioactive glass
	P	26.4	• 100% increase in osteoblast proliferation	—
77 S 80 SiO$_2$– 16CaO–4P$_2$O$_5$	Si	203.11	• Stimulated differentiation of bone marrow cells into osteoblast cells in BG (77 S as well as control 45 SS) conditioned medium	Dissolution kinetics and ion release are not investigated
	Ca	47.06		
	P	6.88	• 77 S NG caused inhibition of osteoblast cell formation	
		—		

Source: Reprinted with permission from Ref. [5]. Copyright 2011, Elsevier.
a) Absolute ion concentration in the cell culture medium after treatment with BG.

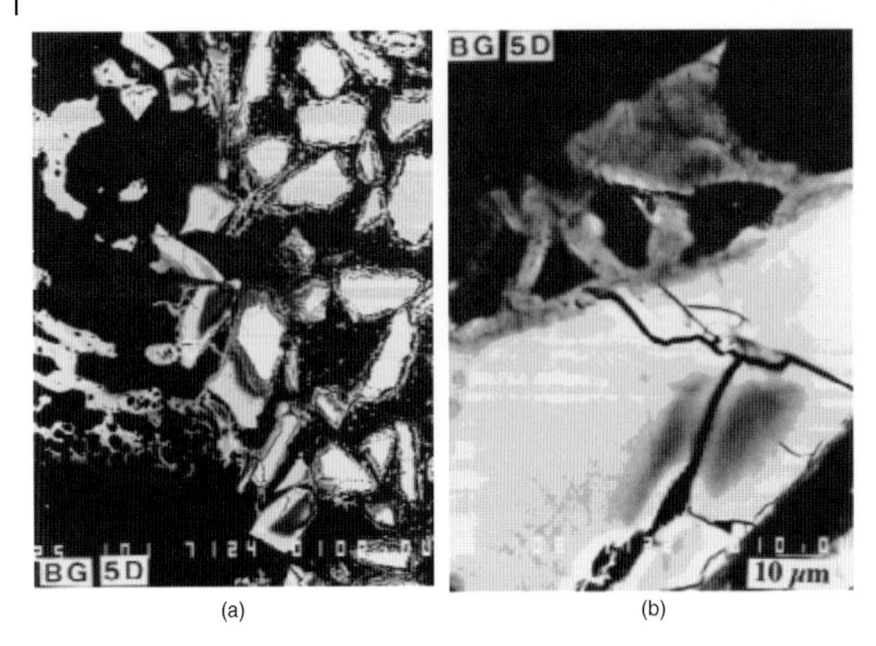

Figure 13.1 Backscattered scanning electron microscopic images of osteoid formation at Bioglass® interfaces at 5 days after implanting femoral condyles of mature rabbits; image (a) shows bone formation in spaces among the particles and image (b) is the associated high-magnification image of (a). (Reprinted with permission from Ref. [15]. Copyright 2000, John Wiley & Sons, Inc.)

hydroxyl and carbonate ions into the amorphous phase from the solution. The solubility and ion-releasing property of glass are important criteria for its bioactivity. Simulated body fluid (SBF), which is an acellular buffer solution constituting similar inorganic concentrations as human blood plasma, however, without proteins, as shown in Table 13.2, is used in *in vitro* investigation of HCA formation on materials surfaces [13]. Conversely, HCA formation and dissolution upon soaking test materials in SBF cannot be used to appropriately estimate the actual bioactivity *in vivo* [14]. Cell culture tests performed *in vitro* allows evaluation of the living cell behaviors on the material surface in a medium with serum proteins. To ensure a more accurate estimation of the bioactivity of a test material, cell studies should be conducted in conjunction with *in vitro* tests in SBF.

Table 13.2 Ion concentrations of SBF and human blood plasma.

Ion	Ca^{2+}	Na^+	Mg^{2+}	K^+	Cl^-	HCO_3^-	HCO_4^-	SO_4^{2-}
SBF (mM)	2.5	142.0	1.5	5.0	148.8	4.2	1.0	0.5
Human blood plasma (mM)	2.5	142.0	1.5	5.0	103.0	27.0	1.0	0.5

13.2 Silicate Glasses

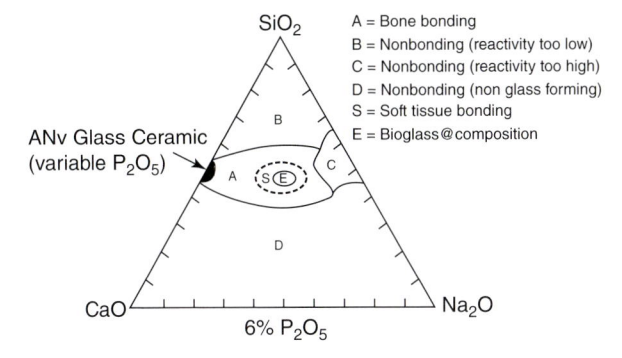

Figure 13.2 Compositional diagram for bone-bonding. (Reprinted with permission from Ref. [15]. Copyright 2006, Springer.)

The phase diagram relating to the bioactivity of SiO_2–Na_2O–CaO system containing 6% P_2O_5 is shown in Figure 13.2 [15]. As observed, changing the chemical composition of the glass system afforded tuning of the HCA formation rate. For example, HCA could be formed within a short period using a glass composition containing 45 mol% silica. In contrast, HCA did not form on the surface of the glass matrix containing more than 60 mol% silica. Thus, HCA formation is greatly influenced by the silica content in the glass matrix. Raman- and nuclear magnetic resonance (NMR) spectroscopy results indicated that glass consisting of predominant Q_{Si}^2 units featured higher bioactivity than that containing large numbers of Q_{Si}^3 units [8]. The Q_{Si}^n unit relates to the concentration of bridging oxygens per tetrahedron, where n is the number of bridging oxygen and ranges from 0 to 4. Thus, Q_{Si}^4 units represent fully linked SiO_4 tetrahedrons that result in the formation of three-dimensional network owing to four bridging oxygens. An isolated orthosilicate is denoted as a Q_{Si}^0 unit. The local structures of three types of glasses simulated by molecular dynamics are shown in Figure 13.3 [16]. The chemical compositions of the glass matrices composed of 45, 55, and 65 wt

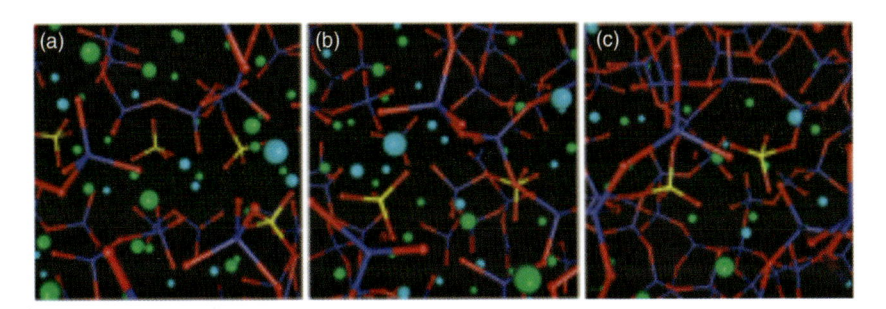

Figure 13.3 Structures of simulated glasses (a) BG45, (b) BG55, and (c) BG65. Red, oxygen; blue, silicon; yellow, phosphorous; dark green, sodium; and cyan, calcium. Sodium and calcium ions are shown as large spheres, and silicon, phosphorous and oxygen are shown as ball-and-stick models. (Reprinted with permission from Ref. [16]. Copyright 2007, American Chemical Society.)

% SiO_2, named as BG45, BG55, and BG65, respectively, and 6 wt% P_2O_5 and equal amounts (24.5, 19.5, and 14.5 wt%) of Na_2O and CaO. The bioactivity of the glass systems decreased in the order of BG45 > BG 55 > BG65. The silica content of the glass matrices increased with increasing amounts of bridging oxygens, resulting in increasing amounts of Q_{Si}^3 and Q_{Si}^4 units and decreasing amounts of Q_{Si}^1 and Q_{Si}^2 units. As noted, trace amounts of Q_{Si}^0 units are detected in BG45 only. Because of the higher release of Q_{Si}^0 unit from the glass structure in solution when compared with that of the other Q_{Si}^n units ($n \geq 1$), silicate units can be released from BG45 in solution without breaking Si—O—Si bonds in the glass network structure, leading to development of high bioactivity. Sodium enrichment of the glass surface, owing to increases in the Na_2O content, also caused rapid hydrolysis, leading to enhanced bioactivity [17]. Substitution of MgO for CaO in S53P4 bioactive glasses slightly decreased the rate of HCA formation [18]. MgO was incorporated into the glass structure as a network former [19]. Accordingly, the increase in MgO content in the glass matrix enabled the construction of a stronger network structure and improvement in the chemical durability of the glass matrix. In the SiO_2–Na_2O–CaO–P_2O_5 glass system, the solubility of the glasses increased with increasing contents of Na_2O substituted for CaO because of weak ionic cross-linking bonds owing to the lower field strength of sodium ion when compared with that of calcium ion [20]. However, an increase in pH owing to ionic exchange reactions of the glasses with higher Na_2O contents induced a cytotoxic response. The dissolution of glass and type of modifier cations should be considered in designing glass systems displaying bioactivity.

The network connectivity, which describes the average number of bridging oxygen atoms per network-forming element, of glasses can be determined by the bridging oxygens in a SiO_4 tetrahedron. The network connectivity of bioactive glasses with a SiO_2–P_2O_5–$M(I)_2O$–$M(II)O$ configuration can be estimated using the follow equation:

$$\text{Network connectivity} = \frac{4[SiO_2] + 6[P_2O_5] - 2\big([M(I)_2O] + [M(II)O]\big)}{[SiO_2]},$$

where $[SiO_2]$ and $[P_2O_5]$ are the molar fractions of silicate and phosphate, respectively, and $[M(I)O_2]$ and $[M(II)O]$ are the molar fractions of the network modifiers [21]. Thus, pure silica glass has a network connectivity of four. The network connectivity decreases with increasing network modifiers contents. The network connectivity influences the dissolution of biocompatible glass [22]. Glass with a poor network connectivity, indicative of a fragmented phospho–silicate network, readily releases soluble silica and phosphate species into solution. Based on molecular dynamics results, bioactive glasses show a silicate network connectivity in the range of 2–3 [16]. Furthermore, experimental results indicated a network connectivity of 2.4 as the cut-off point for HCA formation [23]. The onset time for apatite formation decreases with increasing network connectivity of bioactive glasses, as shown in Figure 13.4 [21].

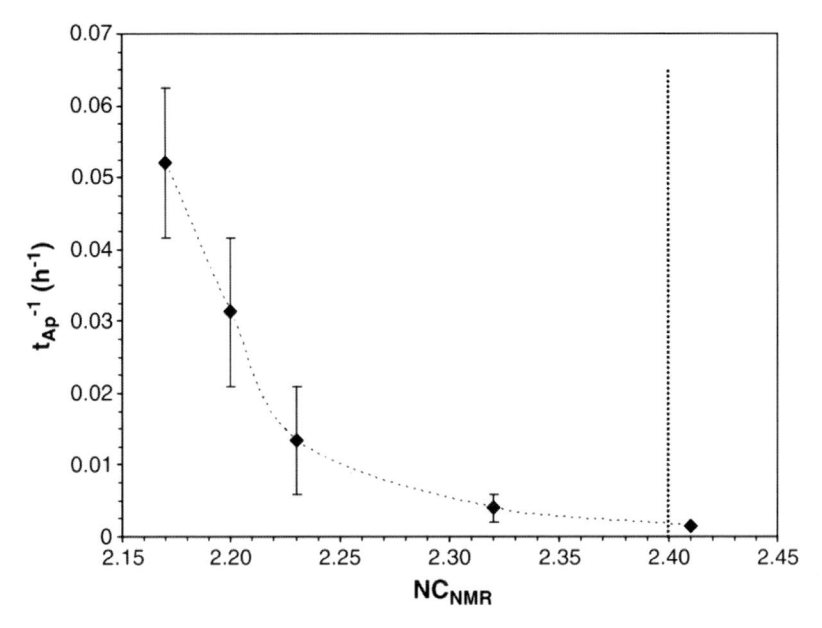

Figure 13.4 Bioactivity (defined as t_{AP}^{-1}, where t_{AP} is the onset time of first apatite formation in SBF as detected by X-ray diffraction) of bioactive glass in ICIE1 with substitution of MgO for CaO as a function of NC_{NMR} (NC = network connectivity) calculated from the proportions of Q_{Si}^2 and Q_{Si}^3 from ^{29}Si magic-angle spinning NMR [19]. The vertical line represents the percolation point (network connectivity = 2.4), that is, the cut-off value for bioactivity as defined by Hill [23]. (Reprinted with permission from Ref. [21]. Copyright 2011, Elsevier.)

Silicate-based bioactive glasses contain a P_2O_5 component as a second network former. The PO_4 tetrahedrons are primarily isolated orthophosphate units in bioactive silicate glasses [24]. The phosphate units are easily released from the glass, leading to rapid HCA formation [25]. The bioactivity of glass with a SiO_2–P_2O_5–CaO–Na_2O configuration increases with increasing phosphate contents greater than 3 mol% at a given ratio of network modifier oxides [26]. In contrast, increasing the phosphate content of the same glass system at varying ratios of network modifier oxides causes increases in the network connectivity of the silicate structure [27], leading to a decrease in bioactivity. The phosphate units have a higher affinity for modifier cations than silicate units in phospho–silicate glasses containing CaO and Na_2O, as observed from molecular dynamics simulations [28]. The orthophosphate units are thought to interact with modifier cations for charge compensation. In a SiO_2–P_2O_5–CaO–Na_2O–CaF_2 glass system, the orthophosphate peak shifted continuously toward a higher magnetic field with increasing $CaO/(CaO + Na_2O)$ ratios [29]. This result was attributed to the difference in the electronegativity of sodium and calcium ions, indicating the interaction of the orthophosphate units with the modifier cations. Furthermore, it was also reported that cluster of PO_4 tetrahedron units could be incorporated into the silicate network, resulting in formation of P—O—Si bonds [30]. The PO_4 units, as the secondary network oxide, influence the bioactivity and dissolution by altering the glass structure though the content of P_2O_5 is small in the glass system.

13.2.2
Sol–Gel-Derived Glasses

Sol–gel is a chemistry-based synthesis involving a solution containing precursors at room temperature. Sol–gel-derived glasses have an inherent unique porous structure, with nanometer features, whereas melt-quenched-derived glasses are dense [31]. Sol–gel-derived glasses feature higher specific surface areas by two orders of magnitude than melt-quenched-derived glasses. HCA formation, which follows dissolution of the glass, is dependent on the specific surface area [31]. Consequently, sol–gel-derived glasses tend to have a higher HCA formation ability. Nanotopography (on the glass surface) influences cellular response [32], thereby indicating that sol–gel-derived bioactive glasses have great potential as bone substitutes. Representative chemical compositions (in mol%) of sol–gel-derived bioactive glasses are $70SiO_2$–$30CaO$ (70S30C), $60SiO_2$–$36CaO$–$4P_2O_5$ (58 S), and $80SiO_2$–$16CaO$–$4P_2O_5$ (77 S) [33,34].

In a typical synthesis, a silicate precursor, such as tetraethylorthosilicate, is hydrolyzed by water under acidic or basic conditions to prepare sol-containing nanoparticles. Subsequently, the nanoparticles agglomerate and bond with each other to form a gel network of assembled nanoparticles. Thermal treatment stabilizes the gel, generating a glass matrix with a nanoporous structure, accompanied with evaporation of water and alcohol produced during the gelation process. The porous structure is derived from the gaps between the aggregated nanoparticles [35]. 70S30C glasses, prepared using tetraethylorthosilicate and calcium nitrate, have unique structures with a homogeneous calcium distribution in the silicate network structure; their fabrication is schematically shown in Figure 13.5 [36]. As observed, there are calcium and nitrate ions, generated upon dissolution of calcium nitrate, in the sol before gelation. During the drying process, these ions are deposited on the rim of the silica secondary particles. Stabilization by heat treatment instigates the decomposition of the nitrate ions and incorporation of some calcium ions into the silica network structure as modifier cations. The remaining calcium ions are used to fuse the silica particles, leading to the formation of a bioactive glass matrix.

13.2.3
Glass-Ceramics

Bioactive glass-ceramics have been successfully prepared by typically heating a MgO–CaO–SiO_2–P_2O_5 glass system of nominal composition of 7.4 mol% MgO, 51.8 mol% CaO, 36.7 mol% SiO_2, 3.7 mol% P_2O_5, and 0.4 mol% CaF_2 [37]. Another example of a glass-ceramics, termed as glass-ceramic A-W, has been prepared from a MgO(22.5 mol%)–CaO(23.7 mol%)–SiO_2(53.8 mol%)

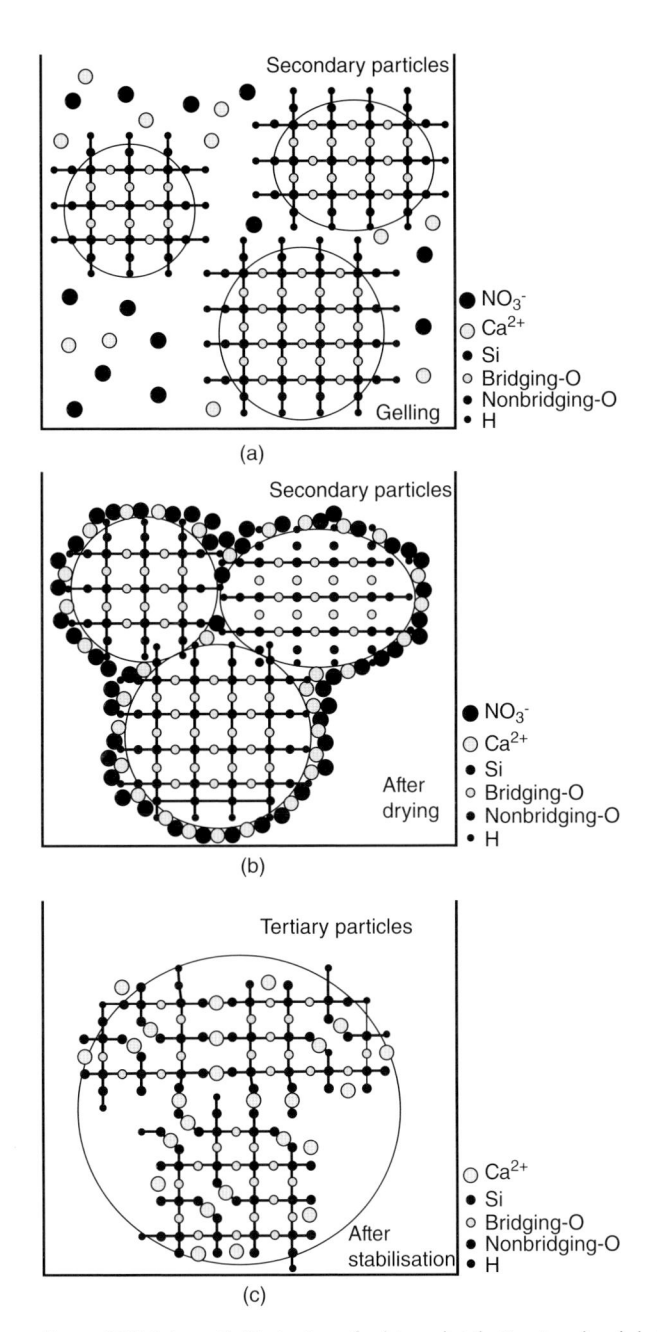

Figure 13.5 Schematic illustration of calcium distribution in sol–gel-derived glasses (a) during the gelling stage, (b) after the drying stage, and (c) the stabilization stage. (Reprinted with permission from Ref. [36]. Copyright 1991, Royal Society of Chemistry.)

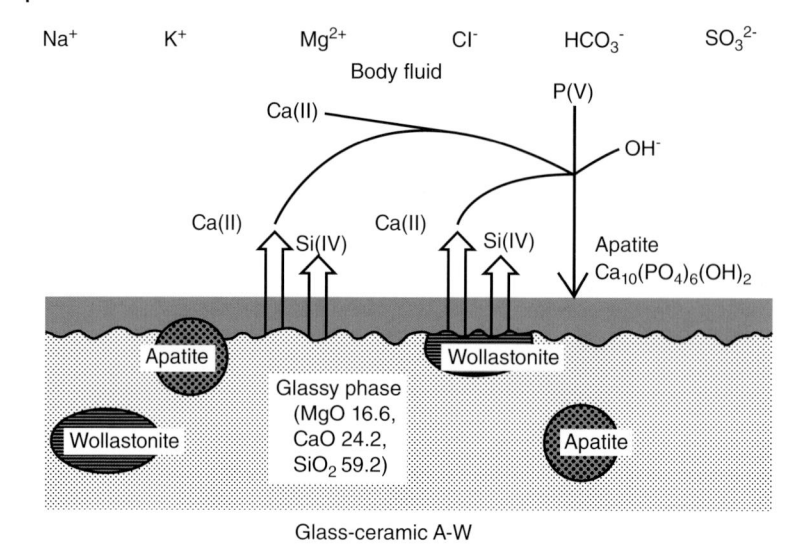

Figure 13.6 Schematic representation of reactions occurring at the surface of glass-ceramic A-W in the body. (Reprinted with permission from Ref. [38]. Copyright 1991, Elsevier.)

system; A-W consists of 38 wt% oxyfluoroapatite ($Ca_{10}(PO_4)_6(O,F)_2$) and 34 wt% ß-wollastonite ($CaSiO_3$) as crystalline phases. A-W not only displays bioactivity but also features high mechanical properties owing to reinforcement effects by precipitation of wollastonite crystals in the matrix. Moreover, it shows machinability, thereby enabling the fabrication of various shapes such as a screw. Figure 13.6 shows a schematic representation of reactions occurring at the surface of A-W. Calcium and silicate ions are released from both wollastonite and the glassy matrix [38]. In contrast, the apatite phase in A-W does not dissolve because the surrounding fluid is supersaturation for apatite. The layer formed at the interface between A-W and the bone contains large amounts of calcium and phosphate with trace amounts of magnesium and silicon. Thus, the dissolution of the calcium and/or silicate ions influences the formation of the layer.

13.2.4
Functionalization of Bioactive Glasses

Third-generation biomaterials are designed to stimulate cellular response as wells as to display bioactivity. The key factor to designing bioactive glasses is the integration of inorganic ions into the chemical composition of bioactive glasses capable of activating cellular response and being released into solution. Doping of different modifier cations, as the inorganic ions capable of exerting stimulating effects, can influence the glass structure because of the different ionic sizes, field strengths, and electronegativities. Bioactive glasses with lower network connectivities show greater cell proliferation than those with higher network connectivities [39]. Thus, evaluation of the stimulating effects exerted by inorganic ions is best performed at a given network connectivity. Examples of some bioactive glasses doped with inorganic ions are given as follows.

Magnesium ion enhances cell attachment and differentiation [40]. The substitution of MgO for CaO in the ICIE1 series increases with increasing polymerization of the silica network structure, indicating reduced ion release [19]. Magnesium displays a unique behavior, whereby it can assume the role of a network modifier or a network former depending on the network structure. Magnesium ions incorporated into bioactive glasses stimulate bone formation despite the reduced ion-releasing property observed in *in vitro* test [41]. Zinc ions enhance alkaline phosphatase activity of osteoblast cells [42]. When zinc ions are incorporated into bioactive glasses, some of them forms Si—O—Zn bonds [43,44]. The zinc-ion-releasing property of the resulting bioactive glass depends on the solution pH: a higher releasing property is observed under acidic conditions owing to acid hydrolysis of the Si—O—Zn bonds, when compared with that observed under neutral pH conditions [45]. Strontium promotes bone formation and replication of osteoblast [46]. Strontium-doped bioactive glasses display both bioactivity and strontium ion-releasing property [47]. Furthermore, bioactive glasses with a CaO–SiO$_2$ system containing SrO dopant featured faster HCA formation kinetics, when compared with glasses without SrO [48]. For activating angiogenesis, which relates the formation of blood vessels by hypoxia, bioactive glasses doped with cobalt, copper, or nickel have been prepared [49–51].

13.3
Borate Glasses

Boron is a trace element required for bone health [52]. Although boron released as borate ions (BO$_3^{3-}$) in solution is toxic, the use of borate glasses promotes cell proliferation and differentiation, and shows no associated toxicity, as determined by *in vitro* tests conducted under dynamic conditions [53,54]. Borate-based glasses are one of the key materials in the design of bioactive glasses. Brink firstly reported the preparation of silicate glasses containing trace amounts of B$_2$O$_3$, thereby expanding the biomedical applications of bioactive glasses [55]. A borate glass matrix was prepared by replacing SiO$_2$ with B$_2$O$_3$, using the same composition as that of 45S5 glass [56]. The weight loss of the glass matrices after soaking in a phosphate solution increased with increasing B$_2$O$_3$ contents. The initial rate of sodium ions released from the glass matrices after soaking was more rapid at the higher B$_2$O$_3$ contents. Increasing the B$_2$O$_3$ content in the glass system enhanced apatite formation after soaking in the phosphate solution. It is well known that the main units of a borate glass network structure are BO$_3$ trihedrons or chains of BO$_3$ triangles. Thus, construction of a complete three-dimensional network using boron is impossible owing to threefold coordination number of boron when compared with silicon, which has a fourfold coordination number. Borate glasses have a lower chemical durability than silicate glasses.

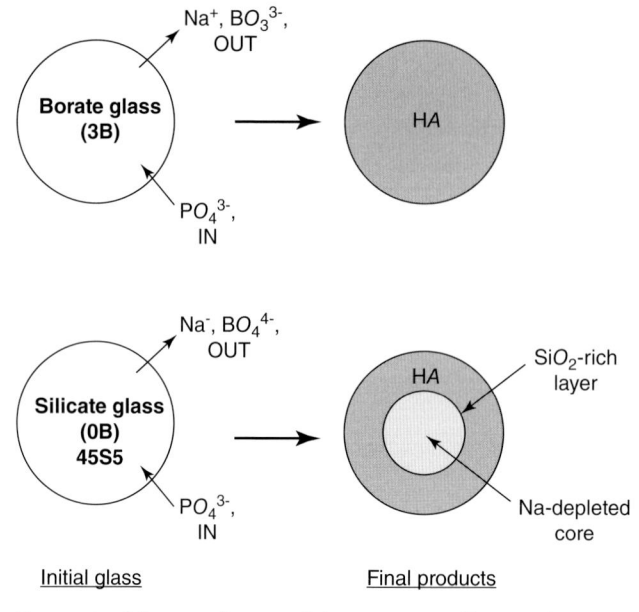

Figure 13.7 Schematic diagram of the conversion of borate glass and 45S5 glass into apatite in a dilute phosphate solution. (Reprinted with permission from Ref. [56]. Copyright 2006, Springer.)

The structural change of the glass upon substitution of SiO_2 with B_2O_3 influenced the solubility and apatite-forming ability of the resulting glass matrix.

The apatite formation mechanism in borate glasses is similar to that occurring in 45S5 glass, however, without the formation of a silica-rich layer. The conversion of borate glass into apatite has been proposed as follows (Figure 13.7) [56]. Sodium and calcium ions are released from the glass. Simultaneously, the BO_3 network structure is attacked by the phosphate solution. Subsequently, the phosphate ions from the solution react with the calcium ions, resulting in nucleation and growth of apatite. The ions, which are needed to form apatite, are provided through the precipitated apatite owing to its highly porous structure, leading to thickening of the apatite layer from the surface of the particle inward. As a result, the borate glass is completely converted into apatite.

13.4
Phosphate Glasses

13.4.1
Conventional Phosphate Glasses

P_2O_5 acts as a network former in phosphate-based glasses. The PO_4 tetrahedron is the basic building unit, and the linkage between PO_4 tetrahedron dominates the glass structure, resulting in a three-dimensional glass network of rings and chains. The network degree of polymerization is denoted as Q_p^n, where n is the number of bridging oxygens and ranges from 0 to 3. Q_p^0 units are orthophosphate units, PO_4^{3-}, with no bridging oxygens to neighbor PO_4 tetrahedron. Q_p^1 units are represented as $P_2O_7^{4-}$ and act as dimers or terminating groups at the end of phosphate chains. Q_p^2 units operate as the middle groups, representing PO_3^-, in the phosphate chains owing to the existence of two bridging oxygens. Q_p^3 units can share three bridging oxygens with neighbor PO_4 tetrahedron units, thus acting as branching units in the network structure of phosphate glasses. The addition of metal oxides as a network modifier to a fully polymerized phosphate network leads to depolymerization of the network with oxygen atoms breaking the P–O–P linkage, resulting in increased amounts of the nonbridging oxygens and reduced amounts of bridging oxygens. The cation of the network modifier coordinates with the nonbridging oxygens of several PO_4 tetrahedrons as shown in Figure 13.8, a process referred as repolymerization. The coordination of the phosphate units with a cation increases with increasing amounts of the network modifier cation, leading to abrupt changes in the glass properties at around 20 mol% network modifier oxides, proposed by Hoppe [57]. The glass network structure was determined by simulation methods in addition to spectroscopy analyses for visualization of the local atomic structure. The simulation methods employed molecular dynamics and reverse Monte Calro modeling. Neutron and X-ray diffraction analyses were additionally used to obtain robust and specific structural information. Combining the simulation with the experimental diffraction data can afford extraction of structural information and avoid misinterpretation of data. Accordingly, the reverse Monte Carlo modeling combined with the X-ray and neutron diffraction results confirmed the short-range structure of calcium metaphosphate, as consistent with Hoppe's model [58].

Phosphorus has a charge of +5, whereas silicon has a charge of +4. Thus, a PO_4 tetrahedron shares only three out of its four oxygen, whereas a SiO_4 tetrahedron shares all four oxygens. Therefore, PO_4 tetrahedron units have a terminal double bond. Phosphate glasses have different functional properties from those displayed by silicate glasses owing to the different network-forming oxide properties. The solubility of phosphate glasses is much higher than that of silicate glasses [59]. The mechanism of the dissolution of conventional phosphate glasses has been suggested [60]. Two processes are required for the dissolution of the glass network as follows. (1) Formation and growth of a hydrated surface layer by permeation of water molecules in the top layer of the glass; (2) chemical reaction between ions or phosphate chains and water molecules inside the hydrated layer.

Phosphate-based glasses with a $CaO-Na_2O-P_2O_5$ system have been investigated for use in biomedical applications because of their high solubility and chemical similarity to the inorganic phase of human bone. The solubility of $CaO-Na_2O-P_2O_5$ glasses can be tuned by changing the CaO/NaO ratio [60]. In the $(Na_2O)_{0.55-x}(CaO)_x(P_2O_5)_{0.45}$ system, glasses with CaO contents in the range of 0.24–0.32 consist of Q_p^1 and Q_p^2 units [61]. The dissolution rate after soaking the glass matrices in distilled water or Hanks buffered saline solution decreased

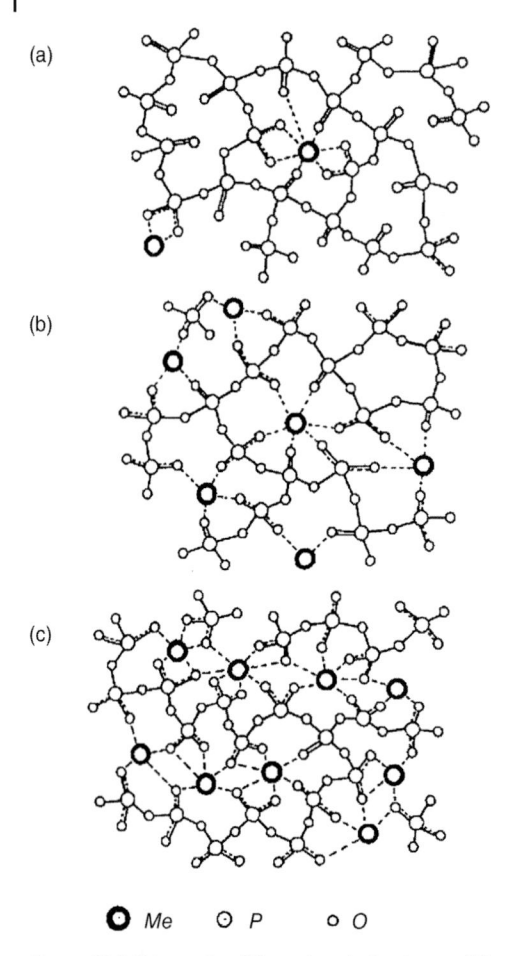

Figure 13.8 Schematic of the network structures of three types of binary phosphate glasses. The MeO contents of the model glasses are (a) 17, (b) 40, and (c) 49 mol%. (Reprinted with permission from Ref. [57]. Copyright 1996, Elsevier.)

with increasing CaO contents, as shown in Figure 13.9. Thus, the solubility of the glass depends on the medium and chemical composition. The CaO content had a great influence on the dissolution rate of the glass. In contrast, sodium ions were released at a higher rate from glasses with a higher solubility owing to the lower ionic strength of sodium when compared with that of calcium [62]. The local- and medium-range structures of $CaO–Na_2O–P_2O_5$ glasses were determined by *ab initio* molecular dynamics simulations [63]. The calcium and sodium ions as cations of the network modifier act as sites for connecting and arranging with different phosphate units as shown in Figure 13.10, enabling control over their folding and interconnectivity. It was clear that the rigidity of the network in phosphate glasses increased with increasing calcium concentrations based on calculation of the pair and angular distribution functions.

13.4 Phosphate Glasses

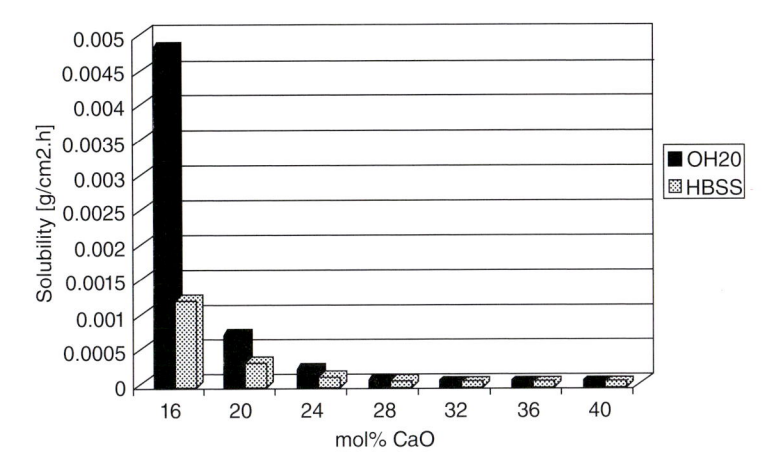

Figure 13.9 Solubility values as a function of glass compositions and aqueous medium. (Reprinted with permission from Ref. [61]. Copyright 2000, Springer.)

Many scientists have reported other glass systems with quaternary and more complex compositions. The incorporation of various oxides, such as K_2O, TiO_2, ZnO, and MgO, into glasses systems toward tuning their solubility has been reported [64–66]. In the case of the $Na_2O-K_2O-CaO-P_2O_5$ system, the solubility of the glasses increased with increasing K_2O contents at a given CaO content. The amount of sodium ions released from the glasses containing K_2O decreased with increasing solubility of the glasses, indicating

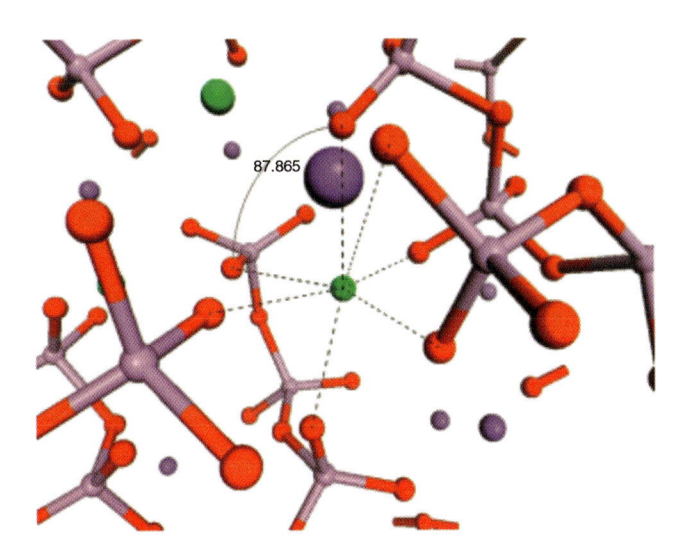

Figure 13.10 Snapshot from the Car–Parrinello molecular dynamics trajectory of glass showing a typical calcium coordination shell. Ca, Na, O, and P are depicted as green, purple, red, and pink spheres, respectively. (Reprinted with permission from Ref. [63]. Copyright 2010, John Wiley & Sons, Inc.)

an anomalous dissolution behavior. This phenomenon could be attributed to the mixed-alkali effect owing to substitution of Na$_2$O with K$_2$O [67]. It is well known that this effect leads to considerable changes in dynamic properties and deviation from linearity. The theoretical network connectivity, which is the number of bridging oxygens per network forming element, was calculated using the following equation:

$$\text{Network connectivity} = \frac{3[P_2O_5] - [M(I)_2O] - [M(II)O] - 2[M(IV)O_2]}{[P_2O_5]},$$

where [P$_2$O$_5$] is the molar fraction of phosphate and [M(I)O$_2$], [M(II)O], and [M(IV)O$_2$] are the molar fractions of the network modifiers [68]. The network connectivity was calculated assuming that magnesium, zinc, and titanium behave as network modifiers. The experimental network connectivity of the glass could also be obtained based on the peak-integrated portions of the deconvoluted ^{31}P magic-angle spinning NMR spectra. Calculation of the network connectivity using the chemical compositions of the glasses afforded prediction of glass structures with complex compositions.

13.4.2
Phosphate Invert Glasses

Metaphosphate glasses (50CaO–50P$_2$O$_5$ in mol%) release phosphate species after soaking in solution, leading to a decrease in the solution pH. As a result, no apatite forms on the surface of metaphosphate glasses after soaking in SBF [69]. Uo *et al.* reported that the cytotoxicity of P$_2$O$_5$–CaO–Na$_2$O glass was related to its degradation, associated pH changes, and the ionic concentration of the medium [70]. For biomedical applications, low releases of phosphate contents from phosphate glass systems are desirable. The fabrication of phosphate invert glasses is one of key approaches to satisfy such a requirement. In the case of invert glass, anionic groups are connected through cations, which act as a network modifier, leading to the glass formation [71]. Kasuga *et al.* reported the synthesis of calcium pyrophosphate invert glass, CaO–P$_2$O$_5$–Na$_2$O–TiO$_2$ system displaying unique structure and bioactivity [72]. The glass consists of orthophosphate and pyrophosphate units and no metaphosphate units. Thus, the phosphate groups are connected to the Ca^{2+} ions [73]. In contrast to metaphosphate glasses, phosphate invert glasses maintained neutral pH in an aqueous solution after soaking. Consequently, an apatite was formed on the glass surface after soaking in SBF [74].

Tuning the chemical composition in calcium phosphate invert glass is one of the key factors to controlling the structure and solubility of the glass and stimulating cellular responses. In the Na$_2$O–Al$_2$O$_3$–TiO$_2$–Nb$_2$O$_5$–P$_2$O$_5$ glass system, it was found that niobium was more effective in improving the chemical durability of the glass than titanium [75]. The dissolution rate of Nb$_2$O$_5$–SrO–P$_2$O$_5$ glass was controlled by increasing the content of niobate [76]. Calcium phosphate invert glasses were prepared by adding Nb$_2$O$_5$ and Na$_2$O [77].

13.4 Phosphate Glasses

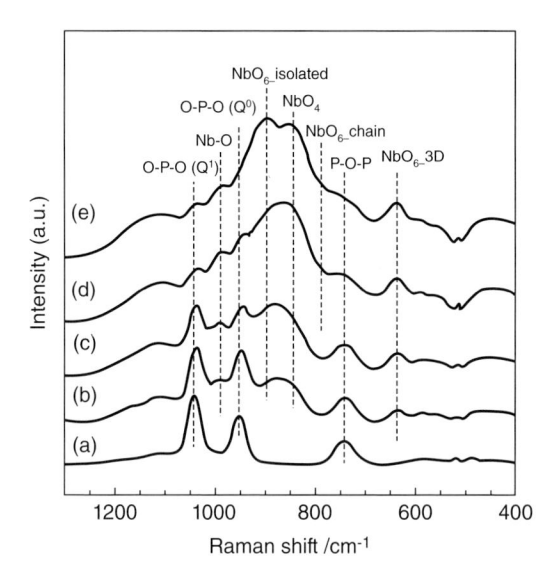

Figure 13.11 Raman spectra of glasses with the nominal compositions of (a) 60CaO–30P$_2$O$_5$–10Na$_2$O, (b) 60CaO–30P$_2$O$_5$–3Nb$_2$O$_5$–7Na$_2$O, (c) 60CaO–30P$_2$O$_5$–5Nb$_2$O$_5$–5Na$_2$O, (d) 60CaO–30P$_2$O$_5$–7Nb$_2$O$_5$–3Na$_2$O, and (e) 60CaO–30P$_2$O$_5$–10Nb$_2$O$_5$. (Reprinted with permission from Ref. [77]. Copyright 2016, Elsevier.)

Raman- and NMR spectra showed that niobate species were present as NbO$_4$ and NbO$_6$ units as a network former and a network modifier, respectively, resulting in the formation of Nb—O—P bonds around the Q_p^0 units in the glass structure (Figure 13.11). The Nb—O—P bond is reported to be stronger than the P—O—P bond [76]. The dissolution test performed on the glasses using a Tris-buffer solution showed that increasing the Nb$_2$O$_5$ content led to decrease in the solubility of the glass matrix. The strong Nb—O—P bonds formed upon incorporation of niobate played an important role on the chemical durability of the glass. *In vitro* evaluation of the bioactivity of glass samples using 1.5 times SBF showed that apatite formation did not occur on the surface of the niobate-containing glasses. Specifically, 60CaO–30P$_2$O$_5$–3TiO$_2$–7Na$_2$O, which has almost the same solubility as 60CaO–30P$_2$O$_5$–3Nb$_2$O$_5$–7Na$_2$O, formed an apatite layer after soaking in SBF, whereas 60CaO–30P$_2$O$_5$–3Nb$_2$O$_5$–7Na$_2$O did not form any apatite layer. It was suggested that the different functional groups, that is, Ti–OH and Nb–OH of the glass matrices influenced the apatite-forming ability of the glass. Conversely, the cell compatibility test showed that 60CaO–30P$_2$O$_5$–3Nb$_2$O$_5$–7Na$_2$O glass promoted differentiation and mineralization of osteoblast-like cells when compared with 60CaO–30P$_2$O$_5$–3TiO$_2$–7Na$_2$O glass [78]. Despite their lower apatite-forming ability, calcium phosphate invert glasses incorporating niobate displayed improved cell activation properties. MgO and SrO has also been incorporated into these types of invert glasses for preparing bioactive glasses with high performance [79–81].

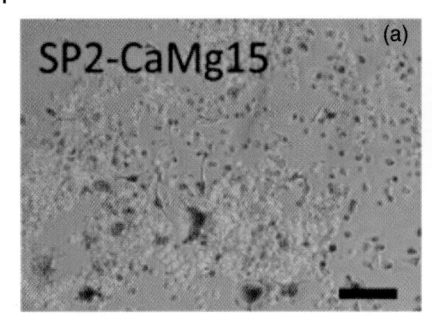

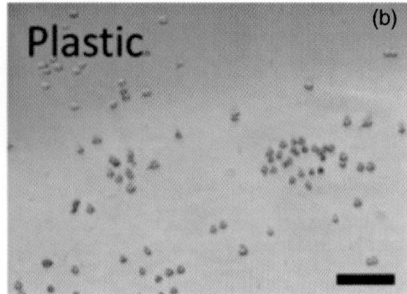

Figure 13.12 Light microscopy images of RAW 267.4 monocyte cells cultured on the surface of glass with a chemical composition of (a) 27.2ZnO–19.8Na$_2$O–14.1SO$_3$–23.9P$_2$O$_5$–7.5CaO–75MgO (SP2-CaMg15) and (b) an adjacent tissue culture plastic (Plastic) following exposure to RANKL and stained for TRAP. The scale bar is 250 μm. (Reprinted with permission from Ref. [84]. Copyright 2013, John Wiley & Sons, Inc.)

13.4.3
Sulfophosphate Glasses

Sulfophosphate glasses with softening temperatures below 400 °C in a SO$_3$–P$_2$O$_5$–ZnO–Na$_2$O system have been successfully prepared with high concentrations of sulfate, which acts as isolated anionic groups, such as SO$_4^{2-}$, and is incorporated into the phosphate matrix [82]. The structure of sulfophosphate glasses typically consists of a pyrophosphate-type network of highly depolymerized Q_p^1, Q_p^0, and SO$_4^{2-}$ groups, and cations located selectively around either the phosphate units or sulfate [83]. The dissolution of such type of glass can be tuned by changing the divalent cations of the SO$_4$–P$_2$O$_5$–MO–Na$_2$O (M = Zn^{2+}, Ca^{2+}, or Mg^{2+}) system [83]. Cell culture tests, using in a direct contact method, showed that monocyte differentiation into osteoclasts was promoted on the sulfophosphate glasses, indicating good bioactivity potential, as shown in Figure 13.12 [84].

13.5
Summary

In this chapter, various bioactive glasses for bone regeneration applications are introduced. The glasses can incorporate inorganic ions suitable for stimulating cellular responses and exhibit bioactivity by tuning the chemical compositions. The solubility of glass material is highly dependent on their chemical composition. Specifically, the types of modifiers and intermediate cations forming the glass structure influence the solubility. The HCA-forming ability depends on the network connectivity and the structure of the phosphate units in silicate-based glasses. The low releases of phosphate species from the glass and the functional groups are key factors to forming HCA on phosphate-based glasses. Altering the chemical composition of the glass material allows tailoring of the bioactivity, solubility, and ion-releasing ability as suitable for biomedical applications.

References

1 Palangkaraya, A. and Yong, J. (2009) Population ageing and its implications on aggregate health care demand: empirical evidence from 22 OECD countries. *Int. J. Health Care Finance Econ.*, **9**, 391–402.

2 Cao, W.P. and Hench, L.L. (1996) Bioactive materials. *Ceram. Inter.*, **22**, 493–507.

3 Hench, L.L., Splinter, R.J., Allen, W.C., and Greenlee, T.K. (1971) Bonding mechanisms at the interface of ceramic prosthetic materials. *J. Biomed. Mater. Res.*, **5**, 117–141.

4 Xynos, I.D., Edgar, A.J., Buttery, L.D.K., Hench, L.L., and Polak, J.M. (2001) Gene-expression profiling of human osteoblasts following treatment with the ionic products of Bioglass® 45S5 dissolution. *J. Biomed. Mater. Res.*, **55**, 151–157.

5 Hoppe, A., Güldal, N.S., and Boccaccini, A.R. (2011) A review of the biological response to ionic dissolution products from bioactive glasses and glass-ceramics. *Biomaterials*, **32**, 2757–2774.

6 Hench, L.L. (2016) Bioglass: 10 milestones from concept to commerce. *J. Non Cryst. Solids*, **432**, 2–8.

7 Andersson, Ö.H., Liu, G., Karlsson, K.H., Niemi, L., Miettinen, J., and Juhanoja, J. (1990) *In vitro* behavior of glasses in the SiO_2–Na_2O–CaO–P_2O_5–Al_2O_3–B_2O_3 system. *J. Mater. Sci. Mater. Med.*, **1**, 219–227.

8 Elgayar, I., Aliev, A.E., Boccaccini, A.R., and Hill, R.G. (2005) Structural analysis of bioactive glasses. *J. Non Cryst. Solids*, **351**, 173–183.

9 Brink, M., Turunen, T., Happonen, R.-P., and Yli-Urpo, A. (1997) Compositional dependence of bioactivity of glasses in the system Na_2O–K_2O–MgO–CaO–B_2O_3–P_2O_5–SiO_2. *J. Biomed. Mater. Res.*, **37**, 114–121.

10 Gomez-Vega, J.M., Saiz, E., Tomsia, A.P., Oku, T., Suganuma, K., Marshall, G.W., and Marshall, S.J. (2000) Novel bioactive functionally graded coatings on Ti_6Al_4V. *Adv. Mater.*, **12**, 894–898.

11 Oonishi, H., Hench, L.L., Wilson, J., Sugihara, F., Tsuji, E., Matsuura, M., Kin, S., Yamamoto, T., and Mizokawa, S. (2000) Quantitative comparison of bone growth behavior in granules of Bioglass®, A-W glass-ceramic, and hydroxyapatite. *J. Biomed. Mater. Res.*, **51**, 37–46.

12 Hench, L.L., Andersson, O.H., and Latorre, G.P. (1991) *Bioceramics*, vol. **4** (eds. W. Bonefield, G.W. Hastings, and K.E. Tanner), Butterworth-Heinemann Ltd, p. 154.

13 Kokubo, T. and Takadama, H. (2006) How useful is SBF in predicting *in vivo* bone bioactivity? *Biomaterials*, **27**, 2907–2915.

14 Bohner, M. and Lemaitre, J. (2009) Can bioactivity be tested *in vitro* with SBF solution? *Biomaterials*, **30**, 2175–2179.

15 Hench, L.L. (2006) The story of bioglass. *J. Mater. Sci. Mater. Med.*, **17**, 967–978.

16 Tilocca, A., Cormack, A.N., and de Leeuw, N.H. (2007) The structure of bioactive silicate glasses: new insight from molecular dynamics simulations. *Chem. Mater.*, **19**, 95–103.

17 Tilocca, A. and Cormack, A.N. (2010) Surface signatures of bioactivity: MD simulations of 45S and 65S silicate glasses. *Langmuir*, **26**, 545–551.

18 Massera, J., Hupa, L., and Hupa, M. (2012) Influence of the partial substitution of CaO with MgO on the thermal properties and *in vitro* reactivity of the bioactive glass S53P4. *J. Non Cryst. Solids*, **358**, 2701–2707.

19 Watts, S.J., Hill, R.G., O'Donnell, M.D., and Law, R.V. (2010) Influence of magnesia on the structure and properties of bioactive glasses. *J. Non Cryst. Solids*, **356**, 517–524.

20 Wallace, K.E., Hill, R.G., Pembroke, J.T., Brown, C.J., and Hatton, P.V. (1999) Influence of sodium oxide content on bioactive glass properties. *J. Mater. Sci. Mater. Med.*, **10**, 697–701.

21 Hill, R.G. and Brauer, D.S. (2011) Predicting the bioactivity of glasses using the network connectivity or split network models. *J. Non Cryst. Solids*, **357**, 3884–3887.
22 Fagerlund, S., Hupa, L., and Hupa, M. (2013) Dissolution patterns of biocompatible glasses in 2-amino-2-hydroxymethyl-propane-1,3-diol (Tris) buffer. *Acta Biomater.*, **9**, 5400–5410.
23 Hill, R. (1996) An alternative view of the degradation of bioglass. *J. Mater. Sci. Lett.*, **15**, 1122–1125.
24 Lockyer, M.W.G., Holland, D., and Dupree, R. (1995) NMR investigation of the structure of some bioactive and related glasses. *J. Non Cryst. Solids*, **188**, 207–219.
25 Peitl, O., Zanotto, E.D., and Hench, L.L. (2001) Highly bioactive P_2O_5–Na_2O–CaO–SiO_2 glass-ceramics. *J. Non Cryst. Solids*, **292**, 115–126.
26 O'Donnell, M.D., Watts, S.J., Hill, R.G., and Law, R.V. (2009) The effect of phosphate content on the bioactivity of soda-lime-phosphosilicate glasses. *J. Mater. Sci. Mater. Med.*, **20**, 1611–1618.
27 O'Donnell, M.D., Watts, S.J., Hill, R.G., and Law, R.V. (2008) Effect of P_2O_5 content in two series of soda lime phosphosilicate glasses on structure and properties-part I: NMR. *J. Non Cryst. Solids*, **354**, 3554–3560.
28 Tilocca, A. and Cormack, A.N. (2007) Structural effects of phosphorus inclusion in bioactive silicate glasses. *J. Phy. Chem. B*, **111**, 14256–14264.
29 Brauer, D.S., Karpukhina, N., Law, R.V., and Hill, R.G. (2009) Structure of fluoride-containing bioactive glasses. *J. Mater. Chem.*, **19**, 5629–5636.
30 Fayon, F., Duée, C., Poumeyrol, T., Allix, M., and Massiot, D. (2013) Evidence of nanometric-sized phosphate clusters in bioactive glasses as revealed by solid-state ^{31}P NMR. *J. Phys. Chem. C*, **117**, 2283–2288.
31 Sepulveda, P., Jones, J.R., and Hench, L.L. (2001) Characterization of melt-derived 45S5 and sol-gel-derived 58S bioactive glasses. *J. Biomed. Mater. Res.*, **58**, 734–740.
32 Chen, W., Villa-Diaz, L.G., Sun, Y., Weng, S., Kim, J.K., Lam, R.H.M., Han, L., Fan, R., Krebsbach, P.H., and Fu, J. (2012) Nanotopography influences adhesion, and self-renewal of human embryonic stem cell. *ACS Nano*, **6**, 4094–4103.
33 Li, R., Clark, A.E., and Hench, L.L. (1991) An investigation of bioactive glass powders by sol–gel processing. *J. Appl. Biomater.*, **2**, 231–239.
34 Saravanapavan, P., Jones, J.R., and Hench, L.L. (2003) Bioactivity of gel-glass powders in the CaO–SiO_2 system: a comparison with ternary (CaO–P_2O_5–SiO_2) and quaternary glasses (SiO_2–CaO–P_2O_5–Na_2O). *J. Biomed. Mater. Res. A*, **66A**, 110–119.
35 Orcel, G., Hench, L.L., Artaki, I., Jonas, J., and Zerda, T.W. (1988) Effect of formamide additive on the chemistry of silica sol-gel II. Gel structure. *J. Non Cryst. Solids*, **105**, 223–231.
36 Lin, S., Ionescu, C., Pike, K.J., Smith, M.E., and Jones, J.R. (2009) Nanostructure evolution and calcium distribution in sol–gel derived bioactive glass. *J. Mater. Chem.*, **19**, 1276–1282.
37 Kokubo, T. (1991) Bioactive glass ceramics: properties and applications. *Biomaterials*, **12**, 155–163.
38 Kokubo, T., Kushitani, H., Ohtsuki, C., and Sakka, S. (1992) Chemical reaction of bioactive glass and glass-ceramics with a simulated body fluid. *J. Mater. Sci. Mater. Med.*, **3**, 79–83.
39 Foppiano, S., Marshall, S.J., Marshall, G.W., Saiz, E., and Tomsia, A.P. (2004) The influence of novel bioactive glasses on *in vitro* osteoblast hebavior. *J. Biomed. Mater. Res. A*, **71A**, 242–249.
40 Zreiqat, H., Howlett, C.R., Zannettino, A., Evans, P., Schulze-Tanzil, G., Knabe, C., and Shakibaei, M. (2002) Mechanisms of magnesium-stimulated adhesion of osteoblastic cells to commonly used orthopaedic implants. *J. Biomed. Mater. Res.*, **62**, 175–184.
41 Diba, M., Tapia, F., Boccaccini, A.R., and Strobel, L.A. (2012) Magnesium-containing bioactive glasses for biomedical applications. *Int. J. Appl. Glass Sci.*, **3**, 221–253.
42 Hall, S.L., Dimai, H.P., and Farley, J.R. (1999) Effects of zinc on human skeletal alkaline phosphatase

activity *in vitro*. *Calcif. Tissue Int.*, **64**, 163–172.

43 Lusvardi, G., Malavasi, G., Menabue, L., Menziani, M.C., Segre, U., Carnasciali, M.M., and Ubaldini, A. (2004) A combined experimental and computational approach to $(Na_2O)_{1-x} \cdot CaO \cdot (ZnO)_x \cdot 2SiO_2$ glasses characterization. *J. Non Cryst. Solids*, **345**, 710–714.

44 Linati, L., Lusvardi, G., Malavasi, G., Menabue, L., Menziani, M.C., Mustarelli, P., and Segre, U. (2005) Qualitative and quantitative structure-property relationships analysis of multicomponent potential bioglasses. *J. Phys. Chem. B*, **109**, 4989–4998.

45 Chen, X., Brauer, D.S., Karpukhina, N., Waite, R.D., Barry, M., McKay, I.J., and Hill, R.G. (2014) 'Smart' acid-degradable zinc-releasing silicate glasses. *Mater. Lett.*, **126**, 278–280.

46 Meunier, P.J., Roux, C., Seeman, E., Ortolani, S., Badurski, J.E., Spector, T.D., Cannata, J., Balogh, A., Lemmel, E.M., and Pors-Nielsen, S. (2004) The effects of strontium ranelate on the risk of vertebral fracture in women with postmenopausal osteoporosis. *N. Engl. J. Med.*, **350**, 459–468.

47 Gentleman, E., Fredholm, Y.C., Jell, G., Lotfibakshaiesh, N., O'Donnell, M.D., Hill, R.G., and Stevens, M.M. (2010) Extracellular matrix-mediated osteogenic differentiation of murine embryonic stem cells. *Biomaterials*, **31**, 3244–3252.

48 Lao, J., Jallot, E., and Nedelec, J.-M. (2008) Strontium-delivering glasses with enhanced bioactivity: a new biomaterial for antiosteoporotic applications? *Chem. Mater.*, **20**, 4969–4973.

49 Hoppe, A., Meszaros, R., Stähli, C., Romeis, S., Schmidt, J., Peukert, J., Marelli, B., Nazhat, S.N., Wondraczek, L., Lao, J., Jallot, E., and Boccaccini, A.R. (2013) *In vitro* reactivity of Cu doped 45S5 Bioglass® derived scaffolds for bone tissue engineering. *J. Mater. Chem. B*, **1**, 5659–5674.

50 Gorustovich, A.A., Roether, J.A., and Boccaccini, A.R. (2010) Effect of bioactive glasses on angiogenesis: a review of *invitro* and *in vivo* evidences. *Tissue Eng. B*, **16**, 199–207.

51 Smith, J.M., Martin, R.A., Cuello, G.J., and Newport, R.J. (2013) Structural characterization of hypoxia-mimicking bioactive glasses. *J. Mater. Chem. B*, **1**, 1296–1303.

52 Nielsen, F.H. (2008) Is boron nutritionally relevant? *Nutr. Rev.*, **66**, 183–191.

53 Marion, N.W., Liang, W., Reilly, G.C., Day, D.E., Rahaman, M.N., and Mao, J.J. (2005) Borate glass supports the *in vitro* osteogen differentiation of human mesenchymal stem cells. *Mech. Adv. Mater. Struct.*, **12**, 239–246.

54 Zhang, X., Jia, W., Gua, Y., Wei, X., Liu, X., Wang, D., Zhang, C., Huang, W., Rahaman, M.N., Day, D.E., and Zhou, N. (2010) Teicoplanin-loaded borate bioactive glass implants for treating chronic bone infection in a rabbit tibia osteomyelitis model. *Biomaterials*, **31**, 5865–5874.

55 Brink, M. (1997) The influence of alkali and alkali earths on the working range for bioactive glasses. *J. Biomed. Mater. Res.*, **36**, 109–117.

56 Huang, W., Day, D.E., Kittiratanapiboon, K., and Rahaman, M.N. (2006) Kinetics and mechanism of the conversion of silicate (45S5) borate, and borosilicate glasses to hydroxyapatite in dilute phosphate solutions. *J. Mater. Sci. Mater. Med.*, **17**, 583–596.

57 Hoppe, U. (1996) A structural model for phosphate glasses. *J. Non Cryst. Solids*, **195**, 138–147.

58 Wetherall, K.M., Pickup, D.M., Newport, R.J., and Mountjoy, G. (2009) The structure of calcium metaphosphate glass obtained from X-ray and neutron diffraction and reverse Monte Carlo modelling. *J. Phys. Condens. Matter*, **21**, 035109.

59 Abou Neel, E.A., Pickup, D.M., Valappil, S.P., Newport, R.J., and Knowles, J.C. (2009) Bioactive functional materials: a perspective on phosphate-based glasses. *J. Mater. Chem.*, **19**, 690–701.

60 Bunker, B.C., Arnold, G.W., and Wilder, J.A. (1984) Phosphate glasses dissolution in aqueous solutions. *J. Non Cryst. Solids*, **64**, 291–316.

61 Abrahams, I., Hawkes, G.E., and Knowles, J. (1997) Phosphorus speciation in

sodium–calcium–phosphate ceramics. *J. Chem. Soc., Dalton Trans.*, 1483–1484.

62 Franks, K., Abrahams, I., and Knowles, J.C. (2000) Development of soluble glasses for biomedical use part I: *in vitro* solubility measurement. *J. Mater. Sci. Mater. Med.*, **11**, 609–614.

63 Tang, E., Di Tommaso, D., and de Leeuw, N.H. (2010) An ab initio molecular dynamics study of bioactive phosphate glasses. *Adv. Eng. Mater.*, **12**, B331–338.

64 Abou Neel, E.A., Mizoguchi, T., Ito, M., Bitar, M., Salih, V., and Knowles, J.C. (2007) *In vitro* bioactivity and gene expression by cells cultured on titanium dioxide doped phosphate-based glasses. *Biomaterials*, **28**, 2967–2977.

65 Salih, V., Patel, A., and Knowles, J.C. (2007) Zinc-containing phosphate-based glasses for tissue engineering. *Biomed. Mater.*, **2**, 11–20.

66 Franks, K., Salih, V., Knowles, J.C., and Olsen, I. (2002) The effect of MgO on the solubility behavior and cell proliferation in a quaternary soluble phosphate based glass system. *J. Mater. Sci. Mater. Med.*, **13**, 549–556.

67 Knowles, J.C., Franks, K., and Abrahams, I. (2001) Investigation of the solubility and ion release in the glass system K_2O–Na_2O–CaO–P_2O_5. *Biomaterials*, **22**, 3091–3096.

68 Brauer, D.S., Karpukhina, N., Law, R.V., and Hill, R.G. (2010) Effect of TiO_2 addition on structure, solubility and crystallization of phosphate invert glasses for biomedical applications. *J. Non Cryst. Solids*, **356**, 2626–2633.

69 Ohtsuki, C., Kokubo, T., Takahara, K., and Yamamuro, T. (1991) Compositional dependence of bioactivity of glasses in the system CaO–SiO_2–P_2O_5: its *in vitro* evaluation. *J. Ceram. Soc. Jpn.*, **99**, 1–6.

70 Uo, M., Mizuno, M., Kuboki, Y., Makishima, A., and Watari, F. (1998) Properties and cytotoxicity for water soluble Na_2O–CaO–P_2O_5 glasses. *Biomaterials*, **19**, 2277–2284.

71 Murthy, M.K., Smith, M.J., and Westman, A.E.R. (1961) Constitution of mixed alkali phosphate glasses: I, constitution of constant lithium variable sodium–potassium phosphate glasses. *J. Am. Ceram. Soc.*, **44**, 97–105.

72 Kasuga, T. (2005) Bioactive calcium pyrophosphate glasses and glass-ceramics. *Acta Biomater.*, **1**, 55–64.

73 Kasuga, T. and Abe, Y. (1999) Calcium phosphate invert glasses with soda and titania. *J. Non Cryst. Solids*, **243**, 70–74.

74 Kasuga, T., Hosoi, Y., Nogami, M., and Niinomi, M. (2001) Apatite formation on calcium phosphate invert glass in simulated body fluid. *J. Am. Ceram. Soc.*, **84**, 450–452.

75 Teixeira, Z., Alves, O.L., and Mazali, I.O. (2007) Structure, thermal behavior, chemical durability, and optical properties of the Na_2O–Al_2O_3–TiO_2–Nb_2O_5–P_2O_5 glass system. *J. Am. Ceram. Soc.*, **90**, 256–263.

76 Hsu, S.M., Wu, J.J., Yung, S.W., Chin, T.S., Zhang, T., Lee, Y.M., Chu, C.M., and Ding, J.Y. (2012) Evaluation of chemical durability, thermal properties and structure characteristics of Nb–Sr-phosphate glasses by Raman and NMR spectroscopy. *J. Non Cryst. Solids*, **358**, 14–19.

77 Maeda, H., Lee, S., Miyajima, T., Obata, A., Ueda, K., Narushima, T., and Kasuga, T. (2016) Structure and physicochemical properties of CaO–P_2O_5–Nb_2O_5–Na_2O glasses. *J. Non Cryst. Solids*, **432**, 60–64.

78 Obata, A., Takahashi, Y., Miyajima, T., Ueda, K., Narushima, T., and Kasuga, T. (2012) Effects of niobium ions released from calcium phosphate invert glasses containing Nb_2O_5 on osteoblasts-like cell functions. *ACS Appl. Mater. Interfaces*, **4**, 5684–5690.

79 Lee, S., Maeda, H., Obata, A., Ueda, K., Narushima, T., and Kasuga, T. (2015) Structures and dissolution behaviors of CaO–P_2O_5–TiO_2/Nb_2O_5 (Ca/P ≥ 1) invert glasses. *J. Non Cryst. Solids*, **426**, 35–42.

80 Lee, S., Obata, A., Brauer, D.S., and Kasuga, T. (2015) Dissolution behavior and cell compatibility of alkali-free MgO–CaO–SrO–TiO_2–P_2O_5 glasses for biomedical applications. *Biomed. Glasses*, **1**, 151–158.

81 Lee, S., Obata, A., and Kasuga, T. (2009) Ion release from SrO–CaO–TiO_2–P_2O_5

glasses in Tris buffer solution. *J. Ceram. Soc. Jpn.*, **117**, 935–938.

82 Da, N., Krolikowski, S., Nielsen, K.H., Kaschta, J., and Wondraczek, L. (2010) Viscosity and softening behavior of alkali zinc sulfophosphate glasses. *J. Am. Ceram. Soc.*, **93**, 2171–2174.

83 Sirotkin, S., Meszaros, R., and Wondraczek, L. (2012) Chemical stability of $ZnO-Na_2O-SO_3-P_2O_5$ glasses. *Inter. J. Appl. Glass Sci.*, **3**, 44–52.

84 Bassett, D.C., Meszaros, R., Orzol, D., Woy, M., Ling Zhang, Y., Tiedemann, K., Wondraczek, L., Komarova, S., and Barralet, J.E. (2014) A new class of bioactive glasses: calcium–magnesium sulfophosphates. *J. Biomed. Mater. Res. A*, **102A**, 2842–2848.

14
Materials for Tissue Engineering

María Vallet-Regí[1,2] and Antonio J. Salinas[1,2]

[1]*Departamento de Química Inorgánica y Bioinorgánica, Universidad Complutense de Madrid, Instituto de Investigación Sanitaria Hospital 12 de Octubre, i + 12, Pza. Ramón y Cajal s/n, 28040 Madrid, Spain*
[2]*Centro de Investigación Biomédica en Red de Bioingeniería, Biomateriales y Nanomedicina (CIBER-BBN), Madrid, Spain*

14.1
Tissue Engineering: General Concepts

Tissue engineering (TE) aims to regenerate damaged tissues, instead of replacing them, by developing biological substitutes that restore, maintain, or improve tissue function [1–4]. This term appeared for the first time in the title of an article in 1984, although the first definition occurred in 1988 in a workshop of the National Science Foundation [5]. Langer and Vacanti included a slight modification of this definition in their seminal article of 1993, often considered the foundation of TE as it is considered currently [6]. This definition is: "TE is an interdisciplinary field that applies the principles of engineering and life sciences toward the development of biological substitutes that restore, maintain, or improve tissue or organ function." Williams proposed another definition of TE in 1998 and considered TE the "persuasion of the body to heal itself, through the delivery to the appropriate sites of molecular signals, cells, and supporting structures" [7]. Both definitions contain the basic elements of TE to be treated in this Chapter.

TE strongly emerged in the last decade of the past century as a meeting point between two, until then, independent disciplines. On one hand, there were the traditional implantable medical devices based on materials engineering principles [8]. The only requisites for these implantable devices were to satisfy certain mechanical requirements, but powerful regenerative forces of the body were not exploited [9]. While these traditional implants worked well in general terms, their properties were further improved through the development of new biomaterials with controlled reactivity, such as degradation or bioactivity [10]. On the other hand, there was the transplant of

Handbook of Solid State Chemistry, First Edition. Edited by Richard Dronskowski, Shinichi Kikkawa, and Andreas Stein.
© 2017 Wiley-VCH Verlag GmbH & Co. KGaA. Published 2017 by Wiley-VCH Verlag GmbH & Co. KGaA.

organs and tissues, the efficacy of which was incremented if only the active components were used [11]. Therefore, TE can be considered as a bridge between both options because it uses biomaterials together with cell components responsible for growth and tissue repair, so that the combination, often called tissue construct [12], helps the patient to regenerate new functional tissue. The material may be biodegradable or not and the biological components can be cells, biomolecules, or both [13]. Therefore, there are many possible combinations as will be discussed below.

Biomaterials are framed within the biomedical engineering field and agglutinate knowledge of areas, including science, engineering, biology, and medicine. Their evolution in the last 50 years has been amazing. The field has switched from the use of inert materials for living tissues substitution to the design of bioactive and/or biodegradable materials for their repair, which has ended at the third generation of biomaterials where the objective is regeneration [14]. In this evolution, very fast in the time, many concepts have changed.

In the 1950s, surgeons and dentists implanted materials with the only requisite that they be tolerated by the organism. The objective was to replace a damaged part without taking into account biology. It was sufficient at this point that they did not react with the organism and could be tolerated by it. In this scenario, materials science monopolized the leading responsibility to look for compatible materials with living tissues among the already existing ones for other technological applications.

In the 1980s, this an important concept is introduced. Researchers in biomaterials realize that the potential chemical reactions of implants with living tissues do not necessarily entail risk, if the reaction outcome is positive for the organism [10]. In this sense, the development, design, and commercialization of bioactive and biodegradable materials started to allow their common use in the repair of diverse living tissues. As a consequence, the basic concept of repair is complemented with the idea of regenerate.

Nowadays replacement and repair processes are chosen whenever it is necessary to place a "spare piece in the human body." Tissue regeneration is now initiating and is field investigation to make reality its application in Medicine in the next future. This change of perception has placed the center of gravity in biology, materials science now serving its demands. Thus, human body repair moves from a bionic approach to a regenerative medicine (RM) approach, which exhibits two faces, TE and cellular therapy, both using third-generation biomaterials, as defined, by Hench and Polak [14].

The objective of TE is the development of biological compounds and biomaterials to implant in the organism, aimed to repair, maintain, or improve the function of tissues and organs. The cells, the signals, and the scaffolds act simultaneously achieve the proposed objective, that is, the reconstruction of certain organs or tissues. A first seed of this discipline was the seed of a collagen matrix with fibroblasts to create neodermis [15]. As we will discuss later, until now the highest success rates of TE are in skin as well as bone and cartilage applications. Figure 14.1 shows the three pillars of TE when applied to regenerate bone, that is, bone TE, BTE (bone tissue engineering) [16]. First, we find the cells responsible for osteogenesis, that is the capacity to produce bone tissue by the action of cells. Second, the biochemical factors, which are the signals and growth factors responsible for the osteoinduction, that is the capacity to promote the bone formation. Third, the scaffolds made with natural and/or synthetic biomaterials, which are responsible for osteoconduction, which is the capacity to allow and favor the growth and organization of the bone tissue.

For treatment of bone defects, the simplest approach is to implant a scaffold, which fulfills the requirements of being biocompatible, porous, biodegradable (or resorbable), and an osteoconductor, regardless of its mechanical properties. The next option is to implant the scaffold in which cells of the own patient were previously seeded. In this second case, we are already within a BTE application. Moreover, there are also other options such as implanting the scaffold functionalized with signals, or to implant the scaffold decorated with both signals and cells. These two options are also BTE approaches as can be observed at the bottom of Figure 14.1.

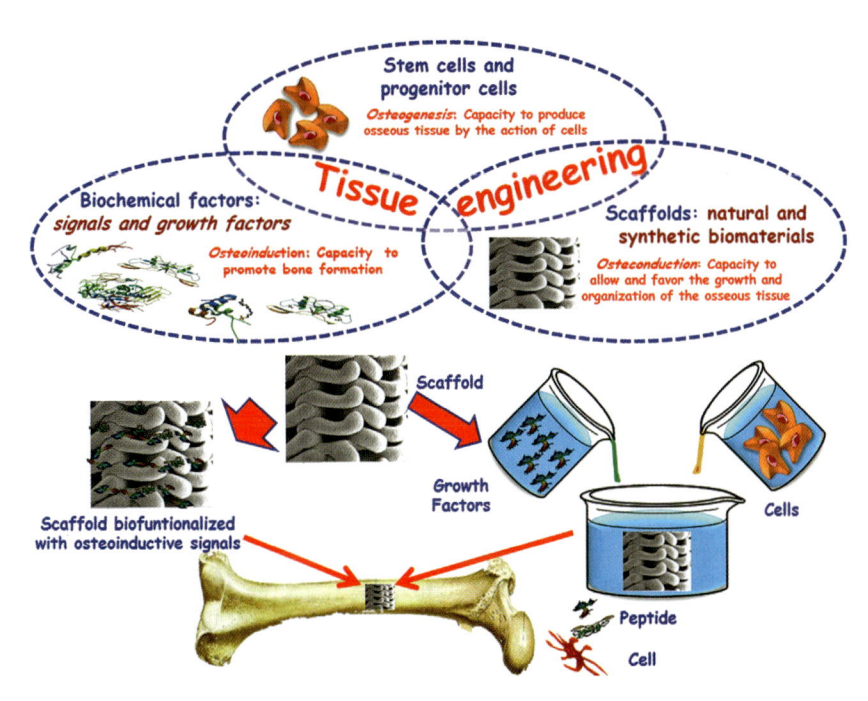

Figure 14.1 Pillars on which TE is sustained and its application to bone tissue engineering.

14.2
What can be Regenerated?

While theoretically in the near future, TE will cover a broad variety of clinical uses [17], at the present this term is associated with applications that repair or replace portions of or whole tissues such as bone, cartilage, blood vessels, bladder, skin, or muscle. Indeed, until now the first clinical successes exhibited for TE have been in

- skin: for the treatment of burns and chronic ulcers;
- cartilage: in the treatment of small injuries; and
- bone regeneration: in nonload-bearing applications.

However, there are a great number of experimental advances with more complex tissues such as nerve and muscle, and in the regeneration of more complex architectures, such as urogenital and cardiovascular systems. Furthermore, TE methods to create blood vessels and heart valves have been successfully used in large animal models.

In contrast to certain animals, humans have a limited ability to repair a tissue once it was formed, and an extremely limited ability to regenerate tissue once we have reached maturity. We have the ability to repair some tissues, for example bone, through the generation of new identical bone at the fracture site, and skin, although the new "skin" is usually more scar tissue than normal dermis and epidermis. With the other tissues, including muscle and nerve, this process of regeneration is difficult if not impossible.

Figure 14.2 shows the types of transplants of organs, tissues, cells, and fluids authorized at the present. At the right of the Figure are included the new opportunities for the treatment of some diseases that will come from the advanced therapies under investigation including cell therapy, gene therapy and TE [18].

The term RM is often used as synonymous to TE, although those procedures involving RM place greater emphasis on the use of stem cells or progenitor cells to produce living tissues [4]. RM applications started with the first transplants of organs in the middle of the twentieth century and today, 60 years later, continue to be characterized by the same limitation: the shortage of organs compatible with the patient. This fact forces us to continue investigating alternative therapies (see Figure 14.3). The latest scientific findings in this field allow us to be optimistic about the future that promises great advances in the field of the biomedicine. No doubt, the twenty first century will be the one of the biomedical revolution.

Together with the TE approaches, thoroughly detailed in this chapter, RM looks for the replacement of functions damaged in the organism mainly in two ways: gene therapy, that is to say, using genes in a therapeutic form and cellular therapy [18,19]. The latter is used to replace the damaged or lost cells by others coming from a donor or obtained in the laboratory from stem cells. That way, they are managed to repair the damages produced in tissues because of disease or trauma. The most common cellular therapy is the use of hematopoietic stem

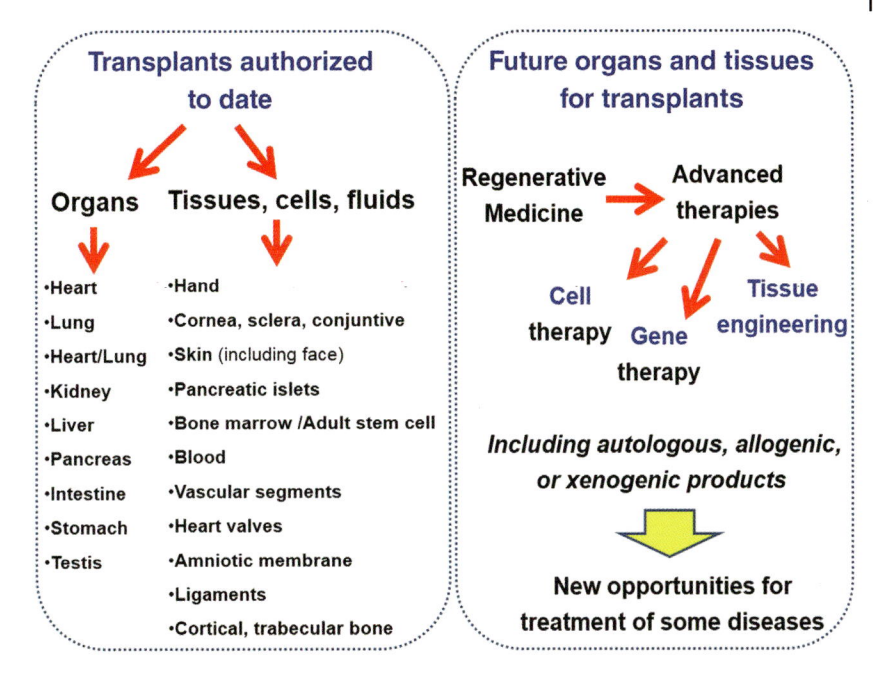

Figure 14.2 Types of transplants of organs, tissues, cells, and fluids authorized at the present and the future of organs and tissues for transplants based on regenerative medicine (RM).

Transplants authorized to date

Organs
- Heart
- Lung
- Heart/Lung
- Kidney
- Liver
- Pancreas
- Intestine
- Stomach
- Testis

Tissues, cells, fluids
- Hand
- Cornea, sclera, conjuntive
- Skin (including face)
- Pancreatic islets
- Bone marrow /Adult stem cell
- Blood
- Vascular segments
- Heart valves
- Amniotic membrane
- Ligaments
- Cortical, trabecular bone

Future organs and tissues for transplants

Regenerative Medicine → Advanced therapies → Cell therapy, Gene therapy, Tissue engineering

Including autologous, allogenic, or xenogenic products

New opportunities for treatment of some diseases

20th century: first transplants of organs → **60 years later:** the same limitation: shortage of organs compatible with patient

Alternative therapies are needed

Regenerative Medicine (RM)

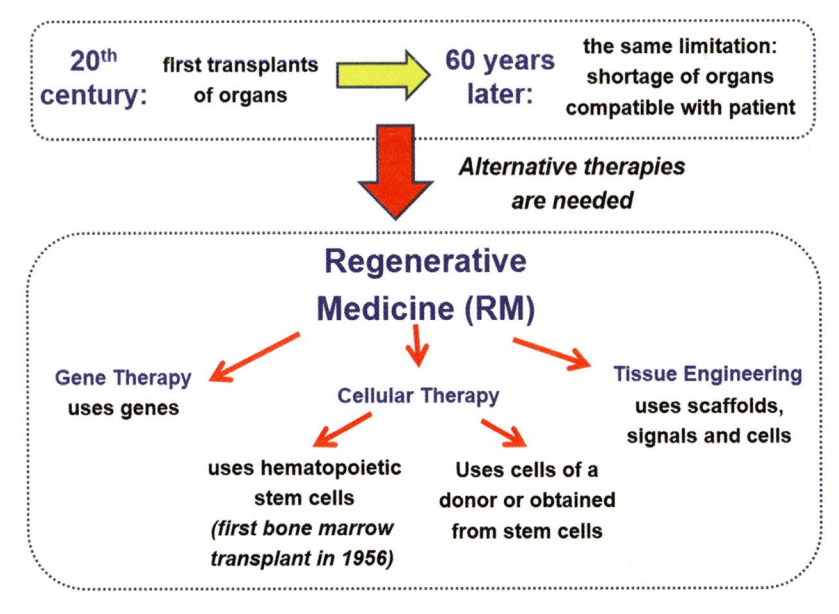

- **Gene Therapy** uses genes
- **Cellular Therapy**
  - uses hematopoietic stem cells (first bone marrow transplant in 1956)
  - Uses cells of a donor or obtained from stem cells
- **Tissue Engineering** uses scaffolds, signals and cells

Figure 14.3 Origins of RM and some important RM approaches.

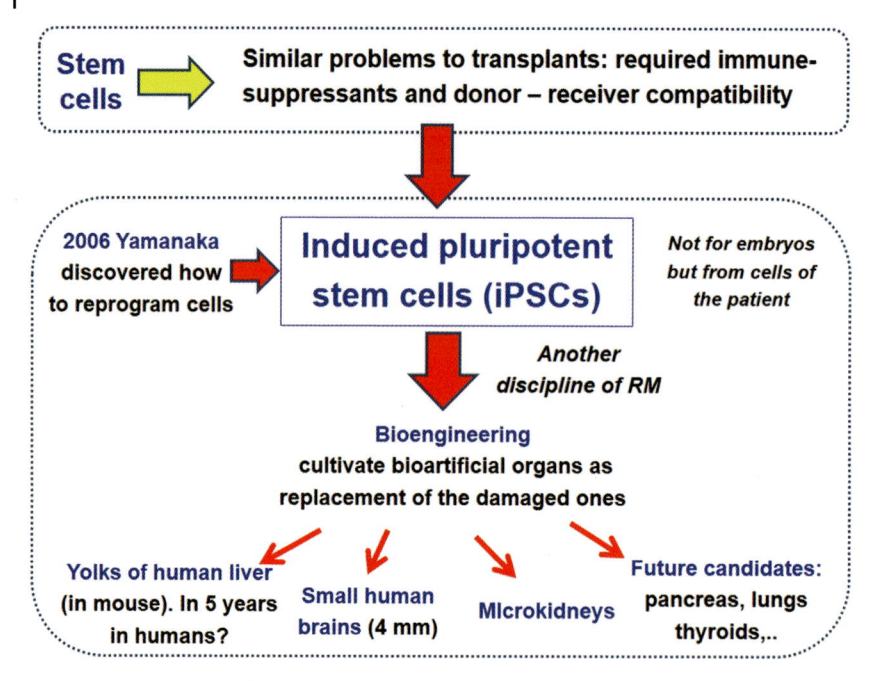

Figure 14.4 Limitations of using stem cells. Discovery and first applications of induced pluripotent stem cells (iPSCs).

cells hosted in bone marrow, which are able to originate all the cellular types of blood. This therapy started in 1956 with the first transplant of bone marrow [20]. Nevertheless, the transplant of stem cells causes problems similar to those of transplants: It requires the administration of immune suppressants in some cases and a certain compatibility between donor and receiver.

As an important development, Takahashi and Yamanaka in 2006 [21] discovered how to reprogram cells and manufacture them from induced pluripotent stem cells (iPSCs) which, besides not coming from embryos (avoiding ethical problems), can be obtained from cells of the patient, eliminating the risk of rejection. In addition, the creation of iPSCs has enabled another discipline of RM, bioengineering, to grow bioartificial organs in the laboratory to replace damaged organs. As Figure 14.4 shows, to date, advances in medical research have allowed scientists to develop yolks of liver [22], small human brains [23], and microkidneys [24]. Yolks of liver developed from iPSCs were transplanted into a mouse. It will take several years before this approach can be applied to humans. However, this is a demonstration of the immense therapeutic potential of these transplants and promises hope for all the patients awaiting the donation of organs. The obtaining of these primordial cells from iPSCs generates hopes to think that someday we will be able to use our own cells to regenerate our ill organs, solving with it the problem of organ shortage for transplants. Until then and, as in the case of the minibrains able to produce advances in neurology or

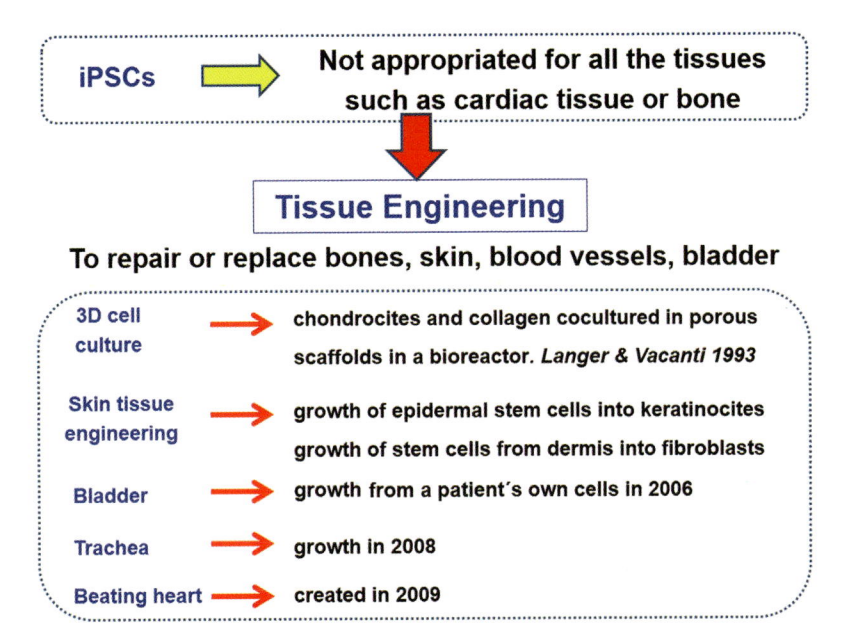

```
┌─────────────────────────────────────────────────────────┐
│  iPSCs   ➡   Not appropriated for all the tissues       │
│              such as cardiac tissue or bone             │
└─────────────────────────────────────────────────────────┘
                           ⬇
                 ┌──────────────────────┐
                 │  Tissue Engineering  │
                 └──────────────────────┘
     To repair or replace bones, skin, blood vessels, bladder

   3D cell        ➡   chondrocites and collagen cocultured in porous
   culture            scaffolds in a bioreactor. Langer & Vacanti 1993

   Skin tissue    ➡   growth of epidermal stem cells into keratinocites
   engineering        growth of stem cells from dermis into fibroblasts

   Bladder        ➡   growth from a patient's own cells in 2006

   Trachea        ➡   growth in 2008

   Beating heart  ➡   created in 2009
```

Figure 14.5 The iPSCs are not equally appropriate for all types of tissue. TE approaches continue to be necessary for the regeneration of many types of tissue. At the bottom, important milestones in TE are included.

the minikidneys in renal diseases, these progresses will facilitate the discovery of new drugs and therapies based on the use of stem cells. More efforts are in progress trying to synthesize in the lab new organs including pancreas, lungs, thyroids, and so on.

Nevertheless iPSCs are yielding lower success rates in the generation of certain tissues like bone or cardiac tissue. Thus, it is necessary to continue investigating TE approaches for repairing or replacing bones, cartilage, skin, blood vessels, or bladder (see Figure 14.5). The lower part of the Figure shows significant advances in TE as the first achievements in TE skin or the first growth of a bladder, a trachea or a beating heart from cells of the patient [25–27]. At present, TE is considered to be one of the most promising biomedical technologies.

14.3
Tissue Engineering in Bone

TE is one of the most active research areas for the treatment of bone tissues [28–31]. Indeed the shift from bone replacement to bone regeneration is often considered the origin of BTE. We can begin by pointing out what can be learned from Nature. In particular, we can start considering the structure and composition of our bones and teeth [32]. The mineral component of hard tissues is an apatite-like phase, nanometric in size, and variable in composition, calcium

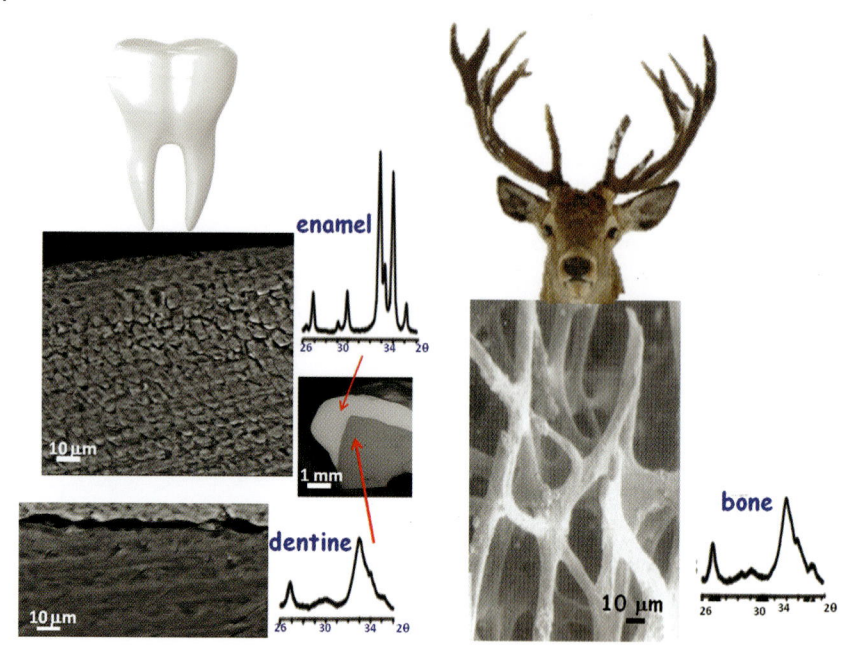

Figure 14.6 Learning from Nature in the search of synthetic biomaterials for BTE.

deficient, and carbonated [33]. These features make it a highly reactive mineral, which is required in order to provide constant and fast continuous resorption and formation processes necessary for bone regeneration. Our bones are natural composites formed by biological apatites, which are located in confined spaces among collagen molecules. This vision of the bones shows their structure at the nanometer scale and to produce bone replacement materials in a laboratory, we must keep these features in mind. Figure 14.6 shows hard natural living tissues that can be used as a source of inspiration for TE applications.

There are two options when looking for materials suitable for bone tissue replacements: (i) to study the possibilities of the living tissues, therefore producing biomaterials of natural origin and (ii) to produce artificial materials in the laboratory. In the design of the materials, we must select several parameters to obtain a material with the appropriate features. Such features include the particle size and potential structural defects of the ceramic component, the conformation method of the piece or implant, and its porosity, and possible surface functionalization.

14.3.1
Scaffolds

A scaffold is a temporary supporting structure for growing cells and tissues where the healing process takes place [34]. The properties of an ideal scaffold for BTE are [34–38] the following: it must provide a biocompatible mechanical support, avoid inducing adverse tissue responses, and temporarily maintain mechanical loads on the tissue where is implanted. Moreover, it must have an appropriate degradation rate, equivalent to the process of regeneration of the living tissue. In addition, three-dimensional (3D) scaffolds must exhibit interconnected porosity where cells can adhere, grow, and proliferate. Furthermore, they must exhibit meso-architecture by designing mesopores able to confine molecules with therapeutic activity. A suitable pore size distribution will promote ingrowths of tissues, blood vessels, and cells, the traffic of metabolites. A high surface area will favor anchoring of cells. Moreover, we must not forget the surface functionalization, both in terms of better biological performance helping bone regeneration, and in terms of drug loading and release. In addition, 3D scaffolds must facilitate the biological recognition, of such form that gives support and promotes adhesion, migration, proliferation, and cellular differentiation. Additionally, it must constitute a place adapted for the development of living tissue that allows capturing and releasing morphogenetic factors.

Figure 14.7 shows the essential role of a scaffold in a typical BTE procedure acting as a support of the cells taken from the patient and cultured *in vitro* before being implanted back into the patient. The scaffolds can be naturally derived materials, for instance bone coming from animals or bone banks in

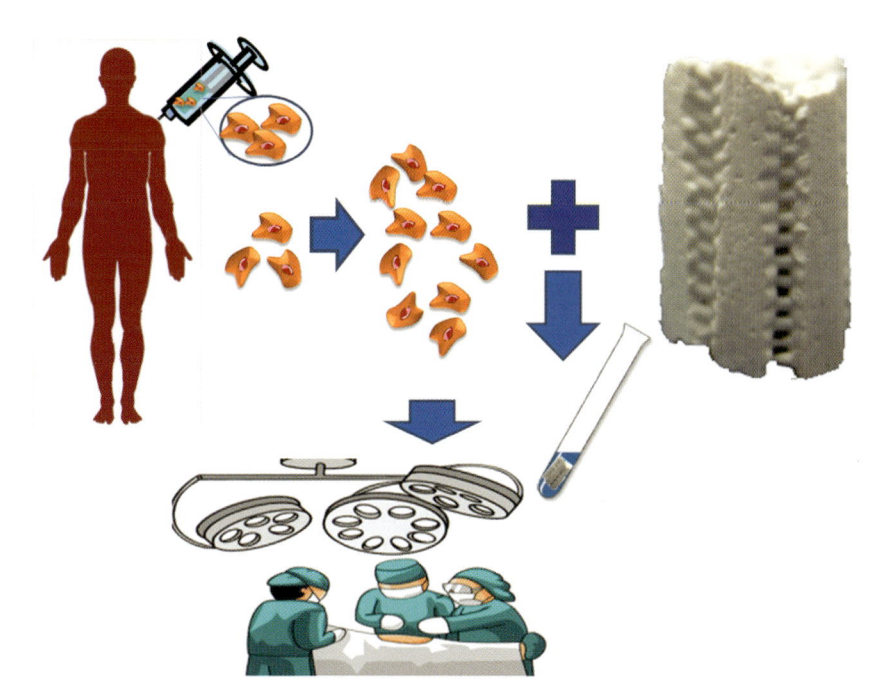

Figure 14.7 A common TE procedure, which starts taking cells from the patient and seeding them onto synthetic scaffolds, cultured under *in vitro* conditions in a bioreactor and implanted back into the patient.

xeno- and allografts or from the patient, or they can be synthetic materials. Because of the limitations of materials of natural origin, including disease transmission, as well as shortage and morbidity in the case of autografts, there is a general trend to investigate the design of new synthetic biomaterials. In the following sections, we will describe the most important features of some scaffolds obtained from bioceramic, polymeric, metallic, and composite materials.

14.3.1.1 Bioceramic Scaffolds

A rather small group of bioceramics has been investigated to design scaffolds for BTE [18]. Most of them are calcium phosphates, including hydroxyapatite, HA, β-calcium phosphate, β–TCP, biphasic calcium phosphate (BCP), and octacalcium phosphate (OCP), as well as bioactive glasses and calcium sulfate. In this section, some representative examples, together with a recently proposed stimuli-responsive system, are described.

Because the mineral component of bones is nanometric carbonate hydroxyapatite (CHA), a logical first approach in this regards is the design of scaffolds based on 3D nanoapatites. The starting point must be the synthesis of apatite with nanometric size. The synthesis method is clearly a wet route method to ensure the nanometric particle size of the bioceramic. Once the material has been synthesized, the next step is to obtain the scaffold with the required size and shape piece conformation. To do so, we must choose methods, which do not need high temperatures or high pressures because they destroy porosity and eventually functionalization. Robocasting devices are a good option. Thus, using computerized axial tomography (CAT) scan and feeding the data obtained to a computer can be used to program the robocasting machine to design an implant with the desired shape, size, and macroporosity [39]. Such an implant can be directly inserted into the patient or can be decorated with signal molecules and cells to obtain a construct.

Functionalization of nanoapatites by bonding molecules with specific features covalently to their surface can modify the interaction of the matrix with the signals included in the scaffolds to improve their osteogenic capabilities. For instance, our group functionalized a bioceramic with amino groups to promote covalent binding with the signal molecule osteostatin [40]. The objective of decorating the scaffold with osteostatin was to inhibit the activity of osteoclasts and improve bone formation. After the loading process, the obtained material is assessed under *in vitro* conditions and then, the construct is subjected to *in vivo* assays to verify if the proposed design fulfills the defined objectives. In this regard, scaffolds with hierarchical porosity based on silicon-doped hydroxyapatite were prepared by rapid prototyping (RP), to be implanted in rabbit tibia. The radiological study after 4 months of implantation showed the partial resorption of the scaffold and the complete repairing of the cortical bone defect [41].

A second example is the production of scaffolds based on bioactive glasses. The desired objectives were identical to describe nanoapatites. Many synthesis features can be controlled from its synthesis at the atomic or molecular level, up to its macrostructure and surface functionalization, and keeping always in mind the nano-

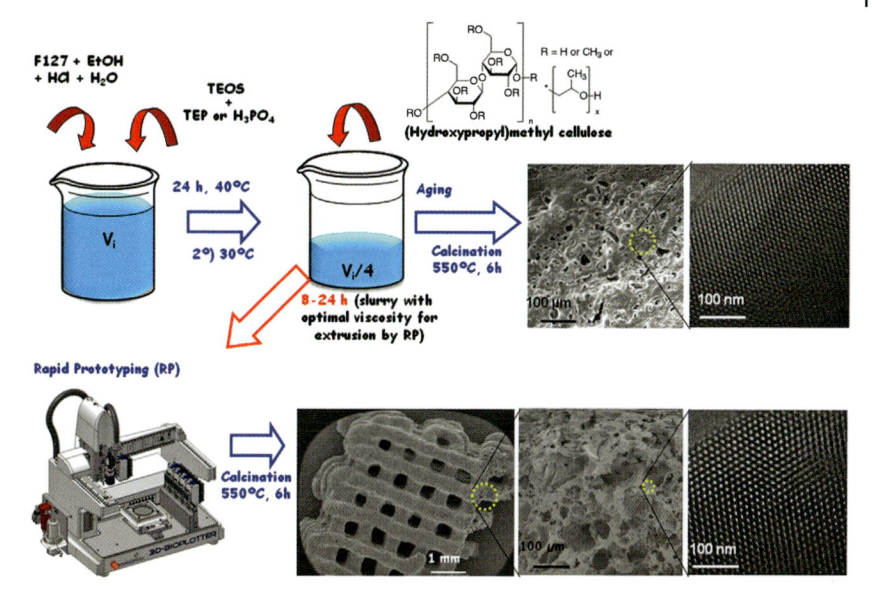

Figure 14.8 After the synthesis of MBGs as powders, they are processed by RP to obtain 3D scaffolds with tailored dimensions and macropores. At the right, SEM and TEM micrographs of the obtained cells.

and microstructures [42]. With the selection of the appropriate compositions it is possible to obtain glasses that exhibit a very quick bioactive response [43]. They can be synthesized by the sol–gel method and during the synthesis amphiphilic molecules can be added to behave as structure-directing agents. In this way, mesoporous bioactive glasses (MBGs) with controlled nanoscale porosity and microstructure are obtained. Figure 14.8 shows the synthesis of the MBGs as powders, by evaporation induced self-assembly (EISA) [44], and then their processing as 3D scaffolds by RP [45,46]. After the synthesis, the obtained materials can be characterized by different experimental techniques, including scanning electron microscopy (SEM) and transmission electron microscopy (TEM) as shown in Figure 14.8. The fabrication of 3D scaffolds based on MBGs by RP yields a hierarchically porous structure with three different levels of porosity at 400 µm, 1–80 µm, and under 10 nm, which exercise different functions in the bone regeneration process [43].

3D scaffolds with designed meso- and macroporosity can be used for BTE applications after cell harvesting from the patient and cell proliferation followed by seeding of the cells on the scaffold, using bioreactors and reaching the desired clinical behavior. These types of scaffolds, fully designed from the starting materials to their final shape and inner porosity, can be a valuable support in BTE.

Moreover, the scaffolds must also have a functionalized surface to achieve the appropriate binding with growth factors, peptides and proteins that can attract cells to stimulate the tissue regeneration process and, as a benefit, the ability to load drugs into them, if needed. Therefore, it is necessary to bear in mind all the different scales because each one fulfils specific functions.

On the other hand, in 2001 our group proposed the application of silica mesoporous materials as matrixes in drug delivery systems [47]. Since then, thousands of articles investigating this property were published [48]. The experience in chemical synthesis of silica mesoporous materials was used to obtain nanoparticles (NPs) of similar composition [28]. Simultaneously, the experience in smart materials can be applied to both bulk and NP materials [49]. With this idea in mind, our research group started to design how to combine scaffolds for hard tissue regeneration with different drugs to obtain efficient controlled drug delivery systems (DDSs) [50].

At this point, we should note that DDSs must exhibit mesoporosity, which ranges between 2 and 10 nm, necessary for applications where biologically active molecules are loaded, and later released. However, scaffolds for BTE must exhibit a hierarchical porosity, similar to bones, to be useful in regeneration properties and, consequently they must exhibit pores in the micrometric scale. It is therefore critical to control pore architecture over wide pore size ranges [51–53]. In the specific application of silica mesoporous materials, weak interactions between this ceramic matrix and different drugs can be induced to obtain efficient controlled DDSs [54]. On the other hand, if strong interactions between the inorganic matrix and biologically active molecules that enable bone tissue regeneration are induced, efficient systems for hard tissue regeneration will be produced. Simultaneously, it is possible to combine weak and strong interactions in the same system, which would then be suitable for both purposes: drug delivery and tissue regeneration.

These objectives can be reached by using a symbiosis between supramolecular chemistry and material science that opens new opportunities. That way, it is possible to combine bioceramic 3D scaffolds with nanotechnology [50]. Figure 14.9

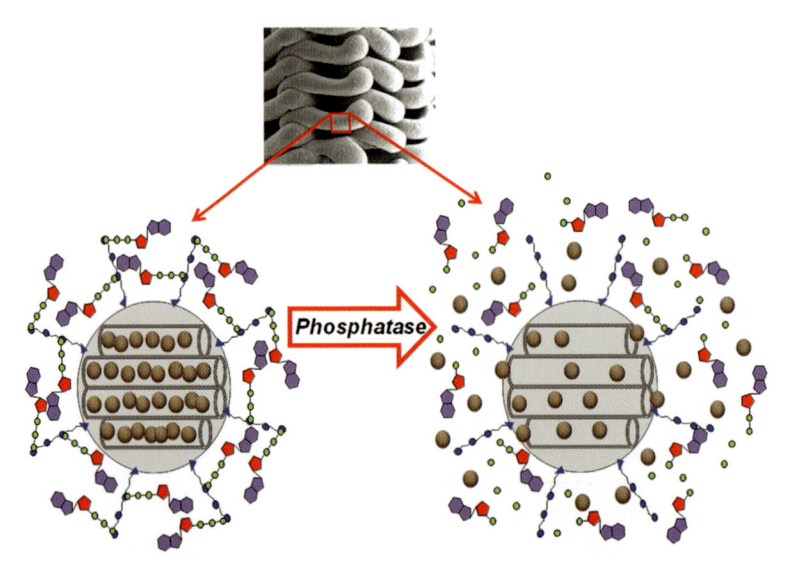

Figure 14.9 A 3D scaffolds containing a stimulus response system based on a mesoporous silica NP able to release the inner drug under the action of the alkaline phosphatase enzyme.

shows an SEM micrograph of a gelatin-based 3D scaffold containing a stimulus responsive system based on mesoporous NPs loaded with a drug that is released to medium with the presence of a specific stimulus. To reach this objective, a design at four length scales is required. Silica NPs must maintain the drug in the pores by molecular nanogates that are opened in response to determined stimuli. That way, scaffolds combined with molecular nanogates, using 3D printing and nanotechnology processes can be obtained.

In going from bottom to top to enclose the drug, it is necessary first to build the drug container, the silica mesoporous NPs, with a gate that is an aminosilane and a system that allows opening or closing of a gate based on ATP chains. One system to open the gate is tartrate-resistant acid phosphatase (TRAP) enzyme. In a study using TRAP, the fluorophore (ruthenium tris(bypyridine)) was used as drug molecule model [50]. That way, when TRAP is in the medium, it hydrolyzes the ATP molecules, hence triggering the drug's release. TRAP was chosen because this enzyme is active in bone metastasis processes and is released by osteoclasts during the bone remodeling processes. Therefore, this system could be used in the treatment of different pathologies of bone.

To obtain the stimuli responsive scaffolds, the NPs loaded with the fluorophore and with the gates closed were imbibed in a suspension of gelatin cross-linked with glutaraldehyde to obtain a paste able to be injected by RP to obtain a 3D scaffold. Obviously, the TRAP must come in contact with the NPs. This aim can be achieved only if the scaffolds exhibit macroporosity. Figure 14.10 contains four micrographs of the scaffolds showing its design at four levels. In Figure 14.10a, the big channels created with the RP equipment are visible to the naked eye. In Figure 14.10b, an SEM image shows the presence of macropores in the walls. In Figure 14.10c, a higher magnification SEM micrograph shows the nanostructured composite material was the spherical particles are clearly visible. Finally, in Figure 14.10d, the TEM image of a silica NP with its corresponding electron diffraction (ED) pattern both clearly show the mesoporous order. This microstructure allows both the external fluid permeability and the drug release. *In vitro*, biocompatibility tests are under progress and the initial results indicate that osteoblasts adhere, extend, and proliferate on the external surface of the scaffolds as well as in the internal macropores. Therefore, this preliminary result opens up possibilities to fabricate scaffolds with a dual purpose; in fact, the combination of bioceramic scaffolds with nanotechnology could allow applications of bone regeneration while simultaneously fighting against scaffold infections.

14.3.1.2 Polymeric Scaffolds

Polymers are widely used for the fabrication of TE scaffolds [55–57]. Indeed, they represent around 70% of the publications of pure biomaterials for BTE, whereas ceramics represent around 25% and the rest are metals [58]. Most polymers used as biomaterials are biodegradable with a clear advantage derived from their ability to support tissue growth during the healing process and to be resorbed later. These materials have been investigated in attempts to grow skin, bone, cartilage, liver, heart valves and arteries, bladder, pancreas, nerves, corneas,

Scaffolds and molecular gates: from macro to nano

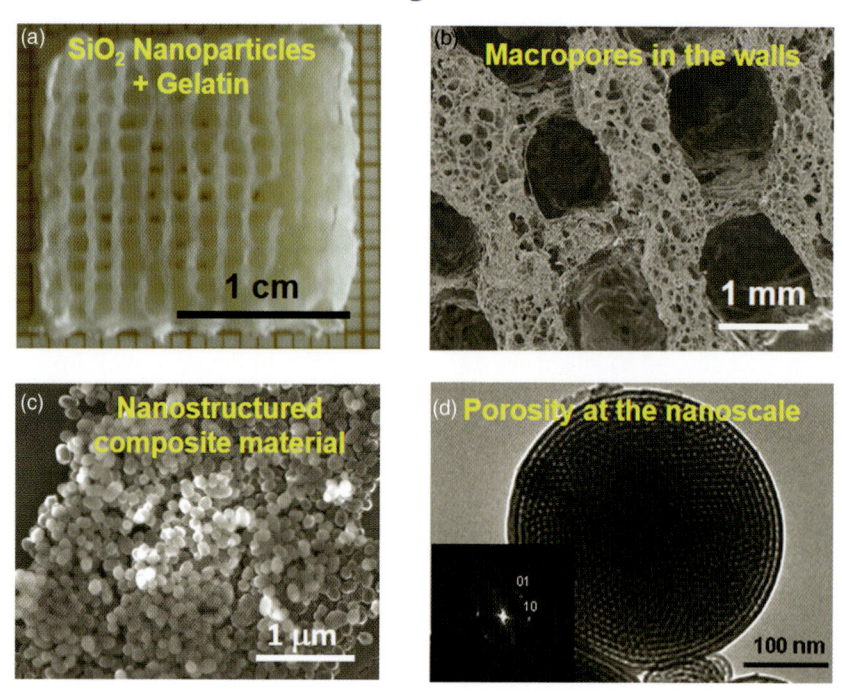

Figure 14.10 Structural features of a scaffold at several length scales.

and other soft tissues. Naturally occurring polymers, synthetic biodegradable, and synthetic nonbiodegradable polymers are the main types of polymers used as biomaterials.

Natural polymers were the first biodegradable materials clinically used. Natural polymers used in BTE are collagen, gelatin, silk, fibrin, alginate, chitosan, and hyaluronic acid. The reason of the use of these polymers is their relatively lower immunogenic potential, which follows the order: polysaccharides < collagen < other proteins, as well as a favorable interaction with the host tissues.

Synthetic biodegradable polymers exhibit more predictable and reproducible mechanical and physical properties, and degradation rates between different lots of the material. In addition, the fabrication of synthetic polymers is easier and the shapes needed for a specific application are more reproducible. Representative examples of polymers of this type used for fabricating porous scaffolding materials are polylactic acid (PLA), polyglycolic acid (PGA), poly(lactic co-glycolic acid) (PLGA), polycaprolactone (PCL), and polyhydroxyalkanoates and polyurethanes. Most of these natural and synthetic polymers have been used to obtain hydrogels, or have been processed into nanofibers for the synthesis of fibrous scaffolds. In both categories, natural or synthetic polymers are excellent candidates to fabricate scaffolds for BTE. Moreover, polymers have been used as matrixes in composites for TE as it will be described later. Particularly, ceramic/polymer composites represent the majority of total.

As in the case of bioceramic scaffolds for BTE, the pore size in polymer scaffolds also plays a crucial role in tissue regeneration. Thus, the optimum pore sizes described are 5 μm for neovascularization, 5–15 μm for fibroblast ingrowth, 20 μm for hepatocytes ingrowth, 200–350 μm for osteoconduction, and 20–125 μm for regeneration of mammalian skin [36,55,59]. Pore interconnectivity is also essential to ensure that cells are within 200 μm of the blood supply to provide transfer of oxygen and nutrients.

There are many methods to fabricate polymer-based 3D porous scaffolds, including solvent casting/salt leaching, ice particle leaching, and gas foaming/salt leaching, all of them used for processing resorbable synthetic or natural polymers. For the fabrication of microspheres, the most used methods are solvent evaporation, particle aggregated scaffold formation, freeze-drying, or thermally induced phase separation. For the fabrication of fibrous scaffolds, the more common methods are nanofiber electrospinning, microfiber wet-spinning, or nonwoven fiber formation by melt-blowing. A complete review of all techniques used for the processing of polymeric scaffolds is in Ref. [55].

The main drawbacks of polymers in BTE applications are limited integration with bone and considerably poorer mechanical compared to bone. In addition, these problems are enlarged if we take into account the porosity required to load the scaffolds with signal molecules and to allow the internal growth of blood vessels and tissues. Thus, the elastic modulus of the majority of the polymers range between 8 MPa and maximum values of 4 GPa for very stiff polymers obtained under special conditions of crystallinity and with very high molecular weight. For comparison, the common units for the elastic modulus of bone or metallic alloys are GPa. Nonetheless, these values are far from the common values of the cortical bone, responsible to support load, which ranges from 10 to 20 GPa. Furthermore, other mechanical properties such as fatigue and creep must be considered when polymeric materials are investigated in the designing of scaffolds for BTE.

Composite materials are investigated for fabrication of scaffolds for BTE to improve the mechanical properties of the polymeric materials. Almost all the composites studied used one of the natural or synthetic polymers mentioned as matrixes. Indeed around 90% of the composites investigated are polymer ceramic, whereas there are only few polymer–metal studies and even fewer studies of ceramic metals for this application [58].

14.3.1.3 Metallic Scaffolds

Whereas soft polymers are mainly investigated for nonload-bearing applications, metallic scaffolds are aimed for load-bearing applications. As in ceramic and polymeric scaffolds, a designed hierarchical and interconnected porosity is required for tissue regeneration. Metals represent the lower proportion of the materials investigated to obtain scaffolds [58]. Metallic scaffolds investigated for BTE applications are usually made of metal alloys such as Ti-6Al-4V or Co-Cr-Mo pristine or coated by polymer or ceramic substances. Moreover, they present the advantage of their structural integrity in long-term applications.

To improve bioactivity and osseointegration of metallic implants, bioactive coatings such as plasma-spayed hydroxyapatite are deposited on the implant. Bioactive glasses are also excellent candidates to improve the bioactivity of metallic scaffolds. In addition, the preparation of sintered glass coatings on metals, or enameling, is another possibility. This technique requires a low trend of the glass to crystallize and a thermal expansion coefficient close to that of the metallic substrate, good adhesion, and obviously a bioactive response.

Important conventional fabrication technologies of metallic scaffolds are sintering metal powders or metal fibres, replication of polymeric sponge, fiber meshes and fiber bonding, spark plasma sintering, gas injection into the metal melt, and decomposition of foaming agents. Furthermore, there are other techniques based on RP such as 3D printing, sacrificial wax template, 3D fiber deposition, electron beam melting, or direct metal deposition [60]. RP complemented with biochemical modification of the surface has undergone significant development in the last years. Thus, porous metallic scaffolds were tested for BTE applications by using cell-based and growth-factor-based strategies. When the metallic scaffolds were coated with proteins such as collagen, RGD-peptide, vibronectin, or fibronectin, accelerated osseointegration and enhanced bone formation took place.

With the exception of degradable magnesium alloys, metallic bone scaffolds must maintain shape, strength, and integrity through the regeneration of damaged bone tissues. Moreover, it must be biocompatibile, radiolucent, easily shaped to fit into the bone defect, nonallergenic, noncarcinogenic, strong enough to endure trauma, stable over time, and osteoconductive. In addition, metallic scaffolds must exhibit the maximum possible porosity with the desired pore sizes and interconnectivity. The optimum pore size for promoting bone ingrowths ranges between 100 and 500 µm.

As part of the scaffold design, numerous surface properties, including topography, surface energy, chemical composition, surface wettability, or surface bioactivity, must be considered. The surface roughness influences the cell morphology and growth. In general, smooth surfaces exhibit less cell adhesion than rough surfaces. Furthermore, surface modifications, such as immobilization of biofunctional polymers and biopolymers, calcium phosphate ceramic coatings, hybridization with biocompatible, and essential biomolecules, are needed to achieve the required osseous tissue induction properties.

Pure Ti and Ti alloys show excellent integration with bone and biocompatibility by the very stable passive layer of TiO_2. Furthermore, their surfaces can be modified by mechanical, physical, chemical, and biochemical methods. The most commonly used mechanical methods are machining, grinding, polishing, and grit-blasting. Numerous physical methods have been attempted, including physical vapor deposition, evaporation, ion plating, sputtering, ion implantation, plasma immersion ion, glow-discharge plasma, thermal spray, or flame spray. Chemical methods that have been used include biomimetics, acid or basic etching, hydrogen peroxide treatment, sol–gel process, electrophoretic deposition, immobilization of functional groups, thermal oxidation, and chemical vapor deposition. Finally, the biochemical methods employed to modify the surface of metallic scaffolds for BTE are modification through polymers, both biological (collagen, fibrin, peptides, alginates, chitosan, or hyaluronic acid) and synthetic (PLA, PGA, PCL), and the incorporation of biochemical factors, inductive signaling molecules, autologous or allogenic bone marrow cells (platelet concentrate), mesenchymal stem cells, or chondrocytes.

Other metals more recently investigated as scaffolds for BTE are tantalum and magnesium. Tantalum is a metal with numerous advantages in terms of biocompatibility and osteoconduction with an elastic modulus close to that of bone. However, at present porous tantalum is in its early stages of development [61]. Some magnesium alloys are bioresorbable, osteoconductive, and do not induce inflammatory response. For these reasons, magnesium was proposed for coronary stents and more recently enriched with extra ions, such as calcium, for BTE [62]. Some concerns about the toxicity of dissolved magnesium have arisen. However, this metal seems to be a good option in the future for trabecular bone regeneration.

Moreover, new developments are under progress for titanium alloys. For instance in the most used alloy Ti–6Al–4 V, elements such as tantalum, niobium, or zirconium have been substituted for aluminum and vanadium with the aim to improve biocompatibility and achieve an elastic modulus closer to that of bone. Ti alloys have an elastic modulus half that of steel or CoCrMo alloys but the value is still five times that of cortical bone. Another important Ti alloy is nickel–titanium alloy, called nitinol that exhibits a shape memory effect, superplasticity, and damping properties. For this reason, it was used in making intramedullar nails and spinal intravertebral spacers [63]. However, the possible toxicity of nickel has restricted the expansion of this alloy. Thus, at present the substitution of nickel by other elements such niobium is under investigation [64].

Most hybrid materials containing metals are in composites reinforcing polymer matrixes, but also titanium scaffolds coated by a ceramic such as tricalcium phosphate and cells have shown osteogenic properties.

The most advanced metallic scaffolds investigated in both preclinical and clinical studies are tantalum, titanium, pristine and loaded with cells, and nitinol. However, magnesium and hybrids titanium with ceramic or polymer and tantalum hybrids have only been subjected to preclinical studies.

14.3.1.4 Composite Scaffolds
Composites of more than one material are formed with the goal to achieve a synergistic effect of their properties. Composites comprise of two phases: the matrix and the disperse phase, the latter being used to increase hardness and stiffness. That way, the main role of the disperse phase as granules, fibres, and so on is to avoid movements in the matrix phase. Mechanical properties of composites can be estimated by the rule of the mixtures, equations that represent the upper and lower bounds of the elastic modulus, respectively. An increase in the

volume percentage of the dispersed phase will result in higher stiffness due to combined elasticity.

Composite materials are used in BTE application to solve several problems of polymers, ceramics, or metallic biomaterials, including failure induced by fluency, fracture toughness, or biocompatibility issues [65,66]. For this application, composite scaffolds are commonly fabricated using a polymer matrix, and for this reason they are called polymer-matrix composites. Then, they are combined with ceramics as disperse phase, producing materials with improved load-bearing capabilities and enhanced host-implant interface interactions with respect to the pure materials. Additionally, the utility of ceramic-matrix composites combined with metal as a reinforcing phase has been found to be physiologically relevant due to the presence of beneficial trace elements within the natural bone tissue. The addition of metal particles or fibres has also shown significant improvements in the compressive strength of the scaffold, osteogenesis, and enhanced bone implant interface interactions. Few investigators have explored the utility of polymer-matrix with metals for BTE, likely due to limitations in fabrication technologies and the inability to retain the desired properties.

Composites used in implants can be classified as nondegradable or biodegradable. The latter sometimes can be only partially biodegradable. The first examples of nondegradable composites designed for orthopedic applications were carbon and glass fiber-reinforced thermoset polymers such as epoxy. Polymer matrixes include poly(sulphone), poly(ether etherketone), and poly(etherimide). An initial example of a composite designed for BTE was that formed by poly(hydroxyethylmethacrylate) and poly(caprolactone) reinforced with polyethylene terephtalate fibers to mimic inter vertebral discs [67].

Due to the importance of the presence of hierarchical porosity in the scaffolds for BTE applications, biodegradable, or at least partially biodegradable composites are more suitable. For instance, a porous poly(L-lactic acid)/apatite composite scaffold demonstrated biocompatibility with hard tissues as well as high osteoconductivity and bioactivity [68].

Composite scaffolds also need to support uniform cell seeding, cell ingrowths, and tissue formation. For this reason, many studies have focused on the major components of natural bone, which consists of an apatite-like phase that reinforces collagen fibers. Thus, numerous promising examples can be found in the literature where bioceramics, including calcium phosphates such as hydroxyapatite or tricalcium phosphate are forming composites with PLLA [69], collagen [70], gelatin [71], or chitosan [72] for use as scaffolding materials for bone repair.

14.3.2
The Cells

As stated above, basic components of TE are cells and the ability to persuade to the cells to regenerate the living tissues [12]. For this purpose, two types of cells can be used: (i) fully differentiated cells of bone, skin, blood vessels, and so on or (ii) mesenchimal stem cells (MSCs) that are precursors of differentiated cells. In adults, MSCs are mainly present in bone marrow and in lower proportion in blood and are able to differentiate into specific cells as required. A rich source of stem cells is the embryo, since embryonic stem cells must be converted into the types of cells required for human growth.

With regards BTE, the traditional approach involves the implantation of osteogenic cells seeded onto appropriate scaffolds composed by some of the biomaterials described in the previous sections. Osteoblastic cells with different differentiation status can be used to drive bone tissue regeneration [73,74]. Moreover, different osteoprogenitor sources were investigated looking for their capacity to regenerate bone [75]. MSCs are multipotent cells capable of differentiating into osteogenic cells [76]. They exhibit interesting activities such as immunosuppressive, cell protective, and angiogenic properties. Some authors indicate that MSCs are pericytes residing on blood vessels able to help regenerate the damaged vascular tissues [77]. MSCs are easily isolated from bone marrow and expanded in cultures for use in different BTE systems. The major drawback of MSCs from various origins, however, lies in their limited quantity and/or differentiation capacity [78,79]. In addition, to elucidate the role of the host contribution to bone tissue regeneration is very important to optimize BTE approaches. Furthermore, MSCs appear to have a different capability to recruit host endothelial cells and, therefore, to increase vascularization [77].

Bone repair is an exceptional process involving the interaction of cellular and molecular events to generate new bone instead of a fibrous scar, which is the outcome in the repairing processes in other connective tissues. In contrast to bone remodeling in adults, which requires a relatively long period to complete at random sites, bone repair occurs in a short period and at a precise location. At the bone injury site, MSCs interact with inflammatory cells [80]. New bone formation starts with the condensation of MSCs, which leads either to the formation of a cartilage template in the bone marrow, endochondral ossification, or to direct differentiation into osteoblasts at the periosteum intramembranous or appositional ossification. The former ossification mechanism predominates in most cases of fracture healing, but rigid fixation of the injury causes primary bone apposition as the major repair mechanism [81]. The general pattern of endochondral ossification after fracture includes several chronological phases. It starts, immediately after injury, with a hematoma formation and inflammation. Osteoprogenitors are then recruited from the periosteum to differentiate into chondrocytes or osteoblastic cells (depending on the oxygen supply). New bone starts to be formed at the borders of the injury site. Simultaneously, a callus mostly made up of hypertrophic cartilage develops and begins to revascularize. Eventually, osteoclasts-mediated remodeling of the newly formed bone leads to the restoration of structural bone integrity [82].

Many of the regulatory factors underlying the control of proliferation and differentiation of osteoblasts from MSCs constitute a matter of investigation as therapies to stimulate bone repair [83]. These factors appear to act through common intracellular signaling pathways, mainly protein kinase (MAPK) and Wnt/β-catenin pathways, to induce proliferation of osteoprogenitors and activation of Runx2, which is an essential transcription factor for osteoblast differentiation [84]. For instance, it was suggested that pretreatment of bone marrow cells with Wnt/β-catenin pathway activators may increase their osteogenic capacity [85]. This approach can be especially important in situations with a low number of osteoprogenitors, such as aging [86]. For this reason, more studies explored the supposed age-related alterations in the pattern of gene expression during bone regeneration in elder humans [87]. In the callus after hip fracture the expression of inflammation-related genes is highest in the earlier stages, and shifts toward bone remodeling genes later on. Interestingly, the expression of Sost, which is a sclerostin-encoding gene modulator of bone remodeling, decreased rapidly in the fracture callus, suggesting that this would allow osteoblasts to escape from its inhibitory effect to promote bone formation.

14.3.3
The Signals

It is possible to achieve osteoproduction by using biochemical signals able to enhance the healing process. Therefore, bone healing is quite often promoted using osteogenic factors that are either directly implanted or previously embedded in a suitable carrier such as a bioceramic scaffold [88]. Osteoinduction properties are conferred to these biomaterials by growth factors that can induce differentiation of MSCs along the bone repair process. Bone morphogenetic proteins (BMPs), specifically BMP-2 and BMP-7, were widely used in animal models and in randomized clinical trials for treating bone fractures and stimulating spine fusion. BMP-2 has proven essential for MSCs recruitment and expansion; it is the only US Food and Drug Administration (FDA) approved molecular therapy for use in the management of fracture repair and spine nonunions. In fact, BMP-2 represented 18% of the US orthopedic biomaterials market in 2009. However, some concerns have arisen due to the poor outcomes of BMP-2 administration in some cases [89].

Recently, our group has started to explore the advantage of the small peptide osteostatin, consisting of the 107–111 domain (Thr–Arg–Ser–Ala–Trp) of PTHrP, to promote bone healing. Osteostatin and BMP-2 present several differences. Therefore, although both substances are osteoinductors, osteostatin is also antiresorptive (a positive effect), while in high doses BMP-2 can be adipogenic (a deleterious effect). An additional advantage of osteostatin is its bone specificity, whereas BMP-2 is pleiotropic exerting its action on many tissues and activating the immune system. Moreover, osteostatin has shown activity at very low concentrations (<nM) and not inhibitors have been described for osteostatin molecule [82]. Finally, due to its smaller size, osteostatin is easier to immobilize in the scaffold. For all these reasons, the local application of osteostatin could bring advantages with respect to BMP-2 in BTE. In this regard, our group recently demonstrated that silica-based ceramic exhibited osteogenic features after being loaded with osteostatin, including stimulation of osteoblastic cells growth and differentiation [90,91]. Moreover, osteostatin has proved to increase

the osteogenic efficacy of fibroblast growth factor-2 (FGF-2) as coated onto Si-doped hydroxyapatite by enhancing its angiogenic potential *in vitro* [92]. These data are consistent with the anabolic features described for the native peptide, PTHrP107–139, which add to the inhibitory effect on bone resorption [93]. The animal model consisted of a cavitary defect in the femoral epiphysis of healthy rabbits. The histological analysis revealed the absence of significant inflammation or bone resorption 8 weeks after implantation [91]. Furthermore, microcomputerized tomography analysis showed that osteostatin-loaded biomaterials were highly osteoconductive and osteoinductive. In addition, these bioceramics-induced revascularization of the defect, related to an increased immunostaining for vascular endothelial growth factor (VEGF) in the healing bone tissue. More recently, these biomaterials were also found to improve the early stage of bone healing in the same bone fracture model in osteoporotic rabbits [94] Thus, these osteostatin-loaded bioceramic scaffolds appear to be an attractive approach for BTE applications even in the setting of osteopenia.

14.4
Achievements in this Area

TE is one of the most exciting multidisciplinary research areas in which scaffold materials and fabrication technologies play a crucial role. TE has undergone huge advances in the last decades, especially with simple tissues (i.e., skin). Moreover important advances in other areas such as BTE have been obtained. This is a potential alternative to the conventional use of bone grafts, due to their unlimited supply and absence of disease transmission risk.

The scaffold fabrication with bioceramics for BTE has allowed producing osteoinductive materials that are able to develop bone in ectopic places [30]. Metallic scaffolds, pure or coated are achieving success for load-bearing applications [60], and a wide range of polymers and techniques of fabrication were used to obtain scaffolds for BTE [55,95,96]. In search of a synergic effect between components, hybrid materials, including copolymers, polymer–polymer blends, and polymer–ceramic composites are also produced. Moreover, advanced hydrogels have been developed for mimicking the extracellular matrix and delivering signals promoting bone tissue regeneration [97]. In addition, numerous techniques are used to modify the scaffolds surface for immunomodulation and shield the biomaterial from protein adsorption. Numerous biodegradable scaffolds have been designed [98]. In this area it is critical to adjust the kinetics of the scaffold degradation and of living tissue formation, and numerous studies have been reported in this regards. Moreover, although there is still a certain controversy regarding the exact values of pore sizes that are optimal for each specific function, there is a general agreement that an interconnected and hierarchical pore network is necessary in a scaffold for TE. The last investigations in this area are focused on designing scaffolds by using nanotechnology principles to be benign to the stem cells that will be hosted.

Embryonic stem cells (ESCs) were discovered in the mid-1990s, but their potential was only recognized in the last decade. They could be used to mend any deteriorated organ from kidney and liver to brain and eye. However, their use brings ethical problems by using human embryos, which has delayed clinical trials. For this reason, investigations are focused on induced pluripotent stem cells, mesenchymal stem cells, adipose-derived stem cells, or stem cells of other origins. Other techniques are now routinely used, such as bone marrow aspirate concentrate, platelet-rich plasma, and bioreactors for *ex-vivo* cultures. On the other hand, the use of osteoblast cells, essential for the development, growth, repair, and maintenance of bone, has been also investigated in BTE [99]. Finally, the inclusion of angiogenic growth factors such as BMPs has already reached the clinical applications.

14.5
Where Are We Going?

TE approaches are reaching the clinical practice very slowly due to several limitations. For instance, BTE aims to induce new functional bone regeneration via the synergistic combination of biomaterials, cells, and factor therapy but it has encountered the problem of the lack of sufficient vascularization at the defect site. Furthermore, many efforts are also looking for more cost-effective systems.

The more significant challenges and limitations in this field are selecting (i) the most effective cell type, (ii) the more suitable scaffolds, which must be mechanically compatible and exhibit appropriate porosity, and (iii) the growth factors, including a combination, to achieve the optimum results including a proper vascularization.

Regarding the scaffolds, most of the scaffolds investigated are biocompatibile and favor new bone formation. However, the most important challenges to be solved in the near future are in the polymeric scaffolds, biocompatibility and biomechanical strength, in the metallic, ion release, limited bioactivity, and biodegradation, and in the ceramics, toughness, reproducible manufacturing techniques, and control of the degradation rate to adjust to the living tissues regeneration. Regarding signals, it is necessary to establish the therapeutic and nontoxic concentrations and the possible side effects in a more effective way.

Other knowledge limitations of this area at present are the donor versus host cell contribution and the possible side effects. Moreover, the appropriate immunomodulatory agents and the most appropriate animal models—including their possible prohibition and substitution by bioreactors—for preclinical approaches need to be defined. Load-bearing large animal models should be used to assess graft functionality because research on small animals (i.e., mice) does not yield relevant results due to major differences in graft size and healing properties. On the other hand, bioreactors may make safer and more effective results obtainable for BTE [100]. Further efforts must also be made to establish efficient intraoperative cell seeding methods to minimize *in vitro* culture of the BTE constructs, and allow for maximized bone tissue regeneration *in vivo*. Finally, the long-evaluation tests in terms of regeneration of living tissue quality and function and the requisites for the regulatory approval of the products obtained need also to be developed.

Future research approaches in BTE will focus on efficient combinations of osteoconductive materials, osteoinductive growth factors, and osteogenic cells ensemble in a construct. The goal will probably be to obtain a functional replacement of the injured hard tissue in a procedure that avoids the step of bone harvesting. However, a perfectly controlled hybrid scaffold remains to be developed.

Acknowledgments

The authors deny any conflicts of interest. This study was supported by research grants from the Ministerio de Economía y Competitividad, project MAT2015-64831-R, European Research Council (ERC-2015-AdG). Advanced Grant Verdi-694160 and Instituto de Salud Carlos III, project PI15/00978.

References

1 ÓBrien, F.J. and ÓBrien, F.J. (2011) Biomaterials & scaffolds for tissue engineering. *Mater. Today*, **14**, 88–95.
2 Bonassar, L.J. and Vacanti, C.A. (1998) Tissue engineering: the first decade and beyond. *J. Cell Biochem. Suppl.*, **29**, 30–31.
3 Atala, A. (2004) Tissue engineering and regenerative medicine: concepts for clinical application. *Rejuvenation Res.*, **7**, 15–31.
4 Kaul, H. and Ventikos, Y. (2015) On the genealogy of tissue engineering and regenerative medicine. *Tissue Eng. Part B*, **21**, 203–217.
5 Heineken, F.G. and Skalak, R. (1991) Tissue engineering: a brief overview. *J. Biomech. Eng.*, **113**, 111–112.
6 Langer, R. and Vacanti, J.P. (1993) Tissue engineering. *Science*, **260**, 920–926.
7 Williams, D.F. (ed.) (1998) *The Williams Dictionary of Biomaterials*, Liverpool University Press, Liverpool, p. 318.
8 Ratner, B.D., Hoffman, A.S., Schoen, F.J., and Lemons, J.E. (2013) *Biomaterials Science: An Evolving, Multidisciplinary Endeavor*, Elsevier Academic Press.
9 Williams, D.F. (2000) *Bone Engineering* (ed. J.E. Davies), Em squared, Toronto.
10 Hench, L.L. (1991) Bioceramics: from concept to clinic. *J. Am. Ceram. Soc.*, **74**, 1487–1510.
11 Russell, P.S. (1985) Selective transplantation. An emerging concept. *Ann. Surg.*, **201**, 255–262.
12 Williams, D.F. (2004) Benefit and risk in tissue engineering. *Mater. Today*, **7**, 24–29.
13 Lysaght, M.J., Nguy, N.A., and Sullivan, K. (1998) An economic survey of the emerging tissue engineering industry. *Tissue Eng.*, **4**, 231–238.
14 Hench, L.L. and Polak, J.M. (2002) Third-generation biomedical materials. *Science*, **295**, 1014–1017.
15 Vacanti, C.A. (2006) The history of tissue engineering. *J. Cell. Mol. Med.*, **10**, 569–576.
16 Moreno-Borchart, A. (2004) Building organs piece by piece. *EMBO Rep.*, **5**, 1025–1028.
17 Griffith, L.G. and Naughton, G. (2002) Tissue engineering – current challenges and expanding opportunities. *Science*, **295**, 1009–1014.
18 Evans, C.H. (2013) Advances in regenerative orthopaedics. *Mayo Clin. Proc.*, **88**, 1323–1339.

19 Evans, C. (2011) Gene therapy for the regeneration of bone. *Injury*, **42**, 599–604.

20 Thomas, E., Storb, R., Clift, R.A., Fefer, A., Johnson, F.L., Neiman, P.E., Lerner, K.G., Glucksberg, H., and Buckner, C.D. (1975) Bone-marrow transplantation (first of two parts). *New Eng. J. Med.*, **292**, 832–843.

21 Takahashi, K. and Yamanaka, S. (2006) Induction of pluripotent stem cells from mouse embryonic and adult fibroblast cultures by defined factors. *Cell*, **126**, 663–676.

22 Takebe, T., Sekine, K., Enomura, M., Koike, H., Kimura, M., Ogaeri, T., Zhang, R.-R., Ueno, Y., Zheng, Y.-W., Koike, N., Aoyama, S., Adachi, Y., and Taniguchi, H. (2013) Vascularized and functional human liver from an iPSC-derived organ bud transplant. *Nature*, **499**, 481–484.

23 Lancaster, M.A., Renner, M., Martin, C.-A., Wenzel, D., Bicknell, L.S., Hurles, M.E., Homfray, T., Penninger, J.M., Jackson, A.P., and Knoblich, J.A. (2013) Cerebral organoids model human brain development and microcephaly. *Nature*, **501**, 373–379.

24 Xia, Y., Nivet, E., Sancho-Martinez, I., Gallegos, T., Suzuki, K., Okamura, D., Wu, M.-Z., Dubova, I., Rodriguez Esteban, C., Montserrat, N., Campistol, J.M., and Izpisua Belmonte, J.C. (2013) Directed differentiation of human pluripotent cells to ureteric bud kidney progenitor-like cells. *Nat. Cell Biol.*, **15**, 1507–1515.

25 Atala, A. (2011) Tissue engineering of human bladder. *Br. Med. Bull.*, **97**, 81–104.

26 Macchiarini, P. (2011) Bioartificial tracheobronchial transplantation. Interview with Paolo Macchiarini. *Regen. Med.*, **6** (6 Supplement), 14–15.

27 Ott, H.C., Matthiesen, T.S., Goh, S.-K., Black, L.D., Kren, S.M., Netoff, T.I., and Taylor, D.A. (2008) Perfusion-decellularized matrix: using nature's platform to engineer a bioartificial heart. *Nat. Med.*, **14**, 213–221.

28 Vallet-Regi, M. and Ruiz-Hernandez, E. (2011) Bioceramics: from bone regeneration to cancer nanomedicine. *Adv. Mater.*, **23**, 5177–5218.

29 Stevens, M.M. (2008) Biomaterials for bone tissue engineering. *Mater. Today*, **11**, 18–25.

30 Amini, A.R., Laurencin, C.T., and Nukavarapu, S.P. (2012) Bone tissue engineering: recent advances and challenges. *Crit. Rev. Biomed. Eng.*, **40**, 363–408.

31 Hutmacher, D.W. (2000) Scaffolds in tissue engineering bone and cartilage. *Biomaterials*, **21**, 2529–2543.

32 Vallet-Regi, M. and Gonzalez-Calbet, J.M. (2004) Calcium phosphates in substitution of bone tissue. *Prog. Solid State Chem.*, **32**, 1–31.

33 Buckwalter, J.A., Glimcher, M.J., Cooper, R.R., and Recker, R. (1996) Bone biology. I: structure, blood supply, cells, matrix, and mineralization. *Instr. Course Lect.*, **45**, 371–386.

34 Murugan, R. and Ramakrishna, S. (2007) Design strategies of tissue engineering scaffolds with controlled fiber orientation. *Tissue Eng.*, **13**, 1845–1866.

35 Bose, S., Roy, M., and Bandyopadhyay, A. (2012) Recent Advances in bone tissue engineering scaffolds. *Tends Biotechnol.*, **30**, 546–554.

36 Salgado, A.J., Coutinho, O.P., and Reis, R.L. (2004) Bone tissue engineering: state of the art and future trends. *Macromol. Biosci.*, **4**, 743–765.

37 Lichte, P., Pape, H.C., Pufe, T., Kobbe, P., and Fischer, H. (2011) Scaffolds for bone healing: concepts material and evidence. *Injury*, **42**, 569–573.

38 Dinesh Kumar, S., Ekanthamoorthy, J., and Senthil Kumar, K. (2015) Study of development and applications of bioactive materials and methods in bone tissue engineering. *Biomed. Res.*, **26**, S53–S59.

39 Ceasarano, J. (1999) A review of robocasting technology. Symposium on Solid Freeform and Additive Fabrication, vol. **542**, pp. 133–139.

40 Manzano, M., Lozano, D., Arcos, D., Portal-Nuñez, S., López, C., Esbrit, P., and Vallet-Regí, M. (2011) Comparison of the osteoblastic activity conferred on Si-doped hydroxyapatite scaffolds by

different osteostatin coatings. *Acta Biomater.*, **7**, 3555–3562.
41 Meseguer-Olmo, L., Vicente-Ortega, V., Alcaraz-Baños, M., Calvo-Guirado, J.L., Vallet-Regi, M., Arcos, D., and Baeza, A. (2013) In-vivo behavior of Si-hydroxyapatite/polycaprolactone/DMB scaffolds fabricated by 3D printing. *J. Biomed. Mater. Res. Part A*, **101**, 2038–2048.
42 Izquierdo-Barba, I., Salinas, A.J., and Vallet-Regi, M. (2013) Bioactive glasses: from macro to nano. *Int. J. Appl. Glass Sci.*, **4**, 149–161.
43 Izquierdo-Barba, I. and Vallet-Regí, M. (2015) Mesoporous bioactive glasses: relevance of their porous structure compared to that of classical bioglasses. *Biomed. Glasses*, **1**, 140–150.
44 Brinker, C.J., Lu, Y., Sellinger, A., and Fan, H. (1999) Evaporation-induced self-assembly: nanostructures made easy. *Adv. Mater.*, **11**, 579–585.
45 Yang, S., Leong, K.F., Du, Z., and Chua, C.K. (2002) The design of scaffolds for use in tissue engineering. Part II. Rapid prototyping techniques. *Tissue Eng.*, **8**, 1–11.
46 Salinas, A.J. and Vallet Regi, M. (2016) Glasses in bone regeneration: a multiscale issue. *J. Non-Cryst. Solids*, **432**, 9–14.
47 Vallet-Regí, M., Ramila, A., del Real, R.P., and Pérez-Pariente, J. (2001) A new property of MCM-41: drug delivery system. *Chem. Mater.*, **13**, 308–311.
48 Doadrio, A.L., Salinas, A.J., Sánchez-Montero, J.M., and Vallet-Regí, M. (2015) Drug release form ordered mesoporous silicas. *Curr. Pharm. Design*, **21**, 6189–6213.
49 Ruiz-Hernández, E., Baeza, A., and Vallet-Regí, M. (2011) Smart drug delivery through DNA/magnetic nanoparticle gates. *ACS Nano*, **5**, 1259–1266.
50 Mas, N., Arcos, D., Aznar, E., Sánchez, S., Sancenón, F., García, A., Marcos, M.D., Baeza, A., Vallet-Regí, M., and Martínez, R. (2014) Towards the development of smart 3D "Gated Scaffolds" for on-command delivery. *Small*, **10**, 4859–4864.
51 Vallet-Regi, M., Izquierdo-Barba, I., and Colilla, M. (2012) Structure and functionalization of mesoporous bioceramics for bone tissue regeneration and local drug delivery. *Phil. Trans. R. Soc. A*, **370**, 1400–1421.
52 Saiz, E., Zimmermann, E.A., Lee, J.S., Wegst, U.G.K., and Tomsia, A. (2013) Perspectives on the role of nanotechnology in bone tissue engineering. *Dent. Mater.*, **29**, 103–115.
53 Gong, T., Xie, J., Liao, J., Zhang, T., Lin, S., and Lin, Y. (2015) Nanomaterials and bone regeneration. *Bone Res.*, **3**, 15029.
54 Vallet-Regí, M., Balas, F., and Arcos, D. (2007) Mesoporous materials for drug delivery. *Angew. Chem,. Int. Ed.*, **46**, 7548–7558.
55 Dhandayuthapani, B. and Yoshida, Y. (2011) Polymeric scaffolds in tissue engineering application: a review. *Int. J. Polym. Sci.*, **2011**, 19.
56 Ji, Y., Ghosh, K., Shu, X.Z., Li, B., Sokolov, J.C., Prestwich, G.D., Clark, R.A.F., and Rafailovich, M.H. (2006) Electrospun threedimensional hyaluronic acid nanofibrous scaffolds. *Biomaterials*, **27**, 3782–3792.
57 Place, E.S., George, J.H., Williams, C.K., and Stevens, M.M. (2009) Synthetic polymer scaffolds for tissue engineering. *Chem. Soc. Rev.*, **38**, 1139–1151.
58 Liu, Y., Lim, J., and T, S.-H. (2013) Development of clinically relevant scaffolds for vascularised bone tissue engineering. *Biotech. Adv.*, **31**, 688–705.
59 Yang, S., Leong, K.F., Du, Z., and Chua, C.K. (2001) The design of scaffolds for use in tissue engineering—part I: traditional factors. *Tissue Eng.*, **7**, 679–689.
60 Alvarez, K. and Nakajima, H. (2009) Metallic scaffolds for bone regeneration. *Materials*, **2**, 790–832.
61 Meneghini, R.M., Lewallen, D.G., and Hansen, A.D. (2008) Use of porous tantalum metaphyseal cones for severe tibial bone loss during revision total knee replacement. *J. Bone joint Surg. Am.*, **90**, 78–84.
62 Li, Z., Gu, X., Lou, S., and Zheng, Y. (2008) The development of binary Mg-Ca

alloys for use as biodegradable materials with bone. *Biomaterials*, **29**, 1329–1344.

63 Bansiddhi, A., Sargeant, T.D., Stupp, S.I., and Dunand, D.C. (2008) Porous NiTi for bone implants: a review. *Acta Biomater.*, **4**, 773–782.

64 Xu, J., Weng, X.-J., Wang, X., Huang, J.-Z., Zhang, C., Muhammad, H., Ma, X., and Liao, Q.-D. (2013) Potential use of porous titanium–niobium alloy in orthopedic implants: preparation and experimental study of its biocompatibility in vitro. *PLoS One*, **8**, e79289.

65 Rezwan, K., Chen, Q.Z., Blaker, J.J., and Boccaccini, A.R. (2006) Biodegradable and bioactive porous polymer/inorganic composite scaffolds for bone tissue engineering. *Biomaterials*, **27**, 3413–3431.

66 De Santis, R., Guarino, V., and Ambrosio, L. (2009) Composite biomaterials for bone repair, in *Bone Repair Biomaterials* (eds J.A. Planell, S.M. Best, D. Lacroix, and A. Merolli.), Woodhead Publishing Limited, Oxford, pp. 252–270.

67 Ambrosio, L., Netti, P.A., Iannace, S., Huang, S.J., and Nicolais, L. (1996) Composite hydrogels for intervertebral disc prosthesies. *J. Mater. Sci. Mater. Med.*, **7**, 252–254.

68 Zhang, R. and Ma, P.X. (1999) Porous poly(L-lactic acid)/apatite composites created by biomimetic process. *J. Biomed. Mater. Res.*, **45**, 285–293.

69 Wei, G. and Ma, P.X. (2004) Structure and properties of nano-hydroxyapatite/polymer composite scaffolds for bone tissue engineering. *Biomaterials*, **25**, 4749–4757.

70 Du, C., Cui, F.Z., Zhu, X.D., and De Groot, K. (1999) Three-dimensional nano-HAp/collagen matrix loading with osteogenic cells in organ culture. *J. Biomed. Mater. Res.*, **44**, 407–415.

71 Bigi, A., Boanini, E., Panzavolta, S., Roveri, N., and Rubini, K. (2002) Bonelike apatite growth on hydroxyapatite-gelatin sponges from simulated body fluid. *J. Biomed. Mater. Res.*, **59**, 709–715.

72 Zhang, Y. and Zhang, M. (2001) Synthesis and characterization of macroporous chitosan/calcium phosphate composite scaffolds for tissue engineering. *J. Biomed. Mater. Res.*, **55**, 304–312.

73 Santos, M.I., Unger, R.E., Sousa, R.A., Reis, R.L., and Kirkpatrick, C.J. (2009) Crosstalk between osteoblasts and endothelial cells co-cultured on a polycaprolactone-starch scaffold and the *in vitro* development of vascularization. *Biomaterials*, **30**, 4407–4415.

74 Arthur, A., Zannettino, A., and Gronthos, S. (2009) The therapeutic applications of multipotential mesenchymal/stromal stem cells in skeletal tissue repair. *J. Cell. Physiol.*, **218**, 237–245.

75 Arvidson, K., Abdallah, B.M., Applegate, L.A., Baldini, N., Cenni, E., Gómez-Barrena, E., Granchi, D., Kassem, M., Konttinen, Y.T., Mustafa, K., Pioletti, D.P., Sillat, T., and Finne-Wistrand, A. (2011) Bone egeneration and stem cells. *J. Cell. Mol. Med.*, **15**, 718–746.

76 Da Silva Meirelles, L., Caplan, A.I., and Nardi, N.B. (2008) In search of the *in vivo* identity of mesenchymal stem cells. *Stem Cells*, **26**, 2287–2299.

77 Sorrell, J.M., Baber, M.A., and Caplan, A.I. (2009) Influence of adult mesenchymal stem cells on *in vitro* vascular formation. *Tissue Eng. Part A*, **15**, 1751–1761.

78 Duplomb, L., Dagouassat, M., Jourdon, P., and Heymann, D. (2007) Concise review: embryonic stem cells: a new tool to study osteoblast and osteoclast differentiation. *Stem Cells*, **25**, 544–552.

79 Chen, Y., Bloemen, V., Impens, S., Mohecen, M., Luyten, F.P., and Schrooten, J. (2011) Characterization and optimization of cell seeding in scaffolds by factorial design: quality by design approach for skeletal tissue engineering. *Tissue Eng. Part C Methods*, **17**, 1211–1221.

80 Claes, L., Recknagel, S., and Ignatius, A. (2012) Fracture healing under healthy and inflammatory conditions. *Nat. Rev. Rheumatol.*, **8**, 133–143.

81 Tortelli, F., Tasso, T., Loiacono, F., and Cancedda, R. (2010) The development of tissue-engineered bone of different origin through endochondral and intramembranous ossification following the implantation of mesenchymal stem

cells and osteoblasts in a murine model. *Biomaterials*, **31**, 242–249.
82 Salinas, A.J., Esbrit, P., and Vallet-Regí, M. (2013) A tissue engineering approach based on the use of bioceramics for bone repair. *Biomater. Sci.*, **1**, 40–51.
83 Deschaseaux, F., Sensébé, L., and Heymann, D. (2009) Mechanisms of bone repair and regeneration. *Trends Mol. Med.*, **15**, 417–429.
84 Vaes, B.L., Ducy, P., Sijbers, A.M., Hendriks, J.M., van Someren, E.P., de Jong, N.G., van den Heuvel, E.R., Olijve, W., van Zoelen, E.J., and Dechering, K.J. (2006) Microarray analysis on Runx2-deficient mouse embryos reveals novel Runx2 functions and target genes during intramembranous and endochondral bone formation. *Bone*, **39**, 724–738.
85 Krause, U., Harris, S., Green, A., Ylostalo, J., Zeitouni, S., Lee, N., and Gregory, C.A. (2010) Pharmaceutical modulation of canonical Wnt signaling in multipotent stromal cells for improved osteoinductive therapy. *Proc. Natl. Acad. Sci. U. S. A.*, **107**, 4147–4152.
86 Roholl, P.J., Blauw, E., Zurcher, C., Dormans, J.A., and Theuns, H.M. (1994) Evidence for a diminished maturation of preosteoblasts into osteoblasts during aging in rats: an ultrastructural analysis. *J. Bone Miner. Res.*, **9**, 355–366.
87 Caetano-Lopes, J., Lopes, A., Rodrigues, A., Fernandes, D., Perpétuo, I.P., Monjardino, T., Lucas, R., Monteiro, J., Konttinen, Y.T., Canhão, H., and Fonseca, J.E. (2011) Upregulation of inflammatory genes and downregulation of sclerostin gene expression are key elements in the early phase of fragility fracture healing. *PLoS One*, **6**, e16947.
88 Manzano, M. and Vallet-Regí, M. (2012) Revisiting bioceramics: bone regenerative and local drug delivery systems. *Prog. Solid State Chem.*, **40**, 17–30.
89 Aro, H.T., Govender, S., Patel, A.D., Hernigou, P., de Gregorio, A.P., Popescu, G.I., Goleen, J.D., Christensen, J., and Valentin, A. (2011) Recombinant human bone morphogenetic protein-2: a randomized trial in open tibial fractures treated with reamed nail fixation. *J. Bone Joint Surg. Am.*, **93**, 801–808.
90 Lozano, D., Manzano, M., Doadrio, J.C., Salinas, A.J., Vallet-Regí, M., Gómez-Barrena, E., and Esbrit, P. (2010) Osteostatin–loaded bioceramics stimulate osteoblastic growth and differenciation. *Acta Biomater.*, **6**, 797–803.
91 Trejo, C.G., Lozano, D., Manzano, M., Doadrio, J.C., Salinas, A.J., Dapía, S., Gómez-Barrena, E., Vallet-Regí, M., García-Honduvilla, N., Buján, J., and Esbrit, P. (2010) The osteoinductive properties of mesoporous silicate coated with osteostatin in a rabbit femur cavity defect model. *Biomaterials*, **31**, 8564–8573.
92 Lozano, D., Feito, M.J., Portal-Núñez, S., Lozano, R.M., Matesanz, M.C., Serrano, M.C., Vallet-Regí, M., Portolés, M.T., and Esbrit, P. (2012) Osteostatin improves the osteogenic activity of fibroblast growth factor-2 immobilized in Si-doped hydroxyapatite in osteoblastic cells. *Acta Biomater.*, **8**, 2770–2777.
93 de Castro, L.F., Lozano, D., Portal-Núñez, S., Maycas, M., De la Fuente, M., Caeiro, J.R., and Esbrit, P. (2012) Comparison of the skeletal effects induced by daily administration of PTHrP (1–36) and PTHrP (107–139) to ovariectomized mice. *J. Cell Physiol.*, **227**, 1752–1760.
94 Lozano, D., Trejo, C.G., Gómez-Barrena, E., Manzano, M., Doadrio, J.C., Salinas, A.J., Vallet-Regí, M., García-Honduvilla, N., Esbrit, P., and Buján, J. (2012) Osteostatin-loaded onto mesoporous ceramics improves the early phase of bone regeneration in a rabbit osteopenia model. *Acta Biomater.*, **8**, 2317–2323.
95 Li, W.J., Laurencin, C.T., Caterson, E.J., Tuan, R.S., and Ko, F.K. (2002) Electrospun nanofibrous structure: a novel scaffold for tissue engineering. *J. Biomed. Mater. Res.*, **60**, 613–621.
96 Collins, M.N. and Birkinshaw, C. (2013) Hyaluronic acid based scaffolds for tissue engineering – a review. *Carbohyd. Polym.*, **92**, 1262–1279.
97 Franco, J., Hunger, P., Launey, M.E., Tomsia, A.P., and Saiz, E. (2010) Direct write assembly of calcium phosphate

scaffolds using a water-based hydrogel. *Acta Biomater.*, **6**, 218–228.

98 Armentano, I., Dottori, M., Fortunati, E., Mattioli, S., and Kenny, J.M. (2010) Biodegradable polymer matrix nanocomposites for tissue engineering: a review. *Polym. Degrad. Stabil.*, **95**, 2126–2146.

99 Jayakumar, P. and Di Silvio, L. (2010) Osteoblasts in bone tissue engineering. *Proc. Inst. Mech. Eng. H*, **224**, 1415–1440.

100 Alman, B.A., Kelley, S.P., and Nam, D. (2011) Heal thyself: using endogenous regeneration to repair bone. *Tissue Eng Part B Rev.*, **17**, 431–436.

Index

a

aba stacking 34
absorption process
– first- order 277
acetic acid 148
acetylene 330
acid treatment 58, 67
activated carbons 122, 165
activation energy 156
adipose-derived stem cells 404
adsorption 144
AFM. *See* atomic force microscopy
AgCl–titanate nanotubes (TNTs) 305
agglomeration 39
agitation 2
Ag nanoparticles 305
AIM, within MOF scaffold 169
ALD, on flat substrate 169
alginate 396
alkali-earth metal oxides 357
alkali-metal aluminosilicate gels 101
alkoxide-derived sol–gel reaction 217
alkoxysilanes 83
– chromophore 320
– modified DR19, 320
alkylamines 172
alkylammonium surfactants 130
alkyl-modified silsesquioxanes 201
alkyltrimethylammonium surfactants 111, 123, 129
allyl glycidyl ether 321
alumina catalysts 122
aluminosilicates 21
– Löwenstein's rule for 104

aluminosilicate zeolites 98
aluminum chloride hexahydrate ($AlCl_3 \cdot 6H_2O$) 218
aluminum hydroxide
– degree of crystallinity 218
– dried 218
– MAS NMR spectra for 220
– monolithic 221
– porous 218
amine-alcohol-silicate hybrid materials 297
amine NH_2 339
3-aminopropyltriethoxysilane (APTES) 172, 283, 329, 331
anhydrous proton conductors 156
anion exchange 72
anthracene monomers 153
antiferromagnetic interactions
– $Cr_3[Cr(CN)_6]_2$ upon O_2 absorption 151
antiferromagnetic material 70
aromatic functional groups 106
arylene-bridged alkoxysilane 229
assembly–disassembly–organization–reassembly strategy (ADOR) 109
atomic force microscopy (AFM) 25, 73
atomic layer deposition 254
Au nanoparticles 253
aurivillius phase 56, 58
Au@SiO_2
– structure of 342
autografts 357
auxiliary reactor 31
$A_xM_yO_z$, general formula 60
azobenzene 330
– chromophore, self-orientation of 322
– molecules 302
azoles 159

Handbook of Solid State Chemistry, First Edition. Edited by Richard Dronskowski, Shinichi Kikkawa, and Andreas Stein.
© 2017 Wiley-VCH Verlag GmbH & Co. KGaA. Published 2017 by Wiley-VCH Verlag GmbH & Co. KGaA.

b

band filling 40
band gap 39
– strain relations 43
– transition 24
base lubricant value 39
basicity 61
BEA zeolite 107
beneficial solid lubrication behavior 43
benzene 152
4,4′,4″-[benzene-1,3,5-triyl-tris(ethyne-2,1-diyl)]tribenzoate (BTE) 144
beryllium 103
binary phosphate glasses, network structures of 372
bioactive glasses 357, 358, 364, 375, 392, 398
– functionalization of 368
bioactivity 357, 365
biocatalysis 260
bioceramic materials 26
bioceramic scaffolds 392, 402
biochemical signals 402
biocompatibility 357
– materials 221
– mechanical support 391
biological grafts 357
biomaterials 384
biomineral 222
biphasic calcium phosphate (BCP) 392
biphenyl-4,4′-dicarboxylate (BPDC) 144
birnessite 67
block-copolymers 243
– silica nanocomposite 294
– surfactants 133
– templates 127
body-centered cubic structure 123, 126, 127, 130
boehmite(γ-AlOOH)-based organic–inorganic hybrid material 283
boiling temperature 70
bonding molecules 392
bone-bonding, compositional diagram 363
bone defects 385
bone formation 401
bone morphogenetic protein (BMP) 402
bone regeneration 402
bone remodeling
– sclerostin-encoding gene modulator 402
bone repair 401
bone tissue engineering (BTE) 45, 385
borate glass 369
– and 45S5 glass, conversion of 370

– matrix 369
boundary lubrication 38
Bragg mirrors 40
branched Azo-POSS conjugates 303
bridged organosilicas 134
Brij family surfactants 202
Brönsted acid sites 61, 100
Brunauer-Emmett-Teller (BET)
– applications 400
– – drawbacks of polymers 397
– constructs
– – in vitro culture of 404
– procedure 391
– scaffold fabrication with bioceramics 403
– surface area 144
– synthetic biomaterials 390
– systems 401
bulk counterparts
– semiconducting behavior 36
bulk crystallites 36
Burstein–Moss shift 39
buserite 67

c

cadmium chloride ($CdCl_2$) 21
$[Ca_2CoO_3]_{0.6}CoO_2$ structure 65
Ca^{2+} ions 143, 374
calcination 123, 130, 132, 224, 229, 247
calcium deficient 390
calcium distribution, in sol–gel-derived glasses 367
calcium ion 357, 364
calcium nitrate 366
calcium phosphate 222, 392
– ceramic coatings 398
– invert glasses 374, 375
camera, visualization of destruction 328
cancer cell lines
– two-photon fluorescence imaging 336
$CaO/(CaO + Na_2O)$ ratios 365
$CaO–Na_2O–P_2O_5$ glasses 371, 372
$CaO–P_2O_5–Na_2O–TiO_2$ system 374
$60CaO–30P_2O_5–3TiO_2–7Na_2O$ glass 375
$CaO–SiO_2$ system 369
capillary condensation 121
carbon 226
carbonate hydroxyapatite (CHA) 392
carbonate ions 362
carbon blacks 122
carbon dot (CD) 283
carbon, fabrication 226
carbon monoliths
– synthesis scheme 227

carbon nanotubes (CNT) 22, 127, 330
carbon, porous 226
catalytic activity 158
cationic surfactants 81, 129, 133, 209
– micelles 125
CdSe quantum dots 259
– ZnS QDs 301
cell compatibility test 375
cell culture tests 362, 376
cell harvesting 393
ceramic component 390
ceramic-matrix composites 400
cetyltrimethylammonium surfactants 128, 129
chalcogen atoms 27
charge density 61, 67
chelating agents 212
chemical catalysis 166
chemical cooling 198
chemical modifications 144
chemical processes 2
– oxidation method 63
chemical reactions 2
chemical reagents
– reaction with 59
chemical separations 165
chemical sol–gel systems 199
chemical vapor deposition 253
chemical vapor transport (CVT) 29, 31
– growth technique 28
chiral stilbazolium cation (CHIDAMS⁺) 323
chitosan 396, 400
chromium-based monoliths 233
chromophores 337
chrysotile 21
– asymmetric structure along c-axis 22
clay galleries 85
clay minerals 81
– ethylene glycol (1,2-ethanediol) use 83
– kaolinite 84
– organophilization of 81
– – organic ligands affecting 83
– – organosilanes used for grafting 85
– palygorskite 84
– sepiolite 84
clay-polymer nanocomposite (CPN) 80, 81
– applications 92
– clay modification 81
– delaminated/exfoliated 85
– elastic modulus 89
– electrical and electrochemical properties 90
– gas permeation 90
– intercalation of polymer 81, 85
– mechanical properties 88

– properties and applications, 88
– – electrical and electrochemical
      properties 90
– – gas permeation 90
– – mechanical properties 88
– – thermal properties and fire retardance 89
– synthesis of, methods 85
– thermal properties and fire retardance 89
– tortuous path in 92
clay sheets 85
click chemistry 144
Cloisite®, 82
closed-cage fullerenes 22
cluster formation 6
CNT. See carbon nanotubes
CO_2 adsorbents 73
cocondensation 208
Co-Cr-Mo pristine 397, 398
Co-doped sol–gel monoliths 344
COF. See covalent organic framework (COF)
collagen 396
colloidal crystals 122
colloidal crystal template (CCT) 243, 268
– bioglasses 267
– colloidal crystals, assembly and examples
    of 248
colloidal particles 243
color intensity 293
color rendering index (CRI) 281
composite
– components 80
– – bulk polymer 80
– – filler/reinforcement 80
– – interfacial polymer 80
– scaffolds 399, 400
composite building unit (CBU) 108
computerized axial tomography (CAT) 392
conductivity 142, 151, 153, 156
– band 41
contact–nanotube interface 42
coordination modes, proposed, of vanadium
    ions to Zr_6-node of V-UiO-66, 181
coordination modulation method 158
coordination networks, strength and stability
    of 159
coordination polymers with interdigitated
    structure (CIDs) 146
copolymers 122
– surfactants 126
copper hydroxide 220
copper sulfoisophthalate 148
core–shell nanotubular structures 33
– inorganic nanotube superstructures 33

core-shell structures 256
coronene tetra-carboxylate (CS) 282
correlated color temperature (CCT) 281
cortical bone 399
cosmology 2
cost-effective systems 404
covalent bonding 308
covalent organic framework (COF) 122
covalent siloxane (Si–O–Si) linkages 134
CPN. *See* clay-polymer nanocomposite (CPN)
crystal-amorphous transition 159
crystal engineering approaches 142, 144, 157
– PCPs/MOFs 157
crystal growth 157
crystalline-amorphous-crystalline transition of ZIF-4 on heating 159
crystalline precursors-layered silicates 108
crystalline sponge method 150
– miyakosyne A guest visualized in pores of a PCP/MOF, 150
crystallization 14, 67
– in noncentrosymmetric space groups 323
crystal
– dielectric properties of 155
– morphology 159
– particle structuring 144
– sizes 158
crystal structure 143, 158
– [Al(OH)(1,4-ndc)] 157
– CPL-1 and regular alignments of C_2H_2 molecules confined in channels of CPL-1 155
– Cu[Ni(pdt)$_2$] incorporating redox active ligand pdt 154
– functionalized MIL-53 156
– of Hofmann-type ST framework [Fe(pz)M(CN)$_4$] (M = Pt, Pd, Ni) 152
– Mn$_2$(DSBDC), and conduction path of infinite metal-sulfur chains 154
– [Zn$_2$(adc)$_2$(dabco)] and changes in fluorescence color 153
CsNi[Cr(CN)$_6$], ferromagnetic interactions between O$_2$ 151
cubic close-packed 243
Cu Lewis acid 148
Cu–TiO$_2$ hybrid nanoparticles 306
CVT. *See* chemical vapor transport
cylindrical arrays 243
cylindrical micelles 123, 124
cytotoxicity
– P$_2$O$_5$–CaO–Na$_2$O glass 374

d

[DAMS]$_4$[M$_2$M′(C$_2$O$_4$)$_6$]·2DAMBA·2H$_2$O
– anionic organic/inorganic and cationic organic layers 324
dangling bonds 29
– annihilation of 28
daughter nanotubes
– growth mechanism 35
daughter WS$_2$ nanotubes
– TEM images 35
Dawson-like POM, 10
deintercalation 63
density functional-based tight binding (DFTB) 28
density functional theory (DFT) 6, 9, 11, 170
desilication technique 111
DFTB. *See* density functional-based tight binding
diblock copolymer
– functionalized with LY dyes and hybrid gold nanoparticles 337
dicalcium phosphate 222
dicalcium phosphate anhydrous (CaHPO$_4$) monolith 222
dielectric properties 142, 150, 151, 154
Diels–Alder reactions 114
diethylenetriamine 172
differential thermal analysis (DTA) 88
diffuse reflectance infrared Fourier transform spectroscopy (DRIFTS) 170
diffuse sunlight 287
diffusion method 143
digital light projector systems 286
dimerization 182
dimethyldimethoxysilane (DMDMS) 210
dimethylsulfoxide (DMSO) 84
2,4-dinitrotoluene 263
Dion–Jacobson phase 55
2,5-dioxidoterephthalate 145
diquaternary phosphonium 106
distributed feedback (DFB) 287
2,5-disulfhydrylbenzene-1,4-dicarboxylic acid (H$_4$DSBDC) 154
dithienylethene (DTE) derivatives 302
DMSO. *See* dimethylsulfoxide (DMSO)
DNA tweezers 338
3DOM Au film 263
3DOM electrodes 261
3DOm-i zeolites 111
3DOM materials
– applications
– – bioactive materials and tissue engineering 267
– – electrochemical sensing 263

-- fuel cells, solid oxide/biofuel cells 264
-- proton-exchange membrane fuel cells 264
-- solar cells 266
  -- dye-sensitized solar cells 267
  -- photovoltaic cells 266
- catalytic applications 260
- electrochemical energy storage 260
- optical applications 258
-- spontaneous emission modification 259
-- photonic crystals 258
-- tunable photonic crystals 258
- synthesis, schematic representation of 244
- structures 244
-- on multiple length scales 245
- synthesis 246, 255
-- assembly from core-shell spheres 256
-- atomic layer deposition 254
-- chemical vapor deposition 253
-- colloidal crystal assembly 247
-- colloidal spheres for CCTs 247
-- double templating 256
-- dual templating 257
-- electrodeposition 252
-- electroless deposition 253
-- electrophoretic deposition 253
-- nanocrystal deposition and sintering 252
-- oxide and salt reduction 251
-- polymerization 251
-- pseudomorphic transformations 255
-- salt precipitation and chemical conversion 250
-- sedimentation and aggregation 254
-- sol-gel chemistry 250
-- spraying techniques 254
-- surface modification 257
-- templating process and synthetic alternatives 249
3DOM metal oxides 250
3DOM NiO-YSZ anodes 265
3DOM oxides 251
3DOM photonic crystals 259
3DOM polyimide matrix 264
3DOM Pt electrodes
- for mass transfer 265
3DOM zeolites 256
- monoliths 257
doping-induced blue shift 40
3D perovskite 58
2D perovskite to 3D perovskite, conversion from 58
dried gel 219
drug delivery system (DDS) 147, 394
- using MIL series compounds 147

drug loading capacity 147
dry gel conversion (DGC) method 104
DTA. *See* differential thermal analysis (DTA)
DTE derivatives 304
DTE-including materials 302
dual photochromic/electrochromic compounds 308
dyes
- bearing thiol groups 335
- bridged hybrids 279
- doped hybrids 279
-- materials 318
- doped organic–inorganic hybrids 286
- encapsulation of 341
- matrix interactions 295
- to nanoparticle distance 335
dynamic light scattering (DLS) 15
dynamic self-assembly 2, 3

e
EDS. *See* energy dispersive X-ray spectroscopy
E_{2g} Raman mode 31
electrical response
- mechanical deformation induced 42
electrical susceptibility 317
electrochemical applications, of 3DOM materials 262
electrochemical capacitor 66
electrodeposition 252
electroluminescent devices 278
electromechanical
- measurements 42
- properties 42
electron carrier concentration 39
electron diffraction (ED) 395
electron doped MoS_2
- Burstein–Moss effect 40
electrospray ionization mass spectroscopy (ESI-MS) 4
electrostatic interactions 61, 63, 308
EM. *See* electromechanical
embryonic stem cell (ESC) 404
emission intensity 289
emission quantum yield 287
enantioselective catalytic reactions 148
energy dispersive X-ray spectroscopy (EDS) 32
energy dissipation 2
enhanced permeability and retention (EPR) 340
enthalpy
- contribution 196
- driven spinodal decomposition 206
enzymatic catalysis 122

epitaxial growth 158
epoxides 216
– mediated sol–gel reaction 217
– mediated sol–gel systems 233
– mediated system 216
ethylacetoacetates 213
ethylenediamine 172, 232
ethylene glycol (1,2-ethanediol) 83
Eu^{3+}-bearing poly(MMA-MA-co-Eu(tta)$_2$phen) copolymer 285
[Eu(btfa)$_3$(MeOH)(bpeta)], 289
Eu^{3+} β-diketonate complexes 288
Eu^{3+}/Tb^{3+} codoped diureasil film
– temperature profile 290
evaporation induced self-assembly (EISA) 124, 393
excitonic transitions 37, 40
external energy source 2

f

fabrication of polymer-based 3D porous scaffolds 397
fabrication technologies 403
face-centered cubic structures 123, 126, 130
fast nanotube-forming reaction 24
Fe$_2$(dobdc)-accommodating hydrocarbons 146
[Fe(pz)Pt(CN)$_4$], guest-induced spin state switching 152
Fermi level 43
ferroelectric properties 57
FET. See field-effect transistor
FIB. See focused ion beam (FIB)
fiber bonding 398
fiber meshes 398
fibroblast growth factor-2 (FGF-2) 403
fibrous clays
– organophilization of 84
fibrous scaffolds 397
fibrous scar 401
field-effect transistor (FET) 26, 41
finite-difference time-domain (FDTD) 344
Flanigen's notation 100
flash-photolysis time-resolved microwave conductivity (FP-TRMC) 154
flexography process
– inkjet, principle of 305
Flory–Huggins formulation 196
flow reactor system approach 14
fluid lubricants 26
fluorescence behavior 335
fluorescence resonance energy transfer (FRET) 301

fluorescent molecular switch (FMS) 293
fluorescent molecular switch Rh-AA-DAE
– chemical structure 294
fluorescent silsesquioxane 285
fluoroalkylsilane 257
fluoropolymer matrix yields 306
Fm3m symmetry 126
– face-centered cubic structures of 124
focused ion beam (FIB) 86
Food and Drug Administration (FDA) 402
formamide (FA) 206
four-probe measurements 42
free radical polymerization 89
freeze-drying 397
fuel cell technology 156
fullerene-like nanoparticles, 25
– doping 31
fused-naphthol[1,2-b]pyran
– photochromic equilibrium 297

g

gallium 103
gas injection 398
gas-phase catalysis 166
gas phase chemical synthesis 33
gas separation 166
gate-opening adsorption behavior, NO gas induced 148
gelatin 396
– based 3D scaffold, SEM micrograph of 395
gelation
– from metal salts and acids 216
gel viscosity 39
gene expression 402
germanium 103
– containing UTL zeolite 108
Gibbs free energy 7, 196
glass
– Car–Parrinello molecular dynamics trajectory of 373
– ceramics 366
– structure 357
glass transition temperature (T_g) 88
glow-discharge plasma 398
glycerol 224
gold doped film
– SEM photo of 343
gold nanocubes 335
gold nanoparticles scattering effect 345
gold nanorods 335
– photosensitizer doped silica shell, synthesis of 340
– polyelectrolytes 338

gold nanoshells 338
gold quantum dots (AuQDs) 340
GONR–Ormosil hybrid materials 332
grain refining 45
graphene-based hybrid materials 332
graphene nanocluster 22
graphene nanoribbons 22
graphene oxide (GO) 330
graphene sheet 21
graphite nanocrystals 21
graphitized carbon monolith 228
Green B, 283
green chemistry 44
green fluorescent protein 145
grit-blasting 398
growth nuclei 27
guest-responsive structural
  transformation 142
gyroidal micellar structure 124

h

HAADF – STEM. *See* high-angle annular dark
  field scanning transmission electron
  microscopy
halogen-metal network 60
heat diffusion mechanisms 325
heat release rate (HRR) 90
hematoma formation 401
heteroanion 4, 8
heteroatom doping 227
heterocyclic organic molecules 156
heterogeneous catalysis 122
heterogeneous catalysts 149
– materials, synthesis of 125
heteropolyanions (HPAs) 4
heteropolybrowns 307
heterotrimer 8
hexagonal mesoporous silica (HMS) 125
hexamethylene tetramine (HMTA) 13, 306
hierarchically porous monoliths 203
high-angle annular dark field scanning
  transmission electron microscopy (HAADF
  – STEM) 32
high energy plasma 34
high-performance liquid chromatography
  (HPLC) 208
high-porosity macroporous silica 205
high-resolution transmission electron
  microscopy (HRTEM) 32, 145
– analysis 27
– micrograph 34
high-throughput systems 143
HMS. *See* hexagonal mesoporous silica (HMS)

Hofmann degradation 105
hollow closed nanostructures 21
homogeneous gelation 216
HOMO-LUMO energy gaps 11
host-guest charge transfer 153
host-guest composite 14, 142, 153
– of $[Cu_3(MTA)_2]$ (HKUST-1) and TCNQ, 153
host-guest interactions 146, 150
host-guest proton conductor 156
HRR. *See* heat release rate (HRR)
hyaluronic acid 396
hybrid aerogels 206
hybrid crystals 157
hybrid luminescent gold nanoshell
  structure 342
hybrid materials 79, 276
– applications 79
– – anticancer drugs 79
– – anticorrosive products 79
– – cosmetics 79
– – fibers 79
– – in medical imaging 79
– – motor vehicle catalysts 79
– for nonlinear optics 317
– – coordination and organometallic
  compounds based 323
– – dye-doped inorganic matrices 318
– – second-order 318
– synthesis of
– – natural clays used in 79
– – synthetic clays used in 79
hybrid nanomaterials 301
hybrid organic–inorganic materials
– applications of 79
– optical properties 275
hybrid PCP/MOF crystals 157
hybrid sol–gel monolithic materials 332
hydrated exchangeable anions 72
hydrogen peroxide treatment 399
hydrogen production 71
hydrophilic clays 81
– organophilic conversion 81
hydrophobic polymers 81
hydrotalcite 72
hydrotalcite-type LDH, schematic illustration
  of 221
hydroxyapatite 392
hydroxycarbonate apatite (HCA) 358
hydroxylated silicon dioxide
  $(SiO_{2-x/2}(OH)_x)$ 121
hydroxyl ions 362
hyperpolarizabilities
– second- and third-order 318

hypoxia 369
hysteresis 152

i

Ia3d symmetry 124, 126
– gyroidal structure of 124
ICP-MS. *See* inductively coupled plasma mass spectrometry
IF/INT
– current-carrying capacity 41
– lubrication mechanism
– – type 36
– nanoparticles
– – applications
– – type 43
– optical properties 36
– tribological properties 36
IF-MoS$_2$ lattice 32
IF-MoS$_2$ nanoparticles 22
– TEM micrographs 23
IF-MoS suspension
– absorbance spectra 37
IF-MoS2 suspension
– extinction spectra 37
IF NPs
– advantageous properties 44
IF-WS$_2$ nanoparticles 25
IF-WS$_2$, posteriori doping 31
imidazolate ligands 148
– 2-nitroimidazole 148
Im3m symmetry 126
– body-centered cubic structure of 124
individual INT
– current-carrying capacity 41
individual WS$_2$ nanotubes
– properties 41
induced pluripotent stem cell (iPSC) 388
inductively coupled plasma mass spectrometry (ICP-MS) 32
inert gas 123
infinite zeolite framework 98
InOMe metallated analog 170
inorganic clay 80
inorganic $Cs_{2.5}H_{0.5}PW_{12}O_{40}$ (CsHPW) cluster systems 264
inorganic 3DOM SiO_2/CsHPW composite membrane 264
inorganic nanotubes (INT) 21, 22
– doping 31
inorganic nodes 166
in situ polymerization 251
in situ Raman spectroscopy 146

insoluble organic ligands 143
insulator 70
INT. *See* inorganic nanotubes
intercalated/exfoliated nanocomposites 86
intercalation 71
– method 60
– properties 66
– reaction 64, 66, 67
interfacial
– energy 200
– polymer 80
– stress 65
interlayer 71
– coupling 41
– polarization forces 28
intermolecular antiferromagnetic interactions 155
International Union of Pure and Applied Chemistry (IUPAC) 110, 291
International Zeolite Association (IZA) 99
interpenetrating networks 144
interplanar spacings 30, 131
intersystem crossing (ISC) 325, 327
INT-WS$_2$
– electromechanical properties 42
– EM measurements 42
– field-effect transistors 41
– a posteriori doping 31
inverse sigma transformation 109
ion concentrations of SBF, and human blood plasma 362
ion conductivity 57
ion exchange 59, 80
– properties 58
– reaction 68
ion exchangers 73
ionic bonding 65
ionic conductivity
– cationic transport mechanism 90
ionic strength 6
ionization potentials 153
ion-releasing ability 377
iPSCs 389
IRMOF-3, with salicylaldehyde/subsequent metallation, with vanadyl *acac*
– bdc-NH$_2$ linker 173
iron hydroxide 220
irradiating multiwall INT-WS$_2$ 34
IR wavelengths 333
isomerism, in POMs 8
isopolyanions (IPAs) 5

j

Jablonski diagram 278

k

kaolinite 21, 84
– intercalation of 84, 87
– – dimethylsulfoxide (DMSO) used for 84
– organophilization of 84
– with tris(hydroxymethyl)aminomethane 84
KDP crystal 323
Keggin and Dawson anions 5
Keggin anions, fully oxidized 9
Keggin heteropolyoxoanion structure, $[\alpha\text{-}XM_{12}O_{40}]^{n-}$ 8
Keggin silicomolybdic acid 8
Keggin-type anion 8
Keggin $[XM_{12}O_{40}]^{3-}$ 7
Keplerate-type structure 5

l

Langmuir–Blodgett 73
– technology 344
lanthanide-bearing organic–inorganic hybrids 284
laponite 85
laser emission 286
laser illumination 286
LaS–TaS$_2$ nanotube 29
LaS–TaS$_2$ tubular crystals
– SEM images 30
– TEM images 30
La$_2$Ti$_2$O$_7$ (A$_2$B$_2$O$_7$) 57
lattice thermal conductivity 65
La$_{2x}$Ba$_x$CuO$_4$, 69
layer-by-layer 73
layered cesium titanate, exfoliation process of 73
layered clays 80
– mineral 80
– – vermiculite 80
layered copper oxides 61, 65, 68
– crystal structure of 69
layered crystalline phosphates 222
layered double hydroxide (LDH) 72, 221
layered manganese oxides 66
– intercalation reaction of 67
layered metal oxides 60
layered perovskite oxides 53
– intercalation properties of 58
(100)-layered perovskite oxides 53
(110)-layered perovskite oxides 57
layered solids 80
– delamination of 80
– exfoliation of 80
– intercalation reactions in 80
– – ion exchange 80
– – surface functionalization 80
layered structures, exfoliation of 73
layered titanium oxide and niobium oxide 71
levynite 101
Lewis acid 148
Li$_x$CoO$_2$ structure 62
LiFePO$_4$/C cathode materials 262
ligand-to-metal charge-transfer (LMCT) 307
light-emitting 283
– hybrid materials 276
– – luminescence 276
– – luminescent solar concentrators 287
– – luminescent thermometers 289
– – random and feedback lasers 286
– – white light emission 280
light emitting diode (LED) 278
light irradiation 141, 147
light scattering by surface plasmon resonance (LSPR) 25
Lindqvist anion 4
Lindqvist $[M_6O_{19}]^{2-}$ 7
Lindqvist-type POMs (hexamolybdate/molybdate anions) 324
liquid-assisted grinding (LAG) 168
liquid chromatography 121, 204
liquid-phase adsorption 122
lithium exchanged low-silica X (Li-LSX) zeolite 100
living tissues 390
Li$_x$CoO$_2$, 62
Ln^{3+} ions 280
load-bearing large animal models 404
localized surface plasmon resonance (LSPR) 302, 333
Löwenstein's rule 104
low-density materials 206
low-spin states 152
low-toxicity compounds 143
low-voltage high-resolution scanning electron microscopy (LV-HRSEM) 145
LSC, schematic representation 288
LSPR. See light scattering by surface plasmon resonance
Lucifer yellow (LY) 337
luminescence 142, 151, 276
luminescent frameworks 152

luminescent hybrid materials 280
luminescent lanthanide (Ln) ions 152
luminescent solar concentrator (LSC) 279, 287
luminescent thermometers 289

m
macropores 393
– control 200
– principle of 198
macroporous arylene-bridged poly(silsesquioxane) gels 229
macroporous silica 195
macroporous titania gels 214
magnetic exchange couplings 151
magnetic $[V^{IV}_{14}E_8O_{50}]^{12-}$ heteropolyoxovanadates (heteroPOVs) 11
magnetic interactions 151
magnetic materials 165
magnetic properties, in frameworks 151
magnetism 142, 151
magnet sponges 151
manganese-based inorganic-layered material (MPS$_3$) 322
manganese oxide 66, 67
– tunnel structure 66, 68
marshmallow-like gels 210, 211
mass transport 2
Materials of Institute Lavoisier (MIL) 147
MBGs synthesis 393
MCF-7 cancer cells
– two-photon imaging of 341
MCM-41 silica 123
mechanical loading images 159
mechanochemical method 143
medical gel (Esracaine) 39
melt-quenched glasses 358
mending effect 39
(3-mercaptopropyl) trimethoxysilane (MPTMS) 283
mesenchymal stem cell (MSC) 400, 404
mesopores
– arrangements of 124
– control 200
– generating agents 111
– supramolecular templating of 204
mesoporous bioactive glasses (MBG) 393
mesoporous bulk silicas 309
mesoporous materials
– synthesis 122–123
mesoporous MOFs 165
mesoporous photochromic (SP) silica shell 300

mesoporous silica 339
mesoporous zeolites 111
metal alkoxides 250
metal-based nodes 173
metal carbides 233
metal-complex components 141
metal-complex properties 141, 151
metal dichalcogenides 41
metal ions 142, 149, 151, 158
– effect 6
metallic biomaterials 400
metallic scaffolds 397
metal nanoparticles
– chromophores, principal strategies to control interactions 334
metal nanostructures
– in situ synthesis 343
metal–organic framework (MOF) 122, 165, 169, 221, 284, 323
– activation 174
– chemistry 166
– containing metal-carboxylate bonds 177
– containing M-N bonds 176
– IRMOF-74-XI, with hexagonal channel structure 145
– kinetic stability 175
– MOF-210 incorporates $Zn_4O(CO_2)_6$ clusters, BTE, and BPDC ligands 145
– MOF-5/IRMOF-2/Zn-JAST-4/Cd-2stp-pyz, rotation of pillar ligands 154
– MOF–PEMA-3.5 hybrid materials
– – photograph of 285
– NU-1000 170
– PCPs/MOFs pores, guest activities 143, 155
– potential applications 179
– – environmental pollution remediation 182
– – gas-phase catalysis with alkenes 180
– – nerve agent degradation 179
– synthesis 166
– – building block replacement 173
– – de novo synthesis 166
– – post-synthesis modification 168
– – – linker modification 172
– – – metal-based node modification 168
– – – organic-based node modification 171
– vanadium(IV) ion incorporation 148
– Zr-based MOFs NU-1000, UiO-66, PCN-225, DUT-67, and MOF-841, 178
metal-organic polyhedron-18, 145
metal oxides 165, 252
– anion 6
– clusters 4
– compounds 6

metal-oxygen anionic clusters 3
metal particles 149
metal phosphate systems 222
– ion conductivity 222
– ion exchange 222
metal polycarboxylates 324
metal salt 217
metal-to-dye interactions 333
metaphosphate glasses 374
(γ-methacrylpropyl)-silsesquioxane 321
methane activation 148
methyl methacrylate 149
5-methyl-4-nitroimidazole 148
methyl orange (MO) 308
methylsilsesquioxane (MSQ)
– aerogels 209
– network formation and pore control in 206
– phase separation of 209
– xerogels 209
methyltriethoxysilane (MTEOS) 330
methyltrimethoxysilane (MTMS) 206
methylviologen cation
– organic–inorganic interface 298
MFI zeolite, schematic of formation 112
Mg-alloys 26
Mg ions 144
$MgO-CaO-SiO_2-P_2O_5$ glass system 366
micelles 123
– expanders 130, 132
micelle-templated materials
– pore diameter of 130
– pore size control in 126
– pore volume of 130
micelle-templated ordered mesoporous materials 123–125
microcomposites
– intercalated/exfoliated nanocomposites, structure difference 86
microcomputerized tomography analysis 403
micromolds 249
micron-sized platelets 38
microporosity 133
microporous adsorbents 122
microporous MOFs 165
microwave methods 143
MIL-101(Cr)/MIL-53(Al)/MIL-125(Ti) structure 178
Miller indices 30
mineralizing agents 102
misfit layered
– compounds (MLC) 24, 27
– lanthanide-based 28
misfit slab 28

misfit strain
– partial relaxation 29
mixed metal oxides 220
miyakosyne A, 151
MLC. See misfit layered, compounds
MLC-NTs, formation mechanism 29
M-N-M angle in ZIFs compared to Si–O–Si angle in zeolites 177
Mo-based POMs 4
Mo-blue and Mo-brown-reduced nanosized POM clusters 5
mode-locked laser 38
MOF. See metal-organic framework (MOF)
M_3O_{13} group 9
molecular conversion 142
molecular dynamics (MD) 6
molecular mechanics (MM) 107
molecular metal oxides 3
molecular mobility 2
molecular vanadium-oxide spin clusters 11
molten halide, capillary imbibition 33
molybdenite (MoS_2) 21
– nanosheets 39
– triatomic layers, with hexagonal symmetry 23
molybdenum 4
molybdenum-blue (MB) 5, 14
molybdenum oxide-based fragments 4
monochromatic LEDs 285
monocyte differentiation 376
– RAW 267.4 monocyte cells cultured on glass surface, microscopy images of 376
monodisperse spherical colloidal crystals 256
monolayer-multilayer adsorption 121
monolithic 233
– capillary column 206
– columns 205
– gels 198, 200
– – formation 206
– macroporous silica gels 203
– pure zirconia 214
– titania columns 214
MoO_3 materials 306
M41S materials 124
– formation of 124
MS_2 nanoparticles, charged-colloidal behavior 37
MSQ aerogels 210
MSQ monolithic gels 208
MSQ xerogel tile, thermal conductivity 211
multichromic hybrid organic–inorganic supramolecular assemblies 307
multicolored systems 302
– combining photochromic dyes 302

multicolor photochromism 302
multicomponent oxides 250
multi-photon absorption (MPA) 318, 325
multiwall carbon nanotubes (MWCNT) 330
multiwall INT-WS$_2$
– syntheses of 24
Wells-Dawson structure $[_2M_{18}O_{54}]^{n-}$ 5
myoglobin 145

n

α-NaMnO$_2$ 67
nanoactuators 43
nanobuilding block (NBB) 295
nanoclay dispersion 92
nanocomposite materials 307
nanocomposite matrix 26
nanocomposites
– applications in electrochemical devices 90
– free radical polymerization, preparation 89
– mechanical properties 44
nanoelectromechanical system (NEMS) 43
nanoelectronics 22
– devices 42
nanometer scale 390
nanoparticle (NP) 80, 394
– properties
– – doping effect 25, 38
nanoplatelets 36
nanorods 338
nanostructures
– templated synthesis of 122
nanotopography 366
nanotubes (NTs) 21
– current transport mechanism 41
– electrical properties 42
– mechanical deformation 42
nanowires 31, 127
Na$_2$O–Al$_2$O$_3$–TiO$_2$–Nb$_2$O$_5$–P$_2$O$_5$ glass system 374
Na$_2$O–K$_2$O–CaO–P$_2$O$_5$ system 373
Na$_3$(2,4,6-trihydroxy-1,3,5-benzenetrisulfonate)-incorporating 1H-1,2,4-triazole 157
natural clays 79
natural gas 166
natural zeolites 97
Na$_x$CoO$_2$ structure 63, 64
Nb$_2$O$_5$–SrO–P$_2$O$_5$ glass 374
near-UV GaN chip 285
NEMS. See nanoelectromechanical system
network modifiers 357, 364, 374
nickel hydroxide 220
NIR fluorophores (IR800) 338
4-nitroaniline, into kaolinite interlayers 323
NLO activity 323
NLO properties 332
N-methylaniline (MA) 153
NMR. See nuclear magnetic resonance (NMR)
N,N-bis-(2-hydroxyethyl)-piperazine (BHEP) 13
N,N-dimethylaniline (DMA) 153
N,N-dimethyl-p-toluidine (DMPT) 153
NO-adsorbing/releasing materials 148
nonaqueous sol–gel process
– scheme of 232
noncrystalline compounds 151
nonfluorescent H-aggregates 279
nonionic triblock copolymer 209
nonlinear materials
– third-order 324
– – dyes dispersion in sol–gel/organic materials 324
– – graphene based hybrid materials for nonlinear absorption 330
– – polysilsesquioxanes hybrids for nonlinear absorption 330
nonlinear optical (NLO) 317, 318
– photoswitching properties 299
nonporous structure 148
nonradiative processes 277
– delayed fluorescence 277
– internal conversion 277
– intersystem crossing 277
nonradiative transitions 277, 278
NO-releasing scaffolds 148
NTs. See nanotubes
n-type conductivity 31
nuclear magnetic resonance (NMR) 86, 363
nucleation mechanisms 6
nylon 6-clay hybrids 80

o

octacalcium phosphate (OCP) 392
OIH photochromic materials 292
oligomeric oxides 204
oligomerization 182
oligosiloxanes, red/green dye-bridged 281
OMM. See ordered mesoporous material (OMM)
open metal sites upon dehydration 148
optical power limiting (OPL) 325
optical tracking 38
order-disorder phase transitions 155
ordered framework-surfactant 125
ordered mesoporous material (OMM) 122

– cubic structures of 124
– nonionic poly(ethylene oxide)-based surfactants as templates for 125
– pore connectivity 133
– pore size control 129–132
– structure control 128
ordered mesoporous silicas 123
– pore size enlargement 130
Oregon Green 488 fluorophore 338
organic dyes (QDs) 287
Organic–inorganic hybrid (OIH) 278
organic ligands 152
organic light-emitting diode (OLED) 279
organic linkers 165, 166
organic structure-directing agent (OSDA) 102
– chemically synthesizable 107
– design of 107
– formation of OSDA–silicate composite 105
– molecular structures and properties of 105
– scope of 105
organoclay 80
organofunctionalized clay
– molecular structures and schematic representation 282
organometallic complexes 149
organophilic bentonites 81
organophilic clays 81
– applications 81
– – agriculture 82
– – – enzyme immobilization 82
– – – herbicide formulations 82
– – antibacterial agents 82
– – biomedical 82
– – – adjuvants in vaccines 82
– – – controlled drug delivery 82
– – – regenerative medicine 82
– – – removal of organic pollutants 82
– minerals 85
– synthesis of 85
organophilization
– agents, tetrasubstituted phosphoniums used as 82
– industrial, tetraalkylammonium cations used for 82
– tetra-substituted ammoniums 82
organosilicas 134
organotrialkoxysilanes 125, 206
organ transplants 387
ORMOSIL (organically modified silanes) hybrid matrices 293
orthophosphate 374
osteoblastic cells 401
osteoclasts 376

– mediated remodeling 401
osteoconduction 399
osteogenic cells 401
osteoid formation, at Bioglass® interfaces 362
osteostatin 402
Ostwald ripening theory 201, 203
oxide sol–gels, polymerization-induced phase separation 196
oxoanions 4, 11
oxyhydroxides 216

p
PA. *See* polyamide-6 (PA)
palygorskite 84
PAO-4 oil 38
particle-packed columns 205
Pauli exclusion principle 39
PbI_2 nanotubes 34
PEO. *See* poly(ethylene oxide) (PEO)
periodic mesoporous organosilica (PMO) 125, 131
perovskites 53, 58
– layer 54
PET. *See* poly(ethylene terephthalate) (PET)
phase diagram 199
phase separation 198
phase transition 32
PHB. *See* poly(3-hydroxybutyrate)
phenylene-bridged poly(silsesquioxane) 229
phosphate glasses 370
– conventional 370
– invert glasses 374
– systems 374
phosphonium clays 82
phosphonium-organophilized smectites 83
phosphonium OSDAs 106
photocatalysis 260
photochemically (P-type photochromism) 291
photochromic hybrid materials 291
– inorganic/organometallic photochromism in hybrids 304
– organic photochromism in hybrids 293
– photochromism 291
photochromic molecules
– micro-/nanostructuring of 293
photochromic polyoxometalates (POMs) 307
photochromism 291
– by proton donors 306
photoelectrochemical cells 40
photon energies 277
photoswitchable materials 297
photoswitching QDs

– coated with amphiphilic photochromic polymer 301
phthalocyanine-based photosensitizers 336
PHTS. *See* plugged hexagonal templated silicas (PHTSs)
physical chemistry 142
– in PCPs/MOFs 159
phytochrome 291
plasma ablation 35
plasma-spayed hydroxyapatite 398
plasmonic device 38
plasmonic hybrid materials 333
– composite materials and thin films 343
– dyes and plasmonic nanostructures 334
– encapsulation of dyes 341
– metal nanoparticles, optical properties of 333
– surface functionalization of metal nanoparticles 335
plasmonic materials 333
plasmonic scattering 38
plastic optical fiber (POF) 288
platinum acetylides doped hybrid silica xerogels
– optical properties of 329
plugged hexagonal templated silicas (PHTSs) 134
pluronics 128, 129, 202
– block copolymers 130
PMMA glasses 332
PMO. *See* periodic mesoporous organosilica (PMO)
polyacrylamide 223
poly(alkyl-thiophenes) 252
poly-alpha-olephin synthetic oil 38
polyamide-6 (PA) 88
polycaprolactone (PCL) 396
polycondensation reactions 198, 200
poly-dimethylsiloxane (PDMS) 212, 249
– oligomers 296
poly(divinylbenzene) (PDVB) gel 226
polyether ether ketone (PEEK) 45
poly(3,4-ethylenedioxythiophene) 252
poly(ethylene glycol) 306
poly(ethylene oxide) (PEO) 80, 125, 202, 218, 223, 225
poly(ethylene terephthalate) (PET) 80
polyglycolic acid (PGA) 396
polyhedral oligomeric silsesquioxane (POSS) 206, 321
– multiwall carbon nanotubes 331
poly(3-hydroxybutyrate) (PHB) 45
poly(2-hydroxyethyl methacrylate) (pHEMA) 286

poly(2-hydroxyethyl methacylate-silica hybrids 281
polylactic acid (PLA) 45, 396
poly(lactic co-glycolic acid) (PLGA) 396
polymer-ceramic composites 403
polymer chemistry in PCP/MOF channels 149
polymeric scaffolds 395, 397
polymeric sponge 398
polymerization
– degree of 196
– of monomers confined in PCP/MOF channels 149
– in restricted channels 150
– of vinyl monomers 149
polymer matrixes 400
polymer nanocomposites 80, 81, 289
– films 36
polymer-polymer blends 403
polymers 80
– poly(ethylene oxide) 80
– polypropylene 80
– polystyrene 80
polymethylmethacrylate (PMMA) 150, 247, 319
poly(methyl methacrylate-butyl acrylate-acrylic acid) P(MMA-BA-AA) 247
polyoxometalates (POMs) 3, 4, 292, 304, 307, 323
– isomerism in 8
polyoxomolybdates 4, 307
– nanocapsule cluster 6
polyoxotungstates 307
poly(phenylene vinylene)-PFO copolymers 283
polypropylene (PP) 80, 88
poly(propylene fumarate) (PPF) 45
poly(propylene oxide) (PPO) 126
polysilsesquioxane 330
polystyrene (PS) 80, 149, 247
poly(styrene-methyl methacrylate-acrylic acid) P(St-MMA-AA) 247
polyurethane tubes 39
polyvinyl alcohol (PVA) 45
polyvinylpyrrolidone 225
POM-based materials 6
POM clusters 4, 5, 8, 12, 13, 14
POM synthesis 13
pores 121
– apertures, of UiO-67, 168
– categories
– – macropores 121
– – mesopores 121
– – micropores 121

– channels in PCPs/MOFs 148
– connectivity control 134
– gas adsorption behavior 121
– interconnectivity 397
– in PCPs/MOFs, usage 144
– surface environment 148
pore size 121, 131, 141, 144, 152, 154, 155, 165, 263
porosity 142, 144, 176
porous
– aluminum hydroxides, SEM images of 219
– arylene-bridged polysilsesquioxane monoliths 229
– calcium phosphate monolith 222
– frameworks 141
– materials 145
– structures 145
– titanium phosphate monolith 225
– zirconium phosphate monolith 223
porous coordination polymer/metal-organic framework (PCP/MOF) 141
– channels 150, 156
– chemistry 142
– compounds 142, 153
– containers 147
– crystals 143, 158
– – engineering of 157
– materials 142
porous monoliths
– XRD patterns of 231
porous poly(L-lactic acid)/apatite composite scaffold 400
porous silica monolith 216
– cumulative and differential pore size distribution of 203
positional isomerism 9
positive electrode material 62
a posteriori method 31
post-gelation aging 204
post-synthesis method 143, 144
post-synthesis modification (PSM) 166, 168
powder X-ray diffraction 170
PP. See polypropylene (PP)
PPO. See poly(propylene oxide) (PPO)
preceramic polymer
– calcination of 232
– route 229
pressure swing adsorption (PSA) 100
Preyssler anion 4
– $[X^{n+}P_5W_{30}O_{110}]^{(15-n)-}$ 5
propylene oxide 216, 217, 218
protonation 6
proton-conducting pathways 156

proton conductivities 156, 157
pseudomorphological replication 221
Pt-Rh catalyst 265
Pt-Ru alloy clusters 265
pure-silica MFI zeolite 102
PVA. See polyvinyl alcohol
pyrochlores 57

q

quantum dot (QD) 279
quantum rattles (QRs) 340
quaternary ammonium cations 105

r

radiative processes
– fluorescence 278
– phosphorescence 278
Raman spectroscopy 155
rapid prototyping (RP) 392
rapid sulfidization 23
Rayleigh limit 290
Re-doped IF-MoS_2
– optical properties 39
– tribological properties 38
redox-inactive anions 9
redox intercalation 61
refractive index 215
regenerative medicine (RM) 384, 386, 387
rehydroxylation 133
reinforced biocomposites 26
renewable production of PET
– reaction pathways for 114
resin matrix
– fracture toughness 36
– hardness 36
resonant tunneling devices 40
resorcinol-formaldehyde (RF)
– polymerization-induced phase separation of 227
reverse Monte Carlo modeling 371
reverse saturable absorption (RSA) 325
– with nonlinear scattering and nonlinear refraction 327
reversible photochromism 291
reversible structural transformation 147
RGD-peptide 398
rhenium atom ionization
– activation barrier 41
Rh6G-bearing diureasils
– normalized emission spectra 286
Rh6G-doped SiO_2 nanoparticles 286
rhodamine 343
rhodamine 610, 342

rhodamine-B (RB) 308
rhodamine doped silica 342
rhodamine 6G (Rh6G) 280, 330
ribosome 2
rim atoms 21
ring-opening metathesis polymerization (ROMP) 302
robocasting devices 392
rock-salt (RS) 53
Rosentsveig–Margolin mechanism 23
rotational isomerism 9
Rothschild–Zak mechanisms 23
RS. See rock-salt (RS)
Ruddlesden–Popper phase 53, 54
Ru(ddphen)$_3^{2+}$
– hybrid polymer thin films 345

S

SAED. See selected area electron diffraction
salicylaldehyde 172
salt encapsulation 33
SBA-15, formation of 127
SBA-16 silica 127
scaffold 390, 404
– 3D, 391, 394
– design 398
– exhibit macroporosity 395
– structural features 396
scanning electron microscopy (SEM) 25, 393
SDS. See sodium dodecyl sulfate (SDS)
second-generation compounds 144
second-generation materials 142
second-harmonic generation (SHG) 300, 319
– enhancement 344
– polymeric layer 345
sedimentation methods 248
selected area electron diffraction (SAED) 29
selective adsorption 142
self-assembly 1, 2, 3
– of metal ions/organic ligands, producing porous frameworks 142
– in molecular systems 2
– reaction networks 6
self-assembly monolayer (SAM) 299
self-organization 2, 3
– tessellated films 44
self-pillared zeolite 111
SEM. See scanning electron microscopy
semiconductors 252
– IF/INT, optical properties 36
– semiconductor nanocrystals, substitutional doping 31

semicrystalline polymers 80
sepiolite 84
shape memory effect 159
shrinkage–re-exapansion 209
SiC monoliths 230
silane matrices, advantages of 293
silanization 83
silanols 122
silica 201
– additional mesopore formation, by aging 202
– applications 205
– block-copolymers 295
– gels 121
– – containing Rh6G-doped SiO$_2$ nanoparticles 286
– mesoporous NPs 395
– pore connectivity in 133
– precursor, cocondensation of 125
– shell thickness, TEM images 339
– synthesis conditions 201
silicalite-1, 102
silicate-based bioactive glasses 365
– biological responses, to inorganic ions released 359
silicate glasses 358
silicate ion 357
silicate/phosphate/borate-based glass systems 358
silicon alkoxides 247
silicon carbide 229
silicon-doped hydroxyapatite 392
silicon oxycarbide 229
silicon precursors 331
siloxane-based organic–inorganic hybrids 279
siloxane bridges 121
siloxane linkages 125
silsesquioxanes 206, 207, 330
– derived from 206
silver halides, photochromic activity 305
simulated glasses, structures of 363
single-crystalline mesostructured MFI nanosheets 107
single-to-triple wall inorganic nanotubes 33
single wall carbon nanotube 24
SiO$_2$–Na$_2$O–CaO–P$_2$O$_5$ glass system 364
SiO$_2$–Na$_2$O–CaO system 363
SiO$_2$–P$_2$O$_5$–CaO–Na$_2$O–CaF$_2$ glass system 365
Si–O–Si bonds 358, 364
Si–O–Zn bonds 369
smectites 82
– alkylammonium cations, arrangements 83
– clay mineral in 82
– – hectorite 82

– – montmorillonite 82
– – saponite 82
– organophilization of 82
– – with neutral molecules 82
– – with organic cations 82
SnS_2
– laser ablation 27
$SnS–SnS_2$ nanoscrolls 28
Sn-substituted polyoxoanions 12
sodium dodecyl sulfate (SDS) 126
soft hybrids 283
soft-ionization 7
soft porous crystals 142
soft templating 122
solar fuels conversion 165
sol-gel chemistry, for preparation of 3DOM materials 250
sol–gel NLO dyes, structures of 320
sol–gel processing 195
– derived glasses 366
– precursors and variations of 197
sol–gel transition 196
– structure formation, in parallel with 199
solid lubrication behavior 43
solid nanolubricants 26
solid-solution-type crystals 158
solid-state lighting (SSL) 279
solid-state materials 165
solid-state reaction 63
– NMR spectroscopy 154, 156
solubility 372, 377
– glass compositions/aqueous medium 373
solvent- and anion-controlled photochromism 302
solvent-assisted ligand incorporation (SALI) 171
solvent-assisted linker exchange (SALE) 174
solvent extraction 123
sonochemical methods 143
SO_3–P_2O_5–ZnO–Na_2O system 376
spark plasma sintering 398
spectroscopic method 302
spherical mesopores 126
– face-centered cubic structure of 131
spherical micelles 123, 124, 127
spin crossover 151
spinodal decomposition 196, 199, 226
– isotropic phase domains, evolution 199
spin–orbit coupling 40
spin transition (ST) 151
spirooxazine compounds, structural changes of 296

spiropyrans (SP) 291
– functional PAA polymer 300
– molecules, schematic illustration 295
– POM self-assembled structures, absorption phenomena 308
spontaneous organization 2
SPR. See surface plasmon resonance
spray pyrolysis 254
spring-back 209
– behavior 210
starting composition and resultant microstructure, relationship between 207
static self-assembly 2, 3
stationary kinetic state 14
steam-assisted crystallization (SAC) 104
stem cells, limitations of 388
steric factors 175
sticking van der Waals forces, role of 28
stilbite 97
strain energy 27
strontium zirconium phosphate 224
styrene 149
suitability of MOFs 165
sulfide nanosheets 28
sulfonated aluminum phtalocyanine 335
sulfophosphate glasses 376
sulforhodamine G (SRG) 282
sulfur vacancies 37
superconductivity 53, 63, 65, 68
– blocks 69
supercritical carbon dioxide ($scCO_2$) 145, 175
superionic conductor (NaSICON)-type metal zirconium phosphates 224
supramolecular templating 204
surface charge 44
surface-emitting lasers 40
surface engineering of PCP/MOF crystals, involving layer-by-layer growth of 158
surface-enhanced Raman spectroscopy (SERS) 334
surface functionalization 36, 80
surface plasmon resonance (SPR) 38
surface roughness, cell morphology and growth 398
surfactant micelles 122, 123
– templating 129
surfactant-micelle-templated OMM, 122
surfactants 122–134, 126, 134, 204, 243
– hydrophilic parts 122
– hydrophobic parts 122
– packing parameter 128
– templating 132, 133
SURMOF technique 145

swelling agents 126, 130
synthetic
– biodegradable polymers 396
– clays 79
– – preparation 85
– hectorite 85
– – laponite as 85
– materials 357
– polymers 397
– tools in MOF chemists' toolbox 167
– tunability 165
– zeolites 97

t

Ta-based microtubes 28
tantalum 399
tartrate-resistant acid phosphatase (TRAP) 395
Tb^{3+} β-diketonate complexes 289
tellurate anion $[TeO_6]^{6-}$ 10
TEM. *See* transmission electron microscopy (TEM)
temperature annealing process 28
templating anions 9
tertiary sulfonium 106
tetraalkoxysilanes, cocondensation of 125
tetraalkylammonium cations 82
tetraethoxysilane (TEOS) 201
tetraethylorthosilicate 366
tetrahedral atoms
– apertures formed from 99
tetramethoxysilane (TMOS) 201
tetramethylammonium (TMA^+) cations 102
tetraphenylethylene derivatives (TPTS) 282
1,3,6,8-tetraphenylpyrene (TPPy)-containing organosilane precursor 280
tetra-substituted ammoniums 82
tetra-substituted phosphoniums 82
TGA. *See* thermogravimetric analysis (TGA)
T_g *See* glass transition temperature (T_g)
thermal expansion 159
– coefficient 398
thermal motion 2
thermal spray 398
thermal transitions 150
thermal treatment 141
thermodynamics, of GD simulant hydrolysis in NU-1000, 180
thermoelectric properties 63
thermogravimetric analysis (TGA) 89
thermoplastic polyurethane (t-PU) 45
thermopower 64

thiol-anchoring group 337
thiol-terminated DNA, 338
third-harmonic generation (THG) 325
three-dimensionally ordered macroporous materials 243
Ti-6Al-4V, 397
TiO_2 nanoparticles 257
TiO_2 photochromic material, using ink-jet and flexography printing processes 305
TIPB. *See* 1,3,5-triisopropylbenzene (TIPB)
tissue engineering (TE) 386, 389
– in bone 389
– concepts 383
– definitions 383
– inert materials for living tissues substitution 384
– objective of 384
– procedure 391
tissue regeneration 394
titania 195, 201, 212
– alkoxides of 212
– application 215
– control over reactivity 214
– nanosheet 74
– starting compounds 212
titania–silica alkoxide systems 216
titanium alloys 399
– niobium oxides structure 71
– preceramic polymer 232
titanium nitride 231
titanium organoclay PVC nanocomposites
– thermal resistance of 91
titanium oxides 231
titanium phosphate 225
titanium scaffolds, coated by ceramic 399
TMOS-PEO-solvent pseudo-ternary system
– relation between starting composition and resultant gel morphology 202
topotactic conversion 108
tortuosity factor 90
T—O—Si bond angles, histograms of 103
trans-4-(4-dimethylaminostyryl)-1-methylpyridinium ($DAMS^+$) cationic ligands 324
transition metal dichalcogenides, doping effects 41
transition metal ions 143
transition metal oxide 60
transition states 7
transition temperature 151
transmetallation at Zn_4O node of MOF-5, 174

transmission electron microscopy (TEM) 21, 86, 123, 393
trialkoxysilanes (RSi(OR′)$_3$) 206
triboluminescence 277
β-tricalcium phosphate 222
triethanol amine (TEA) 13
triethoxysilane (TEOS) 279
1,3,5-triisopropylbenzene (TIPB) 131
1,3,5-trimethylbenzene (TMB) 204
trimethylene oxide 216
tunable photonic crystals 259
tunable surface chemistry 221
tungsten 4
– based structures 4
– oxide nanowhisker 24
– photochromic properties of 306
– suboxide nanowires 23
– suboxide NP, formation 23
two-photon absorption (TPA) 325

u
ultramicroporous MOFs 165
ultraviolet-pumped white phosphors 283
unfeasible zeolites 110
urea glass route 233
USY zeolite, commercial 110
UV irradiation 259
UV-pumped WLEDs 281
UV-vis spectrometer 37

v
vanadium metal oxide clusters 4
van der Waals bonding 65
van der Waals forces 22, 24, 43
van der Waals interactions 24, 33, 107, 146
vapor-phase transport (VPT) 104
vapor solid (VS) 43
vascular endothelial growth factor (VEGF) 403
vasodilatation 148
vermiculite 80, 85
– organophilization of 85
– – tetraalkyl-substituted ammonium salts used in 85
vinylmethyldimethoxysilane (VMDMS) 212
vinyl-modified marshmallow-like gels, preparation of 213
vinyltrimethoxysilane (VTMS) 206
vitamin B12, 145
VS. See vapor solid
V-shaped cavities 249
Vycor®, 195

w
water condensation 6
water-guest systems 156
water-soluble polymer 195
wear resistance 36
Wells–Dawson anion 9
Wells–Dawson structures 10
wet gel 219
– stage 214
WLEDs
– phosphors 283
– photographs of 284
Wnt/β-catenin pathways 401, 402
WO$_3$ nanoparticles, sulfidization 24
WS$_2$ nanotubes 24

x
XM$_9$ moieties 9
XPS. See X-ray photoelectron spectroscopy
X-ray absorption fine structure (XAFS) 32
X-ray diffraction (XRD) 32, 86
X-ray photoelectron spectroscopy (XPS) 32
X-ray structural analyses 150
XRD. See X-ray diffraction

y
YBa$_2$Cu$_3$O$_{7-\delta}$ 70
yttrium aluminum garnet (YAG) 220

z
zeolite 97, 165
– definition of 97
– desilication of 110
– empirical formula of 100
– fabrication of 110
– framework structures of 99
– functionalization of 101
– fundamental aspects of 97
– hydrothermal synthesis of 101
– levels to functionalize 100
– porous 110
– postsynthetic treatments of 110
– properties of 100
– sodium form of 101
– structure/classification 98, 104
– synthesis of 101
– – under hydrothermal conditions 101
– – from layered silicates 108
– – mineralizing agent 102
– units of 98
ZeolitePlus technology 105
zeolitic imidazolate framework (ZIF) 159

zero-dimensional spheres 159
zeta potential (ZP) 37
zinc ions 159, 369
zirconium 195, 201, 212, 399
– alkoxides of 212
– control over reactivity 214
– phosphate 223
– – monolith 224
– starting compounds 212
[ZnBr$_2$(μ-CEbpy).3H$_2$O]
– molecular packing 299
Zn^{II} ions 173
ZnO nanoparticles 253
ZP. *See* zeta potential
Zr$_6$-node of NU-1000 functionalization 171